ISO 9000 Quality Systems Handbook

Using the standards as a framework for business improvement

ISO 9000 Quality Systems Handbook

Using the standards as a framework for business improvement

SIXTH EDITION

David Hoyle

ELSEVIER

AMSTERDAM • BOSTON • HEIDELBERG • LONDON
NEW YORK • OXFORD • PARIS SAN DIEGO
SAN FRANCISCO • SINGAPORE • SYDNEY • TOKYO

Butterworth-Heinemann is an imprint of Elsevier

Butterworth-Heinemann is an imprint of Elsevier
The Boulevard, Langford Lane, Kidlington, Oxford OX5 1GB, UK
30 Corporate Drive, Suite 400, Burlington, MA 01803, USA

First edition 2009
Reprinted 2010

British Library Cataloguing in Publication Data
A catalogue record for this book is available from the British Library

Library of Congress Cataloging-in-Publication Data
A catalog record for this book is available from the Library of Congress

For information on all Butterworth-Heinemann publications
visit our website at www.elsevierdirect.com

ISBN: 978-1-8561-7684-2

Printed and bound in Great Britain

10 11 12 13 14   10 9 8 7 6 5 4 3 2

Contents

Preface to the Sixth Edition

It is now 20 years since the first edition of the ISO 9000 family of standards was published and in that time the popularity and notoriety of the standards (particularly those used for certification purposes) have increased enormously. It is perhaps surprising that what started as a means to eliminate multiple supplier assessments in industries where there was a contractual arrangement between customer and supplier, has had such wide appeal throughout the world.

Over five editions of this handbook its basic purpose has not changed. It is to provide a source of reference not only for those seeking and maintaining ISO 9001 certification but also to provide the reader with the fundamental concepts of quality management so that use of the standard becomes a quest for improving quality in all operations and is not limited to simply getting and keeping a certificate.

ISO standards are reviewed every five years and changes only introduced when there are clear benefits to users. As a result of user feedback it was believed that the clarity of requirements could be improved but no changes to requirements other than editorial were justified on this occasion. Far greater changes are being made to ISO 9004 which breaks the image of a consistent pair and creates a model of an organization in pursuit of sustained success.

Consistent with previous editions this edition provides the reader with an understanding of each requirement of the current version of ISO 9001 through explanation, examples, lists, tables and diagrams. There are over 260 requirements in ISO 9001:2008, and the explanation of each of these forms the major part of the book. With very few exceptions, I have chosen to explain the requirements of the standard in the sequence they are presented in the standard and have added clause numbers to the headings to make it user-friendly. This has presented some difficulty as the clauses in ISO 9001 do not follow a sequence in which a management system might be established, implemented, maintained and improved. Clauses are grouped into some sections because they contain generic requirements and others are dispersed because they are specific to a particular stage of the execution of a contract. To accommodate this approach as best I can, there is considerable cross referencing.

In this edition I have maintained a structured approach with each requirement covered by three basic questions: *What does it mean? Why is it important? How is it demonstrated?*

A NEW STRUCTURE

Feedback from users of this handbook led me to believe that there were significant benefits from changing its structure, clarifying the concepts and providing more practical guidance. As a result instead of structuring the book around the main sections of ISO 9001 and as a consequence creating five large Chapters, the book has been divided into 8 parts with each part containing a number of smaller Chapters thereby making it easier to navigate.

Part 1 contains five Chapters introducing the concepts and principles upon which the ISO 9000 family of standards are based, the growth in certification and addresses some of the important issues to be understood before embarking upon implementation. Part 2 presents a number of approaches to achieving, sustaining and improving quality. Parts 3 to 7, address sections 4 to 8 of ISO 9001:2008 with each part divided into Chapters reflecting the clauses of the standard making 28 Chapters addressing ISO 9001 requirements. Part 8 contains three chapters covering system assessment, certification and continuing development. For convenience of use all the checklists and question-naires of the fifth edition have been consolidated in Part 8.

Each part has an introduction and key messages and each Chapter now has a preview making it easier to choose which to skip or study. Each preview explains for whom the Chapter is primarily written, the subjects covered and in Parts 3–7 there is a diagram showing where the ISO 9001 requirements fit in a managed process, thus placing the requirements in context.

There is an extensive glossary of terms in Appendix C and throughout the book, terms which are included in the glossary are indicated by the superscript symbol$^{\textcircled{\tiny{i}}}$.

NEW CONTENT

New in this edition is a section on EU Directives and more information on other standards in the ISO 9000 family and the vocabulary used. There is a new chapter on stakeholders and their importance in determining organizational objectives. A new chapter that consolidates a number of flawed approaches that have led to ISO 9000 attracting a poor reputation. I included this because I believe that for anyone setting out to use the ISO 9000 family of standards it is important to understand the historical context and avoid repeating the mistakes of the past. There is a new Chapter on the System approach to quality drawing on material from the 5th but adding new material on integrated management systems; a new Chapter on the process approach and a new Chapter on the behavioural approach to quality drawing on material from the fifth edition but expanding and restructuring it to show how human interactions affects quality. A new chapter on using the standards in the ISO 9000 family gives practical guidance and at the end of the book updated material is included from the third edition on preparing for and managing system assessment and certification and what you can do beyond ISO 9001 certification to move towards sustained success.

The interpretations are those of the author and should not be deemed to be those of the International Organization for Standardization, any National Standards Body or Certification Body.

I have retained the direct style of writing referring to the reader as 'you' and me as 'we'. You may be a manager, an auditor, a consultant, an instructor, a member of staff, a student or simply an interested reader. You may not have the power to do what is recommended in this book but may know of someone who does whom you can influence. There will be readers who feel I have laboured topics too much but it never ceases to amaze me how many different ways a certain word, phase or requirement might be interpreted.

I have recognized that although many organizations are using the latest information technology there are some that are not and will continue to use labour intensive ways of generating, maintaining and distributing information. Therefore, if the solutions appear outdated, simply skip over these and remember that more and more of the organizations that are using ISO 9001 are in developing countries.

Whatever your purpose you would benefit from studying the glossary of terms because the meaning given might well differ from that which you may have assumed the term to mean and thus it will affect your judgement.

Companion web site

The following items are available on www.elsevierdirect.com/companions/9781856176842

- Forms from the book in MS Word format
- Flow charts from the books in MS Visio format
- Reading list
- Related web sites from Appendix A
- Maturity Grid from Chapter 38
- Food for thought Questions from Chapter 38
- Requirements checklist from Chapter 38

Acknowledgements

For this sixth edition I am indebted to John Colebrook a director of EOS Ltd, a New Zealand based advisory firm, who provided valuable comment and suggestions for improvement on the fifth edition and new chapters for this edition. John also introduced me to the work of John Bryson, Paul Niven, Peter Senge and Ralph Smith and helped me to refine the relationships between mission, vision and strategy and their link with management system development. The feedback provided from his clients has enabled me to provide more clarity in the concepts and resolve the structural issues with the book.

I am also grateful to my many friends and colleagues in the CQI including Rhian Newton Quality Systems Manager with Flexsys Rubber Chemicals, Wales, UK for her insight that made me see the book more clearly from the users viewpoint and her many comments that improved the structure, explanations and readability; to Tony Brown from London, UK. Chair of CQI Standards Development Group and a UK representative on ISO/TC176 who provided opportunities for me to get closer involvement with the developments in the ISO 9000 family; to Steve Coles quality management consultant in the oil and gas industry for his suggestions on structure and the other standards in the family and finally to Ray Mellett lecturer in quality management at the University of the West of England and Chair of CQI South Western Branch for his comments on the manuscript.

Permission to reproduce extracts from ISO 9001:2008 is granted by BSI on behalf of ISO under licence no. 2008ET0049. British Standards can be obtained in PDF or hard copy formats from the BSI online shop: www.bsigroup.com/Shop or by contacting BSI Customer Services for hard copies only: Tel: +44 (0)20 8996 9001, Email: cservices@ bsigroup.com.

This book is written for those who want to improve the performance of their organization and whether or not certification is a goal or indeed required, it is hoped that the book will continue to provide a source of inspiration in the years ahead.

David Hoyle
Monmouth
E-mail: hoyle@transition-support.com

Part 1

Before You Start

INTRODUCTION TO PART 1

This book goes much further than the requirements of ISO 9001 because it is intended to aid the development of quality management systems that enable organizations to achieve their goals not simply deliver conforming product to customers. For some one of those goals may be ISO 9001 certification but as will become apparent, ISO 9001 is a constraint (rather than a goal) that enables organizations to provide products and services that satisfy their customers every time. An organizations' management system should enable it to go further than satisfy its customers. It should enable it to satisfy all its stakeholders thus creating a enabler for sustained success as reflected in ISO 9004:2009.

All organizations have a way of operating which is intrinsically a management system, whether formalized or not. There are no right or wrong ways of doing things only ways that work for your organization which is why certification should never be the goal. Making this system effective should be a goal of top management and perhaps the most widely recognized tool for developing management systems is ISO 9001 but it can be used in ways that make your system less effective, which is why it is so important that you digest Part 1 of this book first and then "A flawed approach" in Chapter 6 of Part 2 before deciding on your course of action.

There are many views about the value of ISO 9000, some positive and some negative. It has certainly spawned an industry that has not delivered as much as it could have done and even with the release of the 2008 revision there is still much to be done to improve the standards, improve the image, improve the associated infrastructure and improve organizational effectiveness.

This part of the handbook aims to put ISO 9000 as a family of standards in context, define what quality is and why it is important for organizations to make it a high priority and what role the organization's stakeholders play in influencing an organization's approach to the achievement of quality. We take a tour around the standards to appreciate the scope, content and application and then provide a practical guide for using these standards which if adopted will enable you to make your organization more effective by using the ISO 9000 family of standards.

Chapter 1

Putting ISO 9000 in Context

CHAPTER PREVIEW

This chapter is aimed at everyone with an interest in ISO 9000, students, consultants, auditors, quality managers and most importantly the decision makers, CEOs, COOs and Managing Directors. No reader should pass by this chapter without getting an appreciation of what ISO 9000 is all about.

Many people have started their journey towards ISO 9001 certification by reading the standard and trying to understand the requirements. They get so far and then call for help, but they often haven't learnt enough to ask the right questions. The helper might make the assumption that you already know why you are looking at ISO 9001 and therefore may not spend the necessary time for you to understand what it is all about, what pitfalls may lie ahead and whether indeed you need to make this journey at all. This will become clear when you read this chapter.

When you encounter ISO 9001 for the first time, it may be in a conversation, on the Internet, in a leaflet or brochure from your local chamber of commerce or as many have done, from a customer. The source of the message is important as it will influence your choice of strategies:

- Our customers complain about the quality of the products and services we provide.
- We can't demonstrate to our customers' satisfaction that we have the capability of meeting their requirements.
- We need to get ISO 9001 certification as we are losing orders to our competitors that are ISO 9001 registered.
- We need to get ISO 9001 certification to trade within or with Europe.

and if you are busy manager you could be forgiven for either putting it out of your mind or getting someone else to look into it. But you know that as a manager you are either maintaining the status quo or changing it and if you stay with the status quo for too long, your organization will go into decline. So you need to know:

- What the issue is?
- Why this is an issue?
- What this is costing us?
- What you should do about it?
- What the impact of this will be?
- How much it will cost?
- Where the resources are going to come from?

ISO 9000 Quality Systems Handbook
Copyright © 2009, David Hoyle. Published by Elsevier Ltd. All rights reserved.

- When you need to act?
- What the alternatives are and their relative costs?
- What the consequences are of doing nothing?

In this chapter we put ISO 9000 in context by showing the link between ISO 9000 and the fundamental basis for trade and in particular:

- Why customers need confidence and how they go about getting it
- Why organizations need capability and how they go about getting it
- The principles on which the ISO 9000 family of standards has been based.
- The key requirements which underpin the structure of the standards.
- The growth in ISO 9001 certification.
- Where ISO 9001 features in EU Directives

MAKING THE LINK

Since the dawn of civilization the survival of communities has depended on trade. As communities grow they become more dependent on others providing goods and services they are unable to provide from their own resources. Trade continues to this day on the strength of the customer–supplier relationship. The relationship survives through trust and confidence at each stage in the supply chain. A reputation for delivering a product or a service to an agreed specification, at an agreed price on an agreed date is hard to win and organizations will protect their reputation against external threat at all costs. But reputations are often damaged not by those outside but by those inside the organization and by other parties in the supply chain. Broken promises, whatever the cause, harm reputation and promises are broken when an organization does not do what it has committed to do. This can arise either because the organization accepted a commitment it did not have the capability to meet or it had the capability but failed to manage it effectively.

This is what the ISO 9000 family of standards is all about. It is a set of criteria that can be applied to all organizations regardless of type, size and product or service provided. When applied correctly these standards will help organizations develop the capability to create and retain satisfied customers in a manner that satisfies all the other stakeholders. They are not product standards – there are no requirements for specific products or services – they contain criteria that apply to the management of an organization in satisfying customer needs and expectations in a way that satisfies the needs and expectations of other stakeholders.

ISO standards are voluntary and are based on international consensus among the experts in the field. ISO is a non-governmental organization and it has no power to enforce the implementation of the standards it develops. It is a network of the national standards institutes of 160 countries and its aim is to facilitate the international coordination and unification of industrial standards.

By far, the majority of internationally agreed standards apply to specific types of products and services with the aim of ensuring interchangeability, compatibility, interoperability, safety, efficiency and reduction in variation. Mutual recognition of standards between trading organizations and countries increases confidence and decreases the effort spend in verifying that suppliers have shipped acceptable products.

The ISO 9000 family of standards is just one small group of standards among the 17,000 internationally agreed standards and other types of normative documents in ISO's portfolio.

> **In a Nutshell**
>
> The ISO 9000 family of standards will stop you making promises you can't fulfil and help you keep those you can.
>   Broken promises often lead to conflict.

The standards in the ISO 9000 family provide a vehicle for consolidating and communicating concepts in the field of quality management. It is not their purpose to fuel the certification, consulting, training and publishing industries. The primary users of the standards are intended to be organizations acting as either customers or suppliers. Although all ISO standards are voluntary, one of the standards in the ISO 9000 family has become a market requirement. This standard is ISO 9001 which is analysed in detail in Parts 3–7 of this book. However, the primary purpose of these standards is to give confidence to customers that products and services meet the needs and expectations of customers and other stakeholders and improve the capability of organizations to do this.

You don't need to use any of the standards in the ISO 9000 family in order to develop the capability of satisfying your stakeholders, there are other models but none are as detailed or as prescriptive.

A QUEST FOR CONFIDENCE

Customers need confidence that their suppliers can meet their quality, cost and delivery requirements and have a choice as to how they acquire this confidence. They can select their suppliers:

a) Purely on the basis of past performance, reputation or recommendation;
b) By assessing the capability of potential suppliers themselves;
c) On the basis of an assessment of capability performed by a third party.

Most customers select their suppliers using option (a) or (b), but there will be cases where these options are not appropriate either because there is no evidence for using option (a) or resources are not available to use option (b) or it is not economic. It is for these situations that a certification scheme was developed. Organizations submit to a third party audit that is performed by an accredited certification body independent of both customer and supplier. An audit is performed against the requirements of ISO 9001 and if no nonconformities are found, a certificate is awarded. This certificate provides evidence that the organization has the capability to meet customer and regulatory requirements relating to the supply of certain specified good and services. Customers are now able to acquire the confidence they require, simply by establishing whether a supplier holds a current ISO 9001 certificate covering the type of products and services they are seeking. However, the credibility of the certificate rests on the competence of the auditor and the integrity of the certification body, neither of which are guaranteed. (This is addressed further in *System assessment – Chapter 39.*)

Case Study - Trust in the System

China's biggest milk powder manufacturer sold contaminated milk after two brothers who ran a milk collection station added melamine to milk that they sold on from farmers to increase its protein content. Four babies have died and thousands have fallen ill after drinking milk tainted with toxic melamine. Melamine is an industrial chemical used to make plastic cups and saucers.

Providing safe food for children is one of the most basic services of any economy. The toxin was introduced in a ploy by farmers to boost the apparent protein content of the milk that they sold to one of the best-known milk powder manufacturers in the country.

The milk had been rejected several times previously by the manufacturer. Quality controls were ineffective. The milk producer had been told by its joint venture partner in New Zealand to recall the product but the local authorities in China would not do it.

Twelve months prior to the incident the former head of the food and drug agency was executed for taking bribes and as a result the people were told they could trust what they buy.

A QUEST FOR CAPABILITY

Trading organizations need to achieve sustained success in a complex, demanding and ever changing environment. This depends on their capability to:

a) Identify the needs and expectations of their customers and other stakeholders;
b) Convert customer needs and expectations into products and services that will satisfy them;
c) Attract customers to the organization;
d) Supply the products and services that meet customer requirements and deliver the expected benefits;
e) Operate in a manner that satisfies the needs of the other stakeholders.

Food for Thought

Many ISO 9000 registered organizations fail to satisfy their customers but this is largely their own fault – they simply don't do what they say they will do.

If your management is not prepared to change its values, it will always have problems with quality.

Many organizations develop their own ways of working and strive to satisfy their customers in the best way they know how. We will explain this further in more detail but in simple terms the management system is the set of processes that enables the organization to do (a)–(e) and includes both a technical capability and a people capability. Many organizations develop the technical capability but not the people capability and are thus forever struggling to do what they say they will do.

In choosing the best system for them, they can either go through a process of trial and error, select from the vast body of knowledge on management, or utilize one or more management models available that combine proven principles and concepts to develop the organization's capability. ISO 9001 represents one of these models. Others are Business Excellence Model, Six Sigma and Business Process Management (BPM) which we address later in this chapter.

Having given the organization the capability to do (a)–(e), in many business-to-business relationships, organizations are able to give their customers confidence in their capability without becoming registered to ISO 9001. In some market sectors there is a requirement to demonstrate capability through independently regulated conformity assessment procedures before goods and services are purchased. In such cases the organization has no option, but to seek ISO 9001 certification if it wishes to retain business from that particular customer or market sector.

However, it is important to recognize that there is no requirement in the ISO 9000 family of standards for certification. The standards can be used in helping an organization discover the right things to do as well as assess for itself the extent to which its goals and processes meet international standards. Only where customers are imposing ISO 9001 in purchase orders and contracts, would it be necessary to obtain ISO 9001 certification.

THE UNDERLYING PRINCIPLES

The Assurance Principles

The quest for confidence through regulated standards evolved in the defence industry. Defence quality assurance standards (see Chapter 4) were based on the following principles:

- It is essential that products and services be designed, manufactured and provided so as to conform to the purchaser's requirements and this be effected as economically as practicable;
- The quality of products and services depends upon the contractor's control of design, manufacture and other operations that affect quality;
- The contractor needs to institute such control over quality as is necessary to ensure the products and services conform to the purchaser's contractual requirements;
- Contractors need to be prepared to substantiate by objective evidence that they have maintained control over the design, development and manufacturing operations and have performed inspection which demonstrates the acceptability of products and services;
- The purchaser needs to stipulate the assurance required to ascertain that the contractor has control over the operations to be carried out and will ensure that the products and services are properly produced and inspected.

These are the assurance principles, the intent of which is to deliver confidence to customers that the products and services will be or are what they are claimed to be and will be, are being and have been produced under controlled conditions.

By including the phrase "…and this be effected as economically as practicable", there is an implication that contractor's controls were to embrace quality, cost and delivery but other than in Mil-Q-9858A neither the AQAPs nor Def Stans included any requirements for maintaining and using quality cost data. By including the phrase "control of … other operations that affect quality", there is an implication that contractors' controls were to address the processes that created and maintained the working environment but none of the standards included requirements on this topic and it was not until ISO 9001:2000 that requirements for managing the "work environment" were introduced.

The economics of quality was brought out in ISO 9000:1987 in which the intent was stated as:

"to achieve and sustain good economic performance through continual improvement in the customer specification and the organizational system to design and produce the product or service to satisfy the user's needs or requirements"

It was believed that if organizations were able to demonstrate that they were operating a quality system that met international standards, customers would gain greater confidence in the quality of products they purchased. Clearly the intent of these standards was that when implemented, the contractor could be relied upon to meet customer requirements in the most economical manner. The implication being that delivery of nonconforming product or service, late delivery and cost overruns would be an extremely rare event.

The Management Principles

If we ask ourselves, "On what does the achievement of quality depend?" we will find that it rather depends upon our point of view.

In Deming's seminar on Profound Knowledge (c. 1987) he suggested that if you ask people to answer "Yes or No" to the question "Do you believe in quality?" no one would answer "No". They would also know what to do to achieve it and he cited a number of examples which have been put into the cause and effect diagram shown in Fig. 1-1. It is the causes below the labels on the ends of each line that are the determinants. The labels simply categorize the causes.

Deming regarded these factors as all wrong. Either singularly or all together they will not achieve quality. They all require money or learning a new skill and as Lloyd Dobyns (Deming's collaborator on the video library) says "They allow management to duck the issue". However, he tells us that the fact that they won't work does not mean each of them is wrong. Once the processes are predictable and the system is stable, a technique such as Just in Time is a smart thing to do.

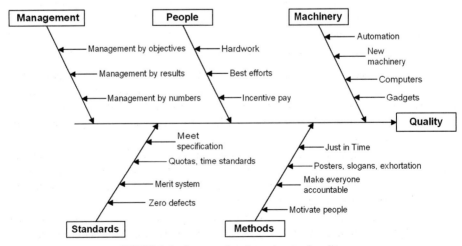

FIGURE 1-1 Inappropriate determinants of quality.

We need principles to help us determine the right things to do and understand why we do what we do. The more prescription we have, the more we get immersed in the detail and lose sight of our objectives – our purpose – our reason for doing what we do. Once we have lost sight of our purpose, our actions and decisions follow the mood of the moment. They are swayed by the political climate or fear of reprisals.

Since the dawn of the industrial revolution, when man came out of the fields into the factories, management became a subject for analysis and synthesis in an attempt to discover some analytical framework upon which to build managerial excellence. There emerged many management principles and indeed they continue to emerge in an attempt to help managers deal with the challenges of management more effectively.

Principles or Rules

A principle is a fundamental law, truth or assumption that is verifiable. Management principles are a guide to action; they are not rules. "No entry to unauthorized personnel" is a rule that is meant to be obeyed without deviation, whereas a principle is flexible, it does not require rigid obedience. A principle may not be useful under all conditions and a violation of a principle under certain conditions may not invalidate the principle for all conditions.

A violation of a principle results in consequences usually by making operations more inefficient or less effective, but that may be a price worth paying under certain circumstances. In order to make this judgement, managers need a full understanding of the consequence of ignoring the principles.

Over the last 20 years a number of principles have been developed that appear to represent the factors upon which the achievement of quality depends:

1. Understanding customer needs and expectations, i.e., a customer focus;
2. Creating a unity of purpose and a quality culture, i.e., leadership;
3. Developing and motivating the people, i.e., involvement of people;
4. Managing processes effectively, i.e., the process approach;
5. Understanding interactions and interdependencies, i.e., the systems approach;
6. Continually seeking better ways of doing things, i.e., continual improvement;
7. Basing decisions on facts, i.e., the factual approach;
8. Realizing that you need others to succeed, i.e., mutual beneficial relationships.

These eight factors represent the causes of quality as shown in the cause and effect diagram of Fig. 1-2. A failure either to understand the nature of any one of these factors or to manage them effectively will invariably lead to a quality failure, the consequences of which may be disastrous for the individual, the customer, the organization, the country and the planet.

A quality management principle is defined by ISO/TC 176 as *a comprehensive and fundamental rule or belief, for leading and operating an organization, aimed at continually improving performance over the long term by focusing on customers while addressing the needs of all other interested parties*. It is a pity that this definition includes the word "rule" because principles are not rules (as this implies inflexibility), but guides to action, implying flexibility and judgement as to their appropriateness.

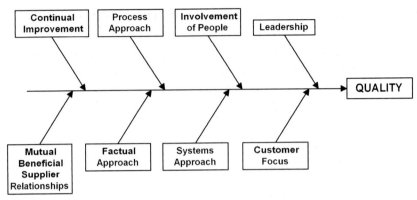

FIGURE 1-2   The eight-quality management principles.

All the requirements of ISO 9001:2008 are related to one or more of these principles. These principles provide the reasons for the requirements and are thus very important. Each of these is addressed below.

Customer Focus

This principle is expressed as follows.

Organizations depend on their customers and therefore should understand current and future customer needs, meet customer requirements and strive to exceed customer expectations.

An organization applying the customer focus principle would be one in which people:

- Understood customer needs and expectations;
- Met customer requirements in a way that met the needs and expectations of all other stakeholders;
- Communicated these needs and expectations throughout the organization;
- Have the knowledge, skills and resources required to satisfy the organization's customers;
- Measured customer satisfaction and acted on results;
- Understood and managed customer relationships;
- Could relate their actions and objectives directly to customer needs and expectations;
- Were sensitive to customer preferences and acted in a way that put the customer first.

Leadership vs Management

Leadership means making sure that the organization is doing the right things. Management means making sure that those things are being done right.

(John M Bryson)

Leadership

This principle is expressed as follows.

Leaders establish unity of purpose and direction for the organization. They should create and maintain the internal environment in which people can become fully involved in achieving the organization's objectives.

An organization applying the leadership principle would be one in which leaders are:

- Establishing and communicating a clear vision of the organization's future;
- Establishing shared values and ethical role models at all levels of the organization;
- Being proactive and leading by example;
- Understanding and responding to changes in the external environment;
- Considering the needs of all stakeholders;
- Building trust and eliminating fear;
- Providing people with the required resources and freedom to act with responsibility and accountability;
- Promoting open and honest communication;
- Educating, training and coaching people;
- Setting challenging goals and targets aligned to the organization's mission and vision;
- Communicating and implementing a strategy to achieve these goals and targets;
- Using performance measures that encourage behaviour consistent with these goals and targets.

Involvement of People

This principle is expressed as follows.

People at all levels are the essence of an organization and their full involvement enables their abilities to be used for the organization's benefit.

An organization applying the involvement of people principle would be one in which people are:

- Accepting ownership and responsibility to solve problems;
- Actively seeking opportunities to make improvements;
- Actively seeking opportunities to enhance their competencies, knowledge and experience;
- Freely sharing knowledge and experience in teams and groups;
- Focusing on the creation of value for customers;
- Being innovative and creative in furthering the organization objectives;
- Better representing the organization to customers, local communities and society at large;
- Deriving satisfaction from their work;
- Enthusiastic and proud to be part of the organization.

Process Approach

This principle is expressed as follows.

A desired result is achieved more efficiently when related resources and activities are managed as a process.

An organization applying the process approach principle would be one in which people:

- Know the objectives they have to achieve and the process that will enable them to achieve these results;
- Know what measures will indicate whether the objectives have been achieved;

- Have clear responsibility, authority and accountability for the results;
- Perform only those activities that are necessary to achieve these objectives and deliver these outputs;
- Assess risks to success and put in place measures that eliminate, reduce or control these risks;
- Know what resources, information and competences are required to achieve the objectives;
- Know whether the process is achieving its objectives as measured;
- Find better ways of achieving the process objectives and of improving process efficiency;
- Regularly confirm that the objectives and targets they are aiming for remain relevant to the needs of the organization.

Systems Approach to Management

This principle is expressed as follows.

Identifying, understanding and managing interrelated processes as a system contributes to the organization's effectiveness and efficiency in achieving its objectives.

An organization applying the system approach principle would be one in which people:

- Are able to visualize the organization as a system of interacting processes;
- Structure the system to achieve the objectives in the most efficient way;
- Understand the interactions and interdependencies between the processes in the system;
- Derive process objectives from system objectives;
- Understand the impact of their actions and decision on other processes with which they interface and on the organization's goals;
- Establish resource constraints prior to action.

Continual Improvement

This principle is expressed as follows.

Continual improvement of the organization's overall performance should be a permanent objective of the organization.

An organization applying the continual improvement principles would be one in which people are:

- Making continual improvement of products, processes and systems - an objective for every individual in the organization;
- Applying the basic improvement concepts of incremental improvement and break-through improvement;
- Using periodic assessments against established criteria of excellence to identify areas for potential improvement;
- Continually improving the efficiency and effectiveness of all processes;
- Promoting prevention-based activities;
- Providing every member of the organization with appropriate education and training, on the methods and tools of continual improvement;

- Establishing measures and goals to guide and track improvements;
- Recognizing improvements.

Factual Approach to Decision Making

This principle is expressed as follows.

Effective decisions are based on the analysis of data and information.

An organization applying the factual approach principle would be one in which people are:

- Defining performance measures that relate to the quality characteristics required for the process, product or service being measured;
- Taking measurements and collecting data and information relevant to the product, process or service objective;
- Ensuring that the data and information are sufficiently accurate, reliable and accessible;
- Analysing the data and information using valid methods;
- Understanding the value of appropriate statistical techniques;
- Making decisions and taking action based on the results of logical analysis balance with experience and intuition.

Mutually Beneficial Supplier Relationships

This principle is expressed as follows.

An organization and its suppliers are interdependent and a mutually beneficial relationship enhances the ability of both to create value.

An organization applying the supplier relationship principle would be one in which people are:

- Identifying and selecting key suppliers on the basis of their ability to meet requirements without compromising quality;
- Establishing supplier relationships that balance short-term gains with long-term considerations for the organization and society at large;
- Creating clear and open communications;
- Initiating joint development and improvement of products and processes;
- Jointly establishing a clear understanding of customers' needs;
- Sharing information and future plans;
- Recognizing supplier improvements and achievements.

This principle has a dual focus even though it refers to suppliers. The organization is a supplier to its customer and is a customer to its suppliers.

Using the Principles

The principles can be used in validating the design of processes, in validating decisions, in auditing system and processes. You look at a process and ask:

- Where is the customer focus in this process?
- Where in this process are there leadership, guiding policies, measurable objectives and the environment that motivate the workforce to achieve these objectives?

- Where in this process is the involvement of people in the design of the process, the making of decisions, the monitoring and measurement of performance and the improvement of performance?
- Where is the process approach to the accomplishment of these objectives?
- Where is the systems' approach to the management of these processes, the optimisation of performance, the elimination of bottlenecks and delays?
- Where in the process are decisions based on fact?
- Where is there continual improvement in performance, efficiency and effectiveness of this process?
- Where is there a mutually beneficial relationship with suppliers in this process?

Also you can review the actual measures used for assessing leadership, customer relationships, personnel, processes, systems, decisions, performance and supplier relationships for alignment with the principles.

THE REQUIREMENTS

The Basis for the Requirements

The requirements of ISO 9001 have been based on the above principles but not derived from them as the principles and the requirements have evolved in parallel over several years. Work is now underway in ISO to bring greater alignment between the principles, concepts, terminology and requirements but this won't be compete until 2013. ISO 9001 contains over 260 requirements spread over five sections but the way the requirements are grouped creates some anomalies and fails to bring clarity to the structure. All the topics addressed by the requirements are listed in Chapter 38.

Purpose of Requirements

The purpose of these requirements is to provide an assurance of product quality as is apparent from ISO 9001 clause 0.1 where it states *"This International Standard can be used by internal and external parties, including certification bodies, to assess the organization's ability to meet customer, statutory and regulatory requirements applicable to the product, and the organization's own requirements."* They are therefore not intended to be for the purpose of developing a quality management system. If we ask of every requirement, "Would confidence in the quality of the product be diminished if this requirement was not met?" We should, in principle, get an affirmative response, but this might not always be the case as it was not the approach taken in the validation of the standard.

There was a change in direction in 2000 when the ISO 9000 family changed its focus from procedures to processes. This change is illustrated in Fig. 1-3. ISO 9001:2008 clearly positions the system of managed processes as the means for generating conforming products with the intent that these create satisfied customers. ISO 9004:2009 goes further and creates a cycle of sustained success, driven from a mission and vision that is influenced by stakeholder needs and expectations through an extended system of managed processes to produce results that satisfy all stakeholders.

FIGURE 1-3 The changing purpose in the ISO 9000 family.

The Basic Management Requirements

Further on, in this book we comment on the structure of ISO 9001 and the 263 requirements in more detail, but we can condense these into the following seven management requirements:

1. **Purpose** – establish the organization's purpose and the needs and expectations of stakeholders relative to this purpose.
2. **Policy** – define, document, maintain and communicate the overall intentions, principles and values related to quality.
3. **Planning** – establish objectives, measures and targets for fulfilling the organization's purpose and its policies, assessing risks and develop plans and processes for achieving the objectives that take due account of these risks.
4. **Implementation** – resource, operate and manage the plans and processes to deliver outputs that achieve the planned results.
5. **Measurement** – monitor, measure and audit processes, the fulfilment of objectives and policies and satisfaction of stakeholders.
6. **Review** – analyse and evaluate the results of measurement, determine performance against objectives and determine changes needed in policies, objectives, measures, targets and processes for the continuing suitability, adequacy and effectiveness of the system.
7. **Improvement** – undertake action to bring about improvement by better control, better utilization of resources and better understanding of stakeholder needs. This might include innovation and learning.

The Basic Assurance Requirements

We can also condense the requirements of ISO 9001 into five assurance requirements.

1. The organization shall demonstrate its commitment to the achievement of quality.
2. The organization shall demonstrate that it has effective policies for creating an environment that will motivate its personnel into satisfying the needs and expectations of its customers and applicable statutory and regulatory requirements.
3. The organization shall demonstrate that it has effectively translated the needs and expectations of its customers and applicable statutory and regulatory requirements into measurable and attainable objectives.
4. The organization shall demonstrate that it has an effective system of interacting processes for enabling the organization to meet these objectives in the most efficient way.
5. The organization shall demonstrate that it is achieving these objectives as measured, that they are being achieved in the best way and that they remain consistent with the needs and expectations of its customers and applicable statutory and regulatory requirements.

POPULARITY OF ISO 9001 CERTIFICATION

ISO 9001 has gained in popularity since 1987 when the UK led the field holding the highest number of ISO 9001, 9002 and 9003 certificates (ISO 9002 and 9003 became obsolete in 2003). Since then certification in the UK has declined from a peak of 66,760 in 2001 to 35,517 by December 2007 pushing the UK into eighth place. This is largely due to the transfer of manufacturing to the Asian economies and why China has held the lead since 2002. The latest year for which there are published figures is 2007, when 18 of 175 countries (10%) held 80% of the total number if certificates issued as detailed in Table 1-1.[1] As quality system standards for automotive and medical devices (ISO/TS 16949 and ISO 13485) include all requirements from ISO 9001, certification to these standards can be added to the numbers of ISO 9001 certificates. The numbers only include data from certification bodies that are accredited by members of the International Accreditation Forum (IAF). The data is for numbers of certificates issued and not for the number of organizations to which certificates are issued which may be less as some organizations register each location.

The 15-year trend since records began in 1992 is shown in Fig. 1-4. The temporary arrest in growth during the transition from ISO 9001:1994 to ISO 9001:2000 is clearly evident. The continued growth is likely to be due to the European Union Directives which are addressed next.

ISO 9001 AND EU DIRECTIVES

With the formation of the European Union (EU) in 1993 there was a need to remove barriers to the free movement of goods across the Union. One part of this was to harmonize standards. At the time each country had its own standards for testing products and for controlling the processes by which they were conceived, developed and produced. This led to a lack of confidence and consequently to the buying organizations

[1] ISO Survey (December 2007) International Organization for Standardisation, Geneva.

TABLE 1-1 Top 10% of Countries with ISO 9001 and Derivative Certificates

Country	Number of certificates issued				
	ISO 9001	ISO/TS 16949	ISO 13485	Total	% of Total
China	210,773	7732	1329	219,834	21.99%
Italy	115,359	1024	1482	117,865	11.79%
Japan	73,176	1106	456	74,738	7.48%
Spain	65,112	928	40	66,080	6.61%
India	46,091	2008	222	48,321	4.83%
Germany	45,195	3068	2204	50,467	5.05%
USA	36,192	4288	2186	42,666	4.27%
UK	35,517	701	589	36,807	3.68%
France	22,981	1165	709	24,855	2.49%
Netherlands	18,922	120	47	19,089	1.91%
Republic of Korea	15,749	3453	6	19,208	1.92%
Brazil	15,384	972	73	16,429	1.64%
Turkey	12,802	504	52	13,358	1.34%
Russian Federation	11,527	78	28	11,633	1.16%
Switzerland	11,077	115	608	11,800	1.18%
Israel	10,846			10,846	1.08%
Hungary	10,473	257	37	10,767	1.08%
Czech Republic	10,458	526	221	11,205	1.12%
Sub-total	**767,634**	**27,262**	**9264**	**805,968**	**80.62%**
Others				193,701	19.38%
Grand total				**999,669**	**100.00%**

undertaking their own product testing and in addition assessment of the seller's quality management systems.

The EU Council was concerned with protecting its citizens with respect to health, safety and environment and therefore decided that it needed a compliance system that regulated the quality of products flowing into and around the EU. The model prevalent in many countries was the conformity assessment regimes operated by procurement agencies where they did the testing, inspection and system assessment.

In May 1985 the EEC Council passed a resolution on a new approach to technical harmonization and standards (85/C 136/01) that was intended to resolve technical barriers to trade and dispel the consequent uncertainty for economic operators. The new approach provided for reference to standards (preferably European but national if

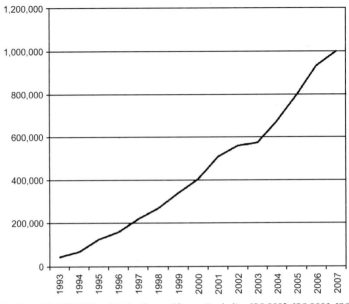

FIGURE 1-4 Growth in ISO 9001 and derivative certificates.*(including ISO 9002, ISO 9003, ISO/TS 16949 & ISO 13485)*

appropriate) for the purposes of defining the technical characteristics of products accompanied by a policy on the assessment of conformity to such standards. These standards would specify the "essential requirements" as regards health protection and safety with which products put on the market must conform. In 1993 the Council adopted 93/465/EEC putting in place the conformity assessment procedures and rules for the affixing and use of the CE conformity marking, that were intended to be used in the technical harmonization directives.

This legislation was repealed in June 2008 when the Council of the European Union adopted a new *Common Framework* for marketing products. This broad package of provisions is intended to remove the remaining obstacles to free circulation of products and represents a major boost for trade in goods between EU Member States. The legislation is intended to bring particular benefits for small and medium sized enterprises (SMEs), who will no longer be discouraged from doing business outside their domestic markets.

ISO 9001 is perceived by the EU Council as ensuring health and safety requirements are met because ISO 9001 now requires organizations to demonstrate that they have the ability to consistently provide product that meets customer and applicable statutory and regulatory requirements. Such requirements would be specified in EU Directives. Where conformity with these directives can be verified by inspection or test of the end product, ISO 9001 is not a requirement for those seeking to supply within and into the EU. Where conformity cannot be ensured without control over design and/or production processes, conformity with ISO 9001 needs to be assessed by a "notified body". The 2008 EU Directives require manufacturers to lodge an application for assessment of their quality system with the notified body of their choice, for the products concerned. A summary of the conformity assessment procedures is shown in Table 1-2. There are rules in Article R17 of the Common Framework regulating "notified bodies".

TABLE 1-2 Conformity Assessment Procedures in EU Legislation

	Module A: Internal production control	Module B: Type examination	Module G: Unit verification	Module H: Full quality assurance
Design	Manufacturer keeps technical documentation at the disposal of national authorities	Manufacturer submits to notified body: • Technical documentation • Supporting evidence for the adequacy of the technical design solution • Specimen(s), representative of the production envisaged, as required Notified body: • Ascertains conformity with essential requirements • Examines technical documentation and supporting evidence to assess adequacy of the technical design • For specimen(s): carries out tests, if necessary Issues EC-type examination certificate	Manufacturer submits technical documentation	EN ISO 9001:2000 (4) Manufacturer: • Operates an approved quality system for design • Submits technical documentation Notified body carries out surveillance of the QMS H1 Notified body: • Verifies conformity of design (1) • Issues EC-design examination certificate (1)

Continued

TABLE 1-2 Conformity Assessment Procedures in EU Legislation—cont'd

Module A: Internal production control	Module C: Conformity to type	Module D: Production quality assurance	Module E: Product quality assurance	Module F: Product verification	Module G: Unit verification	Module H: Full quality assurance
Production A. Manufacturer: • Declares conformity with essential requirements • Affixes required conformity marking	C. Manufacturer: • Declares conformity with approved type • Affixes required conformity marking	EN ISO 9001:2000 (2). Manufacturer: • Operates an approved quality system for production, final inspection and testing • Declares conformity with approved type • Affixes required conformity marking	EN ISO 9001:2000 (3). Manufacturer: • Operates an approved quality system for final inspection and testing • Declares conformity with approved type • Affixes required conformity marking		Manufacturer: • Submits product • Declares conformity • Affixes required conformity marking	Manufacturer: • Operates an approved QMS for production, final inspection and testing • Declares conformity • Affixes required conformity marking

A1. Accredited in-house body or notified body performs tests on specific aspects of the product (1)

A2. Product checks at random intervals (1)

C1. Accredited in-house body or notified body performs tests on specific aspects of the product (1)

C2. Product checks at random intervals (1)

Notified body:
- Approves the QS
- Carries out surveillance of the QMS

D1.
- Declares conformity to essential requirements
- Affixes required conformity marking

Notified body:
- Approves the QS
- Carries out surveillance of the QMS

E1.
- Declares Conformity to essential requirements
- Affixes required conformity marking

Notified body:
- Verifies conformity to essential requirements
- Issues certificate of conformity

Notified body:
- Verifies conformity to essential requirements
- Issues certificate of conformity

Notified body:
Carries out surveillance of the QMS

Notes:

(1) Supplementary requirements which may be used in sectoral legislation
(2) Except for sub-clause 7.3 and requirements relating to customer satisfaction and continual improvement
(3) Except for sub-clauses 7.1, 7.2.3, 7.3, 7.4, 7.5.1, 7.5.2, 7.5.3 and requirements relating to customer satisfaction and continual improvement
(4) Except for requirements relating to customer satisfaction and continual improvement

Chapter 2

Defining and Characterizing Quality

CHAPTER PREVIEW

This chapter is aimed at everyone with an interest in quality, students, consultants, auditors, professionals, and in particular managers formulating and executing strategies that impact the organization's outputs. As the achievement of quality is the raison d'être for the ISO 9000 family, no reader should pass by this chapter without getting an insight into the meaning of quality.

If you are thinking about ISO 9001, you can't get past the front page of the standard without noticing the title in which the first word is *quality* and the second word is *management*. Therefore, it would be unwise to go further without a clear understanding of what quality is and how the achievement of quality is managed. It is also vital that managers have a unified understanding of quality in order to build a coherent strategy for its achievement.

However, in discussions in which the word quality is used, people will differ in their viewpoint either because the word quality has more than one meaning or that they have different perceptions of what the word quality means or because they are drawing conclusions from different premises or concepts. Some of the people are perhaps thinking that quality means goodness or perfection or that quality means adherence to procedure, following the rules etc. or that fewer defects means higher costs or that quality means high class and is expensive. Others might be thinking that controlling quality means rigid systems, inspectors in white coats or that if they push production, quality suffers, or that quality management is what the quality department does.

You may consult ISO 9000 which is invoked in ISO 9001 to gain some appreciation of the concepts and the terms used but this is a rather clinical treatment that does not allow for the wide variation in their application and usage in the real world and this is what this chapter aims to provide.

We examine:

- The different meanings of the term quality in general use; how quality is perceived relative to cost, price, design, reliability and safety and we examine some of the misconceptions that surround the term.
- How differences in design are expressed by class and grade and how they relate to quality
- The characteristics used to measure quality.
- The three dimensions of quality that define a range of meaning through product quality, business quality and enterprise quality.

WHAT IS QUALITY?

Definitions

We are likely to know what quality is when we see or experience it. We are also more likely to ponder the real meaning of the word when we buy something that fails to do what we originally bought it to do. We thus judge quality by making comparisons, based on our own experiences, but defining it in terms that convey the same meaning to others can be difficult. There are dictionary definitions that express how the word *quality* is used but they don't help when we try to take action. When we set out to provide a quality product, formulate a strategy for quality, produce a quality policy, control the quality of something or are faced with an angry customer, we need to know what quality means so that we involve the right people and judge whether the action to be taken is appropriate.

There are a number of definitions in use, each of which is valid when used in a certain context. These are summarized below and then addressed in more detail in the sections that follow.

- A degree of excellence (OED) – The meaning used by the general public.
- Freedom from deficiencies or defects (Juran) – The meaning used by those making a product or delivering a service.
- Conformity to requirements (Crosby) – The meaning used by those designing a product or a service or assessing conformity.
- Fitness for use (Juran) – The meaning used by those accepting a product or service.
- Fitness for purpose (Sales and Supply of Goods Act 1994) – The meaning used by those selling and purchasing goods.
- The degree to which a set of inherent characteristics fulfils requirements (ISO 9000:2005) – The meaning used by those managing or assessing the achievement of quality.
- Sustained satisfaction (Deming) – The meaning used by those in upper management using quality for competitive advantage.

> **Handling Misunderstanding**
>
> It is easy to be deluded into believing there is an understanding when two people use the same words. If you have a disagreement you firstly need to establish what actions and deeds the other person is talking about. Once these are understood, communication can proceed whether or not there is agreement on the meaning of the words. (J. M. Juran)

It therefore becomes important to establish the context of a statement in which the term quality is used, e.g., it would be wrong to say that quality doesn't mean freedom from defects but if the context is a discussion on corporate strategy, it would be foolish to limit ones imagination to that meaning of the word when the purpose of the discussion is to devise a means of gaining a competitive advantage. Even if your products were totally free of deficiencies, you would not gain a competitive advantage if your products lacked the latest features or were not innovative.

Dictionary Definitions

The Oxford English Dictionary (OED) contains 17 meanings, most of which relate to personal characteristics but the following three are relevant to the quality of product:

- A particular class, kind, or grade of something, as determined by its character, especially its excellence. (This meaning is addressed further under Classification of products and services.)
- The standard or nature of something as measured against other things of a similar kind.
- The degree of excellence possessed by a thing.

There are other ways in which we think of quality. Masaaki Imai in his book on Kaizen[1] writes that "when speaking of quality one tends to think first of product quality" and this is indeed the most common context for quality. But Imai goes on to write "when discussed in the context of KAIZEN© strategy the foremost concern is with the quality of people".

Then there is the quality of mercy,[2] the quality of life, the quality of education etc. and in all these cases we are invoking a definition of quality that leans more towards the degree of excellence that is expressed in the OED. It is helpful to remember that dictionaries record common usage and implied meanings not legally correct definitions or definitions resulting from the deliberations of a team of experts. The latter two meanings above are embodied in the more formal definitions that follow.

Freedom from Defects or Deficiencies

The idea that quality means freedom from defects or deficiencies is based on the premise that the fewer the errors, the better the quality so a product with zero defects is a product of superior quality. A defect is *nonconformity with a specified requirement*. Therefore, if the requirement has been agreed with the customer, a defect free product should satisfy the customer. However, at the level where decisions on nonconformity are made, the requirement is likely to be the supplier's own specification and might not address all product characteristics necessary to reflect customer needs; therefore, a defect free product might not be the one with characteristics that satisfy customers.

Juran[3] contrasts two definitions of quality that of freedom from deficiencies and product features which meet customer requirements. He observes that:

- Product features impact sales so higher quality in this sense usually costs more;
- Product deficiencies impact costs so higher quality in this sense usually costs less;

In the eyes of the customer, they see only one kind of quality. The product has to satisfy their needs and expectations and this means that it should possess all the necessary features and be free of deficiencies. It would be foolish to simply focus on reducing defects as a quality strategy because as Deming remarked,[4] reducing defects does not keep the plant open. Innovation is necessary to create new product features to maintain customer loyalty.

Conformity to Requirements or Specification

The idea that quality means conformance or conformity to the requirements is based on the premise that if a product conforms to all the requirements for that product, it is

---

[1] Imai Masaaki (1986) Kaizen The key to Japanese competitive success.

[2] Shakespeare William. Merchant of Venice Act IV Scene 1.

[3] Juran J.M. (1992) Juran on quality by design. The Free Press, Division of Macmillan Inc.

[4] Deming W. Edwards (2000) The New Economics. MIT Press.

a *quality product*. This was the view of the American Quality Guru, Philip B. Cosby[5] in his book 'Quality is Free'. It became one of his four absolutes of quality.[6] This approach depends on the customer or the supplier defining all characteristics that are essential for the product or service to be fit for its use under all conditions it will be used. However, it removes the subjectivity associated with words like goodness, perfection, excellence and eliminated opinions and feelings. It means that no one is in any doubt as to what has to be achieved.

The implication with this definition is that should a product not conform to the specified requirements it will be rejected and deemed poor quality when it might well satisfy the customer. It led Rolls Royce aero engines to declare in the 1980s its Quality Policy as *"Meet the requirements or cause them to be changed"* in order to prevent products being rejected for trivial reasons. There was and still is a tendency with this definition to pursue ever more detailed requirements in an attempt to capture every nuance of customer needs by defining what is and what is not acceptable. Where customer requirements are very detailed it means that the simplest decision on fitness for use has to be deferred to the customer rather than being made locally. However, the specification is often an imperfect definition of what a customer needs. Some needs can be difficult to express clearly and by not conforming, it doesn't mean that the product or service may be unsatisfactory to the customer.

Conformance to the requirements can be an appropriate definition at the operational level where customer needs have been translated into requirements to levels where acceptance decisions are made such as receipt inspection, component test, assembly inspection. Crosby was credited with a 25% reduction in the overall rejection rate and a 30% reduction in scrap costs[7] so understanding quality as conformity to the requirements can bring significant benefits for the supplier and the customer.

It is also possible that a product that conforms to requirements may be unfit for use. It all depends on whose requirements are being met. Companies often define their own requirement as a substitute for conducting in depth market research and misread the market. On the other hand, if the standards are well in excess of what the customer requires, the price may well be much higher than what customers are prepared to pay – there probably isn't a market for a gold-plated mousetrap, except as an ornament perhaps!

The conformance to requirements definition relies on there being requirements with which to conform. The definition does not recognize potential requirements or future needs or wants so as a strategy; it is rooted in the present.

Fitness for Use

The idea that quality means fitness for use is based on the premise that an organization will retain satisfied customers only if it offers for sale products or services that respond to the needs of the user in terms of price, delivery and fitness for use. Juran[8] defined fitness for use as the extent to which the product or service successfully serves the

---

[5] Crosby Philip B. (1979) Quality is Free. McGraw-Hill Inc.

[6] Crosby Philip B. (1986) Quality without tears – The art of hassle-free management. McGraw-Hill Inc.

[7] http://en.wikipedia.org/wiki/Phil_Crosby.

[8] Juran J.M. (1974) Quality Control Handbook 3rd Edition. McGraw-Hill Inc.

purpose of the user during usage (not just at the point of sale) and rather than invent a word for this concept settled on the word 'quality' as being acceptable for this purpose.

Extract from UK Sale and Supply of Good Act 1994, Chapter 35, Section 1

Where the seller sells goods in the course of a business, there is an implied term that the goods supplied under the contract are of satisfactory quality.

For the purposes of this Act, goods are of satisfactory quality if they meet the standard that a reasonable person would regard as satisfactory, taking account of any description of the goods, the price (if relevant) and all the other relevant circumstances.

For the purposes of this Act, the quality of goods includes their state and condition and the following (among others) are in appropriate cases aspects of the quality of goods:

- fitness for all the purposes for which goods of the kind in question are commonly supplied,
- appearance and finish,
- freedom from minor defects,
- safety, and
- durability.

It is interesting to note that Juran did not sit down and ponder on what the word quality meant. He had identified a concept then looked around for a label he could use that would adequately convey his intended meaning. It is only in the ensuing decades that the word quality has been abused and misused.

Juran[9] later recognized that fitness for use definition did not provide the depth for managers to take action and conceived of two branches: product features that meet customer needs and freedom from deficiencies. Nonetheless, as a strategy this definition is also rooted in the present and does not take into account the future needs of customers.

Fitness for Purpose

The UK Sales and Supply of Goods Act 1994, Chapter 35, makes provision as to the terms to be implied in certain agreements for the transfer of property and other trans-actions. An extract from this Act is contained in the boxed text. This definition for quality appears to be based on the premise that quality is a standard that a reasonable person would regard as satisfactory, taking account of any description of the goods, the price (if relevant) and all the other relevant circumstances. The only notion excluded is that of delighting customers but that is where some organizations develop a competitive advantage.

Internationally Agreed Definitions

In 1987, ISO 8402 defined quality as *the totality of characteristics of an entity that bear on its ability to satisfy stated or implied needs.* Although superseded by the definition in ISO 9000:2005 below, in principle it remains relevant even if a little verbose.

---

[9] Juran J.M. (1989) Juran on Leadership for quality. The Free Press, Division of Macmillan Inc.

The problem with it was the term 'entity' which was partially overcome by the new definition in 2000 which is:

The degree to which *a set of inherent characteristics fulfils requirements (ISO 9000:2005).*

This new definition appears to be a retrograde step as it mentions requirements rather than needs, thus arching back to an era where conformity to requirements was the accepted norm. However, we can remove the implied limitation, by combining the definition of the terms *quality* and *requirement* in ISO 9000:2005, and therefore quality can be expressed as *the degree to which a set of inherent characteristics fulfils a need or expectation that is stated, generally implied or obligatory".*

This implies that quality is relative to what something should be and what it is. The something maybe a product, service, decision, document, piece of information or any output from a process.

This means that when we talk of anything using the word quality it simply implies that we are referring to the extent or degree to which a need or expectation is met. It also means that all the principles, methodologies, tools and techniques in the field of quality management serve one purpose, that of enabling organizations to close the gap between the standard required and the standard reached and if desirable, exceed them. In this context, performance, environmental, safety, security and health problems are in fact quality problems because an expectation or a requirement has not been met. If the expectation had been met there would be no problem.

The definition appears to be rooted in the present because it makes no acknowledgement as to whether the 'needs' are present needs or future needs but if we imagine that customers expect continual improvement including innovation, then the definition is sound.

Sustained Satisfaction

Deming wrote that *a product or service possesses quality if it helps somebody and enjoys a good and sustainable market.*[10]

If organizations produce products and service that satisfy their customers and a satisfied customer is deemed as one who does not complain, then the customer may choose a competitor's product next time, not because of dissatisfaction with the previous organization's products but because a more innovative product came on to the market. Even happy customers and loyal customers will switch to suppliers offering innovative products. This does not arise from meeting present customer needs and expectations; it arises from not recognizing that markets change.

Before the age of mobile phones customers were not hammering on the door of the telephone companies demanding mobile phones, before we had video recorders that could pause live TV we were watching, we were not demanding digital video recorders with hard drives; these innovations arose because the designers looked for better and different solutions that would make life easier for their customers. The innovations do not have to involve high technology. It has now become common place in the UK for restaurants to provide chocolate mints after a meal. For a while it delighted customers as they were not expecting it but once it became the norm, its

---

[10] Deming W. Edwards (2000) The New Economics, page 2. MIT Press.

power to delight has diminished and so the restaurant trade has to look to other innovations to keep the customers coming through the door. In business-to-business relationships a quality service is not simply satisfying customers, but enabling your customers to be more successful with their business by using your services. At Lockheed Martin, they say that the core purpose of their corporation is to achieve mission success which they define by saying that *"mission success is when we make our customers successful"*.

Sustained satisfaction therefore takes the meaning of quality beyond the present and attempts to secure the future.

Satisfactory and Unsatisfactory Quality

The definition of quality in ISO 9000:2005 contains the notion of 'degree' implying that quality is not an absolute but a variable. This concept of 'degree' is present in the generally accepted definition of quality in the Oxford English Dictionary and is also implied in the UK Sales and Supply of Goods Act through the phrase 'satisfactory quality'. The concept of 'degree' is illustrated in Fig. 2-1. The diagram expresses several truths:

> **The Customer Decides**
>
> In the final analysis it is the customers who set the standards for quality and they do this by deciding which products to purchase and whom to buy them from.
> (From Kaizen by Masaaki Imai)

- Needs, requirements and expectations are constantly changing;
- Performance needs to be constantly changing to keep pace with the needs;
- Quality is the difference between the standard stated, implied or required and the standard reached;
- Satisfactory quality is where the standard reached is within the range of acceptability defined by the required standard;
- Superior quality is where the standard reached is above the standard required;
- Inferior quality is where the standard reached is below the standard required,

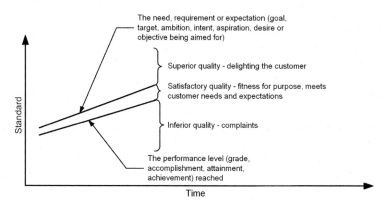

The need, requirement or expectation (goal, target, ambition, intent, aspiration, desire or objective being aimed for)

Superior quality - delighting the customer

Satisfactory quality - fitness for purpose, meets customer needs and expectations

Inferior quality - complaints

The performance level (grade, accomplishment, attainment, achievement) reached

Standard

Time

FIGURE 2-1 The meaning of quality.

We need to express our relative satisfaction with products and services and as a consequence use subjective terms. When a product or service satisfies our needs we are likely to say it is of *good quality* or *satisfactory quality* and likewise when we are dissatisfied we say the product or service is of *poor quality* or of *inferior quality*. When the product or service exceeds our needs we will probably say it is of *high quality* or *superior quality* and likewise if it falls well below our expectations we say it is of *low or unsatisfactory quality*.

Products or services that do not possess the right features and characteristics either by design or by construction are products of poor quality. Those that fail to give customer satisfaction by being uneconomic to use are also products of poor quality, regardless of their conformance to specifications. Often people might claim that a product is of good quality but of poor design, or that a product is of good quality but it has a high maintenance cost. A product may not need to possess defects for it to be regarded as poor quality, for instance it may not possess the features that we would expect, such as access for maintenance. These are design features that give a product its saleability. Products and services that conform to customer requirements are considered to be products of acceptable quality. If an otherwise acceptable product has a blemish – is it now unacceptable? Perhaps not because it may still be far superior to other competing products in those features and characteristics that are acceptable.

For companies supplying products and services, a more precise means of measuring quality is needed. To the supplier, a quality product is the one that meets in full the perceived customer requirements. To the customer, a quality product is one that meets in full the stated customer requirements and it is the supplier's responsibility to ensure that the perceived and stated requirements are within the range of acceptability.

Satisfaction and dissatisfaction are not necessarily opposites as observed by Juran[11] and Deming.[12] There are many products that conformed to requirements and were fit for use and free of defects when produced but no longer satisfy customers because they target a market that has changed, magnetic tape recorders, carburettors, carbon paper, valve radios are a few examples. They did satisfy large numbers of customers at one time but have been replaced by devices offering different functionality and greater satisfaction. Therefore, when judging the quality of a product you need to be sure you are judging competing alternatives (for further discussion see Classification of products and services).

Attainment Levels of Quality

The definitions we have examined all have their place. None of them is entirely incorrect – they can all work but they suggest that there are levels of attainment with respect to quality as shown in the text box.

If we perceive quality as freedom from deficiencies or defects we are limiting our understanding of quality to the current and local requirements. We will lose customers if the local requirements don't align with the customer requirements. We will also reduce costs with this mindset but we will only retain customers for as long as our products are valued.

If we perceive quality as conformity with customer requirements, we recognize that a conforming product is one that is free of deficiencies and meets all local and customer

---

[11] Juran J.M. (1992) Juran on quality by design. The Free Press, Division of Macmillan Inc.
[12] Deming W. Edwards (2000) The New Economics, page 9, MIT Press.

requirements. We are however, limiting our understanding of quality to the current customer requirement and not future needs. With this mindset we will reduce costs and retain more customers but again only as long as our products are valued.

Attainment Levels of Quality

I. Freedom from deficiencies which require better controls and result in lower costs but does not necessarily retain satisfied customers.

II. Conformity with customer requirements which requires capable processes and results in lower costs but does not necessarily retain satisfied customers.

III. Satisfying customer's needs and expectations which requires innovation as well as capable processes and results in lower operating costs and higher development costs but in return creates and retains satisfies customers and leads to sustained success.

If we perceive quality as satisfying customer's needs and expectations, we recognize that a quality product is the one that is free of deficiencies, conforms to customer requirements and satisfies customer needs and expectations. We are not limiting our understanding of quality to current requirements and thus take in future needs and expectations. For example, the customer may not have a requirement to pause live TV but once you make him aware that this is now available, it becomes a customer need and after a month or two, he finds he can't live without it and any other supplier that cannot offer this feature is not even considered. With this mindset we will reduce production costs but increase research and development costs but the bonus is that we will also create and retain more customers for as long as we can continue to innovate.

QUALITY IN CONTEXT

Noun or Adjective

Ordinarily, quality is a noun but is often used as an adjective. The used car dealer displays the placard 'Quality Used Car' on every vehicle to indicate that their condition is of a high standard; the carpet warehouse advertises 'Quality Carpets' indicating that they stock a range of carpets that are suitable for different uses. The seller neither designed nor manufactured the product but nonetheless claim their products to be quality products. These are examples where the word 'quality' comes before the noun and is thus being used as an adjective to give the impression that the products are superior in some way.

Where the word 'quality' comes after a noun, it describes the condition or properties of something. For example, 'air quality' describes the condition of the air in a particular place and time, reflecting the degree to which it is pollution free; 'water quality' described the chemical, physical, and biological characteristics of a particular water-body, usually in relation to its suitability for a particular use.[13]

---

[13] A Dictionary of Environment and Conservation (2007) Oxford University Press.

Classification of Products and Services

If we group products and services by type, category, class and grade, we can use the subdivision to make comparisons on an equitable basis. But when we compare entities we must be careful not to claim one is of better quality than the other unless they are of the same grade. Entities of the same type have at least one attribute in common. Entities of the same grade have been designed for the same functional use and therefore comparisons are valid. Comparisons on quality between entities of different grades, classes, categories or types are invalid because they have been designed for a different use or purpose.

Let us look at some examples to illustrate the point. Food is a type of entity. Transport is another entity. Putting aside the fact that in the food industry the terms *class* and *grade* are used to denote the condition of post-production product, comparison between *types* is like comparing fruit and trucks, i.e., there are no common attributes. Comparisons between *categories* are like comparing fruit and vegetables. Comparisons between *classes* are like comparing apples and oranges. A comparison between grades is like comparing eating apples and cooking apples.

Now let us take another example. Transport is a type of entity. There are different categories of transport such as airliners, ships, automobiles and trains; they are all modes of transport but each has many different attributes. Differences between categories of transport are therefore differences in *modes* of transport. Within each category there are differences in class. For manufactured products, differences between classes imply differences in *purpose*. Luxury cars, large family cars, small family cars, vans, trucks, four-wheel drive vehicles etc. fall within the same category of transport but each was designed for a different purpose. Family cars are in a different class to luxury cars; they were not designed for the same purpose. It is therefore inappropriate to compare a Cadillac with a Chevrolet or a Rolls Royce Silver Shadow with a Ford Mondeo. Entities designed for the same purpose but having different specifications are of different grades. A Ford Mondeo GTX is of a different grade to that of a Mondeo LX. They were both designed for the same purpose but differ in their performance and features and hence comparisons on quality are invalid.

A third example in the service industry would be; accommodation. There are various categories, such as rented, leased and purchased. In the rented category there are hotels, inns, guesthouses, apartments etc. It would be inappropriate to compare hotels with guesthouses or apartments with inns. They are each in a different class. Hotels are a class of accommodation within which are grades such as five stars, four stars, three stars etc. indicating the facilities offered not quality levels. It would therefore be reasonable to expect a one-star hotel to be just as clean as a four-star hotel.

You can legitimately compare the quality of entities if comparing entities of the same grade. If a low-grade product or service meets the needs for which it was designed, it is of the requisite quality. If a high-grade product or service fails to meet the requirements for which it was designed, it is of poor quality, regardless of it still meeting the requirements for the lower grade. There is a market for such differences in products and services but should customer's expectations change then what was once acceptable for a particular grade may no longer be acceptable and regrading may have to occur.

Where manufacturing processes are prone to uncontrollable variation it is not uncommon to grade products as a method of selection. The product that is free of imperfections would be the highest grade and would therefore command the highest price. Any product with imperfections would be downgraded and sold at a correspondingly lower price. Examples of such practice arise in the fruit and vegetables trade and the ceramics, glass and textile industries. In the electronic component industry, grading is a common practice to select devices that operate between certain temperature ranges. In ideal conditions all devices would meet the higher specification but due to variations in the raw material or in the manufacturing process only a few may actually reach full performance. The remainder of the devices has a degraded performance but still offers all the functions of the top-grade components at lower temperatures. To say that these differences are not differences in *quality* would be misleading, because the products were all designed to fulfil the higher specification. As there is a market for such products it is expedient to exploit it. There is a range over which product quality can vary and still create satisfied customers. Outside the lower end of this range, the product is considered to be of poor quality.

Quality and Price

Most of us are attracted to certain products and services by their price. If the price is outside our reach we don't even consider the product or service, whatever its quality, except perhaps to form an opinion about it. We also rely on price as a comparison, hoping that we can obtain the same characteristics at a lower price. In the luxury goods market, a high price is often a mark of quality but occasionally it is a confidence trick aimed at making more profit for the supplier. When certain products and services are rare, the price tends to be high and when plentiful the price is low, regardless of their quality. One can purchase the same item in different stores at different prices, some as much as 50% less and many at 10% less than the highest price. You can also receive a discount for buying in bulk, buying on customer credit card or being a trade customer rather than a retail customer. Often an increase in the price of a product may indicate a better after-sales service, such as free on-site maintenance, free delivery, and free telephone support line. The discount shops may not offer such benefits.

The price label on any product or service, regardless of the inherent features should be for a product or service free of defects. If there are defects the label should say as much, otherwise the supplier may well be in breach of national laws and statutes. Price is therefore not an inherent feature or characteristic of the product. It is not permanent and as shown above varies without any change to the inherent characteristics of the product. Price is also a feature of the service associated with the sale of the product. Price is negotiable for the same quality of product. Some may argue that if you want 'quality' you have to pay for it but what you are paying by a higher price is likely to be a product that is more reliable, more durable and has a longer life or a service providing more comfort, more luxury and greater convenience.

Quality and Cost

Philip Crosby published *Quality is Free* in 1979 and caused a lot of raised eyebrows among executives because they always believed the removal of defects was an in-built cost in running any business. To get quality you had to pay for inspectors to detect the errors! What Crosby told us was that if we could eliminate all the errors and reach zero defects, we would not only reduce our costs but also increase the level of customer satisfaction by several orders of magnitude. In fact there is the cost of doing the right things right first time and the cost of *not* doing the right things right first time. This is often referred to as *quality costs* or the cost incurred because failure is possible.

Using this definition, if failure of a product, a process or a service is not possible, there would be no *quality costs*. It is rather misleading to refer to the cost incurred because failure is possible as *quality costs* because we could classify the costs as avoidable costs and unavoidable costs. We have to pay for labour, materials, facilities, machines, transport etc. To some extent these costs are unavoidable but we are also paying in addition some cost to cover the prevention, detection and removal of errors. Should customers have to pay for the errors made by others? There is a basic cost if failure is not possible and an additional cost in preventing and detecting failures and correcting errors because our prevention and detection programmes are imperfect. We can reduce the basic cost by finding more economical ways of doing things or cheaper materials. However, there is variation in all processes but it is only the variation that exceeds the tolerable limits that incurs a penalty. If you reduce complexity and install failure-prevention measures you will be spending less on failure detection and correction. There is an initial investment to be paid, but in the long term you can meet your customer's requirements at a cost far less than you were spending previously.

Some customers are now forcing their suppliers to reduce internal costs so that they can offer the same products at lower prices. This has the negative effect of forcing suppliers out of business. While the motive is laudable, the method is damaging to industry. There are inefficiencies in industry that need to be reduced but imposing requirements will not solve the problem. Co-operation between customer and supplier would be a better solution and when neither party can identify any further savings the target has been reached. Customers do not benefit by forcing suppliers out of business.

Quality and Design

In examining the terms design and quality, we need to recognize that the word design has different meanings. Here we are not concerned with design as a verb or as the name we give to the process of design or the output of the design process. In this context we are concerned with the term design as an aesthetic characteristic of a product or service rather than a quality characteristic. The quality characteristic embraces the form, fit and function attributes relative to its purpose. The attributes that appeal to the senses are very subjective and cannot be measured with any accuracy, other than by observation and comparison by human senses. So when we talk of quality and design we are not referring to whether the design reflects a product that has the correct features and functions to fulfil its purpose, we are addressing the aesthetic qualities of the product. We could use the word appearance but design goes beyond appearance. It includes all the features that we perceive by sight, touch, smell and hearing.

If the customer requires a product that is aesthetically pleasing to the eye, or is to blend into the environment or appeal to a certain group of people, one way to measure the quality of these subjective characteristics is to present the design to the people concerned and ask them to offer their opinion.

Quality *of* design is a different concept and is *the extent to which the design reflects a product or service that satisfies customer needs and expectations for functionality, cost of ownership and ease of use etc.* All the necessary characteristics should be designed into the product or service at the outset.

Quality, Reliability and Safety

There is a school of thought that distinguishes between quality and reliability and quality and safety. Quality is thought to be a non-time-dependent characteristic and reliability a time-dependent characteristic but the aspect of quality being addressed is the quality of conformity which is the *extent to which the product or service conforms to the design standard*. The design has to be faithfully reproduced in the product or service.

If we take a logical approach to the issue, when a product or service is unreliable, it is clearly unfit for use and therefore of poor quality. If a product is reliable but emits toxic fumes, is too heavy or not transportable when required to be, it is of poor quality. Similarly, if a product is unsafe it is of poor quality even though it may meet its specification in other ways. In such a case the specification is not a true reflection of customer needs. A nuclear plant may meet all the specified safety requirements but if society demands greater safety standards, the plant is not meeting the requirements of society, even though it meets the immediate customer requirements. You therefore need to identify the stakeholders in order to determine the characteristics that need to be satisfied. The needs of all these parties have to be satisfied in order for *quality* to be achieved. But, you can say, "This is a quality product as far as my customer is concerned".

QUALITY CHARACTERISTICS (PRODUCT FEATURES)

There are three fundamental parameters that determine the saleability of products and services; they are price, quality and delivery. Price is a function of cost, profit margin and market forces, and delivery is a function of the organization's efficiency and effectiveness. Price and delivery are easily defined because they can be quantified. Price can be quantified in terms of a number of units of currency and delivery can be quantified in terms of units of time. Quality on the other hand describes the condition or properties of the product which can be quantified in many different ways. Price and delivery are both transient features, whereas the impact of quality is sustained long after the attraction or the pain of price and delivery has subsided.

A product is the output from a process and in describing an output; we express it in terms of its characteristics or features. (In his Quality Control Handbook of 1974 Juran used the term "quality characteristics" but in his later work he abandoned this term as it was only used in the manufacturing industries and preferred the term "product features" as it is more widely used. However, ISO 9000:2005 still defines the term quality characteristic).

To comment on the quality of anything we need a measure of its characteristics and a basis for comparison. Any feature or characteristic of a product or service that is needed to satisfy customer needs or achieve fitness for use is a *quality characteristic*. These characteristics identify the measures of quality, i.e., what we measure to determine that our needs and expectations have been satisfied, fresh bread, hot tea, the promptness of the train service, the softness of a leather chair, the security of an investment etc. When dealing with products the characteristics are almost always

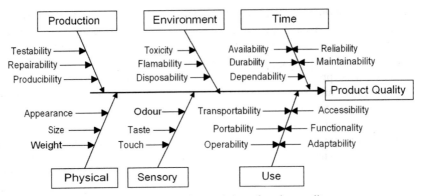

FIGURE 2-2 Some characteristices of product quality.

technical characteristics, whereas service quality characteristics have a human dimension. They may be known as product requirements or service requirements which express the characteristics the product or service needs to exhibit for it to be acceptable. When qualified by units of measure these characteristics become acceptance criteria such as weight 10 kg +/−10 g. Some typical quality characteristics are shown in Fig. 2-2.

Such characteristics need to be specified and their achievement planned, controlled, assured, improved, managed and demonstrated. These characteristics form the subject matter of the product or service requirements referred to in a contract, specification or indeed ISO 9001. When the value of these characteristics is quantified or qualified they are termed *product requirements or service requirements*. We used to use the term *quality requirements* but this caused a division in thinking that resulted in people regarding quality requirements as the domain of the quality personnel and technical requirements being the domain of the technical personnel. All requirements are fundamentally *quality* requirements – they express needs or expectations that are intended to be fulfilled by a process output. We can therefore drop the word *quality*. If a modifying word is needed in front of the word requirements it should be a word that signifies the subject of the requirements. Transportation system requirements would be requirements for a transportation system, audio speaker design requirements would be requirements for the design of an audio speaker, component test requirements would be requirements for testing components, and management training requirements would be requirement for training managers. The requirements of ISO 9001 and its derivatives are often referred to as *quality requirements* as distinct from other types of requirements but this is misleading. ISO 9001 is no more a quality requirement than is ISO 1000 on SI units, ISO 2365 for ammonium nitrate or ISO 246 for rolling bearings. The requirements of ISO 9001 are *quality management system requirements* – requirements for a quality management system.

DIMENSIONS OF QUALITY

There are three *dimensions of quality* two of which extend the perception beyond the concepts outlined previously:

The product quality dimension. This is the extent to which the products and services provided meet the needs of specific customers. Enhancement of product features to satisfy more customers might yield improvement in product quality.

The business quality dimension. This is the extent to which the business serves the needs of all customers (present and future) and is the outward facing view of the organization. Customers are interested in better products and services, delivering greater value and greater benefits. Changes in business strategy, direction or policies might yield improvement in business quality.

The enterprise quality dimension. This is the extent to which the *enterprise* meets the needs of all stakeholders, maximizes its efficiency and effectiveness and is both an inward and outward facing view of the organization. Efficiency is linked with productivity which itself is linked with the motivation of personnel and the capability or processes and utilization of resources. Effectiveness is linked with the utilization of knowledge focusing on the right things to do, taking account of the needs of all stakeholders. The stakeholders are not only interested in the quality of particular products and services but also judge organizations by their potential to create wealth, the continuity of operations, the sustainability of supply, care of the environment, care of people and adherence to health, safety and legal regulations. Seeking best practice might yield improvement in enterprise quality. This directly affects all aspects of quality. Viewing the organization as a system would redefine this dimension as, *the system quality dimension.*

We must separate the three concepts above to avoid confusion. When addressing quality, it is necessary to be specific about the object of our discussion. Is it the quality of products or services, or the quality of the business in which we work, or the enterprise as a whole, about which we are talking? If we only intend that our remarks apply to the quality of products, we should say so.

Many organizations only concentrate on the product quality dimension, but the three are interrelated and interdependent. Deterioration in one eventually leads to deterioration in the others.

Organizations may be able to produce products and services that satisfy their customers under conditions that put employees in fear of losing their jobs, that put suppliers in fear of losing orders and put the local community in fear of losing their quality of life. However, society has a way of dealing with these – through representation in government, laws are passed that regulate the activities of organizations. As we will show in Chapter 3, such organizations are eventually put out of business but there may be a lot of pain all round before this event occurs.

The Importance and Role of Stakeholders

CHAPTER PREVIEW

Although intended for students and those readers whose jobs take them into the territory of strategic management, an understanding of stakeholders is so important to sustained success that no person involved in the achievement of quality should pass it by.

All organizations depend on support from a wide range of other organizations and individuals to achieve their goals; the most obvious being customers but there are others who play a very important role. These contributors are commonly referred to as stakeholders for reasons that will become clear.

We examine the who, what and where of stakeholders and in particular

- The different terms used to describe these contributors and the problems such terms create;
- The idea that all real stakeholders are beneficiaries;
- How we should look upon other stakeholders?
- Just who might these stakeholders be?
- Customers, the most important of all the stakeholder groups;
- Where an organization might find its customers?

We look at why stakeholders are important to all organizations and in particular

- What stakeholders bring to and take from the organization?
- Where the idea of a stakeholder came from?
- Where each of the stakeholders fit in order of importance?
- We dissect the various types of requirements stakeholders impose on organizations and the language used to express these requirements.
- We deal with the tricky problem of balancing the requirements of each of the stakeholders and finally
- We examine the notion of the internal customer and challenge the argument for the continued use of this misnomer.

IDENTIFYING THE STAKEHOLDERS

Multiple Meanings

The concept of stakeholders is complicated by different meanings and uses dependent upon both context and association. In traditional usage a stakeholder is a third party who temporarily holds money or property while the issue of ownership is being resolved between two other parties, e.g., a bet on a race, litigation on ownership of property.

In business usage there are over 20 definitions in Google by simply searching the words 'stakeholder+definition'. These appear to have a number of traits.

- The notion that a 'stake' equates with an 'interest' implying that anyone with an interest in an outcome is a stakeholder, regardless of motive such as a competitor, journalist or politician.
- The notion that a person affecting an outcome is a stakeholder, regardless of motive implying that terrorists and others with a malevolent intent are stakeholders.
- The notion that a person influencing an outcome is a stakeholder, regardless of motive implying that those who abuse their power to exert undue influence are stakeholders.
- The notion that a person being affected by an outcome is a stakeholder, regardless of their contribution, implying that any person in the local or global community affected by an organization's actions such as noise, pollution or business closure is a stakeholder.
- The notion of contributors to an organization's wealth-creation capacity being beneficiaries and risk takers, implying that those who put something into an organization either resources or a commitment are stakeholders (Post, Preston, and Sachs)[1].

> **Food for Thought**
>
> Just consider for a moment that if a stakeholder is anyone affected by the organization, would a burglar who gains access to an organization that makes security systems be one of its stakeholders?

The Significance of Beneficiaries

It is therefore difficult if not impractical in some cases for organizations to set out to satisfy the needs and expectations of all these interested parties. There needs to be some rationalization. The definition postulated by Post, Preston, and Sachs below amalgamates the idea of contributors, beneficiaries, risk takers, voluntary and involuntary parties thus indicating that there is mutuality between stakeholders and organizations.

> "The stakeholders in a corporation are the individuals and constituencies that contribute, either voluntarily or involuntarily, to its wealth-creating capacity and activities, and that are therefore its potential beneficiaries and/or risk bearers"

This accords with the traditional view that businesses exist to create profit for the investors and also recognizes that there is more than one beneficiary. However, Drucker's view was that the purpose of a business is to create customers and this is also true as the business could not exist without customers. That there is more than one beneficiary shows that organizations cannot accomplish their mission without the support of other organizations and individuals who generally want something in return for their support.

---

[1] Redefining the Corporation Stakeholder Management and Organisational Wealth (2002) James E. Post, Lee E. Preston, and Sybille Sachs.

By making a contribution, these beneficiaries have a stake in the performance of the organization. If the organization performs well, they get good value and if it performs poorly, they get a poor value at which point they can withdraw their stake.

Post, Preston, and Sachs make the assumption that organizations do not intentionally destroy wealth, increase risk, or cause harm. However, had organizations not known of the risks, their actions might be deemed unintentional but some decisions are taken knowing that there are risks which are then ignored in the interests of expediency. Enron, WorldCom and Tyco are recent examples where serious fraud was detected and led to prosecutions in the USA. There are many other much smaller organizations deceiving their stakeholders, each day some of which reach the local or national press or are investigated by the Consumer Association, Trading Standards and other independent agencies.

In the drafting of ISO 9000:2000 the term stakeholder was considered because it was a term used with the Excellence Model and the principles on which the Model was based were being incorporated into ISO 9001 at the time. However, the traditional meaning of the term *stakeholder* still pertains in some countries so it was decided to use the term *interested party* instead and define this as:

"a person or group having an interest in the performance or success of an organization"

We have shown that there are problems with this definition as it does not limit these parties to those making a contribution or supporting the organization although the list of examples appending the definition does attempt to do this.

Reconciling the Different Parties

If we settle on stakeholders being the parties that contribute to an organization's wealth-creation capacity and benefit from it, we cannot dismiss all the other interested parties because they don't qualify as a stakeholder as we have defined it above. Their influence needs to be managed. For example,

- It is important to win support from the media as bad press can destroy reputations;
- It is important to keep co-workers on your side as their cooperation might be essential for your success;
- It is important not to alienate other managers as they might limit the availability of resources you need;
- It is important to build rapport with Trade Associations, Unions and other organizations from which you might need help.

Who are Stakeholders?

Following on from the above, organizations depend on support from a wide range of other organizations and individuals. Some are merely different terms for the same thing. These can be placed into five categories as follows:

1. Investors including shareholders, owners, partners, directors, or banks and anyone having a financial stake in the business.
2. Customers including clients, purchasers, consumers and end users. (ISO 9000:2005 also includes beneficiary but this term can apply to any stakeholder. Purchasers also include wholesalers, distributors and retailers.)

3. Employees including temporary and permanent staff and managers. Some texts regard management as a separate stakeholder group but all managers are employees unless they happen to be owners who also manage the organization.
4. Suppliers including manufacturers, service providers, consultants and contract labour.
5. Society including people in the local community, the global community and the various organizations set up to govern, police and regulate the population and its interrelationships.

These labels are not intended to be mutually exclusive. A person might perform the role of customer, supplier, shareholder, employee and member of society all at the same time. An example of this is a bar tender. When serving behind the bar he can be an employee or a supplier if he contracts with an employment agency, but on his day off he might be a customer and since acquiring shares in the brewery he became a shareholder. He also lives in the local community and therefore benefits from the social impact it has on the community. He is a contributor; he affects the outcomes and is affected by those outcomes.

Although stakeholders have the freedom to exert their influence on the organization, there is often little that one individual can do, whether or not that individual is an investor, customer, employee, supplier or simply a citizen. The individual may therefore not be able to exercise their power but they can lobby their parliamentary representatives, consumer groups, trade associations etc. and collectively bring pressure to bear that will change the performance of the organization.

A Closer Look at Customers

The chief current sense of the word customer (according to the OED) is a person who purchases goods or services from a supplier. Turning this around, one might say that a customer is someone who receives goods or services from a supplier and therefore may not be the person who made the purchase. There has to be intent on the part of the supplier to sell the goods and services and intent on the part of the receiver to buy the goods and services (a thief would not be classed as a customer). This customer–supplier relationship is valid even if the goods or services are offered free of charge, but in general the receiver is either charged or offers something in return for the goods or services rendered. There is a transaction between the customer and supplier that has validity in law. Once the sale has been made there is a contract between the supplier and the customer that confers certain rights and obligations on both the parties.

The word 'customer' can be considered as a generic term for the person who buys goods or services from a supplier as other terms are used to convey similar meaning such as:

Client	The customer of a professional service provider such as a law firm, accountant, consultancy practice or architect.
Purchaser	The customer of a supplier that places the order and authorizes payment of the invoice. The purchaser may not be the end-user and so acts on his/her behalf.
Beneficiary	The customer of a charity.
Consumer	The customer of a retailer.

End-user   The person using the goods or services that have been purchased perhaps by someone else. This maybe the person for whom the goods and services were intended but may not be known at the time of purchase.

It is relatively easy to distinguish customers and suppliers in normal trading situations but there are other situations where the transaction is less distinct. In a hospital the patients are customers but so too are the relatives and friends of the patient who might seek information or visit the patient in hospital. In a school the parent is the customer but so too is the pupil unless one considers that the pupil is customer supplied property!

Where are the Customers?

Customers can be buyers in the local community or as far away as the other side of the world. It all depends on the effectiveness of an organization's marketing efforts and the scope of appeal for its products and services.

It also depends on transport technology. Who would have thought 20 years ago that we would find flowers, fruit and vegetables from the far east on western supermarket shelves. The containerization of goods has made it possible to ship almost any product to anywhere in the world within a few hours.

On-line electronic trading has also broadened the field and enabled suppliers to find customers for specialized items in almost every country in the world.

The net result is an increasing demand for transport by land, sea and air with consequences for the environment. For organizations that are environmentally conscious, this means taking a hard look at their customer base and lower their carbon footprint.

Customers are not always the most obvious ones. Products designed for one purpose may well find customers using them for an entirely different purpose and so it pays to continually analyse who is buying what as it may reveal new undiscovered markets and potentially even greater sales.

THE IMPORTANCE OF STAKEHOLDERS

What Do Stakeholders Bring to and Take from the Organization?

Each stakeholder brings something different to an organization in pursuit of their own interests, takes risks in doing so and receives certain benefits in return but is also free to withdraw support when the conditions are no longer favourable.

1. Investors provide financial support in return for increasing value in their investment but may withdraw their support if the actual or projected financial return is no longer profitable. Shares are sold, loans called in, government funding is stopped.
2. Customers provide revenue in return for the benefits that ownership of the product or service brings but may demand refunds if the product does not satisfy the need and are free to withdraw their patronage permanently if they are dissatisfied with the service. The UK Telecoms regulator Offcom stated in the context of broadband services that "competition is only effective where customers can punish 'bad' providers by taking their custom elsewhere, and reward 'good' providers by staying where they are."

3. Employees provide labour in return for good pay and conditions, good leadership and job security but are free to withdraw their labour if they have a legitimate grievance or may seek employment elsewhere if the prospects are more favourable.
4. Suppliers provide products and services in return for payment on time, repeat orders and respect but may refuse to supply or cease supply if the terms and conditions of sale are not honoured or they believe they are being mistreated.
5. Society provides a licence to operate in return for employment and other economic benefits to the community they bring such as support for local projects. However, society can censure the organization's activities through protest and pressure groups. Ultimately, society, through its agents, enacts legislation if the organization's activities are believed to be detrimental to the community such as unjust, unethical, irresponsible and illegal acts by organizations. The organization does not have a choice other than moving to another country where the laws are less onerous.

There is no doubt that customer needs are paramount as without revenue the organization is unable to benefit the other stakeholders. However, as observed by Post, Preston, and Sachs

> "Organizational wealth can be created (or destroyed) through relationships with stakeholders of all kind—resource providers, customers and suppliers, social and political actors. Therefore, effective stakeholder management—that is, managing relationships with stakeholders for mutual benefit—is a critical requirement for corporate success."

A Historical Perspective

As was stated above, no organization can accomplish its purpose and mission without the support of others. However, this belief has not always been so. In the nineteenth century, the only stakeholder of any importance was the owner who ran the business as a wealth-creating machine with himself as chief beneficiary. Workers had no influence and customers bought what was available with little influence over the producer. Suppliers were no more influential than the workers and society was pushed along with the advancing industrial revolution. The successful owners often became philanthropic through guilt and a desire to go to heaven when they died. As a result their wealth was distributed through various trusts and endowments.

Workers were the first to exert influence with the birth of trade unions but it took a century or more for workers rights to be enshrined in law. Customers began to influence decisions of the organization with the increase in competition but it was not until the western industrialists awoke to competition from Japan in the 1970s that a customer revolution emerged. Then slowly in the 1980s and on through the millennium, the green movement began to exert influence resulting in laws and regulations protecting the natural environment. Lastly, the human rights act gave strength to the rights of all citizens including disadvantaged people.

There is now a greater sensitivity to the environmental impact of organizations' actions upon the planet. Workers are more aware of their rights and more confident of censuring their employer when they feel their rights have been abused. The wealth creating organizations now distribute their wealth through their stakeholders rather than through philanthropy but this remains a route for the biggest of corporations as evidenced by Bill Gates and Warren Buffett.

The Relative Importance of Stakeholders

It follows therefore that companies must try to understand better what their stakeholders' needs are and then deal with those needs ahead of time rather than learn about them later.

Case Study – Making Money and Quality

Our CEO is adamant that we are in business to make money and not to satisfy all the stakeholders and won't entertain any proposition that will not make money for the firm. How do I convince him that we have to put satisfying customers as a first priority rather than making money?

Whatever we set out to achieve, we will not get very far unless we are conscious of the factors critical to our success. Many people do this subconsciously – always being aware of waste, relationships, customers etc. But we can be a little more scientific about this. To assess the risks of fulfilling the mission and vision we need to ask

"What affects our ability to fulfil our mission and vision?"

If the mission for the owners were to make money you would ask, "What affects our ability to make money?" There might be a number of answers:

- Maintaining a high profit margin,
- Creating and retaining satisfied customers,
- Employing competent people,
- Recruiting people who share our values,
- Finding a cheap source of raw material,
- Finding capable suppliers,
- Keeping on the right side of the law.

When we examine these answers and ask; "Who benefits if we do these things?" we reveal what are generally referred to as stakeholders.

- Maintaining a high profit margin benefits the shareholders.
- Creating and retaining customers not only benefits customers but also benefits shareholders because it brings in revenue.
- Employing competent people benefits shareholders because competent people make fewer mistakes.
- Recruiting people who share our values benefit shareholders by creating an environment in which any internal threat to probity will be minimised.
- Finding a cheap source of good quality raw material not only benefits customers by keeping prices low but also benefits shareholders by increasing profit margins.
- Finding capable suppliers benefits shareholders because it keeps costs down as a result benefits customers.
- Keeping on the right side of the law benefits shareholders, employees, customers and society.

It will therefore become apparent that the critical factor upon which success depends is satisfying stakeholders. But we know that only the customer brings in revenue therefore making money results from satisfying customers in a way that satisfies the needs of the other stakeholders. "If you take care of the quality, the profits will take of themselves" (Imai Masaaki KAIZEN page 49)

Organizations need to attract, capture and retain the support of those organizations and individuals it depends upon for its success. All are important but some are more important than others.

Investors put up the capital to get the business off the ground. This makes the banks and venture capitalists of prime importance during business start up or regeneration, but once operational it is customers that keeps the business going.

Without customers there is no revenue and without revenue there is no business. Customers are therefore the most important stakeholder after start up but before customers there are:

Employees who provide the human resources that power the engines of marketing and production. Without human resources the business is unlikely to function even if there are investors and potential customers waiting to buy the organization's outputs.

Next in importance are suppliers of products and services upon which the organization depends to produce its outputs. Without suppliers, the engines of marketing and production will run short of fuel.

Finally in importance is society. Society provides the employees and the infrastructure in which the organization operates.

Within any organization, it is important that everyone knows who the customers are so that they can act accordingly. Where individuals come face to face with customers a lack of awareness, attention or respect may lead to such customers taking their business elsewhere. Where individuals are more remote from customers, knowledge of customers in the supply chain will heighten an awareness of the impact they have on customers by what they do.

Organizations cannot survive without customers. Customers are one of the stakeholders but unlike other stakeholders they bring in revenue which is the life blood of every business. Consequently the needs and expectations of customers provide the basis for an organization's objectives, whereas the needs and expectations of the other stakeholders constrain the way in which these objectives are achieved. It follows therefore that the other stakeholders (investors, employees, suppliers and society) should not be regarded as customers as it would introduce conflict by doing so.

However, an organization ignores anyone of these stakeholders at its peril which suggests that there has to be a balancing act but more in this later.

STAKEHOLDER REQUIREMENTS

Organizations are created to achieve a goal, mission or objective but they will only do so if they satisfy the requirements of their stakeholders. Their customers, as one of the stakeholders, will be satisfied only if they provide products and services that meet their needs and expectations, i.e., their requirements. They will retain their customers if they continue to delight them with superior service and convert wants into needs. But they also have to satisfy their customers without harming the interests of the other stakeholders which means giving investors, employees, suppliers and society what they want in return for their contribution to the organization.

Many of these requirements have to be discovered by the organization as they won't all be stated in contracts, orders, regulations and statutes. In addition the organization has an obligation to determine the intent behind these requirements so that they may provide products and services that are fit for purpose.

This creates a language of demands that is expressed using a variety of terms, each signifying something different but collectively we can refer to these as preferences.

Needs

Needs are essential for life, to maintain certain standards, or essential for products and services, to fulfil the purpose for which they have been acquired. For example, a car needs a steering wheel and the wheel needs to withstand the loads put upon it but it does not need to be clad in leather and hand stitched for it to fulfil its purpose.

Everyone's needs will be different and therefore instead of every product and service being different and being prohibitively expensive, we have to accept compromises and live with products and services that in some ways will exceed what we need and in other ways will not quite match our needs. To overcome the diversity of needs, customers define requirements, often selecting existing products because they appear to satisfy the need but might not have been specifically designed to do so.

Wants

By focusing on benefits resulting from products and services, needs can be converted into wants such that a need for food may be converted into a want for a particular brand of chocolate. Sometimes the want is not essential but the higher up the hierarchy of needs we go, (see Figure 9-4) the more a want becomes essential to maintain our social standing, esteem or to realize our personal goals.

In growing their business, organizations create a demand for their products and services but far from the demand arising from a want that is essential to maintain our social standing, it is based on an image created for us by media advertising. We don't need spring vegetables in the winter but because industry has created the organization to supply them, a demand is created that becomes an expectation. Spring vegetables have been available in the winter now for so long that we expect them to be available in the shops and will go elsewhere if they are not. But they are not essential for survival, to safety, to esteem or to realize our potential and their consumption may in fact harm our health because we are no longer absorbing the right chemicals to help us survive the cold winters. We might want it, even need it but it does us harm and regrettably, there are plenty of organizations ready to supply us products that will harm us.

Expectations

Expectations are *implied needs* or *requirements*. They have not been requested because we take them for granted – we regard them to be understood within our particular society as the accepted norm. They may be things to which we are accustomed, based on fashion, style, trends or previous experience. One therefore expects sales staff to be polite and courteous, electronic products to be safe and reliable, policemen to be honest, coffee and soup to be hot etc. One would like businessmen to be honest but in some markets we have come to expect them to be unethical, corruptible and dishonest. As expectations are also born out of experience, after frequent poor service from a train operator, our expectations are that the next time we use that train operator we will once again be disappointed. We would therefore be delighted if, through some well-focused quality initiative, the train operator exceeded our expectations on our next journey.

Requirements

Requirements are stated needs, expectations and wants but often we don't fully realize what we need until after we have stated our requirement. For example, now that we own a mobile telephone we discover we really need hands-free operation when using the phone while driving a vehicle. Another more costly example is with software projects where customers keep on changing the requirements after the architecture has been established. Our requirements at the moment of sale may or may not therefore express all our needs. Requirements may also go beyond needs and include characteristics that are nice to have but not essential. They may also encompass rules and regulations that exist to protect society, prevent harm, fraud and other undesirable situations.

Anything can be expressed as a requirement whether or not it is essential or whether the circumstances it aims to prevent might ever occur or the standards invoked might apply.

Requirements are often an imprecise expression of needs, wants and expectations. Some customers believe that they have to define every characteristic otherwise there is a chance that the product or service will be unsatisfactory. For this reason parameters may be assigned tolerances that are arbitrary simply to provide a basis for acceptance/rejection. It does not follow that a product that fails to meet the requirement will not be fit for use. It simply provides a basis for the customer to use judgement on the failures. The difficulty arises when the producer has no idea of the conditions under which the product will be used. For example, a producer of a power supply may have no knowledge of all the situations in which it might be used. It may be used in domestic, commercial, military or even in equipping a spacecraft. Variations acceptable in domestic equipment might not be acceptable in military equipment but the economics favour selection for use rather than a custom design which would be far more costly.

Desires and Preferences

Customers express their requirements but as we have seen above, these may go beyond what is essential and may include mandatory regulations as well as things that are nice to have – what we can refer to as desires. Sometimes a customer will distinguish between those characteristics that are essential and those that are desirable by using the word 'should'. Desired might also be expressed as preferences, e.g., a customer might have a preference for milk in glass bottles rather than plastic bottles because of the perception that glass is more environmentally friendly than plastic. The determination of stakeholder preferences is an important factor in decision making at the strategic level and in product and service development.

Intent

Behind every want, need, requirement, expectation or desire will be an intent. What the customer is trying to accomplish as a result, the reason for the requirement. In many cases clarifying the intent is not necessary because the requirements express what amounts to common sense or industry practice and norms. But sometimes requirements are expressed in terms that clarify the intent. A good example can be taken from ISO 9001 where in Clause 8.3 it states:

> "The organization shall ensure that product which does not conform to product requirements is identified and controlled to prevent its unintended use or delivery"

The phrase "to prevent its unintended use or delivery" signifies the intent of the requirement but not all requirements are as explicit as this. For example, in Clause 5.5.1 of ISO 9001 it states:

> "Top management shall ensure that responsibilities and authority are defined and communicated within the organization"

There is no expression of intent in this requirement, i.e., it does not clarify why responsibilities and authority need to be defined and communicated. Although in this example it might appear obvious, in the 1994 version of the standard it also required responsibilities and authority to be documented without stating why. Knowing the intent of a requirement is very important when trying to convince someone to change their behaviour.

Demands and Constraints

Requirements become demands at the point when they are imposed on an organization through contract, order, regulation or statute. Until then they simply don't apply. From the outside looking in, all demands imposed on the organization are requirements but from the inside looking out these requirements appear as two categories: one that addresses the objective of the required product or service which we can call *product requirements* and another that addresses the conditions that impact the way in which the required product or service is produced and provided which we can call *constraints*. While customer requirements may be translated into product requirements and constraints, the other stakeholder requirements are only translated into constraints because if the customer requirements were to be removed, there would be no activity upon which to apply the constraint. In other words, requirements and constraints are not mutually exclusive. If we do not have any oil platforms, the safety regulations governing personnel working on oil platforms cannot apply to us.

Is the Customer Always Right?

It might be argued that in theory the customer is always right therefore even if the customer makes a demand that cannot be satisfied without compromising corporate values, the organization has no option but to satisfy that demand.

In reality, organizations can choose not to accept demands that compromise their values or the constraints of other stakeholders particularly those concerning the environment, health, safety and national security.

The product or service may be able to fulfil the requirements without satisfying the constraints. However, in not satisfying the constraints the stakeholders could censure the organization and stop it from continuing production. For example, if the production of the product causes illegal pollution, the Environmental Regulatory authority will sanction or close the production facility. If the organization treats its employees unfairly

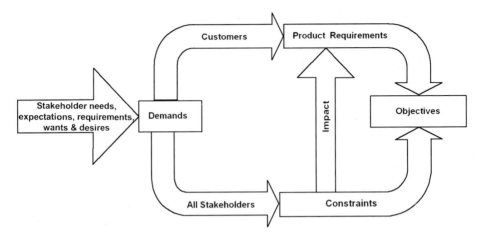

FIGURE 3-1 Relationship between demand requirements and constraints.

and against recognized codes of practices or employment legislation, regulators and employees could litigate against the organization and thus damage its reputation. In other words, Product *Requirements* define the true focus for the organization and *Constraints* define parameters that can modify the focus and influence the organization's process design.

It could be said that fulfilling product requirements is the only true *objective* because all other demands generate constraints on the way the *objective* is to be met (what Bryson refers to as mandates[2]). If the objective was to supply freeze dried coffee to supermarkets, then generating a net profit of 15% using raw materials sourced only under Fair Trade agreements processed without using ozone depleting chemicals are all constraints and not objectives. In practice, however, organizations tend to set objectives based upon both requirements and constraints which often lead to the relationships between requirements and constraints being confused, and consequently the focus on the true objective being lost or forgotten (see also *Quality objectives in Chapter 16*).

The relationship between demands, requirements, constraints and objectives is illustrated in Fig. 3-1.

Handling the Language

The language used differs depending upon whether we are addressing *demands*, *requirements*, *constraints*, *objectives*, *wants*, *intents or desires*, the direction they are coming from and how they are responded to.

In their stated or implied requirements we:

- *place* demands,
- *define* requirements,
- *impose* constraints,
- *set* objectives,
- *state or declare* our intentions,
- *express* our desires or preferences.

---

[2] Bryson J.M. (2004) Creating and implementing your strategic plan. Jossey Bass.

We respond by declaring that:

- demands have been *met*,
- requirements have been *fulfilled*,
- objectives have been *achieved*,
- constraints have been *satisfied*,
- intentions have been *declared*,
- desires have been *satisfied*,
- preferences have been *granted*.

If the direction or response is not clearly understood, what might be perceived and labelled as one of these turns out to be another and as a consequence inappropriate influence and priority are applied.

When the same label is used for two different types of requirements it can result in some people or departments prioritizing actions inappropriately. If it helps to bring about improvement in performance by labelling constraints as objectives, this is not a bad thing provided that people understand they are not trading-off customer satisfaction when doing so.

BALANCING STAKEHOLDER NEEDS

There is a view that the needs of stakeholders have to be balanced because it is virtually impossible to satisfy all of them, all of the time. Managers feel they ought to balance competing objectives when in reality it is not a balancing act.

On face value balancing competing objectives might appear to be a solution but balancing implies that there is some give and take, win/lose, a compromise, a trade-off or reduction in targets so that all needs can be met. Organizations do not reduce customer satisfaction to increase safety, environmental protection or profit. The organization has to satisfy its customers otherwise it would cease to exist, but it needs to do so in a way that satisfies all the other stakeholders as well – hence the cliché 'customer first'. If the organization cannot satisfy the other stakeholders by supplying X, it should negotiate with the customer and reach an agreement whereby the specification of X is modified to allow all stakeholders to be satisfied. If such an agreement cannot be reached the ethical organization will decline to supply X under the conditions specified.

In practice, the attraction of a sale often outweighs any negative impact upon other stakeholders in the short term with managers convinced that future sales will redress the balance. Regrettably, this uncontrolled approach ultimately leads to unrest and destabilization of the business processes as employees, suppliers and eventually customers withdraw their stake. In the worst case scenario, it results in destabilizing the world economy as the credit crisis of 2008 demonstrated. The risks have to be managed effectively for this approach to succeed.

THE INTERNAL CUSTOMER MYTH

It is not uncommon within organizations to find that the people you rely on to get your job done are referred to as suppliers and anyone who counts on you to get their job done is referred to as a customer. On this basis the operator who receives a drawing from the designer would be regarded as a customer in an internal supply chain. But the operator doesn't pay the designer for the drawing, has no contract with the designer, does not pass the output of his work to the designer, does not define the requirements for the drawing and cannot choose to ignore the drawing so is not a customer in the accepted sense but a user of the drawing. The operator may pass his output to another operator in the internal supply chain but this second operator is not a customer of the former, he/she is merely a co-worker in a process.

As internal receivers of product, employees are not stakeholders (they are part of the process) therefore they are not strictly customers but partners or co-workers. Normally the customer is external to the organization supplying the product because to interpret the term customer as either internal or external implies that the internal customer has the same characteristics as an external customer and this is simply not the case as is indicated in Table 3-1.

In Table 3-1 there is only a 30% match between the characteristics of internal and external customers and therefore it would be unwise to use the same term for both parties as misunderstanding could ensue.

The notion of internal customers and suppliers is illustrated in Fig. 3-2. In the upper diagram, requirements are passed along the supply chain and if at each stage there is some embellishment or interpretation by the time the last person in the chain receives the instructions they may well be very much different from what the customer originally required. In reality each stage has to meet the external customer requirement as it applies to the work performed at that stage not as the person performing the previous or subsequent stage wants. This is shown in the lower diagram where at each stage there is an opportunity to verify that the stage output is consistent with the external customer requirement. In a well-designed process, individuals do not impose their own

TABLE 3-1 Internal Versus External Customer

Characteristic	External customer	Internal customer
Places an order for product or service	Yes	No
Has an interest in the performance of the organization	Yes	Yes
Receives product	Yes	Yes
Can reject nonconforming product	Yes	Yes
Provides payment for the output	Yes	No
Is external to the organization supplying the product	Yes	No
Defines requirements the product has to meet	Yes	Sometimes
Can change requirements	Yes	Sometimes
Enters into a legal contract in which there are laws protecting both parties	Yes	No
Is able to take custom elsewhere if not satisfied	Yes	No

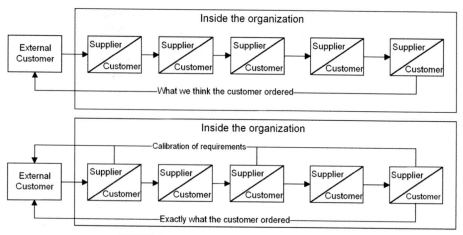

FIGURE 3-2 The internal customer–supplier chain.

requirements on others. The requirements are either all derived from the customer requirements or from the constraints imposed by the other stakeholders.

If organizations want to encourage their employees to behave AS IF they were both customers and suppliers in an internal supply chain then there might be some benefit as it might engender:

- Pride in ones work,
- Greater vigilance,
- Greater accountability,
- Better control,
- Greater cooperation.

But it might also lead to some problems:

- Workers start imposing requirements on their co-workers that are more stringent than the external customer requirements;
- Workers start rejecting input from co-workers for trivial reasons;
- Workers become adversarial in their relationships with co-workers;
- Workers refuse to release resources to co-workers thus causing bottlenecks in the process;
- Departments set up service level agreements that create artificial internal trading conditions;
- Departments begin to compete for resources, bid for the same external contacts and generally waste valuable resources.

If instead of the label internal customers and suppliers, individuals were to regard themselves as **co-workers** in a team that has a common goal, the team would achieve the same intent. In a team, every co-worker is just as important as every other and with each co-worker providing outputs and behaving in a manner that enables the other workers to do their job right first time; the team goal would be achieved. The observation by Phil Crosby that quality is ballet not hockey[3] is very apt. In hockey, the participants do not

---

[3] Crosby Philip. B (1979) Quality is Free. McGraw-Hill Inc

treat each other as customers and suppliers but as team members, each doing their best but the result on most occasions is unpredictable therefore as a metaphor for business, hockey would be inappropriate. In ballet, the participants also do not treat each other as customers and suppliers but as artists playing predetermined roles that are intended to achieve predictable results. Organizational processes are designed to deliver certain outputs and in order to do so individuals need to perform specific roles just like ballet – here the metaphor is more appropriate.

ISO 9000:2005 contains the definitions of terms used in the ISO 9000 family and it defines a customer as an organization or person that receives a product. There is also a note in which it is stated that "A customer can be internal or external to the organization". However, in all cases the term customer is used without qualification. The term internal customer cannot be found in either ISO 9001 or 9004 and therefore one might conclude that wherever the term customer is used either external or internal customer is implied but this would make nonsense of many of the requirements. For example, if every person in the organization was a customer, every error would have to be treated as a nonconformity and the outputs subject to the nonconformity controls of Clause 8.3. Top management would have to ensure that the requirements of internal customers were determined and met with the aim of enhancing internal customer satisfaction – clearly not the intent of ISO 9001 at all.

A much better way than adopting the concept of internal customers is for managers to manage the organization's processes effectively and create conditions in which all employees can be fully involved in achieving the organization's objectives.

Chapter 4

Anatomy of the Standards

CHAPTER PREVIEW

This chapter is aimed more at students, consultants, auditors and quality managers and will also be of interest to managers, in particular, industry sectors where certification schemes other than ISO 9001 are in use.

ISO 9000 is a specific standard but is also the general term for what has become the ISO 9000 phenomenon, meaning not just the single standard but the infrastructure that has grown around ISO 9001 certification. It might therefore be a surprise to learn that there are 18 standards in the ISO 9000 family and while a detailed knowledge of all of them is not essential to meet the intent, in this chapter we provide an overall appreciation of the range of standards and in particular:

- The identity of the standards that make up the ISO 9000 family;
- The characteristics of the four standards in the ISO 9000 series;
- The relationship between the standards to illustrate how they are used;
- The sector specific derivatives such as those produced for the automotive sector and why they exist;
- We also compare ISO 9001 with other models such as the Excellence Model, Six Sigma and Business Process Management (BPM).

GENERIC INTERNATIONAL QUALITY MANAGEMENT AND QUALITY ASSURANCE STANDARDS

All generic international quality management and quality assurance standards are the responsibility of the ISO technical committee ISO/TC176. They include standards commonly referred to as the ISO 9000 series and the ISO 9000 family. Related standards that are sector specific are the responsibility of other ISO technical committees except for ISO/TS 16949 which was administered by ISO/TC 176 and therefore classed as in the ISO 9000 family.

The purpose of these generic standards is to assist organizations operate effective quality management systems. It does this in order to facilitate mutual understanding in national and international trade and help organizations achieve sustained success. This notion of sustained success is brought out in the title of ISO 9004:2009[1] showing clearly

---

[1] Comments on ISO 9004:2009 are made on the draft standard as it had not completed the revision process at the time of going to press.

TABLE 4-1 Generic Quality Management and Quality Assurance Standards (CD means Committee Draft, TS means Technical Specification and TR means Technical Report.)

International standard	Title	Stage of development
ISO 9000:2005	Quality management systems – Fundamentals and vocabulary	Review stage
ISO 9001:2008	Quality management systems – Requirements	Published
ISO 9004:2009	Managing the sustained success of an organization – A quality management approach	Review stage
ISO 10001:2007	Quality management – Customer satisfaction – Guidelines for codes of conduct for organizations	Published
ISO 10002:2004	Quality management – Customer satisfaction – Guidelines for complaints handling in organizations	Review stage
ISO 10003:2007	Quality management – Customer satisfaction – Guidelines for dispute resolution external to organizations	Published
ISO/CD TS 10004	Quality management – Customer satisfaction – Guidelines for monitoring and measuring equipment	Committee stage
ISO 10005:2005	Quality management systems – Guidelines for quality plans	Review stage
ISO 10006:2003	Quality management systems – Guidelines for quality management in projects	Review stage
ISO 10007:2003	Quality management systems – Guidelines for configuration management	Review stage
ISO 10012:2003	Measurement management systems – Requirements for measurement processes and measuring equipment	Review stage
ISO/TR 10013:2001	Guidelines for quality management system documentation	Review stage
ISO 10014:2006	Quality management – Guidelines for realizing financial and economic benefits	Published
ISO 10015:1999	Quality management – Guidelines for training	Review stage
ISO/TR 10017:2003	Guidance on statistical techniques for ISO 9001:2000	Published

Continued

TABLE 4-1 Generic Quality Management and Quality Assurance Standards
(CD means Committee Draft, TS means Technical Specification and TR means
Technical Report.)—cont'd

International standard	Title	Stage of development
ISO/CD 10018	Quality management – Guidelines on people involvement and competences	Committee stage
ISO 10019:2005	Guidelines for the selection of quality management system consultants and use of their services	Review stage
ISO 19011:2002	Guidelines for quality and/or environmental management systems auditing	Review stage

the broad intent but as most organizations are driven towards the ISO 9000 family to gain certification to ISO 9001 rather than to use ISO 9004, this purpose and intent is often overlooked. Whilst ISO 9001 specifies requirements to be met by your management system, it does not dictate how these requirements should be met, that is entirely up to the organization's management. It therefore leaves significant scope for use by different organizations operating in different markets and cultures.

The associated certification schemes (which are not a requirement of any of the standards in the ISO 9000 family) were launched to reduce costs of customer-sponsored audits performed to verify the capability of their suppliers. The schemes were born out of a reticence of customers to trade with organizations that had no credentials in the market place.

There are 18 standards in the ISO 9000 family as shown in Table 4-1 in which the latest versions are identified. (CD means Committee Draft, TS means Technical Specification and TR means Technical Report.)

THE ISO 9000 SERIES

The ISO 9000 series is a subset of the family of ISO/TC 176 standards. It consists of the four standards referenced in Clause 0.1 of ISO 9000:2005 although in this clause they are incorrectly referred to as the ISO 9000 family. Together they form a coherent set of quality management system standards facilitating mutual understanding in national and international trade (*ISO 9000:2005*). Use of these standards is addressed later in this chapter but it is important that each is put in the correct context.

The relationship between the standards in the series is illustrated in Fig. 4-1. At the core is the organization sitting in an environment in which it desires sustained success. To reach this state the fundamental concepts and vocabulary as expressed in ISO 9000:2005 have to be understood, then, if necessary, the organization demonstrates that it has the capability of satisfying customers through assessment against ISO 9001 conducted in accordance with ISO 19011 and finally using ISO 9004 the organization is managed continually as a system of processes focused on delivering sustained success.

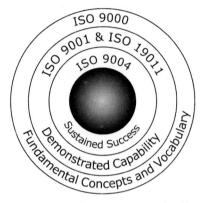

FIGURE 4-1   ISO 9000 family relationships.

Each of these standards has a different purpose, intent, scope and applicability as indicated in Table 4-2.

The terms and definitions in ISO 9000:2005 are invoked in ISO 9001 and thus form part of the requirements and will be the basis on which an auditor can judge the acceptability of something, e.g., the content of a quality manual or the adequacy of a preventive action as these and another 82 terms are defined in ISO 9000:2005.

ISO 9004 is referenced in ISO 9001 as a guide to improvement beyond the requirements of ISO 9001.

THE VOCABULARY OF THE STANDARDS

The ISO 9000 family of standards owes its lineage to the defence standards of the 1960s and 70s and therefore came out of a manufacturing environment. The vocabulary of the first and second editions (1987 and 1994) was certainly more appropriate for the manufacturing sector than the service sector but this changed in 2000 when ISO 9001 and ISO 9004 were completely revised. There are still traces of manufacturing jargon in the standard which shows that it is difficult to find words that have the same meaning in both sectors.

One of the difficulties is that the standards are translated into many languages and the universality of application of the ISO 9000 family of standards requires the use of a technical description without using technical language and a vocabulary that is easily understandable by all potential users of these standards. Thus, ISO 9000:2005 defines the terms used in the family of standards even when some of these terms are words in common use but have acquired a special meaning in the field of quality management.

There are certain terms used in the standards that require further explanation to that given ISO 9000:2005.

Customer

In the context of the ISO 9001, the term customer is always the organization's customer, the client, purchaser, or end user. The customer is the receiver of the product that emerges from the organization's processes. The term customer in ISO 9001 is never an internal customer (a co-worker or groups within the organization).

TABLE 4-2 Overview of the ISO 9000 Series of Standards

Attribute	ISO 9000	ISO 9001	ISO 9004	ISO 19011
Purpose	To facilitate common understanding of the concepts and language used in the family of standards	To provide an equitable basis for assessing the capability of organizations to meet customer and applicable regulatory requirements	To assist organizations achieve sustained success in a complex, demanding, and ever changing, environment	To assist organizations achieve greater consistency and effectiveness in auditing practices
Intent	For use in conjunction with ISO 9001 and ISO 9004. It is invoked in ISO 9001 and therefore forms part of the requirements	This standard is a prescriptive assessment standard used for obtaining an assurance of quality and therefore for contractual and certification purposes only	This standard is a descriptive standard and therefore for guidance only and not intended for certification, regulatory or contractual use	For use in internal and external auditing of management systems
Scope	Defines the principles and fundamental concepts and terms used in the ISO 9000 family	Defines the requirements of a quality management system, the purpose of which is to enable organization to continually satisfy their customers	Describes how organizations can achieve sustained success by application of the quality management principles	Provides guidance on the principles of auditing, managing audit programmes, conducting management system audits and guidance on the competence of management system auditors
Applicability	Applies to all terms used in the ISO 9000 family of standards	Applies where an organization needs to demonstrate its ability to provide products and services that meet customer and regulatory requirements and aims to enhance customer satisfaction	Applies to any organization, regardless of size, type and activity seeking sustained success	Applies to all organizations needing to conduct internal or external audits of quality and/or environmental management systems or to manage an audit programme

Continued

TABLE 4-2 Overview of the ISO 9000 Series of Standards—cont'd

Attribute	ISO 9000	ISO 9001	ISO 9004	ISO 19011
Facts and figures	84 Definitions	8 Sections; 51 Clauses; 263 Requirements	8 Sections; 64 Clauses; No requirements	7 Sections; 45 Clauses; No requirements
Comment	The context and interpretation of the requirements will not be understood without an appreciation of the concepts that underpin the requirements. Also without an understanding of the terms, the standards are prone to misinterpretation	In theory if suppliers satisfy ISO 9001, only conforming product would be shipped. This would reduce the need for customers to verify product on receipt. However, ISO 9001 does not define everything an organization needs to do to satisfy its customers. One omission is the human interaction which influences the pursuit of quality	There are significant benefits in using the standard as a basis for assessing current capability. There is no doubt that if an organization were to follow the guidance given in ISO 9004, it would have no problem in demonstrating it had an effective management system	ISO 19011 expands the requirements of ISO 9001, Clause 8.2.2 on internal auditing. The guidance is equally applicable to any type of management system

Product

In the context of ISO 9001, a product is a result of a process. However, as this result can be intended or unintended the term product in ISO 9001 only refers to intended results. Therefore, product requirements are the requirements for the intended product and not the requirements for unintended product such as waste. Control of nonconforming product is control of intended product that does not conform to requirements. However, the implication in ISO 9001 is that unintended product is outside scope but where customers will be impacted by the unintended product of a process it needs to be controlled, e.g., odour from upholstery.

As results may be given a variety of names such as hardware, software, document, service, material, decision, information etc. it is necessary to use only one generic term for these results and this is the term product. The intention is that for every instance of product in ISO 9001 any one of the other terms could be substituted and the understanding would be preserved. However, product tends to be perceived as something tangible so when the product is intangible such as a pleasant environment for a customer in a restaurant many of the requirements in ISO 9001 simply cannot be applied. For instance, you cannot segregate it, label it, handle it, package it etc. therefore some degree of common sense has to be applied.

Because a product is defined as 'result of a process', it is 'a result' implying that there are other results which a process produces and indeed there are. The results comprise the outputs and the outcomes. Outputs are the effects produced (or emitted) by a process such as a conforming product, nonconforming product, waste material, noise, and odour. These effects may impact customers (see also *A process approach* in Chapter 8).

System

There are many different meanings of the word system but it has only one meaning in the ISO 9000 family and is not used without a qualifier. A system of rules, values or principles is not the type of system that is addressed in the ISO 9000 family. The system addressed by ISO 9004 is the whole organization and the way it functions; it is the system by which the organization accomplishes its mission and vision, i.e., its management system. The system addressed by ISO 9001 is the system that serves the achievement of customer requirements referred to as the quality management system. The system addressed by ISO 14001 is the system that serves the achievement of environmental requirements and referred to as the environmental management system. What this means is that 'a system' is simply a way of looking at an organization as we seek to understand how it functions. It is not a reality. The organization is the reality; the system is a mental construct to enable us to manage the organization effectively. So when we document the quality management system we are only documenting that part of the whole that deals with how we create satisfied customers. However, in this book we have taken the view that it is unwise to limit the quality management system to customer satisfaction when the word quality has a much wider meaning. If all you want is a management system to satisfy customers, you should call it a customer management system (see also *A systems approach* in Chapter 7).

Process

Processes deliver results as expressed above but the term is defined in ISO 9000:2005 as *a set of interrelated or interacting activities which transform inputs into outputs*. This is not strictly correct as there does not have to be a transformation for a process to deliver results. Some processes do change inputs into a different form but it is untrue for all processes, e.g., a purchasing process does not transform an order into goods. It also gives the impression that these processes are somehow limited to the operational aspects of an organization where such transformations occur when in fact processes will be active from the boardroom to the showroom. There is nothing magical about processes, they are all around us. Every result we see has been produced by a process of some type. What ISO 9001 aims to achieve is the effective management of these processes (see also *A process approach* in Chapter 8).

Procedure

Procedures prescribe activities to be performed in the execution of a process but are not a documented process as stated by the ISO 9000:2005 definition. Procedures only deal with the activities people carry out and as will be observed from the above, there will be many interactions that produce the results of a process, only some of which will be carried out by people and only some of these will be prescribed in a procedure. It follows therefore that adherence to procedure is no guarantee of quality.

MISNOMERS

ISO

ISO is not the abbreviation for the International Organization for Standardization because the 'International Organization for Standardization' will have different acronyms in different languages. It was decided to chose '**ISO**', derived from the Greek *isos*, meaning '**equal**'. Whatever the country, or language, the short form of the organization's name is always ISO, but it is often the label used when referring to ISO 9000. The ISO 9000 family has become very well-known outside the community familiar with ISO standards. The ISO 9000 family of standards is just one small group of standards in ISO portfolio of over 17,000 standards, so it would be absurd to use the shorthand 'ISO' as an abbreviation for ISO 9000. Another shorthand is ISO processes, a phrase intended to refer to the processes of an ISO 9001 compliant quality management system but clearly incorrect.

One Standard

Many people are unaware that ISO 9000 is a different standard from the one used for assessment and therefore use the term ISO 9000 in the context of certification when the standard quoted should be ISO 9001 as this is the only standard in the family that is used for certification purposes.

Prior to the year 2000, there were three assessment standards: ISO 9001, ISO 9002 and ISO 9003 and therefore the phrase 'ISO 9000 certification' was excusable if one was referring to certification in general rather than certification to ISO 9001, ISO 9002 or ISO 9003 but ISO 9002 and ISO 9003 became obsolete in 2003 and since then certification should always be with reference to ISO 9001.

Going for ISO 9000?

ISO 9000 is often perceived as a label given to the family of standards and the associated certification scheme. However, certification was never a requirement of any of the standards in the ISO 9000 family – this came from customers. Such notions as 'We are going for ISO 9000' imply that ISO 9000 is a goal like a university degree and like a university degree there are those who pass the exam who will be educated and others who merely pass the exam. You can purchase degrees from unaccredited universities just as you can purchase ISO 9000 certificates from unaccredited certification bodies. The acceptance criteria is the same, it is the means of measurement and therefore the legitimacy of the certificates differs.

Brief Chronology of Management System Standards

The publishers are identified in parenthesis

1956 10CFR 50 Appendix B Quality Assurance Criteria for Nuclear Power Plants and Fuel Reprocessing Plants (US NRC)

1959 Mil-Q-9858, Quality Program Requirements (US DoD)

1968 AQAP 1, NATO Quality Control Requirement for Industry (NATO)

1972 BS 4891, A guide to quality assurance (BSI)

1973 Def Stan 05-21, Quality Control Requirements for Industry (UK MoD)

1974 CSA CAN3 Z299 Quality Assurance Program

1974 BS 5179, Guide to the Operation and Evaluation of Quality Assurance Systems (BSI)

1979 BS 5750, Quality Systems (BSI)

1987 ISO 9000, Quality Management Systems – First series (ISO)

1990 Investors in People Standard (Originally DfEE now IIP)

1992 BS 7750, Specification for environmental management systems (BSI)

1994 ISO 9000, Second series

1995 BS 7799, Information Security management (BSI)

1996 ISO 14001, Environmental management systems – Requirements with guidance for use

1996 BS 8800, Guide to occupational health and safety management systems (BSI)

1997 SA 8000 Social Accountability (Originally the Council on Economic Priorities Accreditation Agency now Social Accountability International)

1999 OHSAS 18001, Occupational Health and Safety Management Systems Specifications (BSI)

2000 ISO 9000, Third series (ISO)

2005 ISO/IEC 27001 Information security management systems requirements (ISO)

2005 ISO 2200 Food Safety Management System – Requirements for any organization in the food chain (ISO)

2008 ISO 9001:2008, Fourth series (ISO)

2009 ISO 9004:2009 Managing for the sustained success of an organization – A quality management approach. Fourth series (ISO)

Similarly, people refer to the ISO 9000 process when what they really mean are processes for managing their organization.

System, Standard or Guide

Some people often think about ISO 9000 as a system. As a group of documents, ISO 9000 is a set of interrelated ideas, principles and rules and could therefore be considered as a system in the same way that we refer to the metric system or the imperial system of measurement.

As many organizations did not perceive they had a quality management system before they embarked on the quest for ISO 9000 certification, the programme, the system and the people were labelled 'ISO 9000' as a kind of shorthand. Soon afterwards, these labels became firmly attached and difficult to shed and consequently that is why people refer to ISO 9000 as a 'system'.

The critics argue that the standards are too open to interpretation to be standards and cannot be more than a framework, model or guide. It is true that application of the ISO 9000 family has produced a wide variation in results from sets of documentation to systems of dynamic processes. Yet if we take a broader view of standards, any set of rules, rituals, requirements, quantities, targets or behaviours that has been agreed by a group of people could be deemed to be a standard. Therefore, by this definition, the documents in the ISO 9000 family are standards.

PUBLISHED INTERPRETATIONS

The ISO technical committee responsible for the ISO 9000 family of standards has a web site (see Appendix B) on which are published sanctioned interpretations of statements in ISO 9001. There are only 37 interpretations numbering RFI 001–052 implying that the process is not well-known. They provide the substance of the request, the background and the interpretation. However, what many fail to do is to provide a complete answer. For example, the request in RFI 001 asks "Does clause 7.4.3 require records of the verification of purchased product?" And the answer given is No, but this is misleading because Clause 8.2.3 requires 'evidence of conformity with the acceptance criteria to be maintained' which translates into records. Other requirements for collecting and analysing data for demonstrating the effectiveness of the quality management system (Clause 8.4) would be difficult to do if there were no records of verification of purchased product. Therefore, a few of the interpretations are questionable and may change in time. These interpretations are addressed by reference number in this book under the appropriate heading.

ISO 9001:2008 COMMENTARY

As mentioned in Preface, ISO 9001:2008 contains no new requirements and therefore it is questionable whether a revision was indeed necessary. Changes can be processed as an amendment in which case the designation would have been ISO 9001:2000 Amd1 or as a new edition and for reasons unknown ISO decided on a new edition even though the degree of change makes this unwarranted. The rules governing the development of management system standards require a Justification Study and this was duly carried out in the period 2003–2004. Feedback was gathered from ISO working groups, user groups and international surveys and this identified the 'need for an amendment, provided that the impact on users would be limited and that changes would only be introduced when there were clear benefits to users.' This therefore ruled out any beneficial changes in requirements that would cause users to change their practices. ISO 9001:2008 is therefore a missed opportunity to raise the requirements to a level commensurate with industry best practice but that may come in the next revision. One of the problems with international standards is that they reflect a consensus. There are many groups with a vested interest that will oppose change if it is too radical no matter how beneficial in the long term.

Although ISO 9001:2008 Annex B does provide an indication of the changes, there is no explanation given as to the impact or the reason for change so the following attempts to fill this gap for some of the more significant changes.

External Influences (Clause 0.1)

Most users of ISO 9001 probably skip the introductory clauses as they contain no requirements. However, a competent auditor would look to these clauses for guidance on what is considered to be important. It is here that there is now recognition that the management system is influenced by forces external to the organization. Although there is no corresponding requirement in the standard, it would not be unreasonable for an auditor to ask "What analysis has been conducted to determine the impact of changes in the business environment on your quality management system?"

Revision of the Process Approach (Clause 0.1)

The description of the process approach has changed. Words have been added in order to clarify that processes should be managed to produce the desired outcome. It would therefore not be unreasonable for an auditor to ask "How do you know that your processes are producing the desired outcome?"

Addition of Statutory Requirements (Clauses 1.1, 1.2, 4.1 & 7.2.1)

The standard now requires organizations to meet customer and applicable statutory and regulatory requirements. There are laws made by statute and regulations for implementing and interpreting these laws which are called statutory regulations. Compliance with a statutory regulation is deemed to be compliance with the law. However, there are regulations other than statutory regulations such as those pertaining to a profession and hence it was necessary to clarify this requirement.

Outsourced Processes (Clause 4.1)

A number of explanatory notes have been added to the requirements on outsourced processes. There has been an increase in the number of organization's outsourcing activities they previously carried out in-house as a way of reducing costs. In principle, using experts to perform activities rather than developing the expertise yourself is sound practice but it does raise a number of problems, especially if you think the only cost to you is what you pay the contractor. These notes are intended to change these perceptions as in reality you may have to do more than what you did when the activities were in-house. Your values were internalized naturally, the internal communication systems reached the people doing the job, work was executed through human interaction not through legal contracts but now these links are broken unless you take action to migrate these informal but vital elements into your contractor's management system.

Record Procedures and External Documents (Clauses 4.2.1 & 4.2.3)

The changes in this requirement rule out the interpretation that

- the only records required were those identified in the standard;
- a procedure has to be a separate document;
- any external documents, whatever their purpose, have to be controlled.

Management Representative (Clause 5.5.2)

The change in this requirement means that you can no longer outsource your management representative unless they have a contract of employment that gives you exclusive use of their services.

Removal of Reference to Product Quality (Clauses 6.2.1 & 6.2.2)

The change in the 2008 edition suggests that work affecting product quality is not the same as work affecting '*conformity to product requirements*'. Work affecting product quality expresses a concept that goes beyond product requirements because until such time in the product development cycle that product requirements have been established, the only basis for judging quality is the organization's perception of customer needs and expectations. In situations where the customer defines the product requirements either in performance terms or in conformance terms, this change has no impact. But where the organization has to translate customer needs and expectations into product requirements, it would appear that the standard no longer requires personnel engaged in such translation work (i.e., developing product requirements) to be competent. How this change brings benefits is unclear as it appears to reduce the value of the standard.

Reversion to Monitoring and Measuring Equipment (Clause 7.6)

ISO 9001:2000, Clause 7.6 referred to measuring devices but this has now been changed to measuring equipment. The advantage of using the term devices was that it reflected any form of measurement and so included non-physical forms such as human senses used in wine tasting or documented criteria as is used in the examination of pupils in the education sector.

ISO 9000:2005 defines measurement equipment as *measuring instrument, software, measurement standard, reference material or auxiliary apparatus or combination thereof necessary to realize a measurement process*. If the measuring instrument can be non-physical, then this change has no impact but the expression 'measuring equipment' will be perceived by many to be physical equipment such as oscilloscopes, micrometers, thermometers etc. How this change brings benefits is unclear as it appears to reduce the value of the standard.

SECTOR SPECIFIC DERIVATIVES

Why Do We Need ISO 9001 Derivatives?

Although the intent of the ISO 9000 family of standards was to reduce the number of international standards in the field of quality management systems by making them generic, a number of sector specific standards have emerged that embody ISO 9001.

The bodies that have developed sector specific standards in this field have all done so because they believed that application of the generic requirements of ISO 9001 would not result in an organization doing the kind of things they expected it to do in order to provide an assurance of product or service quality. A good example is failure modes and effects analysis (FMEA). While there are adequate requirements within ISO 9001 on preventive action, organizations have not been interpreting these requirements such that they would naturally carry out risk assessments let alone FMEA. And more specifically would not perform an FMEA[①] on process design even if they did so on product design. The danger in relying on the generic requirement was that were an auditor to find that FMEA is not being performed, a nonconformity could not be issued against the organization simply because the standard does not stipulate any specific method of preventive action. In the absence of any hard evidence of product or process failures attributable to inadequate risk assessment in product or process design, a competent auditor might challenge the client by questions like "What methods do you use to prevent nonconformity in product and process design?" and if FMEA was not one of the chosen methods, the auditor might ask a follow-up question, "What is considered to be best practice in your industry for risk assessment methods?" If the client had no knowledge of what constitutes best practice in their industry, perhaps a nonconformity could be issued against Clause 5.6.1 on management review as the mechanism for establishing whether the system is effective was clearly inadequate. However, it is much easier for auditors and their clients if the requirements are unambiguous and therefore a simple solution is to impose a series of additional requirements, providing of course they too are unambiguous.

Aerospace Industry

The aerospace industry established the International Aerospace Quality Group (IAQG) in 1998 for the purpose of achieving significant improvements in quality and safety, and reductions in cost, throughout the value stream. This organization includes representation from 57 aerospace companies in the Americas, Asia/Pacific, and Europe including Boeing, Lockheed, BAE Systems, Airbus, EADS, Messier Dowty and Westland, produced a series of standards starting in 2003 with AS 9100. The objectives of IAQG are achieved not only through the basic requirements of AS9100 but also through a family of over 20 standards.

Automotive Industry

In 1996 the International Automotive Task Force (IATF) was established comprising representatives of nine vehicle manufacturers and five national motor trade associations from the Americas and Europe. The nine vehicle manufacturers being BMW Group, Chrysler, Daimler AG, FIAT, FORD, GM, PSA, Renault and VW AG. The five national motor associations being ANFIA, Italy, AIAG, USA, FIEV, France, SMMT, UK and VDA-QMC, Germany. These nations together with representatives from ISO/TC 176 developed a sector standard that became ISO/TS 16949. This technical specification (hence the TS designation) incorporated Section 4 of ISO 9001:1994 and included requirements taken from QS-9000, VDA 6, AVSQ '94 and EAQF '94 and some new requirements, all of which have been agreed by the international members. In March 2002, ISO/TS16949:2002 (aka TS2) was published which embodied ISO 9001:2000 in its entirety. The Japanese Automobile Manufacturers Association (JAMA) was involved as the representative of HONDA, NISSAN and Toyota.

ISO/TS 16949 is not only a technical specification, there is a certification scheme managed on behalf of IATF by five oversight offices located within the national associations that is significantly different from that used for ISO 9001 certifications. The IATF has a closely defined and monitored recognition process for certification bodies as well as a stringent development and qualification criteria for auditors, including a written knowledge and application assessment including a 40-min interview with two qualified evaluators. Oversight offices conduct office assessment and witness audits on a global basis covering all the activities of the recognized certification bodies.

Computer Software

An ISO/IEC technical group published the international standard ISO/IEC 90003 in 2004 containing software engineering guidelines for the application of ISO 9001:2000 to computer software. The standard applies to the acquisition, supply, development, operation and maintenance of computer software and related support services. ISO/IEC 90003:2004 does not add to or otherwise change the requirements of ISO 9001:2000; what it does is to bring unity to what has been an increasingly fragmented approach, given the sheer number of software engineering standards being developed. It cross-refers the many existing discipline-specific standards that already exist to support a software organization's quality programme, e.g., ISO/IEC 12207 Life Cycle Models.

Food and Drink Industry

ISO 15161:2001 on the application of ISO 9001:2000 for the food and drink industry gives information on the possible interactions of the ISO 9000 family of standards and the hazard analysis and critical control points (HACCP) system for food safety requirements. The standard is not intended for certification, regulatory or contractual use.

Medical Devices

In the medical devices sector, ISO 13485:2003 specifies requirements for a quality management system where an organization needs to demonstrate its ability to provide medical devices and related services that consistently meet customer requirements and regulatory requirements applicable to medical devices and related services.

The primary objective of ISO 13485:2003 is to facilitate harmonized medical device regulatory requirements for quality management systems. As a result, it includes some particular requirements for medical devices and excludes some of the requirements of ISO 9001 that are not appropriate as regulatory requirements. Because of these exclusions, organizations whose quality management systems conform to this International Standard cannot claim conformity to ISO 9001 unless their quality management systems conform to all the requirements of ISO 9001. A number of supporting standards in the medical sector have also been published (see Appendix B).

Petroleum, Petrochemical and Natural Gas Industries

ISO/TS 29001 is intended to ensure safe and reliable equipment and services throughout the international oil and gas industries. It provides these sectors with a unique requirements document for quality management (ISO/TS 29001:2006). This standard defines the quality management system requirements for the design, development, production, installation and service of products for the petroleum, petrochemical and natural gas industries. The standard is based on ISO 9001:2000. Limits are addressed on claims of conformity to ISO/TS 29001:2006 if exclusions are made. ISO/TS 29001:2006 does not address competitive or commercial matters such as price, warranties, guarantees, or clauses intended to sustain commercial objectives.

Telecommunications Industry

In the spring of 1996, a group of leading telecommunications service providers initiated an effort to establish better quality requirements for the industry. This group became known as the Quality Excellence for Suppliers of Telecommunications (QuEST) Forum. Its goal was to create a consistent set of quality system requirements and measurements that would apply to the global telecom industry and that, when implemented, will help provide telecommunications customers with faster, better and more cost-effective services. The Quest Forum comprises currently 24 service providers including Verizon, Motorola, British Telecom, France Telecom, China Telecoms, Deutsche Telekom AG, and 65 telecoms suppliers including Lucent, Nortel Networks, Nokia, Corning Cable Systems, Alcatel, SPRINT, Fujitsu, Hitachi and Unisys.

TL 9000 differs from the other ISO 9001 derivatives in that the telecoms sector takes performance measurement very seriously and have added measurement criteria and set up a Measurements Repository System (MRS). The MRS enables customers who are Forum members to access supplier performance data relative to TL 9000 measurement requirements. The measurements are divided into five groups: Common Measurements comprising such things as problem reports, response time and on-time delivery, Outage measurements, Hardware measurements, Software measurements and Service measurements.

Like ISO/TS 16949, TL 9000 is very prescriptive which seems to suggest that customers in these industries are not yet ready to trust suppliers to do the right things right first time without telling them what to do.

Systems Engineering Sector

The first national standard addressing systems engineering emerged in 1969 with Mil Std 499, revised in 1974 as Mil Std 499A and was probably the most widely quoted and used standard in this sector. Systems engineering is an interdisciplinary field of engineering that focuses on how complex engineering projects should be designed and managed throughout the life cycle of the system. Large complex systems often involve multiple contractors in different countries such as aerospace, and defence systems. Recognizing that the systems engineering concepts and techniques used in the aerospace and defence sectors are equally applicable to any industry producing or maintaining complex systems, in 2002 ISO/IEC 15288 was published and offered a portfolio of generic processes for the optimal management of all stages in the life of any product or service, in any sector. This was followed by ISO/IEC TR 90005:2008 which provided guidance for organizations in the application of ISO 9001:2000 for the acquisition, supply, development, operation and maintenance of systems and related support services. It does not add to or otherwise change the requirements of ISO 9001:2000. The guidelines provided are not intended to be used as assessment criteria in quality management system registration or certification. ISO/IEC TR 90005:2008 adopts ISO/IEC 15288 systems life cycle processes as a starting point for system development, operation or maintenance and identifies those equivalent requirements in ISO 9001:2000 that have a bearing on the implementation of ISO/IEC 15288.

COMPARING ISO 9001 WITH OTHER MODELS

The Excellence Model

Business excellence is symbolized by a number of models. These models are based on the premise that excellent results with respect to performance, customers, people and society are achieved through leadership driving policy and strategy that is delivered through people, partnerships and resources, and processes. In Europe excellence is promoted by the European Foundation for Quality Management (EFQM) through the EFQM Excellence Model (see Fig. 8-12). In the USA excellence is promoted by the National Institute of Standards and Technology (NIST) through the Baldrige National Quality Program (BNQP). Around the world, several Quality Awards are being presented for excellence using the same or similar models.

One characteristic that distinguishes these models from ISO 9001 is that the models are a non-prescriptive framework that recognizes there are many approaches to achieve sustainable excellence. Another is the award scheme. Excellence awards are generally annual events where winners collect awards. The award does not have a time limit although an organization can apply again after a suitable lapse, usually five years. However, unlike the ISO 9001 certification scheme there is no continuous assessment and organizations don't lose the award for failing to maintain standards.

There are many similarities between the business excellence concepts and the quality management principles used as a basis for ISO 9000 series of standards as shown in Table 4-3. (Note: the 8QM principles are explained in Chapter 1.)

The differences between the Business Excellence Model and the Quality Management Principles are small enough to be neglected if you take a pragmatic approach. If you take a pedantic approach you can find many gaps but there are more benefits to be gained by looking for synergy rather than for conflict.

Six Sigma

Six sigma is a rigorous and disciplined methodology that uses data and statistical analysis to measure and improve a company's operational performance by identifying and eliminating process 'defects' (see *Six sigma* in Chapter 31). There is nothing in the Six Sigma methodology that should not be included in an ISO 9001 compliant QMS. Table 4-4 shows alignment with the DMAIC technique which is the enabling methodology for six sigma programmes. All the elements of the techniques can be matched with the requirements of ISO 9001.

Other characteristics of the six sigma methodology are a focus on leadership commitment, managing decisions with data, training and cultural change. All of these

TABLE 4-3 Comparison between Business Excellence Principles and Quality Management Principles

Business excellence concepts	Equivalent quality management principles
Customer focus: The customer is the final arbiter of product and service quality and customer loyalty, retention and market share gain are best optimized through a clear focus on the needs of current and potential customers	Customer focus:
Leadership and constancy of purpose: The behaviour of an organization's leaders creates a clarity and unity of purpose within the organization and an environment in which the organization and its people can excel	Leadership:
People development and involvement: The full potential of an organization's people is best released through shared values and a culture of trust and empowerment, which encourages the involvement of everyone	Involvement of people:
Management by processes and facts: Organizations perform more effectively when all interrelated activities are understood and systematically managed and decisions concerning current operations and planned improvements are made using reliable information that includes stakeholder perceptions	Process approach Factual approach to decision making Systems approach
Continuous learning, innovation and improvement: Organizational performance is maximized when it is based on the management and sharing of knowledge within a culture of continuous learning, innovation and improvement	Continual improvement
Partnership development: An organization works more effectively when it has mutually beneficial relationships, built on trust, sharing of knowledge and integration with its partners	Mutually beneficial supplier relationships
Public responsibility: The long-term interest of the organization and its people are best served by adopting an ethical approach and exceeding the expectations and regulations of the community at large	There is no equivalent principle in ISO 9000:2005; however, ISO 9004, Clause 5.2.2 stresses that the success of the organization depends on understanding and considering current and future needs and expectations of the stakeholders
Results orientation: Excellence is dependent on balancing and satisfying the needs of all relevant stakeholders	There is no equivalent principle in ISO 9000:2005; however, ISO 9004, Clause 5.2.2 does recommend that an organization should identify its interested parties and maintain a balanced response to their needs and expectations.

TABLE 4-4 Comparison between Six Sigma Methodology and ISO 9001

	Six sigma methodology[3]	ISO 9001 requirement
Define	Define the Customer and their Critical to Quality (CTQ) issues	Ensure that customer requirements are determined (5.2) Ensure that quality objectives are established to meet requirements for product (5.4.1)
	Define the Core Business Process involved	Plan the QMS in order to meet the quality objectives (5.4.2) Determine the processes needed for the QMS (4.1a)
	Define who customers are, what their requirements are for products and services, and what their expectations are	Identifying customer requirements (7.2.1)
	Define project boundaries – the stop and start of the process	Determine the sequence and interaction of processes (4.1b)
	Define the process to be improved by mapping the process flow	Continual improvement of QMS processes (4.1e)
Measure	Measure the performance of the Core Business Process involved	Monitoring and measurement of processes (8.2.3)
	Develop a data collection plan for the process	Plan the analysis processes needed to continually improve the effectiveness of the quality management system (8.1)
	Collect data from many sources to determine types of defects and metrics	Collect appropriate data to demonstrate the suitability and effectiveness of the quality management system (8.4)
	Compare to customer survey results to determine shortfall	Monitor information relating to customer perception as to whether the organization has fulfilled customer requirements (8.2.1)

Analyse	Analyse the data collected and process map to determine root causes of defects and opportunities for improvement	Analyse appropriate data to demonstrate the suitability and effectiveness of the quality management system (8.4)
	Identify gaps between current performance and goal performance	Evaluate where continual improvement of the quality management system can be made (8.4)
	Prioritize opportunities to improve	Evaluate the need for action to ensure that nonconformities do not recur (8.5.2c)
	Identify sources of variation	Use of statistical techniques (8.1)
Improve	Improve the target process by designing creative solutions to fix and prevent problems	Determine and implement corrective actions needed (8.5.2d)
	Create innovate solutions using technology and discipline	Implement actions necessary to achieve continual improvement of QMS processes (4.1e)
	Develop and deploy implementation plan	Implement actions necessary to achieve planned results and continual improvement of these processes (4.1e)
	Control the improvements to keep the process on the new course	Determine criteria and methods (4.1c) Monitoring and measurement of processes (8.2.3) Monitoring and measurement of product (8.2.4)
Control	Prevent reverting back to the 'old way'	Determine criteria and methods needed to ensure that both the operation and control of these processes are effective (4.1b) Internal audit (8.2.2)
	Require the development, documentation and implementation of an ongoing monitoring plan	Apply suitable methods for monitoring and, where applicable, measurement of the quality management system processes (8.2.3)
	Institutionalize the improvements through the modification of systems and structures (staffing, training, incentives	Ensure the integrity of the QMS is maintained when changes are planned and implemented (5.4.2)

[3]http://www.isixsigma.com/dictionary/DMAIC-57.htm.

should be found in any programme to achieve ISO 9001 certification. Although, six sigma methodology is a problem solving technique, DMAIC could be used to design an effective management system in the same way that using PDCA or the Process Management principles might produce an effective management system.

Business Process Management (BPM)

The concept of business process management has been around for a number of years but like so many management concepts, it has evolved into different forms, thus causing communication difficulties.

Harmon[2] writes of the ambiguity about the phrase business process management and cites two different interpretations.

- *Business process management* (lower case) to refer to aligning processes with the organization's strategic goals (this is the way the term is used in this book).
- Business Process Management (BPM or BPMS) to refer to systems that automate business processes. For example, workflow systems, XML business process languages and packaged Enterprise Resource Planning (ERP) systems. (The 'S' in the acronym is used for 'systems' or 'suites'.)

Harmon also observes that *"BPMS products will play a major role in the development of the corporate use of business processes"*. He advises that *"before a company is ready to automate its processes it first needs to understand them and be confident that the process works well"*. This is good advice and puts BPM (uppercase) in context. Only when you have understood the management system in the way described in Part 2 of this book, modelled the system and its interacting processes as described in Parts 3, put the processes in place as described in Parts 4–6 and have good measurement processes in place as described in Part 7, you will be in any way prepared to contemplate moving to BPMS.

Burlton[3] also puts business process management in context. *"Business process management is a way of thinking and of managing that recognizes business processes as capabilities and hidden assets of the enterprise. They are hidden assets because they aren't found on a balance sheet or annual report. BPM is the discipline that improves measurable business performance for stakeholders, through ongoing optimization and synchronization of enterprise-wide process capabilities"*.

Cleary business process management and BPM have the same purpose as the ISO 9000 family of standards, that of sustained success but go about it in a different way. The application of ISO 9001 can lead to effective business process management when approached in the right way, but applications of ISO 9001 can fall far short of this as we will show in Chapter 6. In Table 4-5, the principles and characteristics of BPM as espoused by Andrew Spanyi in *In Search of BPM Excellence* are contrasted with the equivalent principles in the ISO 9000 series of standards. (Note: the 8QM principles are explained in Chapter 1).

---

[2] Harmon Paul (2007) Business Process Change: A Guide for Business Managers and BPM and Six Sigma Professionals. 2nd edition, Morgan Kaufmann.

[3] Burlton Roger (2005) In Search Of BPM Excellence: Straight From The Thought Leaders, *Meghan-Kiffer Press, Tampa, FL USA.*

TABLE 4-5 Comparing BPM with ISO 9000 Series

Characteristic	BPM	ISO 9000 series
Purpose	Sustained success	ISO 9001 – Meeting customer requirements ISO 9004 – Sustained success
Scope	All processes	ISO 9001 – Those processes that provide the capability to provide product that meets customer and applicable regulatory requirements ISO 9004 – All processes
Focus	First Principle – Successful organizations look at themselves from the outside-in, from the customers perspective as well as the inside out	Customer Focus principle
Strategy	Second principle – Work gets done through cross-functional business processes therefore the strategy should be integrated with the business processes	Process approach principle
Shared vision	Third principle – A successful strategy deployment inspires from the boardroom to the lunchroom and remains front and centre throughout the year	Leadership principle
Process alignment	Fourth principle – Successful business processes are designed to deliver on the organization's strategic goals	Systems approach to management
Organization design	Fifth principle – A successful organization design will enable business process execution • Design stakeholder relationships before processes • Design processes before technological capabilities • Design processes before human competencies • Design processes and human competencies before organizations	ISO 9000:2005 Step 1: determine the needs and expectations of customers and other interested parties Step 2: establish the quality policy and quality objectives of the organization Step 3: determine the processes and responsibilities necessary to attain the quality objectives Step 4: determine and provide the resources necessary to attain the quality objectives Step 5: establish methods to measure the effectiveness and efficiency of each process Step 6: apply these measures to determine the effectiveness and efficiency of each process Step 7: determining means of preventing nonconformities and eliminating their causes Step 8: establish and apply a process for continual improvement of the quality management system

Chapter 5

A Practical Guide to Using these Standards

CHAPTER PREVIEW

This chapter is aimed at students, consultants, auditors, managers and others who will be considering the ISO 9000 family of standards for use as part of a course of study, a programme of improvement or as a basis for assessment and subsequent certification.

There are three ways of using these standards:

- As a source of information on best practice that can be consulted to identify opportunities for improvement in business performance;
- As set of requirements and recommendations that are implemented by the organization;
- As criteria for assessing the capability of a management system or any of its component parts.

In this chapter we address the pros and cons of either consulting, implementing or applying management system standards and in particular

- What you should do before, during and after consulting these standards
- The misnomer of implementing ISO 9001.
- How to go about applying these standards from the point of view of the organization and a customer or third party
- The critical success factors in ISO 9000 programmes and the level of attention because unless attention is pitched at the right level these programmes will fail.

CONSULTING THE STANDARDS

The standards capture what may be regarded as best practice in a particular field. The information has been vetted by those deemed to be experts by ISO member states and therefore one can defer to any of these standards as a legitimate authority in the absence of anything more appropriate. They are, however, but one of several sources of authoritative information.

With this caveat in mind, these standards can be useful in:

- forming ideas,
- settling arguments.

ISO 9000 Quality Systems Handbook

- clarifying terminology, concepts and principles,
- identifying the right things to do,
- identifying the conditions for ensuring things are done right.

Before Consulting the Standards

Before consulting any of the standards, either a need for improvement in performance or a need for demonstration of capability should have been identified and agreed with the senior management.

Ideally the objectives for change and a strategy for change should also have been established in order to indicate the direction and the means of getting there. This will place these standards in the correct context. Consulting the standards before doing this will prejudice the strategy and may result in compliance with the standard becoming the objective thereby changing perceptions as to the motivator for change (further information on making a case for change are included in Chapter 7 in *Quality management essentials*).[1]

The need for improvement might arise from:

- a performance analysis showing a declining market share or significant number of customer complaints either with the product or the associated services;
- a competitor analysis showing that resource utilization needs to be increased to compete on price and delivery;
- a market analysis showing a demand for confidence that operations are being managed effectively. This might arise from EU directives.
- an analysis of the environment identifies opportunities for creating new markets, products or services.

If the organization is currently satisfying its stakeholders but lacks a means of demonstrating its capability to customers or regulators that demand it, certification to assessment standard ISO 9001 may provide a satisfactory solution but it is not the only solution unless given no option by the customer.

While Consulting the Standards

There is no doubt that ISO 9001 is the top selling international standard of all time but other standards in the family have not had similar success, which creates a major problem with the use of these standards.

When consulting these standards, bear in mind the following:

- They reflect the collective wisdom of various organizations and 'experts' that participate in the development of national and international standards.
- They have been produced by different committees and therefore as a group of standards will contain inconsistencies, ambiguities and even conflicting statements. They do not as yet fit together as a system with all prescriptions and descriptions aligned to an overarching purpose and set of principles.
- Compromises often have to be made in order for the standards to be accepted by at least 75% of the voters in the ISO community.

---

[1] Hoyle David (2006) Quality Management Essentials, Butterworth Heinemann.

- What you read is not necessarily the latest thinking on a topic or the result of the latest research primarily because of the review cycle (often five years) being so protracted.
- The standards reflect practices that are well proven and possibly now outdated in some quarters but have stood the test of time and are used universally.
- Common terms may be given an uncommon meaning but terminology is by no means consistent across this class of standards thus making their use more difficult (management system, correction and corrective action being typical examples).
- Some phrases might appear rather unusual in order to preserve meaning when translated into other languages (use of the term interested party in place of stakeholder is one example) and as a result create ambiguities.
- Requirements are not necessarily placed in their true relationship and context due to the constraints of the medium by which the requirement are conveyed. As a result, users and auditors often treat requirements in isolation when in fact they are all interrelated (the misplacement of requirements for measuring equipment[①] being one example see Introduction to Part 6).

Although there is the opportunity for changing these standards, there may not be any desire for change because of the various vested interests. If organizations have based their approach on one or more of these standards they will be reluctant to sponsor any change that might result in additional costs, regardless of the benefits. These organizations might be willing to institute the changes informally rather than to have them imposed through an externally assessed standard. This is particularly true for the 2008 revision of ISO 9001. In the eight years since the previous edition there have been developments in the management sciences such as systems thinking, business process management and Theory Z[①], but no changes in requirements within ISO 9001 were made. It may well take a generation for such changes in management practice to be embedded within the majority of organizations and for ISO 9001 to catch up when there is a consensus for change.

When a family of standards is embraced, studied and applied intelligently, there can be enormous benefits from its use. However, standards of this type can lend themselves to misuse by spreading the information so widely across a number of documents and by not translating the concepts into requirements with a clarity that removes any ambiguity.

> **Axiom**
>
> 1. Understand the intent
> 2. Understand the impact
> 3. Understand how to make it happen
> 4. Understand how to make it a habit

The most important factor is that whatever the statement in these standards, it is necessary to understand the intent, i.e., what it is designed to achieve. There is simply no point in following advice unless you fully understand the consequences (i.e., what the impact will be) and have a good idea of what you might have to do to make it happen and to sustain the benefits it will bring. Sustained levels of performance will only arise when the new practices become ingrained in the culture and become habitual. This makes it imperative that you do not limit your reading to ISO 9001 alone but also include the guidance standards and other relevant literature.

After Consulting the Standards

Having consulted the standards you need to:

- put your findings in context as not everything you read will be applicable in your organization.
- assess the impact (benefits, drawbacks) on the organization of applicable provisions.
- validate your findings with other sources (books, articles, peers etc.).

If it seems like what is expressed in the standards accords with best practice and offers practical benefits then by all means follow the advice given.

Before You Change Anything

At some stage after you have obtained your ISO 9001 certificate, people will ask you about the benefits it has brought to the organization. Unless you capture the state of the organization and its performance beforehand you can only provide subjective opinion. It is therefore highly advisable to record a series of benchmarks that you can use later to determine how far you have progressed. A simple model is provided in *Self assessment* in Chapter 38 relative to the quality management principles but you also need measurements against your key performance indicators[①] such as:

- Time to market (time it takes to get a new product into the market);
- Customer satisfaction (customer perception of your organization and its products);
- Conformity (measure of conformity, e.g., ratio of the number of products returned to those shipped or system availability if you are a service provider such as phone company, energy supplier etc.);
- Supplier relationships (supplier perception of your organization and the way you deal with them);
- On-time delivery;
- Processing delays (the impact of shortages, bottlenecks, down time);
- Employee satisfaction (employee perception of your organization and the way you attend to their needs);
- External failure costs (costs of correcting failures after product delivery);
- Internal failure costs (costs of correcting errors detected before product delivery);
- Appraisal costs (costs of detecting errors).

Remember to use the same measurement process after certification otherwise the results will be invalidated.

IMPLEMENTING MANAGEMENT SYSTEM STANDARDS

Some ways in which these standards have been promoted have not helped their cause because they have been perceived as addressing issues separate from the business of managing the organization. Invariably organizations are being told to implement ISO 9000 or some other standard but implementation is often not the best approach to take. Hence, in response, some organizations have set up new systems of documentation that run in parallel to the operating systems in place.

Regrettably, certification has followed implementation and it is certification that has driven the rate of adoption rather than a quest for economic performance.

When we implement something we put it into effect, we fulfil an obligation. In fact many organizations have implemented these standards because they have put it into effect and fulfilled an obligation to do as required and recommended by the standard.

Implementation implies we pick up the standard and do what it requires. As the

> **Stop What You are Doing**
>
> The biggest mistake many make is in following the ritual; *Document what you do, do what you document and prove it* and continue to pursue activities and behaviours that adversely affect performance.
>
> This approach is like taking medicine but continuing the lifestyle that prompted the medication in the first place.

standards don't tell us to stop doing those things that adversely affect performance, these things continue. If the culture is not conducive for the pursuit of quality, these things will not only continue but also make any implementation of standards ineffective.

Doing as the standards require will not necessarily result in improved performance. A far better way is to consult the standards (as described above), establish a management system that enables the organization to fulfil its goals (as described in *Establishing a quality management system* in Chapter 10), then assess the system by applying the standards as described below.

APPLYING MANAGEMENT SYSTEM STANDARDS

By the Organization

If we apply these standards instead of implementing them, we design a system that enables us to achieve our goals then use ISO 9001 to assess whether this system conforms to the requirements. The guides may help you consider various options, even find the right things to do, but it is your system, your organization so only you know what is relevant.

In applying these standards you should not create a separate system but look at the organization as a system of processes and look for alignment with the requirements and recommendations of the various standards. This is self assessment and is addressed further in Chapter 38. Only change the organization's processes to bring about an improvement in its performance, utilization of resources or alignment with stakeholder needs and expectations. Where there is no alignment:

- verify that the requirement is really applicable in your circumstances;
- change the organization's processes only if it will yield a business benefit.

Changing a process simply to meet the requirements of a standard is absurd, there has to be a real benefit to the organization. If you can't conceive any benefit, take advice from experts who should be able to explain what benefits your organization will get from a change.

By the Customer or a Third Party (Conformity Assessment)

Customers and third party certification bodies use the assessment standards such as ISO 9001 and ISO 14001 to determine the capability of other organizations to satisfy certain requirements (customer, environment, security etc.). This is called conformity asses-sment[©] which refers to a variety of processes whereby goods and/or services are deter-mined to meet voluntary or mandatory standards or specifications. Conformity assessment is therefore limited to the scope of the standard being used and thus (unlike the excellence model or the self assessment criteria in ISO 9004) it is not intended to grade organizations on their capability. An organization either conforms or it doesn't conform.

ISO 9001 was primarily intended for situations where customers and suppliers were in a contractual relationship. It was not intended for use where there were no contractual relationships. It was therefore surprising that schools, hospitals, local authorities and many other organizations not having a contractual relationship with their 'customer' would seek ISO 9001 certification. However, since publication of the 1994 version, ISO 9001 has been applied in non-contractual situations with the result that organizations created over complicated systems with far more documentation than they needed. In non-contractual situations there is usually no need to demonstrate potential capability. Customers normally purchase on the basis of recommendation or prior knowledge. Even in contractual situations, demonstration of capability is often only necessary when the customer cannot verify the quality of the products or services after delivery (see *EU directives* in Chapter 1). The customer may not have any way of knowing that the product or service meets the agreed requirements until it is put into service by which time it is costly in time, resource and reputation to make corrections. In cases where the customer has the capability to verify conformity, the time and effort required are added burden and their elimination help to reduce costs to the end user.

ISO 9001 was a neat solution to this problem as it embodied most of the requirements customers needed to obtain an assurance of quality. Any additional requirements could be put into the contract. Standardization in this case improved efficiency in getting orders out. However, in the mad rush to use ISO 9001, the buyers in the purchasing departments overlooked a vital step. Having determined the product or service to be procured and the specification of its characteristics, they should have asked themselves a key question:

> "Are the consequences of failure such that we need the supplier to demonstrate it has the capability to meet our requirements or do we have sufficient confidence that we are willing to compensate for any problems that might arise?"

In many cases, using ISO 9001 as a contractual requirement was like using a sledge-hammer to crack a nut – it was totally unnecessary and much simpler models should have been used.

LEVEL OF ATTENTION TO QUALITY

In the first section of the Introduction to ISO 9001 there is a statement that might appear progressive but depending on how it is interpreted, it could be regressive. The statement is: *"The adoption of a quality management system should be a strategic decision of an organization"*. What would top management be doing if they did this? Would they be:

- Agreeing to implement the requirements of ISO 9001 and subject the organization to periodic third party audit as evidence of commitment to quality?

- Agreeing to document the approach they take for the management of product quality and to subsequently do what they have documented?
- Agreeing to manage the organization as a system of interacting processes delivering stakeholder satisfaction?

It all comes down to their understanding of the word *quality* and this is what will determine the level of attention to quality.

Whilst the decision to make the *management of quality* a strategic issue will be an executive decision, the attention it is given at each level in the organization will have a bearing on the degree of success attained. There are three primary organization levels: the *enterprise level, the business level* and the *operations level.*[2] Between each level there are barriers.

At the enterprise level, the executive management responds to the 'voice' of the stakeholders and on one level is concerned with profit, return on capital employed, market share etc. and on another level with care of the environment, its people and the community. At the business level, the managers are concerned with products and services and so respond to the 'voice' of the customer. At the operational level, the middle managers, supervisors, operators etc. focus on processes that produce products and services and so respond to the 'voice' of the processes carried out within their own function.

In reality, these levels overlap, particularly in small organizations. The Chief Executive Officer (CEO) of a small company will be involved at all three levels whereas in the large multinational, the CEO spends all of the time at the enterprise level, barely touching the business level, except when major deals with potential customers are being negotiated. Once the contract is won, the CEO of the multinational may confine his or her involvement to monitoring performance through metrics and goals.

Quality should be a strategic issue that involves the owners because it delivers fiscal performance. Low quality will ultimately cause a decline in fiscal performance.

The typical focus for a quality management system is at the operations level. ISO 9001 is seen as an initiative for work process improvement. The documentation is often

TABLE 5-1 Attention Levels

Organizational level	Principle process focus	Basic team structure	Performance issue focus	Typical quality system focus	Ideal quality system focus
Enterprise	Strategic	Cross-business	Ownership	Market	Strategic
Business	Business	Cross-functional	Customer	Administrative	Business process
Operations	Work	Departmental	Process	Task process	Work process

---

[2] Watson, Gregory H., (1994). *Business Systems Engineering*, Wiley.

developed at the work process level and focused on functions. Much of the effort is focused on the processes within the functions rather than across the functions and only involves the business level at the customer interface, as illustrated in Table 5-1. For the application of ISO 9001 to be successful, quality has to be a strategic issue with every function of the organization embraced by the management system that is focused on satisfying the needs of all stakeholders.

Key Messages from Part 1

1. ISO 9001 serves two fundamental needs, a quest for confidence by customers and a quest for capability by suppliers.
2. Customers need to be able to trust the products and services they use, a concept that is as old as trade itself but where they cannot judge the quality for themselves, they use other methods and the ISO 9000 family of standards was developed for this purpose.
3. A quest for confidence translates into a need for an assurance of quality and this resulted in the adoption of assurance principles. The quest for capability translates into a need for effective management and this resulted in the adoption of eight quality management principles.
4. The standards changed direction in 2000 away from a belief that adherence to procedures produces products that satisfy customers to the belief that satisfied customers result from a system of managed processes, the objectives of which have been derived from customer needs.
5. Certification to ISO 9001 is not a requirement of any standards in the ISO 9000 family but may be a requirement of customers and entry into certain markets.
6. ISO 9001 certification is invoked in EU Directives because it infers that certified organizations have a system that ensures their products comply with applicable statutory and regulatory requirements.
7. There are three attainment levels of quality each having equal merit in the right context. There is the understanding that quality means (1) freedom from deficiencies, (2) conformity with customer requirements, (3) satisfying customer needs and expectations. As we move through these levels we build a stronger foundation for sustained success. By confining our understanding to freedom of deficiencies we may be deluding ourselves that we are competitive on quality but we will lose our customers to other organizations that rise through levels 1 and 2.
8. Quality does not mean high price, high class or high grade but whatever the need, it's the extent to which that need is fulfilled.
9. Quality is defined by measurable features or characteristics of a product or service such as reliability, appearance or comfort, courtesy and responsiveness.
10. All organizations have stakeholders that contribute to its wealth-creating capacity and benefit from it. It is therefore inconceivable that managers can ignore any one of these stakeholders and sustain success.
11. All stakeholders can be placed in one of five stakeholder groups, namely, customers, investors, employees, suppliers and society; offend any one of these

stakeholders and pressure will be brought to bear that will force a change in direction eventually.

12. For an organization to succeed it needs to derive its objectives from customers needs and to derive the measures for determining how well these objectives are being achieved from the constraints imposed by all the stakeholders.

13. Quality is like ballet not hockey. The result of a game of hockey is unpredictable. In an organization individuals need to perform predetermined roles that are intended to achieve predictable results in the same way as ballet; right first time and every time.

14. An effective way to use the ISO 9000 family of standards is to use ISO 9000:2005 to get to grips with the concepts and vocabulary. Then use ISO 9001 and ISO 19011 to demonstrate the capability of your management system to meet customer and legal requirements and finally you use ISO 9004 to develop your capability to sustain success Don't use ISO 9001 to develop your system.

15. It is important to use the right language in connection with ISO 9000 so as to transmit the correct messages to the workforce, to customers and suppliers. Abbreviated language can send out the wrong message entirely and result in systems that add no value.

16. The eight quality management principles on which ISO 9001 has been based compare well with the European Excellence Model, Six Sigma Methodology and Business Process Management (BPM) but are deficient on critical to quality (CTQ) issues, public responsibility and the needs of all stakeholders being as they are, limited to a customer focus.

17. Before consulting the standards set a benchmark that you can refer back to after you have used the ISO 9000 family of standards to measure the benefits any improvement may bring.

18. Understand the intent and the impact of the requirements before proceeding and understand that if you expect improvement in performance you will have to change your habits. Documenting processes will not change anything except your understanding of what you do and how you do it.

19. Don't implement ISO 9001 as this is not the best use of this standard. ISO 9001 is an assessment standard not a design standard. The best approach is to design a system of managed processes that enables your organization to achieve its goals and then use ISO 9001 to assess its readiness for certification. If there are gaps, only change your processes if by doing so it brings a business benefit.

20. Your success in using the ISO 9000 family of standards to assist your pursuit of quality depends on management commitment. The adoption of a quality management system should be a strategic decision. Management should therefore be committed to make quality a strategic issue and the development of the system of processes for delivering quality products and services a strategic decision.

Approaches to Achieving, Sustaining and Improving Quality

INTRODUCTION TO PART 2

The title to this part of the Handbook can be paraphrased as getting there, staying there and getting better – three common aspirations of all organizations. Where 'there' is, is the level of quality that will lead to sustained success. The work involved in achieving the desired level of quality will be different from the work involved in sustaining a level of quality and different again from improving the level of quality.

As explained previously, quality is a result produced when a need, expectation, requirement or demand is met or satisfied. This result (quality) is what most people try to achieve in whatever they do, in all organizations and families. Most of us want the result of our efforts to satisfy the need, expectation, requirement or demand however it is expressed. ISO 9001 does not define how you should develop a management system; it does not provide any guidance apart from emphasizing the importance of managing processes in the introduction. So how do we go about it?

A TASK BASED APPROACH

One approach is to prescribe what has to happen then supervise adherence to these rules or procedures. This was the Taylor System of management conceived by Frederick Winslow Taylor (1856–1915). Taylor formulated four principles of scientific management which were:

1. Develop a science for each element of a man's work. This is almost equivalent to the ISO 9001 requirement for documented procedures.
2. Scientifically select and then train, teach and develop the workman. This is equivalent to the ISO 9001 requirement for training and competence.
3. Heartily cooperate with the men so as to ensure all of the work is being done in accordance with the principles of the science that have been developed. This is equivalent to the ISO 9001 requirements for verification and audit.
4. The management takes over all of the work for which it is better fitted than the worker. This is not prescribed by ISO 9001 except for requiring certain activities of top management and a management representative.

Taylor clearly recognized that in an industrial age, work needed to be managed as a system and that management and workers are partners within it and not adversaries. But it was, as Taylor admitted, a "task based system". Taylor is credited with the idea of separating decision making from work that at one extreme is interpreted as 'leaving your brain at the door'. In simple terms this leads to a separation between planners and doers but as workers became more educated they could undertake more of the planning and see planning and doing as two roles rather than two jobs.

A RISK BASED APPROACH

Another approach is to identify the risks to achieving quality and then manage these risks effectively. This was how ISO 9001 evolved. It was a compilation of the measures taken to remove the risk of shipping defective product to customers so they focused on the prevention, detection and correction of defects.

This started alongside Taylor's task based system by introducing inspection as a means of sorting good products from bad products. In an attempt to reduce end of line rejects, in-process inspection was introduced and eventually defect investigation cells were created to discover the cause of the rejects and put in place measures to prevent recurrence. The concept developed to an extent where there were

- Final inspections to reduce the risk of shipping defective product to customers;
- In-process inspections to reduce the risk of passing defective product to the next stage in the process;
- Receipt inspections to reduce the risk of passing defective product into the process;
- Supplier appraisals to reduce the risk of receiving defective product from suppliers;
- Design reviews to reduce the risk of releasing deficient designs into production;
- Reliability analysis to reduce the risk of in-service failures;
- Hazards analysis to reduce the risk of harming people producing, using, maintaining or disposing of the product;
- Contract reviews to reduce the risk of entering into contracts, the organization is unable to fulfil.

There are many other risks we could enter into this list but we can characterize the risk-based approach by seeking answers to five questions

1. What could jeopardize our ability to achieve our goal?
2. What measures can we take to contain these risks?
3. How will we know these risks have been contained?
4. How will we ensure the integrity of these checks?
5. How will we know these failures won't recur?

However, this approach can lead to a dependence on inspection to detect problems before they enter the next process which is counter to what Deming advocated in the third of his 14 points when he said *"Cease dependence on inspection to achieve quality. Eliminate the need for inspection on a mass basis by building quality into the product in the first place"*.

GOAL BASED APPROACHES

Quality does not appear by chance, or if it does, it may not be repeated and as Deming advocated one has to design quality into the products and services. The risk based approach depends on our ability to predict the effect that our decisions have on others and we may go over the top as is often the case with health and safety measures taken by local authorities or we might not have sufficient imagination to identify the consequence of our actions. We therefore need other means to deliver quality products – we have to adopt practices that enable us to achieve our objectives while preventing failures from occurring.

Complementing the task and risk based approaches are three other approaches which are addressed through Chapters 7, 8 and 9.

A system approach: This views the organization as a system of processes and it is the effective management of the interactions between these processes that will enable the organization to achieve its goals.

A process approach: This recognizes that work is performed through a process to achieve an objective and it is the effective management of the activities and resources within this process that will deliver outputs that achieve the objective.

A behavioural approach: This recognizes that all work is performed by people and that it is the effective management of the interactions between people that will enable the organization to achieve and sustain success.

But first we need to examine some of the approaches taken to the implementation of ISO 9001 in the belief that these were valid approaches to achieving, sustaining and improving quality. This is the subject of Chapter 6 which is given the title "A flawed approach" because the approaches taken have not yielded the anticipated benefits.

A Flawed Approach

CHAPTER PREVIEW

This chapter is aimed at everyone with an interest in the ISO 9000 family of standards, students, consultants, auditors, managers and most importantly the decision makers, CEOs, COOs and Managing Directors. Without an appreciation of the various perceptions and misconception and flawed approaches over the last 20 years any attempt at using or even talking about ISO 9000 may fail to represent it in the right context – this is an essential reading so that you do not repeat the mistakes of the past.

To the advocate, ISO 9000 is simply a family of standards that embodies common sense and all the negative comments have nothing to do with the standard but the way it has been interpreted by organizations, consultants and auditors. To the critics, ISO 9000 is what it is perceived to be and this tends to be the standard and its support infrastructure. John Seddon, industrial psychologist, author and critic of ISO 9000 and command and control strategies, has challenged business leaders, government and ISO on many occasions to steer them away from using this flawed approach and we use some of the arguments from his books to illustrate the misconceptions and reassure users that the 2008 version of ISO 9001 is not as flawed as the 1994 version on which John Seddon bases most of his arguments. One of the problems in assessing the validity of the pros and cons of the debate is the very term ISO 9000 because it means different things to different people. This makes any discussion on the subject difficult and inevitably leads to disagreement.

After 20 years, since its inception one would have thought that flawed perceptions and misconceptions about ISO 9000 and flawed approaches to the requirements would have dwindled but that is not so. As the standards have changed over the last 20 years, different approaches, perceptions and misconceptions have been spawned but some of the old ones remain.

In this chapter, we address the many ideas that have led to ineffective applications of the standards including:

- The argument that ISO 9000 is a flawed approach to the assurance of quality as it makes certification and not customer satisfaction the goal;
- The approaches to ISO 9001 that made us separate the quality management system from the business;
- The approaches to ISO 9001 that led to systems of documentation which made us lose sight of the objectives;
- The requirements that encouraged us to measure conformance rather than performance;

91

- The way third party auditors have influenced the organization's approach to ISO 9001;
- The way the requirements have misconstrued the responsibility for quality and led to departments set up to maintain ISO 9001 registration.

APPROACH TO QUALITY ASSURANCE

A Requirement for Doing Business

A number of ISO standards – mainly those concerned with quality, health, safety or the environment – have been adopted in some countries as part of their regulatory frame-work, or are referred to in legislation for which they serve as the technical basis. However, such adoptions are sovereign decisions by the regulatory authorities or governments of the countries concerned. ISO itself does not regulate or legislate. Although voluntary, ISO standards may become a market requirement, as has happened in the case of ISO 9001 and this has led to the perception that ISO 9001 is a requirement for doing business.

ISO 9001 was designed for use by customers to gain an assurance of quality. It replaced a multitude of customer specific requirements which suppliers had to meet and thus made it easier for them to bid for work. Coupled with the certification scheme it enabled suppliers to demonstrate that they had the ability to consistently meet customer requirements and thus reduced multiple assessments and therefore reduced costs.

It is not that this approach to quality assurance is flawed for it goes back centuries to when traders joined guilds to prove their competence and keep charlatans out of their market. What is flawed is the approach of using ISO 9001 for situations where it is simply inappropriate.

ISO 9001 was designed for situations where there was a contractual relationship between customer and supplier and even then it is only applicable where an organization needs to demonstrate its ability to consistently provide product that meets customer and statutory requirements. This is expressed at the front of the standard but many customers have invoked it in contracts, regardless of the need.

ISO 9001 does not require purchasers to impose ISO 9001 on their suppliers, but what it *does* require is for purchasers to determine the controls necessary to ensure whether purchased product meets their requirements. ISO 9001 is now being used through the supply chain as a means of passing customer requirement down the line and saving the purchaser from having to assess for themselves the capability of suppliers and this has led to certification becoming the goal.

In *A quest for confidence* in Chapter 1 we cited three ways in which customers can select their suppliers.

a) Purely on the basis of past performance, reputation or recommendation. This option is often selected for general services, inexpensive or non-critical products coupled with some basic receipt or service completion checks.

b) By assessing the capability of potential suppliers themselves. This option is often selected for bespoke services and products where quality verification by the purchaser is possible.

c) On the basis of an assessment of capability performed by a third party. This option is often selected for professional services and complex or critical products where the quality cannot be verified by external examination of the output alone.

ISO 9001 is a solution for case (c) only. It can be used in the other cases where customer intervention is not economic but not without assessing the risks (see also *Purchasing* in Chapter 26).

Making Certification the Goal

Many organizations have been driven to seek ISO 9001 certification by pressure from customers rather than as an incentive to improve business performance and therefore have sought the quickest route.

Seddon[1] calls this coercion and argues that it does not foster learning. In order to be able to tender for business Seddon claims that *"people cheat, they do what they need to do to avoid the feared consequence of not being registered"*. This is unfortunately a consequence of any separate inspection regime. As Seddon observes, it is happening in schools, in social work, in fact in any sector where there is regulation by inspection – or what Seddon calls "command and control". The relationship between the inspector and the inspected is one in which conformity is the standard and nonconformity is a black mark that can result in lost business and lost reputation that are very difficult to regain. Therefore, it is not surprising that some organizations will play the game to win at all costs.

What was out of character was that suppliers that were well known to customers were made to jump through this hoop in order to get a tick in a box in a list of approved suppliers. Customers were at fault by imposing ISO 9001 in situations where it was unnecessary.

The flaw in the approach was that customers were led to believe that imposing ISO 9001 would improve quality. To top it all, the organizations themselves believed that by getting the certificate they have somehow, overnight become a champion of quality. Putting the badge on the wall made them feel 'World Class' but in reality, not very much had changed. The processes were not being managed any more effectively and the process outputs were not showing significant improvement in performance.

To achieve anything in our society we inevitably have to impose rules and regulations – what the critics regard as *command and control* – but unfortunately, any progress we make masks the disadvantages of this strategy and because we only do what we are required to do, few people learn. When people make errors, more rules are imposed until we are put in a straightjacket and productivity plummets. There is a need for regulations to keep sharks out of the bathing area, but if the regulations prevent bathing we defeat the objective, as did many of the customers that imposed ISO 9001.

The Acceptance Criteria

The flaw in the certification process is that the standard used as the acceptance criteria (ISO 9001) is so prescriptive that it is easy to find nonconformity. If the acceptance

---

[1] Seddon John (2000) The case against ISO 9000. Oak Tree Press 2000.

criteria[①] were based on performance relative to stakeholder needs and expectations, it would level the playing field, make the certificate worth having and make it more easy to determine capability (see *The basic assurance requirements* in Chapter 1).

It is highly unlikely that ISO 9001 will be reduced to the requirements stated in Chapter 1 but if users were to keep these in mind as they use ISO 9001, they might not be persuaded into doing things that add no value for their organization or their clients. An existing alternative is to use the eight quality management principles (see Chapter 1). Simply filter every nonconformity through the eight quality management principles to determine whether it violates any one of these principles. If it doesn't it can't be a valid nonconformity. This might not please the external auditors as they would claim that the principles are not requirements of the standard but it is worth trying.

APPROACH TO SYSTEM DEVELOPMENT

Designed for Auditors Not for the Business

Seddon argues that "the typical method of implementation is bound to cause sub-optimisation of performance as it assumes that if properly implemented ISO 9001 will have a beneficial impact on performance and this is not proven nor theoretically sound". Invariably, ISO 9001 is implemented incorrectly. It is an assessment standard but has been used as a design standard resulting in new systems of documentation that exist for the benefit of auditors and not the business. By focusing only on the assurance requirements as interpreted by external auditors (see below), the management systems have been designed to pass the scrutiny of the third party auditors rather than the scrutiny of top management. In some cases the standard has been used wisely by looking at what it requires that is not done and assessing the benefits of change, but this is quite rare. Also Seddon is right to question the link between cause and effect because ISO 9001 does not address all of the factors upon which the achievement of quality depends. It omits the human factor. It is claimed in the Introduction to ISO 9001 that the eight quality management principles have been taken into account in its development but if this was the case there would be far more emphasis on the human factors, supplier relationships and leadership as they are so important in the achievement of quality. Such factors are addressed in ISO 9004 which is one of the reasons why this would be a better model to use for management system development than ISO 9001 (see also *A behavioural approach* in Chapter 9).

Neglecting Variation

Seddon argues that ISO 9000 has discouraged managers from learning about the theory of variation. He claims that ISO 9001 has encouraged managers to believe that adherence to procedures will reduce variation. If by variation he means variation in practices then by documenting best practice and getting everyone to follow those practices, variation in practice is reduced. This is not new. Frederick Taylor observed in the latter part of the nineteenth century that *"each worker did a day's work with great variation in output between the workers"*. He found this was because instead of there being one standard way of doing a task there were 50–100 ways of doing it. Taylor developed what he called the task system (see Part 2 Introduction). Where Taylor's approach differs from ISO 9001 is that Taylor talks about a science for each element of a man's work and ISO 9001:1994 reduced this to a procedure which is not the same thing at all. The science of which Taylor spoke included the tools and everything needed to produce the required outputs. It was under Taylor's system that Walter Shewart (1891–1967) developed statistical quality control thus introducing the theory of variation into the management of quality. There is no doubt that the early versions of ISO 9001 encouraged systems of documentation rather than documented systems but all that was to change with the introduction of ISO 9000:2000. Procedures were replaced by processes and this should have encouraged managers to believe that effectively managed processes will reduce variation. Regrettably the standard described a process as simply something that transforms inputs into outputs and a procedure as a documented process and failed to define what an effectively managed process should look like so managers may still not be managing variation (see Chapter 8 for further guidance on managing processes effectively).

The Organization as a System

Seddon argues that ISO 9000 has discouraged managers from learning about the theory of a system. Although ISO 9001:1994 defined what the quality system was required to achieve, that was to *"ensure product met specified requirements"*, it did not recognize the dynamic behaviour of systems.

Managers often think of their organization as the people and if they think of the organization as a system, it will be as a system of people not a system of processes. They fail to recognize the interactions and will change one function or process without considering the effects on another.

There was nothing in the 1994 version to suggest that the system being referred to is the organization. In fact it appeared to present a system as a set of documents but again this changed with the complete revision of the ISO 9000 family of standards. There are a few instances in ISO 9000 family where systems thinking is now recognized.

- A system is now defined as a set of interrelated or interacting elements (Ref. ISO 9000:2005, Clause 3.2.1).
- A management system is now defined as a system to establish policy and objectives and to achieve those objectives (Ref. ISO 9000:2005, Clause 3.2.4), thus expressing the dynamic relationship between the management system and the organization's outputs. This should encourage managers to think of the management system as the enabler of results rather than as a set of policies and procedures.
- In applying the systems approach it states that this typically leads to structuring a system to achieve the organization's objectives in the most effective and efficient way (Ref. ISO 9004:2009, Annex B). By using the phrase "typically leads to" the authors are expressing what the outcome of applying the systems approach should be rather than what the outcome might be. An auditor can therefore use this to test the effectiveness of the system.
- The quality management principles stated in ISO 9000 and ISO 9004 have been taken into consideration during the development of ISO 9001:2008 (Ref. ISO 9001:2008, Clause 0.1). These principles include a systems approach and a process

approach but the way these two approaches were expressed was not sufficiently robust to convey the real meaning behind them as the standard still carries over a significant number of requirements that were present in the 1994 version.

- The process approach now refers to processes and *their management to produce the desired outcome* (Ref. ISO 9001:2008, Clause 0.2). This change should encourage managers to manage the organization as if it were a system of processes.
- The sequence and interaction of the processes needed for the quality management system are now required to be determined (Ref. ISO 9001:2008, Clause 4.1b). By emphasising 'interaction', the standard recognizes the dynamics of processes and thus a key attribute of systems thinking.
- Top management is now required to ensure that the integrity of the system is maintained whenever changes are made, thus recognizing that there is interaction between elements (Ref. ISO 9001:2008, Clause 5.4.2b).

This is all good news as the family is now proceeding in the right direction but there is little guidance as to what all this means in the context of a quality management system that had been established as a system of documentation. The ISO TC 176 guide to the process approach[2] came out too late for 1994–2000 transition and as it is not mandated in ISO 9001:2008, it probably won't be used. One of the aims of this book is to fill this gap and provide guidance on making the transition from a system of documentation to a system of interacting processes.

Separate from the Business

ISO 9001 requires organizations to establish a quality management system as a means of ensuring that customer requirements are met. The misconception here is that many organizations failed to appreciate that they already had a management system – a way of doing things and

> **Food for Thought**
>
> Is our management system the way we run our business or is it simply a set of documents we used to show compliance with ISO 9001?

because the language used in ISO 9001 was not consistent with the language of their business, many people did not see the connection between what they did already and what the standard required. So instead of mapping the requirements of ISO 9001 onto the business they started to create a paper system that responded to the requirements of ISO 9001 thus separating this 'system' from the business as shown in Fig. 6-1.

An unintended consequence of ISO 9001 was the formalization of only those parts of the system that served the achievement of product quality – often diverting resources away from the other parts of the system. Activities were only documented and performed because the standard required it. It isolated parts of the organization and made them less efficient. When ISO 14001 came along this resulted in the formalization of another part of their management system to create an Environmental Management System (EMS). The danger is that as more and more management system standards emerge, more and more management systems will be created separating more parts from the business.

---

[2] Guidance on the concept and use of the process approach for management systems: *ISO/TC 176/SC 2/N544R3 October 2008.*

This is not the way to approach these standards and a more effective approach is addressed by *A Systems approach* in Chapter 7.

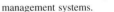

All organizations have a way of doing things. For some it rests in the mind of the leaders, for others it is translated onto paper and for most it is a mixture of both. Before ISO 9001 came on the radar of an organization they had inevitably found ways of doing things that worked for them. They were faced with meeting all manner of rules and regulations. Government inspectors and financial auditors frequently examined the books and practices for evidence of wrong-

FIGURE 6-1 Separate management systems.

doing but none of these resulted in organizations' creating something that was not integrated within the routines they applied to manage the business. We seem to forget that before ISO 9001, we had built the pyramids, created the mass production of consumer goods, broken the sound barrier, put a man on the moon and brought him safely back to earth. It was organizational systems that made these achievements possible. Systems, with all their inadequacies and inefficiencies, enabled mankind to achieve objectives that until 1987 had completely revolutionized society. The ISO 9000 family of standards simply consolidated the principles and practices that had enabled us to reach these goals so in fact it was not telling us anything new. It merely brought together proven practices.

The next logical step was to apply these principles and practices where they were not already being applied and improve systems making them more predictable, more efficient and more effective – optimizing performance across the whole organization – not focusing on particular parts at the expense of the others.

Misunderstanding in Professional Services

There has also been a perception in the service industries that ISO 9000 quality systems only deal with the procedural aspects of a service and not the professional aspects. For instance in a medical practice, the ISO 9000 quality system is often used only for processing patients and not for the medical treatment. In legal practices, the quality system again has been focused only on the administrative aspects and not on the legal issues. The argument for this is that there are professional bodies that deal with the professional side of the business. In other words, the quality system only addresses the non-technical issues, leaving the profession to address the technical issues. This is not *quality management*. The quality of the service depends on both the technical and non-technical aspects of the service. Patients who are given the wrong advice would remain dissatisfied even if their papers were in order or even if they were given courteous attention and advised promptly. To achieve quality, one has to consider both the product and the service. A faulty product delivered on time, within budget and with a smile remains a faulty product!

The Exclusive or Inclusive System

Seddon argues that ISO 9000 has failed to foster good customer–supplier relations as it obliges suppliers to show they are registered. He claims that this reinforces an "arms-length" view of management whereas the Japanese learnt to see their organizations as

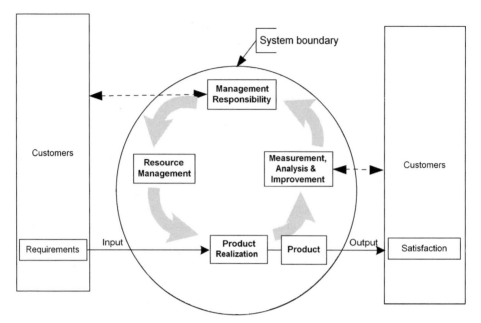

FIGURE 6-2   ISO 9001 based system with stakeholders outside the system.

systems that included their suppliers and customers. While ISO 9001:2008 does take a far different approach than earlier versions and does embrace some aspects of systems thinking, it does not go so far as to include the customers and the suppliers within the system. This could be due to what the authors believe a system is. As stated previously, the pre-2000 quality management systems were more likely to be systems of documentation than documented systems, so when the standard embraced the systems approach the idea that the system included suppliers and customers would have been 'off the radar'. The concepts and principles developed by Peter Senge in the Fifth Discipline[3] appear to have not been well known among the representatives who formulated ISO 9001:2008 even though his book was first published in 1990. The ISO 9001:2008 model of a process-based quality management system redrawn in Fig. 6-2 shows the suppliers and customer outside the system and yet the behaviour of customers and suppliers influences the organization's outputs so they ought to be inside the system as shown in Fig. 6-3.

APPROACH TO DOCUMENTATION

The Document What You Do Approach

An approach to ISO 9000 that found favour was that of 'Document what you do, do what you document and prove it'. It sounded so simple and it appeared to match the expectations of third party auditors who often asked questions such as:

---

[3] Senge Peter M (2006) The Fifth Discipline, The Art and Practice of the Learning Organization. Random House.

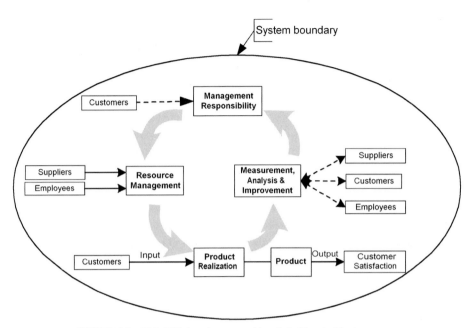

FIGURE 6-3 ISO 9001 based system with stakeholders inside the system.

- What do you do?
- In which procedure is it documented?
- Can you show me evidence of conformity with this procedure?

This approach was described by Jack Small of IBM.[4] Although Small explains the approach slightly differently, the explanatory statements he made were often overlooked by those who adopted the approach:

- Say what you do, i.e., establish appropriate quality controls and systems;
- Do what you say, i.e., ensure that everyone involved follows the established processes;
- Show me, i.e., demonstrate compliance of your quality system to an external auditor.

Although it may appear as though this was the tenet of ISO 9001 prior to the 2000 version, this was not in fact what the

SAY–DO

When Admiral Rickover stepped down as head of US Navy nuclear power programs in the early 1980s he addressed a joint session of Congress in which he mentioned a trend he had observed in the Navy that concerned him greatly. This consisted of a leader devising a plan to address a problem and then just simply not executing it. He called it SAY–DO meaning "SAYing" that something would be done, but not actually "DOing" whatever it was that was said would be done.

This was a flaw in the implementation of ISO 9001 a decade or so after Rickover's observations as firms issued their quality policies and procedures and then failed to implement them.

---

[4] Small Jack E Dr. (1997) ISO 900 for Executives. Lanchester Press Inc Sunnyvale, CA, USA.

standard required. The flaw in this approach is that the standard required the organization *"to establish, document and maintain a quality system as a means of <u>ensuring</u> that product conforms to specified requirements."* It was not recognized then that it needed far more than documented procedures to ensure that product conforms to specified requirements. In subsequent versions it requires the organization *"to establish, document, implement and maintain a quality management system … to ensure that customer requirements are determined and are met with the aim of enhancing customer satisfaction"*. Therefore, if after documenting what you do and doing what you documented and proving it to third party auditors, your quality management system failed to:

- ensure that product conforms to specified requirements,
- ensure customer requirements were determined,
- ensure customer requirements were met,
- enhance customer satisfaction,

… then clearly your quality management system should be deemed ineffective.

By 'documenting what you do' you overlook the possibility that what you are doing is not consistent with the above criteria. You may in fact be doing things that result in delivery of nonconforming product, that result in customer complaints, therefore why would you want to document these? This approach also has

> **Food for Thought**
>
> If we document what we do, can we be sure that we will be documenting everything that affects our ability to satisfy our customers?

a tendency to focus only on tangible activities and overlooks the way people think, the informal network that makes things happen, the values that shape behaviour and lead to action and so the result of 'documenting what you do' creates an imperfect representation of how the organization is managed.

By proving only that you do what you have documented, you overlook the objectives of the system and the results it is delivering. If you test products before shipment and document you do this, then demonstrate that you are testing products before shipment you have 'documented what you do, you have done what you documented and have proven it'. However, if the people doing these tests are not customer focused, they might skip some tests to avoid the tedium and go home early. If all you have is a record that the test had been performed or a tick in the appropriate box you would be none the wiser. To be confident that what was done was what was supposed to be done, you need confidence in the people. This requires a different approach (see *A behavioural approach* in Chapter 9).

Documentation for the One in a Million Event

The persistence of the auditors to require documentation led to situations where documentation only existed in case something went wrong – in case someone was knocked down by a bus. The flaw in this approach is that while the unexpected can result in disaster for an organization, it needs to be based on a risk assessment. There was often no assessment of the risks or the consequences. This could have been avoided simply by asking the question 'so what?' So there are no written instructions for someone to take over the job but even if there were, would it guarantee there were no hiccups? Would it *ensure* product quality? Often the new person sees improvements that the previous person missed or deliberately chose not to make – often the written instructions are of no use without training and often the written instructions are of no value whatsoever because they were written by people who were not doing the job. Requiring documented instructions for every activity would be sensible if what we were creating was a computer program because the instructions were needed to make the computer function as intended. People don't need written instructions to make them function; a management system is not a computer program. Those people who have been brought into the organization to accomplish an objective will seize the opportunity and begin to work without waiting for written instructions.

Management-led or Customer-led Approach

Primarily, ISO 9001 is to be used "to assess the organization's ability to meet customer, statutory and regulatory requirements applicable to the product, and the organization's own requirements" (Clause 0.1). It is not designed to be used as a design specification for management systems. It is suggested that, *"beginning with ISO 9000:2000, you adopt ISO 9001:2000 to achieve a first level of perfor-mance. The practices described in ISO 9004:2000 may then be implemented to make your quality management system increasingly effective in achieving your own business goals".*[5] The flaw in this approach is that management systems are being established to meet the requirements of ISO 9001 on the demand of customers or the market and therefore this is a market- or customer-led approach. It may not result in outcomes which will satisfy all stakeholders. In such a documented system there are likely to be no processes beyond those specified in ISO 9001 and within those processes no activities that could not be traced to a requirement in ISO 9001. Invariably, users go no further and do not embrace ISO 9004. Had ISO 9004 been promoted and used as a system design requirement, the management system would be designed to enable the organization to deliver outcomes that satisfied all stakeholders. ISO 9001 could then be used to assess the organization's ability to meet customer requirements and if necessary, ISO 14001 could be used to assess the organization's ability to meet environmental requirements and so on for health, safety, security etc.

A Misunderstood Purpose

It was believed that by operating in accordance with documented procedures, errors would be reduced and consistency of output would be ensured. If you find the best way of achieving a result, put in place measures to prevent variation, document it and train others to apply it, it follows that the results produced should be consistently good. The flaw in this argument is that you can build a system from a set of procedures as though a management system is just a pile of paper. If it were a pile of paper it wouldn't do very

---

[5] Selection and Use of the ISO 9000:2000 family of standards. Available from http://www.iso.ch/iso/en/ iso9000-14000/iso9000/selection_use/selection_use.html.

much on its own, there has to be some energizing force for the system to ensure customer requirements are met.

In the third edition of ISO 9001 in 2000, only six documented procedures were required and the emphasis place upon processes. Some organizations going through the transition from previous versions only produced six documented procedures and converted the other procedures into work instructions or renamed them as processes which largely missed the point. The tragedy was that certification body auditors accepted this approach. They misunderstood the difference between procedures and processes and continued to prescribe activities as though by doing so they were describing processes. Those starting afresh were not so constrained and had the opportunity to take a process approach but invariably this has resulted in procedures presented as flow charts instead of text thus again exhibiting a misunderstanding between procedures and processes.

Seddon argues that ISO 9000 encourages organizations to act in ways which make things worse for their customers; it adds cost, demoralises staff and prevents improvement in performance. It forces organizations to write procedures and follow them regardless and lose sight of the objective. This claim is hard to refute and it is true not only for ISO 9000 registered organizations but also for any organization that places adherence to procedures above achievement of objectives. In ISO 9001:2008 there is less emphasis on writing procedures but the standard still misses the point. Look at the first requirement in Clause 4.1 where it requires the organization to establish, document, implement and maintain a quality management system and continually improve its effectiveness <u>in accordance with the requirements of this International Standard</u>. This emphasizes that the purpose of the system is to meet the requirements of ISO 9001 and not to enable the organization to satisfy its stakeholders. A simple amendment would have changed the focus considerably. If we look at Clause 8.2.2 on internal audit we see that it requires the organization to conduct internal audits to determine whether the quality management system conforms to the planned arrangements and the requirements of ISO 9001; again a misunderstanding of purpose. There is no requirement to audit the system to establish how effective it is in enabling the organization to satisfy its customers. Again, a simple amendment would make internal audits add value instead of a being a box ticking exercise.

APPROACH TO MEASUREMENT

Measure of Effectiveness

ISO 9001 requires top management to review the management system at defined intervals to ensure its continuing suitability, adequacy and effectiveness, where effectiveness was deemed to be the extent to which the system was implemented. The flaw in this approach is that it led to quality being thought of as conformity with procedures. This preoccupation with documentation has alienated upper management so that internal auditors have great difficulty in getting commitment from managers to undertake corrective action. Where auditors do discover serious breaches of company policy or non-adherence to procedure, the managers might commit to take action but when the majority of the audit findings focus on what they might regard as trivia, the auditor loses the confidence of management.

If auditors apply the process approach they would firstly look at what results were being achieved and whether they were consistent with the intent of ISO 9001, then discover what processes were delivering these results and only after doing this, establish whether these processes complied with stated policies, procedures and standards. The management reviews were fuelled by customer complaints and nonconformities from audits and product and process inspections which when resolved maintained the status quo but did not measure the effectiveness of the system to achieve the organization's objectives. But as the system was not considered to be the way the organization achieved its results, it was not surprising that these totally inadequate management reviews continued in the name of keeping the badge on the wall. Had top management understood that the system was simply the way the organization functioned and that reviewing the system was synonymous with reviewing the effectiveness of the organization in meeting its goals, they might have held a different perception of management reviews and committed more time and energy into making them effective.

Measuring Conformity with Procedures

Seddon argues that ISO 9000 starts from the flawed presumption that work is best controlled by specifying and controlling procedures. Controlling performance by controlling people's activity makes performance worse. Seddon also argues that when people are subjected to external controls, they will be inclined to pay attention only to those things which are affected by the controls. He makes a valid point that people do what you count and not what counts. Again these claims are hard to refute and ISO 9001 has indeed encouraged the notion that following the correct procedures was all that was needed to provide a quality product or service. This approach was one of the factors that led to the death of a baby in Haringey, North London, UK in August 2007 as social workers stuck by the rules and their supervisor defended their position. The belief that following the correct procedures produces quality may not be the case in top management but in a large organization, managers at lower levels are often judged on their ability to play the game, stick to the rules, and adhere to the policy and procedures. Under an authoritarian management style, people don't step out of line for fear of losing their jobs. Therefore, Seddon is right to question the efficacy of the approach. It is true in most organizations and particularly within those where targets are set for every conceivable variable.

Many quality managers feel obliged to take external auditors seriously because their boss would not be pleased to receive reports of nonconformity. Instead of appointing a person with a wealth of experience in quality management who might expect a salary appropriate to their experience, organizations sometimes chose for their quality manager, a less qualified person who was at an immediate disadvantage with the third party auditor. Sometimes they select a person with many years of experience with a certification body. This can have the desired effect of facing like with like, but a third party auditor might not have sufficient experience in developing quality management systems. It is more important for a quality manager to understand the factors on which the achievement of quality depends and know how to influence them rather than understand the requirements of ISO 9001 because such a person will be able to prevent the organization from being adversely influenced by an external auditor.

Regrettably, ISO 9001:2008 does not eliminate this perception. Look at Clause 8.1 where it requires the organization to implement the measurement processes needed to ensure conformity of the quality management system. Such a requirement leads people to measure conformance and not performance. If this requirement had led organizations to implement measurement processes needed to establish that the management system was effective in enabling the organization to achieve its objectives, there would have been more focus on results and less on following procedures.

APPROACH TO EXTERNAL AUDITORS

Ticking Boxes Not Achieving Objectives

Unfortunately a preoccupation with conformity assessment has resulted in a 'tick in the box approach'. This has led to the misconception that if an auditor finds sufficient evidence of compliance against each clause of the standard[①] he/she can declare that the organization satisfies ISO 9001 and thus has an effective management system. Auditors are focusing on conformity and not on effectiveness. There are over 260 requirements in ISO 9001:2008 and invariably auditors will not check all of them in all parts of an organization they might apply – this would be a mammoth task and too costly. What they attempt to do is to take samples and test for conformity in the belief that if conformity is found the system is effective. The flaw in this approach is that by and large it is checking inputs not outputs. For example, it is checking that a quality policy exists (the input) that a few people can recite it but not that the policy is driving the outputs that are being achieved (the results). The auditor looks for customer complaints to see if they have been closed and ticks the box if they have. If the system was designed to ensure that customer requirements were met, any complaint would be indicative of a system failure but in general if the auditor is satisfied that people are following the documented procedure no action will be taken.

> **Food for Thought**
>
> What is more important for us to sustain success, is it to achieve objectives or get ticks in boxes?

This approach has led to auditors demanding that things be done when they add no value. The assertive manager would ask, "Why would I want to do that?" and if the auditor or consultant could not give a sound business case for doing it, the manager would rightly take no action. By focusing on the detail the auditor loses sight of the objective.

After all the boxes have been ticked, the auditor and the manager should establish whether sufficient evidence has been gathered to demonstrate that the objectives are being achieved. If there isn't then the audit should continue until evidence is found indicative of the weakness in the system. Of course, it would make sense if the auditor were to start by looking for evidence that objectives have been achieved but this requires a different set of competences – something that might come about as a result of the application of ISO 17021.

Auditor Competence

Seddon argues that ISO 9001 relies too much on interpretation of the requirements. He questions the competence of auditors and the training they receive and asks if the audit process has been tested. The audit process is tested but against standards that are in the same mould as ISO 9001. The accreditation bodies receive their income from the certification bodies so are unlikely to be too radical. The certification bodies receive their income from their client so they too are not going to make too many demands for fear of losing a customer.

External auditing has become a regular job and one where there is little accountability which gives the auditor significant power without responsibility. Obvious malpractice and misconduct will be revealed quickly but other than repeating an audit there is little one can do to ensure

> **Food for Thought**
>
> Do we want our staff to learn new skills that will help us improve our performance or do we want them to learn how to catch us out?

that the results obtained are a true reflection of the state of conformity. Witness audits may reveal ineffective methods but witness audits affect what they are measuring. As with most professional services, the results of the audit are entirely dependent on the competence of the auditor. The five-day lead auditor courses were designed for people who were already experienced in some aspect of quality management. It is therefore not surprising that the course does not turn a novice into a competent auditor in five days. In fact many finish the course having learnt more about ISO 9001 than auditing.

Training is doing things over and over until a skill is acquired and mastered but rules have forced training bodies to cover certain topics in a certain time. Commercial pressure has resulted in training bodies cutting costs to keep the courses running. Delegates were being subjected to endless slide shows of ISO 9001, the attributes of auditors, the procedures for planning, conducting and reporting audits but few gained experience practicing auditing in a realistic environment while being trained. The auditing skills are tested in classroom conditions and only a few courses provide the opportunity for the trainees to carry out a real audit in a company whilst being observed; but it is not training. Even the examinations had no practical components in which a person's competence to carry out audits was tested. Customers would not pay for more than they thought they needed but they did not know what they needed. Tell them what is required to convert a novice into a competent auditor and they wince at the cost! The misconception here is that listening is mistaken for training. When there are providers only too willing to relieve them of their cash, customers opt for the cheaper solution. Had customers of training courses been purchasing a product that failed to function there would have been an outcry, but the results of training were less likely to be measured. The training auditors received focused on auditing for conformity and led to auditors learning to catch people doing things not prescribed by the procedures. It did not lead to imparting the skills necessary for them to determine whether the organization had an effective system for producing satisfied customers. Had the training been designed for this objective, the courses might have been completely different.

The policy of the accreditation bodies issued before the launch of ISO 9000:2000 was intended to make big changes in the way that audits were conducted. However, eight years later, we see that it has not changed in any significant way as another new standard is being prepared that is intended to improve auditor competence. This latest standard (ISO 17021) neither prescribes a process approach to auditing nor does it prescribe any method of gathering the objective evidence, what to look at and what to look for. All this is assumed to be addressed by the training but the competence requirements in the draft of ISO 17021–2 only list the characteristics and not the acceptance criteria. We are therefore not moving forward. Once again the authorities are attempting to use the status quo to change the status quo and with the probable effect that very little will change.

Surviving the Audit

Quality managers scurried around before and after the auditor and in doing so led everyone else to believe that the only thing of real importance to the auditor was documentation. The misconception here

> **Food for Thought**
>
> Is our goal to survive the audit or to improve our performance?

was that this led others in the organization to focus on the things the auditor looked for not on the things that mattered – they became so focussed on satisfying the auditor that they lost sight of their objectives. They focused on surviving the audit and not on improving the performance. It has the same effect as the student who crams for an examination. The certificate may be won but an education is lost. What would the organization rather have – a certificate or an effective management system? Organizations had it in their power to terminate the contract with their certification body if they did not like the way the audits were being handled. They had it in their power to complain to the Accreditation Body if they were not satisfied with the service rendered but on both counts they often failed to take any action. Certification Bodies are suppliers, not regulators. What went wrong with ISO 9001 assessments is that the auditors lost sight of the objective of the audit which was to find opportunities to improve the quality of products and services. They failed to ask themselves whether the discrepancies they found had or would have any bearing on the quality of the product or on customer satisfaction. Many of the nonconformities were only classified as such because the organization had chosen to document what it did, regardless of its impact on quality. Auditors often held the view that if an organization took the trouble to document *it*, *it* must be essential for product quality and therefore by not doing *it*, product quality must be affected! But this was often not the case and as a result procedures were rewritten, removing anything that was not essential on which an auditor could pin nonconformity. So out went the guidance and what remained was a skeleton of a procedure that failed to guide the user into doing the job in the best way.

Validity of Audit Conclusions

Certification bodies are in competition and this leads to auditors spending less time conducting audits than is *really* needed. They focus on the easy things to spot and not on whether the system is effective. This is where there is a misconception that in engaging the services of a certification body to determine whether the organization has a management system that satisfies ISO 9001, a process is initiated which will determine whether the system is effective. When one examines the report after the audit and notices what has been checked and what was found nonconforming one is left wondering how the auditors could have reached this conclusion without talking with the CEO, the Marketing Director and the Financial Director and finding out how the organization was performing relative to its objectives, but they didn't. The audit report will be carefully worded. It will more than likely declare that the quality management system meets or continues to meet the requirements of ISO 9001. It won't say that *sufficient objective evidence has been found to demonstrate that the quality management system is effective, i.e., it*

> **Food for Thought**
>
> Do we want our third party auditors to prove to us that our system is effective or simply tell us if it's compliant with ISO 9001?

provides the organization with the ability to consistently provide product that meets customer and applicable statutory and regulatory requirements. But this is the objective of ISO 9001 as stated in the clause 1.1. So you are getting an objective opinion that the requirements have been met but not that the objective is being achieved which is perhaps what you thought you were paying for – its another example of ticking boxes not achieving objectives.

These inconsistencies and anomalies have not gone unnoticed by ISO and in 2001 a *Joint Working Group on Image and Integrity of Conformity Assessment* was established to discuss what contribution IAF, ILAC and CASCO can collaboratively make for improving the integrity of conformity assessment practices. Among the problems identified by the working group were malpractices and unethical and dishonest practices of conformity assessment bodies. The result of their deliberations is ISO 17021.

APPROACH TO RESPONSIBILITY FOR QUALITY

A Department with Responsibility for Quality

As attention to quality increased, organizations began to build inspection departments and then quality departments. Most organizations structure the division of labour on a functional basis – a function being a group of specialists needed for the organization to fulfil its purpose. The organization structures were distinguished

> **Shifting Priority**
>
> When moving responsibility for quality from the quality department to the production department, don't expect any change in performance unless you make quality the first priority.

by having quality managers who led a team of specialists whose mission was to remove the risk of shipping defective product to customers. What these departments did was act as a regulator with the authority to stop release of product. However, this is control of supply and not of control of quality as authority to change product remained in the hands of the producing departments.

By putting the word quality in the title of a department and a manager, it sent a signal to the other managers that this department was responsible for quality. As the design, marketing, purchasing and production managers took responsibility for what they were managing, it followed that the quality manager would be responsible for managing quality.

It was a flawed approach because it appeared to take the responsibility for quality away from those who created the product and because responsibility for quality cannot rest in a single department. Juran highlighted the anomaly both in his Quality Control Handbook of 1974 and again in Juran on Leadership for quality in 1989. Here he writes that *"the question 'Who is responsible for quality?' is unanswerable as the question needs to be broken down into responsibilities for specific actions and decisions".* The question also needs a context to identify the subject for which we are trying to determine responsibilities. People can only be responsible for what they do and the actions of the people under their control. Many people are involved in producing a product or service ranging from those who determine the requirements to those who prepare the product for shipment or deliver the service.

Organizational Freedom

Mil-Q-9858A of 1963 required *"that personnel performing quality functions have sufficient, well-defined responsibility, authority and the organizational freedom to identify and evaluate quality problems and to initiate, recommend or provide solutions".* It also stated that *"It does not mean that the fulfilment of the requirements of this specification is the responsibility of any single contractor's organization, function or person".* The term "personnel performing quality functions" was not therefore intended to mean the personnel in the quality department but all personnel throughout the organization whose work affects the quality of the product. This requirement was included in BS 5750:1979 without the requirement for organizational freedom but with the same intent and also carried over into ISO 9001:1987.

The requirement arises as a solution to the problem of self-control. Many organizations operate along the military command and control lines where directives are issued from management that are to be obeyed. Under such conditions, employees performing quality functions may be given directives that conflict with what they perceive to be their authority and responsibility, e.g., releasing substandard product, not undertaking certain tests, reducing safety factors etc. all with the aim of putting profits, delivery or other interests above quality. In this respect it turns out that the early versions of ISO 9001 were not encouraging command and control thinking as much as some critics would argue. However, the requirement for management responsibility creates a conflict and the organizational freedom concept has been omitted from ISO 9001:2000 and 2008 editions. It is not clear whether this is as a result of introducing requirements for top management commitment which should remove the possibility of directives that conflict with the declared quality policy. However, a piece of paper has not stopped a manager from acting in his/her own interests in the past so there is no doubt that it won't prevent problems in the future.

Independent Inspection

It was often thought that the standard required review, approval, inspection and audit activities to be performed by personnel independent of the work. Critics argue that as a consequence both worker and inspector assumed the other would find the errors. The misconception here is that ISO 9001 does not require independent inspection. There is no requirement that prohibits a worker from inspecting his or her own work or approving his or her own documents. Seddon was mistaken[6] when he claimed that ISO 9001 required independent inspection. It is the management that chooses a policy of not delegating authority for accepting results to those who produce them. There will be circumstances when independent inspection is necessary either as a blind check or when

---

[6] Seddon John (2000) The case against ISO 9000 Oak Second Edition, Tree Press.

safety, cost, reputation or national security could be compromised by inadvertent errors or where a motive for misdemeanour exists. What organizations could have done, and this would have met ISO 9001 requirements, is to let the worker decide on the need for independent inspection except in special cases. The worker remains in control but delegates the measurement task to others with the necessary capability.

All work is a process and results in outputs and those producing these outputs should be placed in a state of self-control so that they can be held accountable for the results. Therefore, when we cause workers to pass their work to inspectors to determine conformity, we are removing the worker's right to self-control. However, it is never as simple as that. It has often been said that one cannot inspect quality into a product. A product remains the same after inspection as it did before, so no amount of inspection will change the quality of the product. However, what inspection does is to measure quality in a way that allows us to make decisions on whether or not to release a piece of work. Work that passes inspection should be quality work but inspection is unfortunately not 100% reliable. Most inspection relies on human judgement and this can be affected by many factors, some of which are outside our control (such as the private life, health or mood of the inspector). A balanced approach is to carry out a risk assessment and impose independent or additional inspection where the risks warrant it. However, inspection is no substitute for getting it right first time. ISO 9001 simply emphases the prudent practice of verifying product before use, processing or delivery.

> **Rickover on Inspection**
>
> "All work should be checked through an independent and impartial review. Even the most dedicated individual makes mistakes – and many workers are less than dedicated. I have seen much poor work and sheer nonsense generated in government and in industry because it was not checked properly."
>
> (From Doing a Job by Admiral Hyman G. Rickover delivered to *Columbia University School of Engineering* Nov 1981)

The resources needed to determine conformity might be considerable; gauges, test equipment, specialist skills and knowledge and therefore rather than equip every worker with the means to determine conformity, the task is managed by a separate group of dedicated specialists. In effect these specialists support the worker and allow the worker to make the judgement on acceptability. But when this judgement remains with the specialists it again removes the worker's right to self-control. One of the significant benefits to arise from computerization is its ability to enable the worker to control his output. Skills and tools, once the preserve of specialists in the quality department, are now available to everyone. However, someone needs to manage these resources so that the worker can depend on them being capable and available when needed. This role might be handed to a reformed quality department.

The Management Representative

The idea of a management representative did not come out of the first national standard on quality management but in BS 5750:1979 where it required *"the supplier to appoint a management representative, preferably independent of other functions, with the necessary authority and the responsibility for ensuring that the requirements of this standard are implemented and maintained"*. The requirement was carried over into ISO 9001:1987 but it was an approach that was flawed.

Requiring a person to ensure the requirements of the standard were met is advocating a command and control approach because meeting the requirements is not the correct objective. What this did was to put pressure on managers to comply with the standard, regardless of whether it was beneficial to the organization to do so. It imposed solutions that were not always appropriate and made conformity the measure of success rather than customer satisfaction.

These lessons were learnt in the 2000 revision when the requirement for a management representative was changed. However, the requirement is still flawed because it still focuses on ensuring things happen rather than enabling them to happen and overlooks the whole purpose of the system.

Quality Management System Specialists

ISO 9001 has led organizations creating a position in their management structure that responded to the standard. It often starts off with the appointment of an ISO 9000 Project Manager or Coordinator who works with the consultant to document the system through to certification. Thereafter this person manages the system and possibly acts as the management representative or works for such a person. There is indeed a need to develop a management system, maintain and improve it but that job is the responsibility of the whole management team. If they choose to assign this responsibility to another, that is their choice but it is an approach that is flawed because the management system is the way the organization functions and to make anyone other than the CEO responsible would be illogical.

What this person often results in doing is maintaining the manuals, processing change requests on documents and managing the quality audit programme. This tends to place all the system documents under the control of one person or department which is not healthy because it inevitably leads to situations where the documentation is always lagging behind actual practices. The internal audits pick up these issues and the Quality System Department then spend most of their time chasing paper and not attending to real quality problems. With the advent of electronic communications it is now possible for managers to access the server where the documents describing their processes are located and change them, bypassing a central quality system department.

There is a role at the centre of an organization for a systems specialist who assists managers in developing, maintaining and improving processes so that they interact in ways that achieve the desired outcomes. The role would be enhanced by the addition of the system audit function thus providing data with which to identify opportunities for improvement without removing the responsibility for performance, efficiency and effectiveness from the individual managers.

Chapter 7

A Systems Approach

CHAPTER PREVIEW

This chapter is aimed at those with responsibility for the performance of an organization and those charged with formalizing, managing and improving the systems that enable the organization to fulfil its goals. It will interest students, consultants, auditors, managers and most importantly top management who often look upon management systems as a bureaucratic necessity to qualify the organization as a supplier in a particular market. What should become clear is that far from being a bureaucratic necessity, when the organization becomes a system of managed processes, a management system is just another name for an organization. It is an essential reading for those setting out to take a systems approach to management and develop a process based management system. More detail on this latter point can be found in *Establishing a quality management system* in Chapter 10 but first it is important for you to have an understanding of management systems or systems of managed processes.

As revealed in Fig. 1-3, the approach taken by ISO 9001:1994 was that establishing and implementing a system of documented procedures which was periodically audited for compliance would ensure the supply of conforming products and services. This changed dramatically in 2000 with two new approaches to ensure customer satisfaction – the systems approach and process approach. ISO 9004:2009 now takes this one step closer to a holistic approach by focusing on sustained success and emphasising the importance of mission, vision, policy, strategy and a system of managed processes in achieving sustained success.

In this chapter, we examine the systems approach to the management of quality and look at:

- The relationships between systems and quality;
- The differences between the systems approach and the process approach;
- The nature of management systems and explore several different perceptions;
- The question of integrated management systems and several misconceptions that have grown up about this subject;
- The factors that characterize management systems and make them what they are;
- System models that explain the relationship and interaction between the processes in the system.

THE RELATIONSHIP BETWEEN SYSTEMS AND QUALITY

There are three primary factors upon which the achievement of product quality depends: quality of management, quality of design and quality of conformity. There are eight quality management principles as another set of factors upon which the achievement of quality depends. We could take another perspective and produce even more factors such as, People, Machines, Methods and Materials and we could add Measurement and Money. These factors are often used as categories for grouping causes in cause and effect diagrams. What this illustrates is that there are many dependencies; many things have to be right for the organization to produce quality products. It is not sufficient to simply focus on the production or delivery processes as these depend on other processes for their inputs. For example, in a fast food outlet, speed is of the essence and service quality depends on being able to serve the customer quickly. But if the process for maintaining the cookers breaks down, orders cannot be completed and service delivery fails. If the people from the maintenance process cleaning the entrance fail to understand the values which the organization stands for, they will commence cleaning the front doors just when there is a rush of customers that get in their way. This will leave a bad impression and clearly indicate that the organization is not demonstrating it is customer focused. If quality is perceived as sustained satisfaction any number of things can cause a failure to deliver on this promise from the boardroom to the boiler-room. Thus, the organization depends on a system of managed processes to produce the desired outcomes if it has to satisfy its customers and other stakeholders.

SYSTEMS APPROACH VERSUS PROCESS APPROACH

We will explain more about systems further on and processes are explained in detail in Chapter 8 but there is a distinction to be made between systems and processes because they both produce results but the results are produced in different ways. A process produces results through work done

> **Systems Approach**
>
> Viewing the organization as a system process and managing their interactions to produce desired outcomes.

in the process. A system produces results through the interaction of processes. We are not claiming this relationship to be true in any other context than organizational systems and organizational processes.

There are therefore two quite different types of management. There is process management which will manage the achievement of results by planning, organizing, controlling and continually improving the work required to produce them. There is system management which will manage a system of interacting processes that function together to achieve certain goals.

The boundary between process and system is where the output fulfils a system objective. For example, if we treat the series of steps in frying an egg as a process, we will find that frying the egg is one stage of a meal preparation process. The meal preparation process is one stage in the service delivery process. Any of these processes can interact with the outputs from a process that manages the resources used in meal preparation process. Cut off the supply of electricity and the cook can't fry the egg. We therefore ascend through a hierarchy of processes to a system of processes which function together to deliver an experience that will delight the customer – the objective of frying the egg, preparing the meal and delivering the service.

The system manager is obviously the manager of the business as no other person would have the authority and responsibility to bring together the process that attracts the customer to the food outlet, the process that ensures all the facilities needed are in place and operational and the process that delivers the food in a way that lives up to the fast food outlet's mission statement.

Organizations can be simple or complex but all strive to achieve goals through the combined efforts of the people, for without people there is no organization. We can view the organization as an arrangement of people and we depict this arrangement through the organization chart. But the chart is a static two dimensional image of the organization so it cannot achieve the goals. We can look beyond the chart, walk around the building watching people at work and machines operating, things happening. This is more like the organization because it is dynamic and looks like it might be achieving the goals. One thing we won't see is some people working on quality, others working on cost and others working on delivery. Everyone will be trying to achieve standards, on time and in the most economical way for the organization. However, just like the organization chart is one perspective, we can take another perspective and view the organization as a system of processes, all working together to achieve the organizational goals. The organization itself is the assemblage of the people and machines etc. the organization structure is the way all these pieces are put together and how they interact and it requires more than an organization chart to show these relationships and interactions.

When we think of the organization as a system we take a holistic approach. This looks at the whole rather than the parts and examines the relationship between the parts because it is the interplay between the parts that produce the outcomes. Therefore, the performance of the whole results from the interactions between the parts and cannot be predicted by analysing each part separately. However, it is not always clear what the whole is and what the parts are because a whole may be a part of another whole like a person might be considered a whole but is part of a department which itself is a whole but also part of an organization which is itself a whole but also part of an industry etc. This introduces the concept of system boundaries that will be dealt with later.

A common way of looking at organizations is to study individual parts and draw conclusions about a group based on the analysis of its constituent parts. This may lead to changes being made in the parts which have unintended consequences as a result of ignoring the influence that individual parts have on each other. This approach is often referred to as reductionism[1]. To understand and predict the behaviour of systems, we have to look at and analyse the whole and not its parts – this is the systems approach.

THE NATURE OF MANAGEMENT SYSTEMS

Defining a Management System

If the organization is to be managed as a system, what might a management system be? We should approach the answer to this question from a different direction, primarily because of the term's origin and the various misconceptions.

Is it a Set of Rules?

There is a meaning that expresses a system as a ordered set of principles or rules like the metric system and this might be where the idea that a management system is a set of rules and requirements originated. The use of standards to define requirements for management systems has led to the belief that the standards themselves are management systems. This was illustrated on the BSI – Global Web site a few years ago where a training course on 'Implementing an Integrated Management System', was prefaced with the following "The business challenge today is to manage activities more holistically rather than the traditional approach of having ISO 9001, ISO 14001, OHSAS 18001 and other management systems as peripheral arrangements". This phrase is no longer on the BSI web site but has found its way into other web sites implying it is believed to be true. In reality ISO 9001, ISO 14001, OHSAS 18001 are documents not systems. To say that a management system is a set of requirements would be like saying that a specification for a car is the car itself. The requirements within these standards may characterize the system, but they are not the system.

Is it a Set of Documents?

One way of looking at a management system is as a system of documentation. There is no doubt that when ISO 9000 was launched in 1987, organizations received the message that in order to meet this standards it was simply a case of documenting what you do, doing what you document and proving to external auditors that the documented system was in place. Throughout the world this resulted in look-alike documentation. All had a Quality Manual, procedures, work instructions and files of records. It mirrored the pyramid illustrated in Fig. 7-1 so often put forward as how the quality system should be structured. The procedures were documents that defined *who* had the responsibility for doing *what* and *when* with the *how* described in Work Instructions.

FIGURE 7-1 Pre-ISO 9001:2000 system documentation structure.

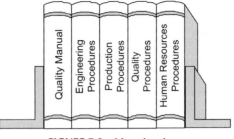

FIGURE 7-2   Manuals galore.

These were often compiled into Departmental Procedure Manuals as illustrated in Fig. 7-2. The Quality Procedures were sometimes procedures that were limited to the activities of the Quality Department but often they would include procedures written to meet specific requirements in ISO 9001 such as Nonconforming Material Control, Contract Review and Document Control etc.

The misconception was fuelled further by:

- Auditors asking to look at the 'quality system' implying that it is a set of documents;
- Comments that it is a 'nice little system' implying again that it is a set of documents;
- Adverts for software that claim to provide an 'electronic management system' implying that it is a tool;
- Requests to update 'the system' implying that it is documentation;
- Reviews that are limited to nonconformities and document changes, implying that it is about following procedures;
- The management representative or quality manager being responsible for 'the system' implying that other managers are not responsible for it;
- A statement that employees had to meet the requirements of the QMS, implying that it is a set of rules. This misconception is displayed in Clause 8.2.2 where it requires audits to be conducted to verify that the quality management system conforms to the quality management system requirements.

A search on Google for the definition of a quality management system will reveal many definitions that imply it is a set of documents. All are overcomplicated and worth a look to see how bizarre they can get.

Is it a Set of Tools?

An extension of the notion of a system of documentation is that some of the methods prescribed within the documents refer to tools and indeed as many paper-based procedures are computerized, it is the tool that attracts the label quality management system. Document control has evolved from being paper-based to computer-based so the software tools to control changes form part of the system. Databases for capturing measurement data, nonconformity data, corrective and preventive action reports are now automated such that apart from policies, many procedures are implemented through software tools. By the click of a mouse, reports can be produced and distributed world wide. They are accessible from the workstation together with all the documentation so the temptation is to call what people see when looking at the computer screen, the quality management system. The server on which sit all these documents and tools is certainly part of the system but it is not the whole system because without human interaction and other resources it is passive. Systems are dynamic; they produce effects like satisfied and dissatisfied customers and disgruntled employees.

Is it a Set of Processes?

The new definition in ISO 9000:2005 states that a management system is "a set of interrelated or interacting elements to establish policy and objectives and to achieve those objectives".

This definition is weak because it provides no clue as to the type of objectives referred to but the fact that it includes the phrase "to achieve those objectives" must make the system dynamic. It is weak also because there is no clue as to what these interacting elements might be. What might

> **Purpose & Objectives**
>
> The words 'objective' and 'purpose' are often interchangeable. The word purpose might be used to express a permanent state such as the reason for existence whereas the word objective might be used to express a transient state such as something that is aimed for. Things with a purpose are sometimes used by other things to achieve an objective.

be implied is that these elements are documents rather than processes. Indeed, the objective might be simply to deploy communication policies to the workforce and therefore a set of hyperlinked documents posted on an Intranet promulgating communication policies on this basis could be regarded as a management system. It is a set of documents and hence is interrelated. It is hyperlinked so it is interacting. It is structured into sections so it has got elements and it establishes communication policy and it achieves the communication objectives by being available to all through the intranet – but it is not the type of system that carries the title 'management system'. If the definition had included the phrase *management policy and objectives* it might have made more sense. Another anomaly in the definition is that objectives are established and achieved but policies are only established by the system. For completeness, one would have thought that policies would also be implemented.

Is it clear from the definition of a process approach in ISO 9001, Clause 0.2 that what is being addressed is a system of processes therefore these interrelated and interacting elements in the ISO 9000:2005 definition are processes.

Isaac Newton tells us that for every action there is an equal and opposite reaction, so when we bring together people, equipment and activities there will be a series of actions and reactions that cause a number of results. This is what we call *process*.

When we observe the human anatomy we see a collection of organs. If we are able to take a walk around the human body as we did for an organization previously, we would observe that there are not only organs but also processes at work: the digestive system, the respiratory system, the reproductive system, the nervous system etc. (We refer to these as systems because we have traditionally drawn a boundary around them.)

When we create a central heating system we bring together various components but until energized the system is dormant. When we open the valves, turn on the gas supply and light the boiler, the system springs into life and the processes begin to operate.

It therefore appears that in both the above cases, it is not the components themselves that form a system – simply connecting them together only connects the pathways or channels. The components need to be energized or triggered, blood needs to flow, water needs to pass through the pipes for any kind of result to be produced and that result is going to depend on how well each of the components performs its function. If the heater does not raise the temperature of the water to the required level, the radiators will not put out the heat required to raise the temperature in the room. If the heart does not pump the blood around the body, the nervous system will cease to function.

When we observe the people interacting within an organization, the outcome will not be simply the product of a person performing an activity using a tool. The activity does not take place in a vacuum – the environment in which the activity is performed influences the behaviour of the person and the impact of that person's actions upon others either directly or indirectly influences the results. By introducing the concept of processes, we capture all the forces that interact to generate results. The results from one process will be used by another process in order to deliver its outputs, thereby making the system dynamic. The outputs from another process might not be used by other processes but may influence their behaviour. A process can be thought of as a set of interrelated activities and resources that produce results. Therefore, in order to accomplish a specific purpose we need to design and manage a series of processes to deliver results that will fulfil our purpose. A management system can therefore be formed from set of interacting processes designed to function together to fulfil a specific purpose.

Multiple Systems

The word 'management' in the term 'management system' is intended to tell us what type of system it is and as systems achieve objectives or fulfil a purpose, it becomes evident that management systems achieve management objectives just as clearly as security systems achieve security objectives, storage systems achieve storage objectives and communication systems achieve communication objectives.

But the word management can be applied to anything that needs to be managed so we get database management systems, information management systems, content management systems, learning management systems, identity management systems and of course

> **Management System Objectives**
>
> **Management systems** achieve management objectives just as a security system achieves security objectives.
>
> A **quality management system** therefore achieves management objectives for quality just as a financial management system achieves management objectives for finance.
>
> A **business management system** must therefore achieve management objectives for the business which will include all the former management objectives and more besides.

environmental management systems and safety management systems. The objective in each case is management's objective for that aspect of performance (learning, identity, content, information, safety etc.). It would therefore appear that the word qualifying the term 'management system' is the subject of the management system and hence the focus for the system's objectives. Therefore:

- A quality management system is a set of interacting processes designed to function together to fulfil quality objectives;

- An environmental management system is a set of interacting processes designed to function together to fulfil environmental objectives;
- A financial management system is a set of interacting processes designed to function together to fulfil financial objectives.

Whatever the objective we could develop a management system to achieve it, which might result in a free for all as illustrated in Fig. 7-3. It follows therefore that if we want to focus on the whole organization, we should either refer to its management system as a *business* management system or an *enterprise* management system but we need to tread carefully. An Internet search will reveal that Enterprise Management Systems are software driven systems that speed up transactions between customer and supplier through the supply chain using the latest technology. Even an Internet search on Business Management Systems will produce a similar result although some of these do come up with ISO 9001 based management systems. It looks therefore that outside the world of ISO 9001 and its derivatives, these terms are used for software solutions rather than a description of how the business is managed.

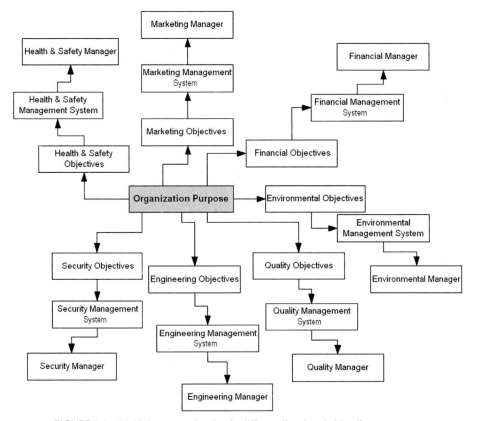

FIGURE 7-3   Multiple systems heading in different directions led by disparate teams.

Towards a Single System

It therefore becomes difficult to use any established terms to describe the set of inter-acting processes that achieve the business objectives, because the most appropriate ones like 'business' and 'enterprise' have been hijacked by the software industry. But we will not be daunted by this because our logic is sound – the system that enables the business to achieve its objectives, we will call a ***business management system***. The notion of qualifying terms to clarify purpose and scope is a good one and should be used more widely. Most quality management systems are limited in scope to products so they should strictly be called product management systems.

The business management system is therefore a system for managing the business and not a set of procedures for making widgets. It will deliver procedures to the places where they are needed. It will also deliver products but primarily the system is the enabler of business results. If it were only concerned with producing product, it would not be a management system but a production system. If it were only concerned with emissions, it would be an emission control system. If it were only concerned with product safety, it would be a product safety programme. These important systems and programmes are not the management system but are the product of the management system. It's all to do with context. The 'management system' that concerns us here should be thought of as a system of managed processes and its place in the cycle of sustained success is illustrated in Fig. 7-4.

The cycle of sustained success shows that:

- Stakeholders place demands upon the organization and these are fundamental in determining its mission and vision.
- The organization's mission, vision and values reflect what the organization is trying to do, where it is going and what principles will drive it towards satisfying stake-holder needs and expectations.
- The organization accomplishes its mission and vision through a set of interacting processes that collectively form the business management system focused on the mission. In this respect each process will comprise the activities, resources and behaviours needed to produce the outputs necessary to accomplish the mission and vision.
- The business management system delivers the organization's results that produce satisfied stakeholders.
- The stakeholders consider whether their needs continue to be satisfied and through one means or another, redefine the demands they place upon the organization.

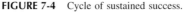

FIGURE 7-4   Cycle of sustained success.

This approach to a business management system has implications because for some it is fundamentally different. For such a management system to achieve the organization's desired results it has to include more than a set of documents. It must include resources because all work needs to be resourced to deliver results. It must also include behaviours because the attitudes, beliefs, motivation and other aspects of the organization's culture affect the way in which work is performed. Work can be performed quickly or slowly with enthusiasm or with a grudge, precisely or shoddily – it is a consequence of how people behave in given situations. Unlike machines, people are prone to emotions that impact their work.

SYSTEMS INTEGRATION

In the field of management systems there is a belief that organizations have multiple systems which create inefficiencies and there are therefore business benefits from integrating these systems, but we will show that this logic is flawed, for organizations only have one management system.

> **Integration**
>
> Integration combines parts to provide a function greater than the sum of the individual parts.

Quite literally, to integrate means to combine parts into a whole, bringing parts together or amalgamating parts to make complete, to desegregate or to incorporate into a larger unit.[1] In the context of management, integration might be putting all the internal management practices into one system or bringing together separate disciplines to work on a problem, or joining the processes that serve a particular objective. Think of the opposite word: disintegrate. If something disintegrates it shatters into tiny pieces. However, it was once whole and therefore for something to be integrated, it does not just sit next to the other components, it has to interact with the other components so as to make a whole. If the integrated whole is energized, all the parts will be energized or will provide a platform for the energized parts. There will be no part that does not have a function within the whole.

In a concert hall, there is the orchestra and the audience. Both are groups of people. The orchestra can be said to be integrated – a whole. If a section of the orchestra is missing, the orchestra cannot perform the piece they had intended. The audience is not integrated. It is simply a collection of individuals who are related by their interest in music. There is a relationship between the orchestra and the audience; one plays and other listens. If a few people fail to turn up for the concert, it has no effect on the audience. The audience can still perform its function even when numbers are reduced to single figures – hence the distinction between a 'collection of parts' and 'combination of parts'. The orchestra forms part of the entertainment system. The audience are part of that system as their interaction with the orchestra will determine the effectiveness of this entertainment system. The owners, managers and maintainers of the auditorium are also part of the system for any failure on their part will again impact the effectiveness of this entertainment system. The audience is not part of the orchestra but can effect its performance; no applause and the orchestra attempts to do better. The air-conditioning fails and the whole experience becomes unbearable; the system fails. There are therefore

---

[1] Concise Oxford English Dictionary.

active parts of a system that need to be integrated (the orchestra, the auditorium etc.) and some that need to be related (the audience).

The Misconceptions of Integration

If the organization is thought of as a system, the two terms would be synonymous and hence there would be no question about integration. However, because many do not see it that way and organizations are invariably not managed effectively, we need to address a number of perceptions. There is much confusion about what is being integrated. In fact one can read articles on the subject and get to the end without any clue as to what the writer means by the two words 'system' and 'integration'. They write of the benefits, advantages and the disadvantages but fail to explain just what is being integrated. So what is it that we are integrating?

Many organizations have created separate systems of documentation in response to management system standards and therefore they may perceive that integration is about *integrating documentation*. Those advocates of management system integration have the idea of managing activities more holistically which suggests we might be *integrating management*. There is also a notion that the structure of management system standards has been a barrier to integration which suggests we might be *integrating standards*. Another view is that organizations have a tendency to create functional silos or departments that focus only on departmental goals with the attendant disadvantages for the enterprise as a whole; therefore, we might be *integrating functions*. Finally, the management system standards are perceived by some to be a way of reducing risk so another view might be the *integration of risk management systems*. We will deal with each of these in more detail to discover just what is it that we are combining, amalgamating, incorporating or making complete. This is the issue that sits at the centre of the argument.

Are we integrating standards such as ISO 9001, ISO 14001 and OHSAS 18001? Is the goal to produce one management system standard as illustrated in Fig. 7-5? There might be a case for integrating these standards but we really need to examine their purpose before we pursue this argument.

These standards contain management system requirements and are used contractually by purchasers as a means of obtaining confidence in the capability of their suppliers. They are also used by certification bodies as criteria for determining the capability of an organization's management system. On demonstrating conformity with the requirements of one of these standards to a certification body, an organization will receive a certificate. By having three standards, it provides organizations with a choice of certification. They may not need certification at all and therefore would not use any of these standards. They might only need certification to ISO 9001 and would therefore ignore the other two standards or they

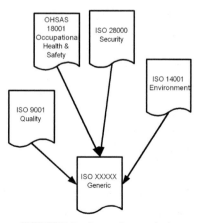

FIGURE 7-5   Integrating standards.

might need certification to all three standards and would therefore receive three certificates.

There is therefore a case for <u>not</u> integrating these standards as it provides organizations with a free choice. This is the situation when we perceive these standards as assessment tools.

If we perceive these standards as design tools the picture is completely different. Quality, security, environment and, occupational health and safety might be considered to be three aspects of management requiring three separate systems. If organizations respond to the standards with a manual and a series of procedures and work instructions, they might well have several systems of documentation and this presents another candidate for integration.

Integrating Documentation

Before national standards for different types of management systems emerged, a company would have one system that had many functions. Some of these were documented and some were not. Often these were based upon departmental or functional practices. On paper it wasn't a system just a collection of practices. But in reality, there was a system

> **Management System Standards**
>
> Management System Standards are criteria for assessing the capability of organizations to meet specified requirements not requirements to be achieved.

that consisted of custom and practice. Most of it was not written down but it worked because of the skills, knowledge and working relationships the people developed and put into the business. Management style meant that it was quite fragile in some organizations and strong in others. Change the people, the processes or indeed change anything and stability could not be guaranteed. The learning that had been acquired by one generation was passed on by word of mouth. Today's world is changing at a much faster rate than 50 years ago and hence we cannot rely on informal systems to reach our goals except in very small organizations.

Although all organizations had a management system or in other words, a system for managing the business, there was no consistency. Each department might or might not have defined its working practices and it was not until we started to formalize these practices in response to external standards such as ISO 9001 and ISO 14001 that they began to take shape as several different systems of documentation as illustrated in Fig. 7-6.

Getting companies to formalize their practices all in one go would not have been successful. Pressure from government procurement agencies and corporate purchasing for improvement in product quality came first and this resulted in an increase in Quality Management Systems being created and documented. This was followed by pressure from the environmental lobby, the Rio Conference, Kyoto etc. that resulted in Environmental Management Systems. Although the Health and Safety legislation has been around for some time it was only after the authorities realized that imposing rules did not necessarily improve performance that the notion of Occupational health and safety systems came about. All of these systems have been driven by standards. Take away the standards and the 'systems' cease to exist as formal systems. Had the movement been launched with one management system standard, organizations would have created one

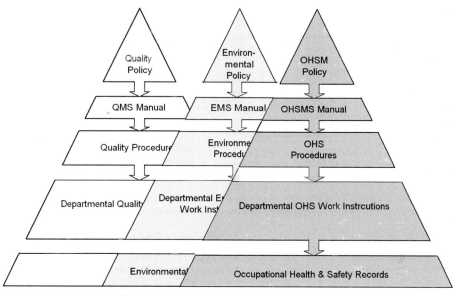

FIGURE 7-6 Separate systems of documentation.

system. It was therefore the piecemeal publication of standards that led to the creation of separate systems.

Organizations have always had their own way of working so taking away the standards does not remove all formality in management. It might remove the motivation to define and document management practices which is why the systems created were systems of documentation rather than documented systems.

If the manuals produced in response to the various standards are integrated to form one composite manual – there are perhaps savings in paper or storage and it may simplify navigation through the documents. For instance there are many common elements:

- Management responsibility,
- Management review,
- Corrective action,
- Preventive action,
- Internal audit,
- Document control,
- Records control,
- Resource management,
- Continual improvement.

This is not surprising as they are ISO 9001 derivatives. The procedures developed for dealing with these issues might well be common for quality, environmental, health, safety, security etc. but putting the financial system documentation, the quality system documentation, the environmental system documentation etc. into one book of policies and procedures is not integrating management systems – it is merely assembling the documentation that describes such systems in a more holistic manner and eliminating unnecessary duplication. Sometimes such action can be detrimental when it removes flexibility.

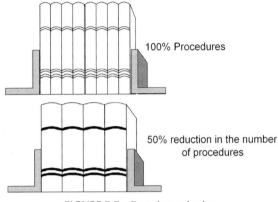

100% Procedures

50% reduction in the number
of procedures

FIGURE 7-7   Procedure reduction.

One of the claims made for adopting an Integrated Management System is that the numbers of procedures can be reduced by up to 50% – a great saving! But if the volume is no different as illustrated in Fig. 7-7, the reduction in quantity has no beneficial impact.

- Will reducing the number of documented procedures improve the bottom line?
- Will reducing the number of documented procedures increase the number of customers?
- Will reducing the number of documented procedures reduce variation in performance?

It is very doubtful that it will do any of the above if the objective is measured in terms of the number of procedures. It would be absurd to use size as a measure of effectiveness. Who really cares how many documents there are or how big they are? A positive outcome might be a reduction in the variation of methods thus instead of having 20 ways of doing a job, you reduce it to one effective and economical way of doing the same job.

Some organizations have installed software packages that handle quality, safety and environmental documentation. Some of these software packages retain menus for quality, environment, health and safety so all they do is to put the documents in one place. For these systems to be an integral part of the company's management system there have to be linkages so that you don't notice the transition between disciplines. The categories of quality, environment, health and safety disappear. The notion of separate systems disappears; even the terms disappear so that it becomes the management system that covers quality, environment, health, safety, finance, security and everything else we do.

Integrating Disciplines or Functions

Putting the quality manager, safety manager and environmental manager in one office or one department is another candidate for integration. It might reduce resources, office space for example and it might improve communication between these people, but it is not system integration. This is bringing separate disciplines under one roof but it is not a function of the organization unless one conceives of it as corporate governance when in which case it might be considered as a function because it makes a unique contribution to organizational performance.

There is no management system function in an organization. There might be a person or group of people whose primary concern is the maintenance and improvement of working practices. They might maintain the system descriptions as a means to convey best practice within the organization but these groups are not separate functions in the organization. Again if the management system is a risk management system, one might conceive of those people concerned with that system to be part of a corporate governance function.

Another example is a matrix organization where staff are loaned to a project manager responsible for the successful completion of the project. Although people come together to discuss common themes, issues etc. by returning to their *parent function* afterwards they can slip back into functional mode because the influence of the team has been only temporary. Putting people together in a team and removing them from their parent function will have the advantage of removing influences from the parent function.

If one perceives health, safety, environment, quality, security etc. to be disciplines, it is conceivable that they might work together but whether the outcomes are integrated will largely depend how they are managed. In this context, health, safety, environment and security and quality are all constraints.

There is no separate quality function in an organization because quality is not a function but an outcome. There may well be people with the word 'quality' in their job title and they may work in a quality department but such people do not perform all activities necessary to produce products and service that satisfy customer requirements. They have a limited role and perform some of these activities, primarily those concerned with setting up systems that enable others to meet standards, verifying compliance with standards and coordinating improvement. If this is what is understood as 'quality' then as a discipline it can be combined with safety, health and environment. But if 'quality' is perceived as meeting requirements, not simply product and service requirements, anybody and everybody in an organization is responsible for quality and has a role in its achievement and control.

Integrating Risk Management Systems

As health and safety management systems as well as environmental management systems are risk management systems, joining risk management systems together can constitute an integrated (Risk) Management System. This is a view taken by IOSH and therefore the Quality Management System, Security Management System and Information Technology Management Systems can also be integrated with the other risk management systems. But that is what it is, an integrated risk management system not an integrated management system.

In their guide on the integration of management systems,[2] the UK Institute of Occupational Health and Safety (IOSH) states clearly that on the subject of integration they are referring to the integration of such matters as organizational structures, strategic decision-making, resource allocation and the processes of auditing and reviewing performance. Regarding the integration of organization structures, there is no doubt that IOSH is addressing the separate disciplines of heath, safety, environment and quality rather than the wider aspect of the complete organization. On the integration of decision-making IOSH is referring to decisions concerning health, safety, quality and

---

[2] Joined-up working: an introduction to integrated management systems published by the Institute of Occupational Health and Safety 2006.

environment, where quality along with the other topics is perceived as a constraint. On resource allocation, IOSH is concerned about the proper allocation of resource to each discipline so that integration does not compromise safety, environment etc. by resulting in fewer resources. Regarding auditing processes, there is no doubt that IOSH is addressing the benefits from combined quality, health, safety and environmental audits. There is nothing in their arguments about organizational purpose and the processes needed for the organization to fulfil that purpose. The systems that IOSH is referring to are subsets of an all-embracing risk management system which appears to be a system of documentation and tools.

Integrating Management

There are several forms of management: functional management, project management, product management etc. each of which makes a specific impact on the organization structure. In functional management, the structure is composed of functional groups, each making a unique contribution to the organization's goals but each being a collection of specialists, e.g., sales, marketing, engineering, production, purchasing, quality etc. Under functional structures there also tend to be professional institutions or societies that support the specialism with the attendant disadvantage that there is sub-optimization of performance. Each group strives to maximize its performance often at the expense of the performance of the whole. This results in those outside the organization claiming that it lacks 'joined up thinking'. One group issues directives that are contradicted by other groups. Another group carries out activities that are undone by the activities of other groups. A classic example is in maintaining a county's infrastructure. No sooner has a hole been filled in a road than another utility comes along to dig it up again – no joined up thinking!

With a project organization several specialists from line functions are seconded to the project to serve a project objective with the kind of division that suits the project not the functional structure. The work of the project team is centrally coordinated so you don't get people undoing work recently completed by other team members. Product management follows the same pattern and in both these cases the group serves the objectives of the group with the distinct advantage that there is no sub-optimization of performance. If such a model could be made to work for the whole organization, everyone would focus on the organization's goals.

The CQI definition of integrated management seems to sum it up. *"Integrated Management is the understanding and effective direction of every aspect of an organization so that the needs and expectations of all stakeholders are equitably satisfied by the best use of all resources."*

The clues in the definition are the words 'understanding' and 'direction'. Integrated management is not about changing the structure of the organization. There are performance advantages in grouping people together by speciality, discipline or common interest so there is no need to change this. Integrated management is about understanding and directing activities to achieve common objectives.

The classic approach is to deploy the organization's objectives to each functional group but this often results in some functions being allocated objectives that can only be achieved through the participation of other functions. Examples of this are financial objectives, quality objectives, environmental objectives, and safety objectives. Their achievement requires the collective participation of all employees, whereas, a product objective might only require the participation of those in the engineering and production function. Engineering has to design it like the customer wanted and production has to make it like designer designed it.

Integrating Systems

If we perceive that quality, health, safety, environment etc. are objectives for which systems need to be established to achieve them as we illustrated in Fig. 7-3, the integration of these systems would result in an Integrated Management System (IMS) addressing all the objectives. This is illustrated in Fig. 7-8 but it is still not a Business Management System or BMS.

As with separate systems they sit outside the organization although they are now joined together to fulfil the requirements of the various standards. In Fig. 7-8 the IMS is a management system that comprises only those systems that are the subject of national or international standards such as QMS – ISO 9001, EMS – ISO 14001, OHSMS – OHSAS 18001, ISMS – BS 7799. We call it an IMS because the standards caused disintegration and therefore putting the pieces back together might be termed integration. But what is being integrated are systems designed around measurement tools for obtaining assurance. They are not design tools for designing enabling processes. However, it remains separate from the business because it excludes the result producing activities. It only includes the activities that satisfy the constraints.

Recognizing that these 'requirements' are not objectives but constraints that comprise value adding and non-value adding 'requirements', the integration of these systems will therefore look like that illustrated in Fig. 7-9. But, the grey disc is much smaller than the white disc so even if all requirements added value there would still be some white showing. The IMS would not have eclipsed the Organization. The system therefore needs to be wider in scope.

If we now embrace all activities within an organization, regardless of whether there are external standards governing their management we will eliminate the artificial boundaries created by these standards and once again treat the organization as a system. There will remain some constraints that add no value to the organization per sé but need to be satisfied if the organization desires to continue trading in its chosen markets. The net result is illustrated in Fig. 7-10. The alignment of the arrows is significant because it shows that the constraints are being filtered to align value adding constraints with the objectives.

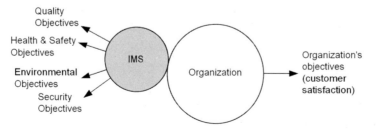

FIGURE 7-8   Integrated requirements.

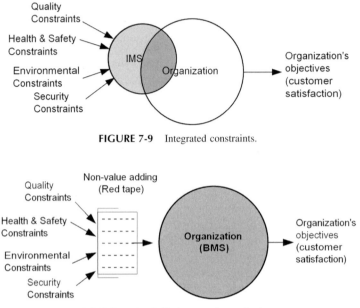

FIGURE 7-9   Integrated constraints.

Quality
Constraints

Non-value adding
(Red tape)

Health & Safety
Constraints

Environmental
Constraints

Security
Constraints

Organization
(BMS)

Organization's
objectives
(customer
satisfaction)

FIGURE 7-10   Fully integrated organization.

The organization depicted in Fig. 7-10 is a system that consists of a set of managed processes arranged in such a way as to deliver the organization's objectives day after day, year after year. *The organization is the BMS.* When the objectives need to change, the processes are configured in such a way that the need for change will be recognized and the processes reconfigured to achieve these new objectives. Instead of deploying objectives to functions, they are deployed to processes that are designed to achieve them. The activities that need to be carried out to achieve the objectives are assigned to people with the necessary competence and authority from which the roles are determined and people from the appropriate functions are assigned to perform these roles. Performance is reviewed against process objectives rather than functional objectives so that outputs are aligned and optimized not sub-optimized. This is the systems approach to management, an approach that enables the organization to develop the capability to satisfy the needs and expectations of all its stakeholders.

Now we have a different perception of the organization we can redraw the Cycle of sustained success of Fig. 7-4 and depict it as shown in Fig. 7-11.

SYSTEM CHARACTERISTICS

System Boundaries

Senge describes a key principle,[3] that of the system boundary which is that "the interactions that must be examined are those most important to the issue at hand

---

[3] Senge Peter M (2006) The Fifth Discipline, The Art and Practice of the Learning Organization. Random House.

FIGURE 7-11 Revised cycle of sustained success.

regardless of parochial organizational boundaries". There are two important interactions in any quality management system and one of these is between the organization and its suppliers, therefore the supplier must be part of the system. And the other is between the organization and its customers. Certification bodies are suppliers and therefore they should be perceived as part of the system. By seeing customers and suppliers as part of the system they can each be developed so that instead of being at arms-length they are brought into an environment which fosters learning.

Vulnerability

Just as the solar system or a command and control system can be adversely affected by an external force so powerful that it creates instability, an organization can be affected by the economic climate, competition, terrorists, regulations, fire, flood and earthquake etc. all of which are external to it and have the power to destabilize it, even destroy it. Two such forces are mergers and acquisitions.

> **Vulnerability**
>
> All systems are subject to external influences that have a potential for causing a change in state.

When two organizations merge or one is acquired by another, the quality management system is often the last thing considered by top management and yet both organizations declare a commitment to maintaining it. There maybe pressure to merge the two systems but this should not be attempted without first revisiting the definition of a system[①] and considering the following:

- if there is no interaction between the processes brought together by the merger, there remains two systems;
- if the mission or objectives of the two organizations remain unchanged and the different, there remains two systems;
- if the merger or acquisition creates two separate business units under one management, there remains two systems.

Interaction between processes may be found at a strategic level rather than at operational level. It may be that one of the reasons for the merger or acquisition was a synergy in the two organizations that increases their leverage in a shared market. In such circumstances, one might expect interaction between the demand creation processes in each organization so that they could be unified into one process delivering outputs of greater value then two separate processes.

A downside of acquisitions is that the dominant party strips the assets from the other party leaving a different landscape and a dysfunctional management system. In such cases you have no option but to start again and develop a new system from scratch. The mission, strategy and resource base have changed so the processes will be very different in practice but perhaps not so different in principle.

Connections and Interconnections

Within a management system connections and interconnections exist on paper but rarely in practice. On paper we can depict processes by boxes and depict the channels along which information and product flow as lines connecting two or more boxes, but these lines do not exist in reality. Unlike a physical system that might have wires and pipes for connecting the various system components, the connection between processes is through interrelationships and interactions.

Relations and Interrelations

When we examine human society we observe that there are relationships between people. One might say that we are all related, having descended from the same gene pool but when we say we are related to a specific individual we imply that this individual appears in our family tree. This person could be dead and even if alive it does not mean that we have anything to do with them. A relationship simply expresses an attribute or feature that two or more things have in common. An interrelationship expresses how two or more things relate to each other.

The processes in a system are interrelated because they function together. When processes within a system don't function together but serve different masters, we say the system is broken. All processes within a system are related by having a common master, the system objectives or organizational goals. We show process relationships through a system model (see Fig. 7-15) and the hierarchy of processes (see Chapter 8).

The people who manage and operate the processes are also related by speciality or function and this relationship is usually depicted in an organization chart.

Interdependencies

All processes within a system are interdependent. They depend on each other to provide something or to take something. Even when we place the customer and supplier within the system, the customer depends upon the supplier honouring commitments for it to meet its obligations to its customer in a supply chain.

> **Utility**
>
> The force exerted by a component upon a system should always be consistent with the purpose of that system otherwise it may cause instability.

Utility

Systems function on the utilitarian principle which means judging each action by its utility, that is to say its usefulness in bringing about consequences of a certain kind. The consequences in the case of a system are its aims, purpose or objectives. Every action and decision taken by a component of the system should serve the aims, purpose or objectives of the system – to do otherwise puts that component in conflict or competition with other components or makes it superfluous. This would result in system breakdown or destruction at worst and system inefficiency at best.

Interactions

If we want to express how one thing affects another we look at the interactions not the relationships or interrelationships. Processes within a system interact to produce the system outcomes as illustrated in Fig. 7-12. All processes don't interact with each other as the processes may form a chain in which interfacing processes interact and others are simply interrelated.

An example of interaction is where the Purchasing process may have as its objective the minimization of costs and select suppliers on lowest price not realizing or even ignoring the fact that product quality is lower and as a consequence the Production process cannot meet its objectives for product quality.

The Finance management process uses cash flow as the measure of performance and as a consequence delays paying suppliers on time. This has a knock-on effect on production because in retaliation, the suppliers withhold further deliveries until outstanding invoices have been paid. Another example is where the Packer gets

> **Optimization**
>
> Getting the best from your function's resources may not result in the organization getting the best from its total resources.

a deduction from his wages if there is a customer complaint because the Packer is the last person to check the product before delivery. As a consequence the Packer won't let anything through the gate until everything has been checked, including the most trivial

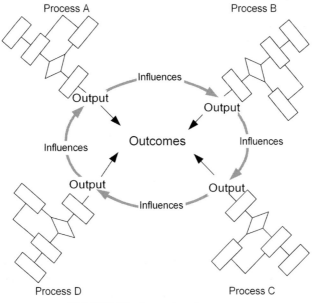

FIGURE 7-12 Interaction of processes.

of issues. A component left to its own devices will attempt to kill off other components in a competitive environment whether inside or outside an organization.

When system performance is running at an optimum, the performance of its components may not necessarily be at an optimum but the interactions are being balanced. It is therefore incumbent upon managers to manage the interactions between processes to achieve the organization's goals and not unilaterally change their objectives, practices or performance thus destabilizing the organization.

Value Chains

Within a system demand and cash flow from the customer. This is called a *value chain* or *demand chain* and is a flow from the customer through the organization to suppliers. The concept of value chains was developed by Michael Porter in the 1980s[4] as a competitive strategy and it is interesting to note that value is a variable and dependent on context. What a customer values one day may be different the next as the environment in which operate changes. Value is an experience and because it is derived from customer needs, activities that do not contribute to meeting these needs are 'non-value-added'. Not all work done add value and not all time taken is time adding value. There is the obvious downtime such as coffee breaks, lunch breaks, meetings then time spend doing work over instead of right first time. An analysis technique called value stream mapping[①] is used to identify these non-value added activities and eliminate them.

Supply Chains

Within a system products and services flow to the customer. This is called a *supply chain* from supplier to supplier and into the organization and out the customer. Supply chain is a term "now commonly used internationally – to encompass every effort involved in producing and delivering a final product or service, from the supplier's supplier to the customer's customer".[5] In a global market, the supply chain can become very complex and span many countries.

The integrity of the supply chain depends upon each party honouring their commitments and this depends upon each supplier having processes that have the capability to deliver quality product on time. Once products begin to flow along the supply chain, any disruptions either due to poor quality or late delivery cause costs to rise further along the chain that are irrecoverable. The end customer will only pay for product that meets requirements therefore if buffer stocks have to be held and staff paid waiting time as a result of supply chain unreliability, these costs have to be born by the producers. Process capability and product and service quality along the supply chain becomes the most vital factors in delivering outputs that satisfy the end customer requirement.

Delays

Within any system there will be delays. Some may be planned others unplanned. Planned delays are where inputs are awaited from external sources outside the

---

[4] M. Porter (1985) Competitive Advantage, Creating and Sustaining Superior Performance, The Free Press, New York.

[5] Supply-Chain Council (2005), available at: www.supply-chain.org.

organization's control. Unplanned delays are due to technical problems, absence, shortages, underestimates etc. Delays in a process output reaching its destination can have significant impact as Peter Senge illustrates with the Beer Game. Briefly, the Publican orders crates of beer monthly but between deliveries there is a special promotion that boosts sales. The publican starts to run low on beer so he doubles the next order. However, the delays in the ordering system result in him not receiving the increase for another month. Meanwhile demand increases but he cannot speed up delivery and by the time he receives the beer he has ordered, the customers are no longer buying the special brew so he is left with crate upon crate that he cannot sell. The advertising people were oblivious to the time it would take for the system to react to a special promotion and there was no provision for ordering outside the monthly schedule thus illustrating the way processes interact within a system.

Reserves

Holding reserves can be a consequence of previous delays but it not uncommon for people to order more than they need 'just in case' we need them. If every person was empowered to manage their own processes, we might find that everyone bought what they needed without reference to anyone else and the place was awash with pens, paper, light bulbs etc. Interventions that empower people and do other things like this can create instability in the system. They might be full of good intentions but a systems analysis is always necessary before authorizing the change. Reserves are also part of the contingency plan in the system for when something does go wrong and much of these reserves are in deposits or investments to pay out in times of lay-offs or bring in external assistance.

Overproduction

Some systems are designed to implement the 'sell what we can make' policy which results in making for stock. Western automotive industry has used this approach for decades and in late 2008 realized that a transition to a 'make what we can sell' policy was long overdue. This is the basis of mass production. It was the only way to get the price of a car down to what the customer could afford. It was the economy of scale that brought affordable transport to the masses but it was very wasteful. It depended upon there being a constant stream of buyers and in the post-millennium recession, with thousands of cars waiting to be sold, and no buyers, it was pointless continuing to make cars. After WWII the Japanese were aware that their productivity was about one tenth that in America and so believing that they must be wasting something, they set out to eliminate waste and this idea marked the start of the Toyota Production System which was based on a 'sell what we can make' policy. They eliminated the warehouse and rather than make for stock they make to order. Every car coming off the production line has a customer.

SYSTEM MODELS

When we move our thinking from rules and documents to processes we are turning a system from something tangible as a set of documents to something that is a representation of a dynamic entity (the organization) but which itself is intangible except as a model. The system exists only in our imagination and as a description on paper or other media. The reality is partially what we see when we walk around an organization, the rest is exhibited through actions, interactions and the tangible outputs. All we can do is to model the structure and behaviour of the organization as best we can. We cannot see management processes. All we see are their effects. We can see people doing things that are connected with a particular process but unlike industrial processes there are no conveyor belts carrying the information from one stage to another. The connections are more than likely invisible as information is conveyed from person to person in a variety of ways.

On reading ISO 9001:2008 we can interpret the requirements as a customer satisfaction cycle as shown in Fig. 7-13. This model shows the interaction between the elements. Here the quality policy and objectives drive the system of managed processes to deliver conforming products that satisfy customers who influence the policy and objectives and so on.

If we look at ISO 9004 we get a different perspective as we showed previously in Fig. 7-4 and reproduced in Fig. 7-14. Here the driver is the Mission and vision which has been crafted in response to the needs and expectations of stakeholders. The system of managed processes now covers the whole organization and the results are the outputs and outcomes of the organization which should satisfy all stakeholders and thus lead to sustained success. If we analyse all of the organization's outputs, we are likely to find that they can be placed into one of four processes.

1. The group of outputs that create demand will be delivered by a Demand Creation Process.
2. The group of outputs that satisfy a demand will be delivered by a Demand Fulfilment Process.
3. The group of outputs that provide resources will be delivered by a Resource Management Process.
4. The group of outputs that establish goals and strategy etc. will be delivered by a Mission Management Process.

These are generic names and each organization may choose different names to suite its culture and operating environment. The system of managed processes in Fig. 7-14 can therefore be modelled using these processes as shown in Fig. 7-15.

A further derivation of these processes is provided in Chapter 8 and further definition of these processes is addressed in Chapter 10.

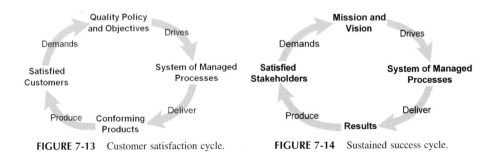

FIGURE 7-13   Customer satisfaction cycle.   **FIGURE 7-14**   Sustained success cycle.

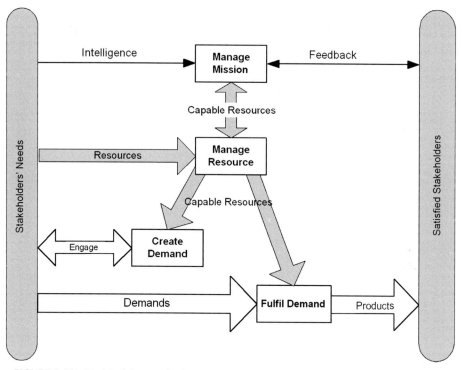

FIGURE 7-15   Model of the organization as a system of managed processes (generic system model).

Chapter 8

A Process Approach

CHAPTER PREVIEW

This chapter is aimed at those who want to understand and manage processes and details what the process approach is all about. It will interest, students, consultants, auditors, managers and most importantly top management who tend to opt for reorganization where performance is not good enough. It is an essential reading for those setting out to take a process approach to management and develop a process-based management system. More detail on this latter point can be found in *Establishing a quality management system* in Chapter 10, but first it is important for you to have an understanding of processes.

It's worth saying many times that 'all work is a process' because the use of ISO 9001 has placed in some people's minds the idea that all work can be described by procedures. In this chapter we hope to change this perception. The process approach was defined in ISO 9001:2000 but the definition has been revised to correct a misunderstanding. The new definition is in the text box. The additional words are 'to produce the desired outcome'. Thus, the

> **The Process Approach**
>
> The application of a system of processes within an organization, together with the identification and interactions of these processes, and their management to produce the desired outcome, can be referred to as the "process approach".
> (Ref. ISO 9001:2008)

purpose of taking a process approach is clarified. These outcomes are those expected by the organization's stakeholders. Also note that it is a system of processes, not a system of rules, policies or procedures and that it is the interaction of these processes that needs to be managed, and not the interaction within the processes. This definition is therefore more appropriate as a definition of the systems approach as it addresses a system of processes rather than reveal what it takes to manage an individual process indicating that these terms are still evolving.

In this chapter we examine process in depth and look at:

- The relationship between processes and quality;
- The differences between a functional approach and a process approach;
- The nature of processes, how they are defined, what types of processes there are and how they are classified;
- Perceptions of processes from different viewpoints;
- Several different ways of representing a process through process models;

- Business process re-engineering and the Excellence Model;
- The principles that underpin the process approach;
- A range of process characteristics.

THE RELATIONSHIP BETWEEN PROCESSES AND QUALITY

It has been stated previously that all work is a process but what do we mean by this? When we undertake work there is a series of actions we take, tools and equipment that we use, energy and materials we consume, information we need and decisions we make from the beginning to the end. These are the variables. The work progresses until completion. This progression is a process; it has a beginning and an end. It begins with an event or when we receive the command or reach a date and ends when we are satisfied with the resultant output. This output may be a tangible product or a service we provide to someone else. The product or service will possess certain features or characteristics, some of them needed, some of them not needed. The process we use determines these features or characteristics. Therefore, we can design or manipulate this process to produce any features or characteristics we so desire by altering the variables. If we start work with a determination to create an output possessing certain desired features or characteristics, we will produce outputs with those qualities. It follows therefore that if one manages processes effectively they will consistently and continually produce outputs of the desired quality.

Quality does not happen by chance, it has to be designed into a product or service and any amount of inspection will not change its quality. It is therefore within the process from conception to final delivery that holds the most potential for creating products or services of the utmost quality. We can manage product quality by sorting good products from bad products but that is wasteful. Sometimes we have no choice if we don't own the process but when we have control over all the variables that influence the quality of the outputs, we can guarantee the quality of the product or service. Some of the variables have a greater influence on output quality than others and although we may not be able to control all the variables all the time, we should aim to control those variables which have greatest impact so that customer satisfaction is assured. In some industries they call these variables Critical to Quality characteristics (CTQs).

FUNCTION APPROACH VERSUS PROCESS APPROACH

Most organizations are structured into functions that are collections of specialists performing tasks. The functions are like silos into which work is passed and executed under the directive of a function manager before being passed into another silo. In the next silo the work waits its turn because the people in that silo have different priorities and were not lucky enough to receive the resources they requested. Each function competes for scarce resources and completes a part of what is needed to deliver product to customers. This approach to work came out of the industrial revolution influenced firstly by Adam Smith and later by Frederick Winslow Taylor, Henry Fayol and others. When Smith and Taylor made their observations and formulated their theories, workers were not as educated as they are today. Technology was not as available and machines not as portable. Transportation of goods and information in the eighteenth and nineteenth centuries was totally different from today. As a means to transform a domestic economy to an industrial economy, the theory was right for the time. Mass production would not have been possible under the domestic systems used at that time.

Drucker defined a function as a collection of activities that make a common and unique contribution to the purpose and mission of the business.[1] Functional structures often include marketing, finance, research & development and production, each divided into departmental structures that include design, manufacturing, tooling, maintenance, purchasing, quality, personnel, accounting etc. In some cases the function is carried out by a single department and in other cases it is split among several departments.

The marketing function in a business generates revenue and the people contributing to marketing may possess many different skills, e.g., planning, organizing, selling, negotiating, data analysis etc. It is quite common to group work by its contribution to the business and to refer to these groupings as functions so that there is a marketing function, a design function a production function etc. However, it should not be assumed that all those who contribute to a function reside in one department. The marketing department may contain many staff with many skills, but often the design staff contributes to marketing. Likewise, the design function may have the major contribution from the design department but may also have contributors from research, test laboratory, trials and customer support. Therefore, the organization chart may in fact not define functions at all but a collection of departments that provide a mixture of contributions. In a simple structure the functions will be clear but in a complex organization, there could be many departments concerned with the marketing function, the design function, the production function etc. For example, the Reliability Engineering Department may be located in the Quality Department for reasons of independence but contributes to the design function.

However, the combined expertise of all these departments is needed to fulfil a customer's requirement. It is rare to find one department or function that fulfils an organizational objective without the support of other departments or function. However, the functional structure has proved to be very successful primarily because it develops core competences and hence attracts individuals who want to have a career in a particular discipline. This is the strength of the functional structure but because work is always executed as a process it passes through a variety of functions before the desired results are achieved. This causes bottlenecks, conflicts and sub-optimization. A functional approach tends to create gaps between functions and does not optimize overall performance. One department will optimize its activities around its objectives at the expense of other departments. We gave examples of this sub-optimization in *Interactions* in Chapter 7.

One approach that aims to avoid these conflicts is what is referred to as 'balancing objectives'. On face value, this might appear to be a solution but balancing implies that there is some give and take, a compromise or reduction in targets so that all objectives can be met. The result is often arrived at by negotiation implying that quality is negotiable when in reality it is not. Customers require products that meet their requirements not products that more or less meet their requirements.

A typical functional structure of an organization developing computer systems for military and civil applications of the 1980s is illustrated in Fig. 8-1.

---

[1] Drucker, Peter F. (1977). *Management: Tasks, Responsibilities, Practices.* Pan Business Management.

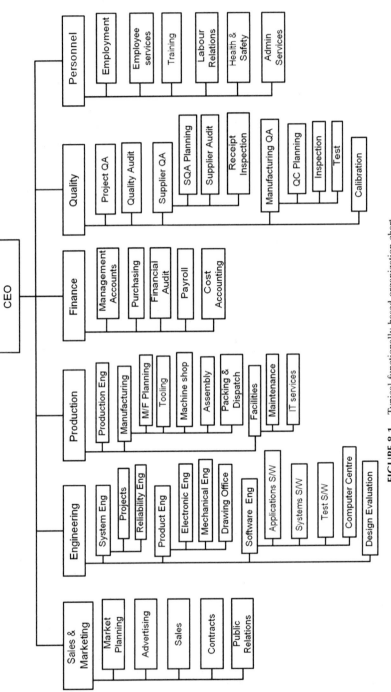

FIGURE 8-1   Typical functionally based organization chart.

In this structure you will observe that the Quality Department includes inspection and test rather than Production implying that the Production Department cannot be trusted to put quality first and not put delivery first. Also note that Purchasing is under the control of the Finance Department where money is king, thus implying that Production is not to be trusted to buy what it needs. All design engineering is in the Engineering Department including the Project Managers thus giving Engineering dominance. A project would have staff assigned from other engineering functions, production and quality functions to form the project team. Note that financial audit and quality audit are separate and that IT wasn't of high importance as most information systems were paper-based. With advances in technology and management theory, functional structures change. IT would become more dominant, production staff would have greater access to the tools needed for operators to exercise self-control over the processes they used and thus inspection and test activities could be transferred into the manufacturing department. The Quality Department may be renamed Quality Assurance or in an attempt to remove the obstacles that the word quality creates, a new name of Assurance Technology would emerge thus signalling to the organization that its business is provision of assurance and not decisions on conformity as had been the case previously.

When objectives are derived from stakeholder needs, internal negotiation is not a viable approach. The only negotiation is with the customer as explained in Chapter 3 in *Balancing stakeholder needs*.

Some of the other differences are indicated in Table 8-1.

Business outputs are generated by the combined efforts of all departments so processes tend to be cross-functional. Rarely does a single department produce a business output entirely without support from others. The interfaces for the same organization as Fig. 8-1 are shown in Fig. 8-2. (For simplicity, not all outputs are shown.) In principle the grey arrows indicate direction in which the process flows across departments. In reality it is probably not as simple as this because there will be transactions

TABLE 8-1 Function Versus Process

Attribute	Functional approach	Process approach
Objectives focus	Satisfying departmental ambitions	Satisfying stakeholder needs
Inputs	From other functions	From other processes
Outputs	To other functions	To other processes
Work	Task focused	Result focused
Teams	Departmental	Cross-functional
Resources	Territorial	Shared
Ownership	Departmental manager	Shared
Procedures	Departmental based	Task based
Performance review	Departmental	Process

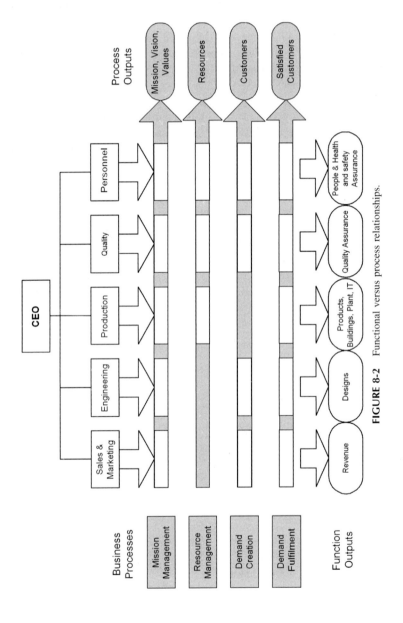

FIGURE 8-2   Functional versus process relationships.

that flow black and forth between departments. The organization structure shows that functional outputs are indeed different from process outputs and obviously make an important contribution, but it is the output from business processes that is important to the business.

When we organize work functionally, the hierarchy can be represented by the waterfall diagram of Fig. 8-3.

In this diagram the top-level description of the way work is managed will probably be contained in a Quality Manual with supporting Department Manuals. A common mistake when converting to a process approach is to simply group activities together and call them processes but retaining the Function/Department division. This perpetuates the practice of separating organization objectives into Departmental objectives and then into process objectives. This is not strictly managing work as a process at an organizational level. Another anomaly is that it makes the assumption that the manufacturing department provides everything needed to perform the identified activities when in fact other departments are involved, such as the Quality Department providing inspection and test as shown Fig. 8-3. A more effective approach ignores functional and departmental boundaries as represented by Fig. 8-4.

Superficially it may appear as though all we have done is to change some words but it is more profound than that. By positioning the Business process at the top level we are changing the way work is managed, instead of managing results by the contributions made by separate functions and departments, we manage the process which delivers the results regardless of which function or departments does the work. This does not mean we disband the functions/departments; they still have a role in the organization of work. Work can be organized in three ways.[2] By stages in a process, by moving work to where the skill or tool is located or assembling a multi-skilled team and moving it to where the work is. In all of these cases we can still manage the work as a process or as a function. It comes down to what we declare as the objectives, how these were derived and how we intend to measure performance. If we ask three questions, "What are we trying to do, how will we make it happen and how will we know it's right?" we can either decide to make it happen through a process or through a number of functions/departments and measure performance accordingly. By 'making it happen' through a process, we overcome the disadvantages of the functional approach. A complementary view of the process – function debate is provided by Jeston and Nelis.[3]

> **People and Process**
>
> If a process is supposed to transform inputs into outputs, it can't produce an output without the people and other resources which is why the people and other resources form part of the process.
>
> It is what people do that influence process dynamics. Their actions or inactions create flow, sequence, delay, breakdown, stability and many other process attributes.

---

[2] Drucker, Peter F. (1977). *Management: Tasks, Responsibilities, Practices.* Pan Business Management.

[3] John Jeston and Johan Nelis (2008) Business Process Management: Practical Guidelines to Successful Implementations. Butterworth Heinemann.

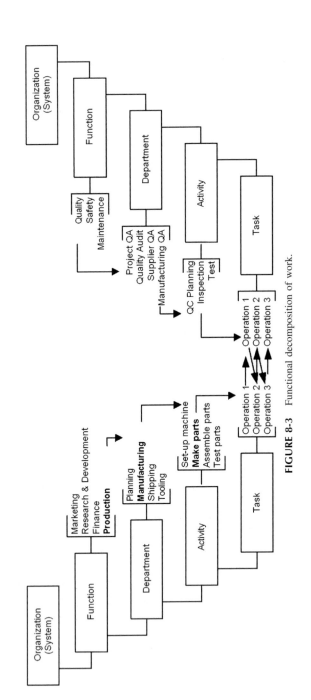

FIGURE 8-3 Functional decomposition of work.

FIGURE 8-4   Process decomposition of work.

THE NATURE OF PROCESSES

Finding a Definition

There are different schools of thought on what constitutes a process.

A process is defined in ISO 9000:2005 as *a set of interrelated or interacting activities which transform inputs into outputs* and goes on to state that *processes in an organization are generally planned and carried out under controlled conditions to add value.* The inclusion of the word generally tends to suggest that organizations may have processes that are not planned, not carried out under controlled conditions and do not add value and indeed they do therefore this explanation only goes to confuse the definition.

Juran defines a process[4] *as a systematic series of actions directed to the achievement of a goal.* In Juran's model the inputs are the goals and required product features and the outputs are products possessing the features required to meet customer needs. The ISO 9000:2005 definition does not refer to goals or objectives.

Hammer defines a process[5] as *a collection of activities that takes one or more kinds of inputs and creates an output that is of value to the customer.*

Davenport defines a process[6] as *a structured measured set of activities designed to produce a specified output for a particular customer or market.*

The concept of adding value and the party receiving the added value is seen as important in these definitions. This distinguishes processes from procedures and places customer value as a criterion for a process unlike the ISO 9000:2005 definition.

---

[4] Juran J.M. (1992) Juran on quality by design. The Free Press, Division of Macmillan Inc.

[5] Hammer, Michael and Champy, James, (1993). Reengineering the corporation, Harper Business.

[6] Davenport, T. H., (1993). Process Innovation: Reengineering work through Information Technology, Harvard Business School Press.

However, the ISO 9000:2005 definition is for the term process and although not stated, Hammer and Davenport are clearly talking about business processes because not every type of process creates value.

It is easy to see how these definitions can be misinterpreted but it doesn't explain why for many it results in flowcharts they call processes. They may describe the process flow but they are not in themselves processes because they simply define trans-actions. A series of transactions can repre-sent a chain from input to output but it does not cause things to happen. Add the resources, the behaviours, the constraints

> **Documented Processes**
>
> A procedure is not a documented process; it generally only documents the activities to be carried out, not the resources and behaviours required and the methods of managing the process – this is the role of the process description.

and make the necessary connections and you might have a process that will cause things to happen. Therefore, any process description that does not connect the activities and resources with the objectives and results is invalid. In fact any attempt to justify the charted activities with causing the outputs becomes futile.

Processes Versus Procedures

The procedural approach is about doing a task, conforming to the rules, doing what we are told to do, whereas, *the process approach* is about, understanding needs, finding the best way of fulfilling these needs, checking whether the needs are being satisfied and in the best way and checking whether our understanding of these needs remains valid. Some differences between processes and procedures are indicated in Table 8.2.

A view in the literature supporting ISO 9001:2008 is that the procedural approach is about how you do things and processes are about what you do. This is misleading as it places the person outside the process when in fact the person is part of the process. It also sends out a signal that processes are just a set of instructions rather than a dynamic mechanism for achieving results. This message could jeopardize the benefits to be gained from using the process approach.

Types of Processes

In thermodynamics we have the isobaric process of constant pressure, the isothermal process of constant temperature and the isochoric process of constant volume or indeed the adiabatic process where there is no heat transfer. But we are not discussing thermodynamic processes, chemical processes, computer processes or organic processes such as the digestive process. We are focused on organization processes where there are also different types of processes but the attributes that characterize them are not physical attributes.

Process Characterization by Purpose

All organization processes are feedback processes, i.e., processes where

a) a sensor measures output and feeds information to a comparator;
b) a comparator transmits a signal to the action component;

TABLE 8-2 Processes Versus Procedures

Procedures	Processes
Procedures are driven by completion of the task	Processes are driven by achievement of a desired outcome
Procedures are implemented	Processes are operated
Procedures steps are completed by different people in different departments with different objectives	Process stages are completed by different people with the same objectives – departments do not matter
Procedures are discontinuous	Processes flow to conclusion
Procedures focus on satisfying the rules	Processes focus on satisfying the customer
Procedures define the sequence of steps to execute a task	Processes generate results through use of resources
Procedures are used by people to carry out a task	Processes use people to achieve an objective
Procedures exist, they are static	Processes behave, they are dynamic
Procedures only cause people to take actions and decisions	Processes make things to happen, regardless of people following procedures
Procedures prescribe actions to be taken	Processes function through the actions and decisions that are taken
Procedures identify the tasks to be carried out	Processes select the procedures to be followed

c) an action component adjusts a parameter if necessary so that the output remains on target.

There are two types of feedback processes. Senge[7] refers to these as reinforcing or amplifying processes and balancing or stabilizing processes. The reinforcing processes are engines of growth and the balancing processes are engines of stability. If the target is to grow, increase or decrease the amount of something or widen the gap between two levels, reinforcing processes are being used. If the target is to maintain a certain level of performance, a certain speed, maintain cash flow, balancing processes are being used. The terminology varies in this regard. Juran[8] refers to these as Breakthrough and Control processes where breakthrough is reaching new levels of performance and control is maintaining an existing level of performance.

---

[7] Senge Peter M. (2006) The Fifth Discipline, The Art and Practice of the Learning Organization. Random House.

[8] Juran J.M. (1964) Managerial Breakthrough, McGraw-Hill Inc.

There are many balancing processes in organizations, in fact most of the organization's processes are balancing processes as their aim is to maintain the status quo, keep revenues flowing, keep customers happy, keep to the production quotas, keep to the performance targets etc. A few processes are reinforcing processes such as research and development processes, the process for expanding markets, building new factories etc. All these place a burden on the balancing processes until they can no longer handle the capacity and something has to change. Likewise, the reinforcing processes decrease orders, innovation and as a consequence the balancing processes have surplus capacity and again something has to change.

Process Characterization by Class

Processes are also characterized by class. As stated previously, all work is a process and all processes produce outputs therefore if we look at the organization as whole and ask, "What outputs will our stakeholders look for as evidence that their needs are being met?" we identify the organization's outputs. There must be processes producing these outputs and we call these macro-processes. These processes are multi-functional in nature consisting of numerous micro-processes. Macro-processes deliver business outputs and are commonly referred to as *Business Processes.* For processes to be classed as business processes they need to be in a chain of processes having the same stakeholder at each end of the chain. The input is an input to the business and the output is an output from the business.

Some people classify business processes into core processes and support processes but this distinction has little value, in fact it may create in people's minds the perception that core processes are more important than support processes. All processes have equal value in the system as all are dependent upon each other to achieve the organization's goals.

If we ask of each of these business processes "What affects our ability to deliver the business process outputs?" we identify the critical activities which at this level are processes because they deliver outputs upon which delivery of the business output depends. These processes are the micro-processes and they deliver departmental outputs and are task oriented. In this book these are referred to as *Work Processes.* A management system is not just a collection of work processes, but also the interaction of business processes. The relationship between these two types of processes is addressed in Table 8-3.[9] Some people call work processes, sub-processes.

The American Quality and Productivity Centre published a Process Classification framework in 1995 to encourage organizations to see their activities from a cross-industry process viewpoint instead of from a narrow functional viewpoint. The main classifications as revised in 2006 are as follows:

1. Develop vision and strategy,
2. Design and develop products and services,
3. Market and sell products and services,

---

[9] Juran, J. M. (1992). Juran on Quality by design, The Free Press. Based on Figure 11-1.

4. Deliver products and services,
5. Manage customer service,
6. Develop and manage human capital,
7. Manage information technology,
8. Manage financial resources,
9. Acquire, construct and manage property,
10. Manage environmental health and safety,
11. Manage external relationships,
12. Manage knowledge, improvement and change.

TABLE 8-3 Relationship of Business Process to Work Processes

Scope	Business process	Work process
Relationship to organization hierarchy	Unrelated	Closely related
Ownership of process	No natural owner	Departmental head or supervisor
Level of attention	Executive level	Supervisory or operator level
Relationship to business goals	Directly related	Indirectly related and sometimes (incorrectly) unrelated
Responsibility	Multi-functional	Invariably single function (but not exclusively)
Customers	Generally external or other business processes	Other departments or personnel in same department
Suppliers	Generally external or other business processes	Other departments or personnel in same department
Measures	Quality, cost delivery	Errors, quantities, response time
Units of measure	Customer satisfaction, shareholder value, cycle time	% Defective, % Sales cancelled, % Throughput

This classification was conceived out of a need for organizations to make comparisons when benchmarking[©] their processes. It was not intended as a basis for designing management systems. We can see from this list that several processes have similar outputs. For example, there are a group of processes with resources as the output. Also, some of these processes are not core processes but themes running through core processes. For example, the process for executing an environmental management program has a process design element but its implementation will be embodied in other result producing processes as it does not on its own form part of a chain of processes. Similarly with managing external relations, there will be many processes that have external interfaces so rather than one process there should be objectives for external relationships that are achieved by all processes with external interfaces.

Deriving the Business Processes

Taking the view that a business process has the same stakeholder at each end, we would conclude that product design is not a business process because the stakeholders are different at each end. On the input end could be sales and the output end could be produce and deliver. Under this logic, produce and deliver would not be a business process because on the input could be product design and the output could be the customer. Therefore, the business process flow is: customer to sales, sales to product design, product design to produce and deliver, produce and deliver to customer and customer to bank. On this basis the business process is 'order to cash'. The important point here is that the measure of success is not whether a design is completed on time, or a product meets its specification but whether the products designed, produced and delivered satisfy customer requirements to the extent that the invoice is paid in full. With the above approach, there would be one process that creates a demand for the organization's products and services. This is often referred to as marketing but this is also the label given to a department therefore we need a different term to avoid confusion. A suitable name might be Demand creation process.

Having created a demand, there must be a process that fulfils this demand. This might be production but if the customer requirement is detailed in performance terms rather than in terms of a solution, it might also include product design. There are many other ways of satisfying a demand and once again to avoid using labels that are also names of departments, a suitable name might be a Demand fulfilment process.

Both these processes need capable resources and clearly the planning, acquisition, maintenance and disposal of these resources would not be part of demand creation or fulfilment as resources are not an output of these processes. There is therefore a need for a process that manages the organization's resources and so we might call this the Resource management process.

Lastly, all the work involved in determining stakeholder needs, determining the mission, the vision and strategy, the business outputs and designing the processes to deliver these outputs is clearly a separate process. It is also important that the performance of the organization is subject to continual review and improvement and this is clearly a process. But neither can exist in isolation, they are in fact a continuum and when brought together would have the same stakeholder at each end. We have a choice of names for this process. We could call it a business management process but we might call the system the business management system so this could cause confusion. As the process plans the direction of the business and reviews performance against plan, we could call this process the vision, mission or goal management process. The precise title is not important, provided it conveys the right meaning to those who use and manage the process. If we take the AQPC processes and ask a simple question: *"What contribution do these activity groups make to the business?"* We can reduce the number of processes to four. If we now ask: *"What name should we give to the process that makes this contribution?"* We will identify four processes into which we could place all organization's activities as shown in Table 8-4

These business processes and the purpose of each process explained as follows with Table 8-5 showing the stakeholders.

TABLE 8-4 Process Classification Alignment

AQPC[i]	Process classification framework (main classifications)	Contribution	Business process
1	Develop vision and strategy	These set the goals and enable us to achieve them	Mission Management
12	Manage knowledge, improvement and change		
10	Manage environmental health and safety		
11	Manage external relationships		
2	Design and develop products and services	These create a demand	Demand Creation
3	Market and sell products and services		
4	Deliver products and services	These fulfil a demand	Demand Fulfilment
5	Manage customer service		
6	Develop and manage human capital	These provide capable resources	Resource management
7	Manage information technology		
8	Manage financial resources		
9	Acquire, construct and manage property		

TABLE 8-5 Business Process Stakeholders

Business process	Input stakeholder (inputs)	Output stakeholder (outputs)
Mission management	Investors, owners (vision)	Investors, owners (mission accomplished)
Demand creation	Customer (need)	Customer (demand)
Demand fulfilment	Customer (demand)	Customer (demand satisfied)
Resource management	Resource user (resource need)	Resource user (resource satisfies need)

Mission management process	Determines the direction of the business, continually confirms that the business is proceeding in the right direction and makes course corrections to keep the business focused on its mission. The business processes are developed within mission management as the enabling mechanism by which the mission is accomplished.
Resource management process	Specifies, acquires and maintains the resources required by the business to fulfil the mission and disposes off any resources that are no longer required.
Demand creation process	Penetrates new markets and exploits existing markets with products and a promotional strategy that influences decision makers and attracts potential customers to the organization. New product development would form part of this process if the business were market driven.

Demand fulfilment process	Converts customer requirements into products and services in a manner that satisfies all stakeholders. New product development would form part of this process if the business were order driven (i.e., the order contained performance requirements for which a new product or service had to be designed).

Perceptions of Process

Previously we said that all work is a process but as we have seen there are macro-processes and micro-processes, it all depends on our perception. If we were to ask the same question to three workers cutting stone on a building site, we might be surprised to get three different answers.

We approach the first stone cutter and ask, "What are you doing?"
"Breaking stone" he replies rather abruptly
This stone cutter has no vision of what he is doing beyond the task and will therefore be blind to its impact

We approach the second stone cutter and ask, "What are you doing?"
"I'm making a window" he replies with enthusiasm
This stone cutter sees beyond the task to a useful output but not where this outputs fits in the great scheme of things

We approach the third stone cutter and ask, "What are you doing?"
"I am building a Cathedral" he replies with considerable pride.
This stone cutter sees himself as part of a process and has a vision of what he is trying to achieve that will influence what he does.

If we allow ourselves to be persuaded that a single task is a process, we might well deduce that our organization has several thousand processes. If we go further and try to manage each of these nano-processes (they are smaller than micro-processes), we will lose sight of our objective very quickly. By seeing where the task fits in the activity, the activity fits within a process and the process fits within a system, we create a line of sight to the overall objective. By managing the system we manage the processes and in doing this we manage the activities. However, system design is crucial. If the processes are not designed to function together to fulfil the organizational goals, they can't be made to do so by tinkering with the activities.

Process Models

In the context of organizational analysis, a simple model of a process is shown in Fig. 8-5. This appeared in ISO 9000:1994 but clearly assumes everything other than inputs and outputs are contained in the process. The process transforms the inputs into outputs but the diagram does not in itself indicate whether these outputs are of added value or where the resources come from.

FIGURE 8-5   Simple process model.

Figure 8-6 reminds us that processes can produce outputs that are not wanted, therefore, if we want to model an effective process we should modify the information displayed.

FIGURE 8-6   Unwanted process outputs.

Another model, Fig. 8-7 taken from BS 7850:1992 shows resources and controls to be external to the process implying that they are drawn into the process when needed and yet without either a process cannot function. So can a process be a process without them? If it can't, the label on the box should either be 'activities' or these inputs should be shown as outputs from another process. Some controls might also be an output of another process but controls would be built-in to the process during process design and resources would be acquired when building a process other than any output specific resources. Therefore, in this respect the diagram is misleading but it has been around for many years.

The process model adopted by ISO/TC 176 did show procedures as an external input to a process but the updated version in 2003[10] (Fig. 8-8) shows resources as inputs but does qualify the outputs as being 'Requirements Satisfied' which is a far cry from

---

[10] ISO/TC 176/SC2 N544R2, available on www.iso.org.

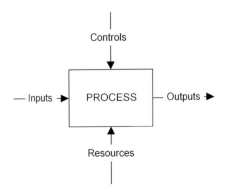

FIGURE 8-7   BS 7850 process model.

simply outputs. The central box is also different. The process label has now changed to activities which is more accurate. However, if we apply the ISO 9000:2005 definition to this model, it implies that as the resources are inputs they are all transformed into outputs or consumed by the process which clearly cannot be the case. People and facilities are resources and are not transformed or consumed by the process (assuming the process is functioning correctly!)

Therefore, there would appear to be a difference between a process that transforms inputs into outputs and one that takes a requirement and produces a result that satisfies this requirement. If we accept that a process is a series of activities that uses resources to

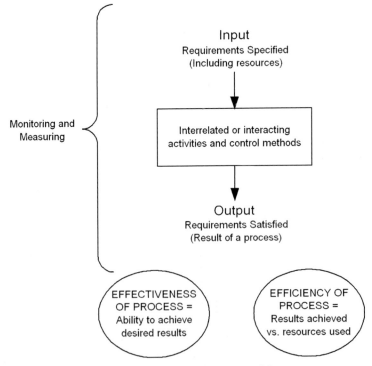

FIGURE 8-8   ISO 9000 process model.

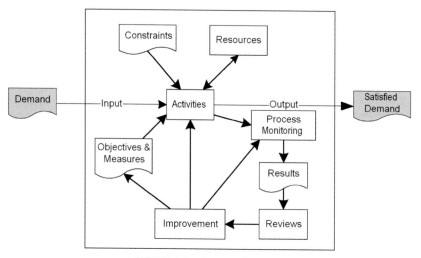

FIGURE 8-9   A managed process.

deliver a result and an effective process as being the one that achieves an objective, a more useful model might be that of Fig. 8-9. This model shows that the process is resourced to receive a demand and when a demand is placed upon the process, a number of predetermined activities are carried out using the available resources and constrained in a manner that will produce an output that satisfies the demand as well as the other stakeholders. These activities have been deemed as those necessary to achieve a defined objective and the results are reviewed and action taken were appropriate to:

- Improve the results by better control,
- Improve the way the activities are carried out,
- Improve alignment of the objectives and measures with current and future demands.

If we go inside the box labelled 'Activities' we would find planning, doing and checking activities and feedback loops designed to control the outputs. These can be represented as a generic control model (see Fig,. 8-10).

If we now replace the Activities box in Fig. 8-9 with Fig. 8-10, rearrange and expand the other elements, we can provide a version of Fig. 8-9 that we can use to map the clauses of ISO 9001:2008 and thus provide a context to the requirements. This model is shown in Fig. 8-11 and will be replicated in each chapter preview in Parts 3–7 indicating where the requirements apply.

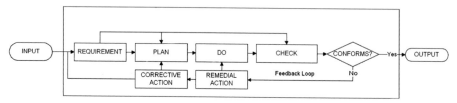

FIGURE 8-10   Generic control model.

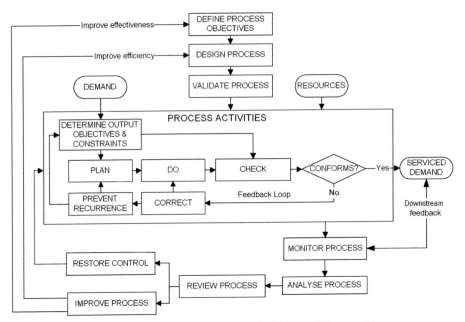

FIGURE 8-11   A managed process that is ISO 9001:2008 compatible.

Business Process Re-engineering

Re-engineering is the fundamental rethinking and radical redesign of business processes to achieve dramatic improvements in critical contemporary measures of performance such as costs, quality, service and speed.[11] Business process re-engineering is about turning the organization on its head. Abandoning the old traditional way of organizing work as a set of tasks to organizing it as a process. According to Hammer, re-engineering means scrapping the organization charts and starting again. But this does not need to happen. Process Management is principally about managing processes that involve people. A functional organization structure might well reflect the best way to develop the talents, skills and competence of the people but not the best way of managing stakeholder needs and expectations.

Processes in the Excellence Model

The introduction of national quality awards such as the Malcolm Baldrige Award in the US (MBNQA), European Quality Award, UK Business Excellence Award and many others across the world has brought the notion of Process Management into Quality Management.

All of the 'excellence' models are based upon a number of common, underlying principles, namely Leadership including organizational culture; Planning including strategy, policies, stakeholder expectation, resources; Process and Knowledge Management including innovation and problem solving, and finally Performance Results

---

[11] Hammer, Michael and Champy, James, (1993). Re-engineering the corporation, Harper Business.

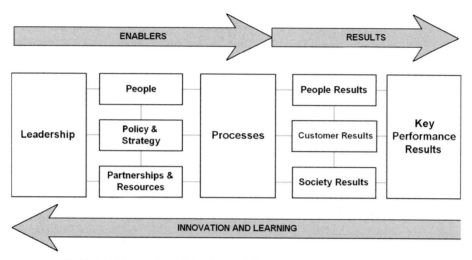

FIGURE 8-12 The EFQM Excellence Model *(Copyright © 1999-2003EFQM).*

covering all stakeholder expectations. Pivotal to organizational success is effective and efficient process management. The EFQM Excellence Model® in Fig. 8-12 clearly illustrates these principles and the importance of processes as an enabler of results.

On first encounter, the EFQM Excellence Model in Fig. 8-12 appears to suggest that processes are separate from Leadership, People, Policy and Strategy, Partnerships and Resources because Processes are placed in a box with these factors shown as 'inputs'. This also suggests that the processes are more concerned with the 'engine room' than with the 'boardroom'. In reality, there are processes in the boardroom as well as in the engine room and the showroom. Clearly there must be strategic planning processes, policy-making processes, resource management processes, processes for building and maintaining partnerships and above all processes for leading the organization towards its goals. The model therefore must be viewed as representing an organization as a system with each of the factors continually interacting with each other to achieve the goals. Whatever an organization desires to do, it does it through a system of processes. However, we must not forget that fundamentally the EFQM Excellence Model® is an assessment tool. It was not intended to be a design tool. Also it was not designed as a diagnostic tool although it can be a helpful input to a diagnosis. It is used in assessing an organization's commitment to the excellence principles and to allow comparison of such commitment and performance between organisations.

PROCESS MANAGEMENT PRINCIPLES

A set of seven principles has begun to emerge on which effective process management is based.

Consistency of Purpose

Processes will deliver the required outputs when there is consistency between the process purpose and the external stakeholders. When this principle is applied the process objectives, measures, targets, activities, resources and reviews would have been derived from the needs and expectations of the stakeholders.

Clarity of Purpose

Clear measurable objectives with defined targets establish a clear focus for all actions and decisions and enable the degree of achievement to be measured relative to stakeholder satisfaction. When this principle is applied people know what they are trying to do and how their performance will be measured.

Connectivity with Objectives

The actions and decisions that are undertaken in any process will be those necessary to achieve the objectives and hence there will be demonstrable connectivity between the two. When this principle is applied, the actions and decisions that people take will be those necessary to deliver the outputs needed to achieve the process objectives and no others.

Competence and Capability

The quality of process outputs is directly proportional to the competence of the people, including their behaviour, and is also directly proportional to the capability of the equipment used by these people. When this principle is applied, personnel will be assigned on the basis of their competence to deliver the required outputs and equipment will be selected on the basis of its capability to produce the required results.

Certainty of Results

Desired results are more certain when they are measured frequently using soundly based methods and the results are reviewed against the agreed targets. When this principle is applied, people will know how the process is performing.

Conformity to Best Practice

Process performance reaches an optimum when actions and decisions conform to best practice. When this principle is applied work is performed in the manner intended and there is confidence that it is being performed in the most efficient and effective way.

Clear Line of Sight

The process outputs are more likely to satisfy stakeholder expectations when periodic reviews verify whether there is a clear line of site between objectives, measures and targets and the needs and expectations of stakeholders. When this principle is applied, the process objectives, measures and targets will periodically change causing realignment of activities and resources thus ensuring continual improvement.

Using the Principles

Whether you design, manage, operate, or evaluate a process you can apply these principles to verify whether the process is being managed effectively and is robust. You simply take one of the principles and look for evidence that it is being properly applied.

You will note that each principle has two parts. There is the principle and a statement of its application. So if we wanted to know whether there was consistency of purpose in a particular process, we review the principle, note what it says about its application and then examine in this case the process objectives, measures, targets, activities, resources and reviews to find evidence that they have been derived from the needs and expectations of the stakeholders. Clearly we would need to discover what the process designers established as the needs and expectations of the stakeholders. It would not be sensible for *us* to define stakeholder needs and expectations as this would more than likely yield different results. We are more interested in what data the process designers used. It is therefore more effective if questioning is used as the investigatory technique rather than a desk study.

PROCESS CHARACTERISTICS

Process Name

Processes can be named using the verb-focused convention or the noun-focused convention. An example of the verb-focused convention is a process with the name 'Create demand'. This communicates very clearly what the process objective is: to create a demand. Conversely with the noun-focused convention the action or purpose could be obscured. It may also be confused with department names such as naming a process 'Marketing'. However, with the Create Demand process, even if we reverse the words to Demand Creation process it remains just as clear. Examples of the alternate naming conventions are shown in Fig. 8-13.

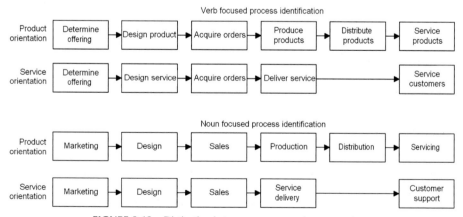

FIGURE 8-13 Distinction between process naming conventions.

Purpose

From the definitions of a process it is clear that every process needs a purpose for it to add value. The purpose provides a reason for its existence. It is the first thing to agree upon and is determined by the desired results. Process analysis does not begin by analysing activities or operations. It begins by defining the results to be achieved. Those who commence process analysis by determining the results to be achieved will soon find themselves asking, "Why do we do this, and why do we do that?" There may be no answer other than "We have always done it that way".

The particular question to ask to reveal the purpose of a process will vary depending on the answers you get. It might be "What are you trying to do, what results are you trying to achieve, what is the end product or what is the output of this process?"

The purpose statement should be expressed in terms of what the process does and in doing so identify what if anything is to be converted or transformed. The purpose of a sales process may be to convert prospects into orders for the organization's products. The results we are looking for is orders for the organization's products and therefore we need a process to deliver that result. Instead of calling the process a sales process you could call it the *prospect to order process*. Similarly the purpose of a design process may be to convert customer needs into product features that satisfy these needs.

Outputs

The outputs of a process are considered to be the direct effects produced by a process. They are tangible or intangible results such as a conforming product, a nonconforming product, noise and odour in the workplace or ambiance in a restaurant. The principal process outputs will be the same as the process objectives. Note that the output is not simply product but conforming product thus indicating its quality.

However, for a process output to be an objective it has to be predefined – in other

> **Results**
>
> **Results** are the outcomes and outputs of a process and both have impacts on stakeholders.
> **Outputs** are the direct effects produced by a process. These may have a direct effect upon a stakeholder.
> **Outcomes** are the indirect effects of a process upon a stakeholder.

words it has to be something you are aiming for, not necessarily something you are currently achieving. An example may clarify this. A current process output might be 50 units/week but this does not mean that 50 units/week is the objective. The objective might be to produce only 20 conforming units/week so of the 50 produced, how many are conforming? If all are conforming, the process is producing surplus output. If less than 20 are conforming, the process is out of control. Therefore, doing what you are currently doing may not be achieving what you are trying to do.

The outputs from business processes should be the same as the business outputs and these should arise out of an analysis of stakeholder needs and expectations. If we ask *"What will the stakeholders be looking for as evidence that their needs and expectations are being satisfied?"* the answers constitute the outputs that the business needs to produce. From this we ask, *"Which process will deliver these outputs?"* and we have now defined the required process outputs for each business process but we may need to pass through an intermediate stage as explained in Chapter 10.

Outcomes

In addition to outputs, processes have outcomes. There is an indirect effect that the process has on its surroundings. An outcome of a process may be a detrimental affect on the environment. Satisfaction of either customers or employees is an outcome not an output. However, processes are designed to deliver outputs because the outputs are measured before they emerge from the process, whereas, outcomes arise long after the process has delivered its outputs and therefore cannot be used to control process performance. Any attempt to do so would induce an erratic performance. (See process measures.) Outcomes are controlled by process design – you design the process to deliver the outputs that will produce the desired outcomes.

Objectives

As the objective of any process is to deliver the required results, it follows that we can discover the process objectives from an analysis of its required outputs. All that is required is to construct a sentence out of the output statement. For example, if the output is *growth in the number of enquires* the process objective is *to grow the number of enquiries*. Clearly the output is not simply 'enquiries' as is so often depicted on process flow charts. The process measurements should determine growth not simply whether or not there were enquiries.

In some cases the wording might need to be different whilst retaining the same intent. For example, a measure of employee satisfaction might be staff turnover and management style may be considered a critical success factor. The output the employee is looking for as evidence that management has adopted an appropriate style is a motivated workforce. Motivation is a result but there is no process that produces motivation. It is an outcome not an output. Instead of expressing the objective of the process as *to motivate the workforce*, it becomes 'To maintain conditions that sustain worker motivation' and the process thus becomes one of developing, maintaining and refreshing these conditions.

Measures

Measures are the characteristics used to judge performance. They are the characteristics that need to be controlled in order that an objective will be achieved. Juran refers to these as the control subjects.

There are two types of measures – stakeholder measures and process measures. Stakeholder measures respond to the question *"What measures will the stakeholders use to reveal whether their needs and expectations have been met?"* Some call these key performance indicators[1]. Process measures respond to the question: *"What measures will reveal whether the process objectives have been met?"* Profit is a stakeholder measure of performance (specifically the investors or stockholders) but would be of no use as a process measure because it is a lagging measure. Lagging measures indicate an aspect of performance long after the conditions that created it have changed. To control a process we need leading measures. Leading measures indicate an aspect of performance while the conditions that created it still prevail (e.g., response time, conformity).

There are also output driven measures and input driven measures. Measures defined in verbs are more likely to be input driven. Those defined by nouns are more likely to be output driven. For example, in an office cleaning process we can either measure performance by whether the office has been cleaned when required or by whether the office is clean. The supervisor asks, "Have you cleaned the office?" The answer might be yes because you dragged a brush around the floor an hour ago. This is an input driven measure because it is focused on a task. But if the supervisor asks, "Is the office clean? You need some criteria to judge cleanliness – this is an output driven measure because it is focused on the purpose of the process. Governments often use input measures to claim that their policies are successful. For example, the success of a policy of investment in the health service is measured by how much money has been pumped in and not by how much service quality has improved.

The word 'measures' does have different meanings. It can also refer to activities being undertaken to implement a policy or objective. For example, a Government minister says "You will begin to see a distinct reduction in traffic congestion as a result of the measures we are taking".

Process measures are not the same as stakeholder measures. Process measures need to be derived from stakeholder measures. A typical example of where they are not was the case in the UK National Health Service in 2005. Performance of hospitals was measured by waiting time for operations but the patient cares more about total unwell time. Even if the hospital operation waiting time was zero, it still might take two years getting through the system from when the symptoms first appear to when the problem is finally resolved. There are so many other waiting periods in the process that to only measure one of them (no matter how important) is totally misleading. Other delays started to be addressed once the waiting time for operations fell below the upper limit set by Government but in the interim period time was lost by not addressing other bottle-necks. Response is a performance measure in the UK Emergency Medical Service in 2005. The target was limited to a measure of response time. There were no targets for whether a life was saved by the crew's actions. There were also no targets for the number of instances where an ill-equipped ambulance got to the location on time and as a consequence failed to save a life (see also Chapter 9 in *Measurements*).

Targets

Measurements will produce data but not information. And not all information is knowledge. Managers need to know whether the result is good or bad. So when someone asks "What is the response time? and you tell them its "10 minutes" they ask, "Is this good or bad?" You need a target value to convey a meaningful answer. The target obviously needs to be related to what is being measured which is why the targets are set only after determining the measurement method. Setting targets without any idea of the capability of the process is futile. Setting targets without any idea what process will deliver them is incompetence – but it is not uncommon for targets to be set

Unrealistic Targets

Managers can only expect average results from average people and perhaps there are not enough extraordinary people to go round to produce the extraordinary results they demand! This principle was expressed in another way by Sir Peter Spencer, Chief of Defence Procurement, UK MoD in the Evening Post, 30 July 2004 when he said:

"A culture of unrealistic expectations leads to the setting of unachievable targets. We've got to change and stop being over-optimistic. We must set goals that we can be confident of achieving and then do just that".

without any thought being given to the process that will achieve them. Staff might be reprimanded for results over which they have no control; staff might suffer frustration and stress trying to achieve an unachievable target.

A realistic method for setting targets is to monitor what the process currently achieves, observe the variation, then set a target that lies outside the upper and lower limits of variation, then you know the process will meet the target. There is clearly no point in setting a target well above current performance unless you are prepared to redesign the whole process. However, performance measurement should be iterative.

Inputs

There are different types of process inputs. Within the process depicted in Fig. 8-10 the demand is an input into the planning stage of the process whereas material that needs to be transformed would be an input into the doing stage of the process. People and equipment might input into any stage of a process and none of them would be transformed. Other than demand specific controls, constraints and resources all others are built into the process design.

Activators

Processes need to be activated in order to produce results. The activator or trigger can be event based, time based or input based. With an event activated process operations commence when something occurs, e.g., a disaster recovery process. With a time activated process operations commence when a date is reached, e.g., an annual review process. With an input activated process operations commence on receipt of a prescribed input, e.g., printed books are received into the binding process.

The concept of process activators enables us to see more clearly how processes operate and better understand the realities of process management.

Hierarchy

There is a hierarchy of processes from the business processes to individual operations such as 'measure dimension' as illustrated in Fig. 8-14. Depending on the level within the process hierarchy, an activity might be as grand as 'Design product' or as small as 'Verify drawing'.

There are several Activity Levels. If we examine this hierarchy in a Demand Creation Process the result might be as follows:

- A Level 1 Activity might be 'Develop new product'. If we view this activity as a process we can conceive a series of activities that together produce a new product design. These we will call Level 2 Activities.
- A Level 2 Activity might be 'Plan new product development'. If we view this activity as a process we can conceive a further series of activities that together produce a new product development plan. These we will call Level 3 Activities.
- A Level 3 Activity might be 'Verify new product development plan'. If we view this activity as a process we can conceive a further series of activities that together

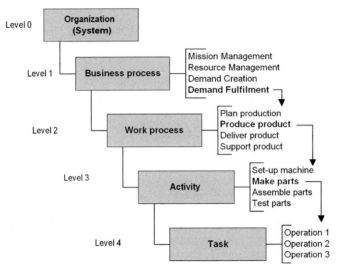

FIGURE 8-14   Process hierarchy.

produce a record of new product development plan verification. These we will call Level 4 Activities.

- A Level 4 Activity might be 'Select verification record blank'. Now if we were to go any further in the hierarchy we would be in danger of noting arm movements. Therefore, in this example we have reached the limit of activities at Level 4.

If we now examine these series of activities and look for those having an output that serves a stakeholder's needs, we will find that there are only two. The Demand creation process has 'demand' as its output. This serves the customer and the New product development process has 'Product design' as its output and this also serves the customer. The series of New product development planning activities has an output which is only used by its parent process so remains a series of activities. The activity of 'Verify product design plan' and 'Select verification record blank' only has any meaning within the context of a specific process so cannot be classed as processes.

In some processes we might decompose to seven or eight levels before we reach this limit. In such cases the hierarchy would be inadequate and so we could use the convention of:

- Business process (Level 1), e.g., Demand Creation
   - Level 2 process, e.g., Develop New Product/Service
      - Level 3 process, e.g., Plan Project
         - Level 4 process, e.g., Prepare Project Plan
            - Activity, e.g., Develop Gantt Chart
               - Task, e.g., Determine project milestones

Another convention might be to simply categorize the levels as process, sub-process, sub-sub-process, sub-sub-sub-process etc. but it might create some confusion in conversations.

Activities

Process activities are the actions and decisions that collectively deliver the process outputs. They include all the activities in the PDCA cycle. The Plan, Do, Check, Act cycle is a good model with which to check the activities but don't expect every process to have a PDCA sequence; it won't. The activities are not determined by PDCA. Activities are determined by answering the questions "What affects our ability to deliver these process outputs?" The sequence of these activities will be determined by simply asking "What do we/should we do next?"

At a high level the sequence might be as follows:

- On receipt of a demand there will be planning activities to establish how the deliverables will be produced and delivered;
- There will be doing activities that implement the plans;
- There will be checking activities to verify that the plans have been implemented as intended and to verify that outputs meet requirements;
- There will be activities resulting from the checking in order to correct mistakes or modify the plans.

In principle it should be possible to place all activities needed to achieve an objective into one of these categories. In reality there may be some processes where the best way of doing something does not follow exactly in this sequence.

Process Flow

Sequential Flow

A process is often depicted as a flow chart representing a sequence of activities with an input at one end and an output at the other. When the process activator is an input this might well be the case but it is by no means always the case. If we examine the Demand Creation process for a typical 'Business to Business' organization illustrated in Fig. 8-15, we find that while the activity of converting enquiries follows that of promoting product, by presenting these activities as a flow it implies not only that one follows the other but also the later does not commence until the former has been

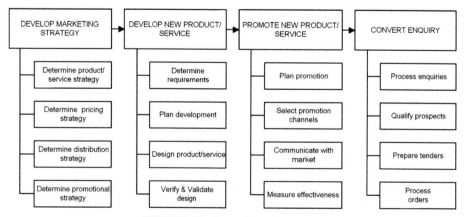

FIGURE 8-15 Demand creation process.

completed. This is clearly not the case. Product promotion continues well after the first enquires are received and enquires may well come in before the first promotion activity has started. In fact all parts of this process may well be active at the same time but if we take one specific product and one specific customer, clearly when the customer makes enquiries, the promotion activities have been effective in attracting this customer to the organization.

Where the output depends upon work being executed in a defined sequence then it can be represented as a flow chart but when activities are activated by events or by time as opposed to inputs, there may be no flow between them as is the case with the work processes in the first stage of the Demand creation process.

Parallel Flow

More than one activity can happen at the same time. An activity or decision may be followed by two or more other activities (depending on the outputs from the previous activity) performed in parallel and creating a new sequential flow that returns to the main flow at some point. A typical example is where product is sorted and there are three routes, one for good product, one for product that can be reworked and one for product that is scrapped and recycled. Two of these flows are all part of the same process rather than being separate processes because they have the same objective.

Feedback Loops

Wherever there is a sensor or measuring station, detecting conformity, variation or change there should be a comparator where the data from the sensor is compared with the target value, standard or acceptance criteria, analysed and converted into a form that facilitates interpretation by decision makers. If the result transmitted to the decision maker is acceptable the output passes through the process to its destination. If the result is not acceptable a decision is made as to what to do about it. This is the feedback loop. There might be two decision makers: one for dealing with results on target (usually the process operator) and other for dealing with results that are not on target. This may also be the process operator depending on the nature of the output and the magnitude and significance of the result. Managers often set alert limits so that they are informed of exceptions thus leaving day-to-day running to the process operators.

Preventive Measures

If failure was not possible we would need no precautions but the potential for failure of one sort or another is always present due to variation and uncertainty. In order that a process delivers conforming output every time, it is necessary to build protection into the process. A way of doing that that has been used successfully in many industries is the failure mode and effects analysis (FMEA). More detail is provided under *Determining potential nonconformities* in Chapter 37 but the basic approach is to answer the questions:

"How could this process fail to achieve its objectives?" This results in the identification of failure modes. Next ask for each failure mode:

"What effect would this failure have on the performance of this process?" This results in identifying the effects of failure. Then ask:

"What effect would this failure have on the performance of this process?

"What could cause this failure?" This identifies the causes of failure of which they may be several so you need to get to the root cause by using the *five Whys technique*[①]. Now ask:

"How likely is it that this will occur regardless of any controls being in place?" This results in a list of priorities.

Now examine the process and establish what provisions are currently in place to prevent, control or reduce the risk of failure. Some failure modes might be removed by process redesign. In other cases, review or inspection measures might be needed and finally, strengthening of the routines, process instructions, training or other provisions may reduce the probability of failure into the realms of the unlikely.

This methodology can be used to test processes for compliance with requirements of national or international standards. Even if no provision has been made to remove the risk, it could be that the risk is so unlikely that no action is needed.

When performing risk assessment, the failure modes should be realistic. They should be based on experience of what has happened – possibly not in your organization but somewhere else. A potential failure is not the one that might never happen otherwise you will never get out of bed in a morning, but it is one that either has happened previously or the laws of science suggest it will happen when certain conditions are met.

The changes that you make as a result of the risk assessment should reduce the probability of process failure within manageable limits.

Resources

The resources in a process are the supplies that can be drawn on when needed by the process. Resources are classified into human, physical and financial resources. The physical resources include materials, equipment, buildings, land, plant and machinery. You can also include information and knowledge as a resource. Time is also a resource that can be planned, acquired and used but it is being used continually whether the process runs or not unlike other resources. Time is also a measure of performance. Human resources include managers and staff including employees, contractors, volunteers and partners. The financial resources include money, credit and sponsorship. Resources are used or consumed by a process. There is a view that resources to a process are used (not consumed) and are those things that don't change during the process. People and machinery are resources that are used (not consumed) because they are the same at the start of the process as they are at the end, i.e., they don't lose anything to the process. Whereas materials, components and money are either lost to the process, converted or transformed and could therefore be classed as process inputs. People would be inputs not resources if the process transforms them.

For a process to be deemed operational it must be resourced. A process that has not been resourced remains in development or moribund. There is a view that resources are acquired by the process when required and indeed, input specific resources are, but resources that are independent of the inputs such as energy, tooling, machinery, people etc. and the channel along which they flow will have been established during process development. Resources are often shared and depleted and have to be replenished but the idea that a process can exist on paper is not credible. A process exists when it is ready to be activated. Those processes that are activated infrequently need to be resourced otherwise they will not be capable of delivering the desired outputs on demand. In such cases resources might be stored for use when required. For instance, you would not set out to acquire back-up software after there had been a computer failure unless you were managing by the seat of your pants!

The resources for a process will be those needed to achieve the process objectives and therefore must include not only the physical resources and number of people but also the capability of the physical resources and the competence of the people including their behaviours.

Responsibilities

The number and competence of the people engaged in a process are human resource matters. Where these resources are located and who is responsible for developing and providing them are organizational matters.

The questions of who undertakes which activity and who makes which decision are important ones in making things happen. When a process is activated something or someone must undertake the first step. Unless responsibilities for the actions have been assigned and authority for decisions delegated nothing will happen. Even if the first step is undertaken by a machine, someone will have responsibility for ensuring that it is able to operate when called upon to do so. In some respects it matters not who does what provided they are competent and indeed, in an emergency managers may undertake tasks normally performed by others if they are competent to do so. However, work and labour are divided in organizations in order to make work productive and the worker achieving. Activities are often grouped by speciality or discipline rather than by process. For example, all the quality engineers are situated in the same department irrespective of some working in the product development process, others in the purchasing process and others in the manufacturing process.

Constraints

The constraints on a process are the things that limit its freedom. Legislation, policies, procedures, codes of practices etc. all constrain how the activities are carried out. Actions should be performed within the boundaries of the law and regulations impose conditions on such aspects as hygiene, emissions and the internal and external environment. They may constrain resources (including time), effects, methods, decisions and many other factors depending on the type of process, the risks and its significance with respect to the business and society. Constraints may also arise out of a PEST[①] and SWOT[①] analysis carried out to determine the Critical Success Factors. Values, principles and guidelines are also constraints that limit freedom for the benefit of the organization. After all it wouldn't do for everyone to have his or her own way! Some people call these things controls rather than constraints but include among them, the customer requirements that trigger the process and these could just as well be inputs. Customer requirements for the most part are objectives not constraints but they may include constraints over how those objectives are to be achieved. For instance they may impose sustainability requirements that constrain the options open to the designer.

To determine the objectives and constraints pass the mission through the stakeholders and ask *"What are this stakeholder's needs and expectations relative to our mission?"* The result will be a series of needs and expectations that can be classified as objectives or constraints. The objectives arise from the outputs the customer requires and the constraints arise from the conditions the other stakeholders impose relative to these outputs as illustrated in Fig. 8-16. The important thing to remember about constraints is that they only apply when relevant to the business. For exampl, if you don't use substances hazardous to health in your organization, the regulations regarding their acquisition, storage, use and disposal are not applicable to your organization.

> **Results**
>
> **Results** imply any results, good or bad.
> **Specified results** imply results that are communicated.
> **Required results** imply results that are demanded by stakeholders.
> **Desired results** imply results that are wanted by stakeholders.

Views differ and whilst a purist might argue that requirements are controls not inputs, and materials are inputs not resources, it matters not in the management of quality. All it might affect is the manner in which the process is described diagrammatically. The requirements would enter the process from above and not from the side if you drew the chart as a horizontal flow.

Results

Results comprise outcomes and outputs and impact on stakeholders either directly or indirectly.

The results of a process arise out of measuring performance using the planned methods for the defined measures, at the planned frequency and against the planned targets.

If the planned measurement methods have been implemented there should be sufficient objective evidence with which to compare current performance against the agreed targets. Therefore, one would expect the results to be presented as graphs, charts,

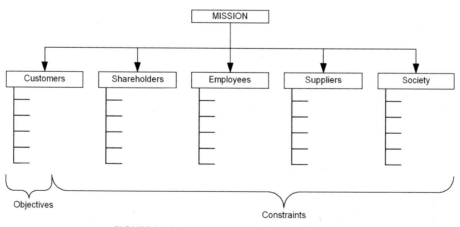
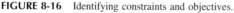

FIGURE 8-16   Identifying constraints and objectives.

and figures. These might show performance to be improving, declining or remaining unchanged relative to a particular process parameter. The scale is important as are trends over time so that decision makers can see the whole picture and not be led into a knee-jerk reaction.

Reviews

There are three dimensions of process performance that can be expressed by three questions:

- How are we doing against the plan?
- Are we doing it in the best way?
- How do we know it's the right thing to do?

The first question establishes whether the objectives are being achieved in the way we planned to achieve them. This means that not only are planned outputs being produced but when you examine the process throughput over the last week, month, year or even longer, the level and quality of the output are consistent and these outputs are being produced in the way you said you would, i.e., you are adhering to the specified policies and procedures etc.

The second question establishes whether the ways in which the planned results are being achieved are the best ways. For example, optimizing resources (time, finance, people, space, materials etc.) such that they are utilized more efficiently and effectively. This would mean that you are satisfied in achieving your objective by using no more than the allocated resources but can reduce the operating cost by optimizing the resources or using more appropriate resources such as new technologies, new materials, new working practices. These improvements arise out of doing things better not by removing waste. If planned outputs were being achieved there would be no unavoidable waste.

The third question establishes whether the planned outputs are still appropriate and relevant to meeting stakeholder needs and expectations. This would mean that irrespective of the planned results being achieved and irrespective of utilizing best practices we could be wasting our time if the goal posts have moved. Maybe the needs and expectations of stakeholders have changed. Maybe they no longer measure our performance in the same way. Some objectives remain unchanged for years, others change rapidly. As all our outputs are derived from stakeholder needs and expectations it is vital to establish that the outputs remain continually relevant and appropriate.

The answers to each of these questions require a different approach because the purpose, method and timing of these reviews are different. This results in there being three specific and independent process reviews.

Process Capability

A process is as capable of producing rubbish as it is of producing the required outputs. It is all a matter of how well it is designed and managed. The inherent capability of manufacturing processes can often be determined because they are built to produce one product. It is therefore possible to study the conformity of the products, remove special cause variation[①], bring common cause variation[①] well within the target for every critical to quality characteristic[①] and engineer the process so that it produces conforming product during multiple cycles of operation (see also Taking action on process variation in Chapter 31). With business processes the situation is somewhat different. Every output although being of the same type, will be different. These processes need to be designed to handle varying and often irregular demand. Some of the work processes may only be activated annually but nonetheless they need to be capable of delivering the right results every time. The equivalent of critical to quality characteristics in a business process are generally the success measures for the process.

PROCESS EFFECTIVENESS

The ISO 9000 process model defines Process Effectiveness as an ability to achieve desired results. We showed previously that results, outcomes and outputs impact on stakeholders, therefore process effectiveness is not simply delivering required outputs but that the outputs delivered the desired outcomes. In the definition of a process approach in ISO 9001, it refers to the process producing the desired outcome. Therefore, if the process delivers the required outputs it is only effective if these outputs produce the designed outcomes.

If in producing the process output the laws of the land are breached, the process is clearly not effective. If in producing the output, the producers are exploited, are forced to work under appalling conditions or become de-motivated and only deliver the goods when stimulated by fear, the process is again not effective. So we could fix all these factors and deliver the required output and have satisfied employees.

Employees are but one of the stakeholders and customers are the most important but although an output may be of good quality to its producers, it may not be a product that satisfies customers. The costs of operating the process may not yield a profit for the organization and its investors, and even if in compliance with current environmental laws, it may waste natural resources, dissatisfy the community and place unreasonable constraints on suppliers such that they decline to supply the process's material inputs. There is therefore only one measure of process effectiveness – that the process outcomes satisfy all stakeholders.

Chapter 9

A Behavioural Approach

CHAPTER PREVIEW

In the North of England there is a saying that "There's nowt so funny as folk" ('Nowt' is dialect for 'nothing'). This is because we are all different and unless you are a student of human behaviour, the things that people do, don't do and say or don't say will never cease to amaze you. You would think that people whose job it is to get results through other people would understand human behaviour, but invariably they don't. Not only do they expect extraordinary results from ordinary people, they continually impose policies and make decisions without any thought as to their effect on the people they expect to implement them and those who will be affected by them. Deming advocated[1] that "a manager of people needs to understand that all people are different and the performance of anyone is largely governed by the system in which they work." A technical approach to management places all the emphasis on the goal and getting the job done, regardless of the human cost. A behavioural approach to management places the emphasis on the interaction between the people so that they are motivated to achieve the gaol of their own volition.

In this chapter we examine aspects of a behavioural approach and look at:

- The relationship between behaviour and quality;
- How the behavioural approach differs from other approaches;
- Customer and supplier relationships and the issues that arise managing these relationships;
- Employer and employee relationships and the issues that arise in managing these relationships.

THE RELATIONSHIP BETWEEN BEHAVIOUR AND QUALITY

We could have two processes, each with the same sequence of steps and therefore the same flow chart. Process A performs well while Process B constantly under performs, reject rates are high, morale is low etc. The procedures, equipment and controls are identical. What causes such a difference in performance? In Process B the people have not been trained, the supervisors are in conflict and ruled by fear and there is high absenteeism. There is poor leadership and because of the time spent on correcting mistakes,

---

[1] Deming W. Edwards (1994) The New Economics for industry, government and education. Second Edition, MIT Press.

there is no time for maintaining equipment, cleanliness and documentation. In Process A, the supervisors spend time building relationships before launching the process. They share their vision for the process, how it should perform and what it should deliver. They plan ahead, train and empower their staff. Everyone engaged in the process is constantly looking for opportunities for improvement and because the process runs smoothly, they have time for maintaining equipment, documentation and cleanliness – morale is high so there is no absenteeism. Therefore, to describe a process in a way that will show how the process objectives are achieved, it is necessary to describe the features and characteristics of the process that cause success and this warrants more than can be depicted on a flow chart. It is not intended that behaviours should be documented but the activities that reflect appropriate behaviours such as planning, preparing, checking, communicating, mentoring can be reflected in the process description.

BEHAVIOURAL APPROACH VERSUS OTHER APPROACHES

Although we have labelled this chapter 'A behavioural approach' we are not proposing that it is an alternative to either the systems or process approach. These two approaches are very much technical; they deal with hard issues and not soft issues even though the management of people is central to their effectiveness. In every process there are actions and decisions taken by people as illustrated above. Choosing what to do, when to do it and how to do it can to some degree be prescribed in procedures but procedures are only effective where judgment is no longer required. Judgement is made by a person at the scene of the action taking in all the signals and reaching what that person considers is the right way forward. Whether this person performs an action or takes a decision that is consistent with how the organization wishes to be perceived by its stakeholders rather depends on his/her commitment to the organization. It is neither healthy nor profitable to manipulate or exploit other people because people perform effectively when they understand and are understood not because they are ordered to comply with directives from their paymaster. We pointed out that the interaction between processes is a key characteristic of the system approach but neither the process approach nor the systems approach will be effective unless the interaction between the people in the system is managed effectively. This is what we mean by the behavioural approach.

Many managers are well-equipped to deal with technical issues but not as equipped to deal with human relationships and yet to get anything done requires people. People who get results are often the ones promoted into management but success in a technical discipline does not necessarily mean that the individual is equipped to manage human relationships. An organization needs to be productive and indeed technical expertise is necessary but it is only one side of the equation. You might be able to get results in the short term by telling people what to do and punishing them when they fail but in the long term the frustrations and resentment built up will cause your downfall as a manager. You need to develop a capability in handling people as well as technical capability to be able to achieve sustained success.

We can use the bicycle metaphor to illustrate these two capabilities[2] as shown in Fig. 9-1. Technical and people capability can be thought of as the two wheels of a bicycle.

---

[2] Hunsaker Philip L & Alessandra Anthony J (1980) The Art of Managing People. Simon & Shuster Inc.

The back wheel is the technical capability, providing the motive power and the front wheel is the people capability taking the bicycle where the rider needs to go. This shows that you can have all the technical capability in the world but if the people won't cooperate or don't understand where to go, you won't get very far. This metaphor marries the systems and process approach with the behavioural approach to propel the organization in the desired direction but it would be counterproductive to think of these as three separate approaches when the behavioural approach must pervade both the systems and process approach for either to be effective.

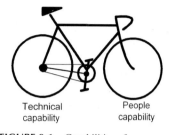

Technical
capability

People
capability

FIGURE 9-1 Capabilities of a manager.

Jon Choppin wrote in Quality Through People[3] that "Any organization can become the best, but only with the full co-operation and participation of each and every individual contributor". A similar point was made by Steven Covey[4] "A cardinal principle of total quality escapes many managers: you cannot continually improve interdependent systems and processes until you progressively perfect interdependent, interpersonal relationships".

If we view the organization as a system in which there is the organization and its customers, suppliers and members of the community with which it interacts, these individuals that Choppin refers to need to include the individuals within the organization, its customer and supplier organizations and the community with which members of the organization communicate. Creating anything but a harmonious relationship with its employees, customers, suppliers and the community can only be detrimental to the organization's success.

CUSTOMER–SUPPLIER RELATIONSHIPS

Engaging

It is important that the people you assign to engage with your customers project a positive image of the organization. If you want your organization to be perceived as being customer focused you would not want to employ the type of sales personnel who pride themselves on cheating customers – the person who wins the sale by deception, lies and bullying. Stephen Covey's fourth Habit advocates "Think win–win" as a principle of seeking mutual benefit. You would therefore want sales personnel to aim for a win–win situation, whereby the customer wins a product or service that will satisfy their needs and expectations, providing good value for money and the company wins an honest sale and a loyal customer (a combination of customer focus and mutually beneficial relationships).

The relationship between customer and supplier has often been adversarial and at arms length bonded by a legal contract that imposes conditions, most of which are only used if one or other party is in breach of contract. Such contracts are devoid of flexibility but there is another way, one of building partnerships of mutual understanding where the

---

[3] Choppin Jon (1997) Total Quality Through People, Rushmere Wynne, England.
[4] Covey Stephen R (1992) Principled Centred Leadership. Simon & Shuster.

customer sees the supplier as an extension of the organization and develops the supplier's capability to meet their requirements.

Mutually beneficial supplier relationships is one of the eight quality management principles but it is amazing that ISO 9001:2008 conveys no requirements that apply this principle. As we found in Chapter 2, organizations depend on their suppliers to enable them to meet their customer requirements therefore the interactions between a supplier and its customer need to be managed if the quality management system is to be effective.

Selection of suppliers has to include an alignment of values to be successful. Deming advocated with the fourth of his 14 points when he said *"End the practice of awarding business on the basis of price tag. Instead, minimize total cost. Move toward a single supplier for any one item, on a long-term relationship of loyalty and trust"*. Ford Motors Q1 award was given to suppliers on the basis of process capability and a demonstrated commitment to Dr Deming's principles and the wide scale use of statistical methods. In Nissan it takes 6–10 months to approve a supplier for similar reasons.

A supplier is more inclined to keep its promises if its relationship with its customers secures future orders. Where there is more empathy, the customer sees the supplier's point of view and vice versa. When a relationship of mutual respect prevails, there is more give and take that binds the two organizations closer together and ultimately there is trust that holds the partnership together. Absent will be adversarial relationships and one-off transactions when either party can walk away from the deal. The partnerships will also encourage better after sales care and more customer focus throughout the organization (everyone knows their customers because there are fewer of them).

Disputes

When a customer is dissatisfied with an organization, they will boycott its products and services and if enough customers are dissatisfied this will force down sales. Customers may take legal action if they have incurred unnecessary costs or hardship as a result of the organization's failure. When a supplier has a dispute with its customers the supplier might cease to supply vital resources and this may force plant closure or work stoppage.

An approach should be adopted whereby conflict is resolved without having to resort to litigation. It should enable people to work collaboratively together to develop and consider alternatives that can lead to a mutually satisfying resolution of their issues. If the conflict is one of ethics, ethical sensitivity is a precursor to moral judgment in that a person must recognize the existence of an ethical problem before such a problem can be resolved. Some organizations undertake a conflict assessment to identify issues that give rise to serious concerns and need to be dealt with as a matter of urgency.

EMPLOYER–EMPLOYEE RELATIONSHIPS

Engaging

Employers cannot employ a part of a person – they take the whole person or none at all. Every person has knowledge and experience beyond the job they have been assigned to perform. Some are leaders in the community; some are architects of social events, building projects and expeditions. No one is limited in knowledge and experience to the current job they do.

Closed-door management leads to distrust among the workforce. It is therefore not uncommon for those affected by decisions to be absent from the discussions with decision makers. Decisions that stand the test of time are more likely to be made when those affected by them have been involved. Managers should be seen to operate with integrity and this means involving the people.

Because we all differ in our individual abilities, some types of people are better at some types of jobs than others, regardless of their technical capability. This requires managers to appreciate the different behavioural styles that people exhibit. Place an amiable person in a job that requires a person to take decisions quickly and confidently and you will be disappointed by their performance.

Unless the recruitment process recognizes the importance of matching people with the culture, mavericks may well enter the organization and either cause havoc in the work environment or be totally ineffective due to a lack of cultural awareness.

Behavioural Styles

Hunsaker and Alessandra adapted David W Merrill's original research in 1964 and postulated four behavioural styles, Amiable, Analytical, Expressive and Driving styles.

- The Amiable person tends to be slow in taking action and decisions, likes close relationships but works well with others and therefore has good counselling skills.
- The Analytical person tends to be cautious in taking actions and decisions, likes structure, prefers working alone but good problem solving skills.
- The Expressive person tends to be spontaneous in taking actions and decisions, like involvement, works well with others and has good persuasive skills.
- The Driving person tends to take firm actions and decisions, likes control, works quickly by him/herself and has good administrative skills.

From The Art of Managing People by Philip L Hunsaker & Anthony J Alessandra.

Leading

Leadership is one of the eight quality management principles but there are very few requirements in ISO 9001:2008 that apply this principle. Leadership without customer-focus will drive organizations towards profit for its own sake. Leadership without involving people will leave behind those who do not share the same vision. If the workforce is unhappy, de-motivated and dissatisfied, it is the fault of the leaders. The vision, culture and motivation in an organization arise from leadership. It is the leaders in an organization who through their actions and decision create the vision and either create or destroy the culture and motivate or de-motivate the workforce thus making the organization's vision and culture key to the achievement of quality.

Vision

A vision is not the mission but the two are related. Mission expresses the purpose of the organization whereas a vision expresses what success will look like, i.e., what the organization will be doing, how it will be performing, what position it will have in the market and how it will be regarded in the community when fulfilling its mission and implementing its strategy. Deming[5] expresses a shared vision in the first of his 14 points for management as *"Create constancy of purpose toward improvement of product and service, with the aim to become competitive and to stay in business, and to provide jobs"*. Deming criticized management for chasing short-term profits, budgets, forecasts and getting entangled in the problems of the day rather than securing the future. A shared vision was the third of Peter Senge's five disciplines[6] for building learning organizations. Establishing a vision for the organization is the eighth of John Bryson's 10-step strategic planning cycle (see page 321).[7]

A shared vision is not a vision statement we all agree with, it is a force that binds people together with a common aspiration. It is the answer to the question "What do we want to create or what do we want success to look like?" Although the vision may be written down, it should reside in people's heads so that it is something that matters deeply to them. When people share a vision, the work they do together becomes focused on a common purpose and is not adversarial or competitive. The individual pursues a vision because they want to not because they have to.

The difficulty in creating a shared vision is that whilst it is an expression of a future state, it will inevitably be created using the knowledge of the present. Those people creating a vision for their organization in 2004 would probably not have imagined the economic crisis we face in 2009. Hence visions need to be regularly reviewed and updated. Both Senge and Bryson provide useful guidance with developing a shared vision

---

Case Study – Rules Based Culture

A hotel has a procedure for maintaining the facilities and this requires annual maintenance of the air conditioning system. Maintenance staff perform exactly as required by the procedure but take no account of the effect of their actions on customers using the conference facilities. The conference department performs exactly in accordance with the conference management procedure but takes no account of facility maintenance. One possible outcome is that the air-conditioning is out of action due to maintenance at the same time a conference is held in the summer just when air conditioning is needed.

The maintenance department will claim they followed the procedures and therefore did nothing wrong – except of course use their common sense; something that is often missing from procedures. Had there been an effective maintenance process in place there would have been process measures that aligned with stakeholder needs and this would cause the maintenance crew to schedule their work when customers of the hotel would not be adversely affected by their actions.

Culture

If we ask people to describe what it is like to work for a particular organization, they often reply in terms of their feelings and emotions that are their perceptions of the essential

---

[5] Deming W. Edwards (1986) Out of the crisis, MIT Press.

[6] Senge Peter M (2006) The Fifth Discipline, The Art and Practice of the Learning Organization. Random House

[7] Bryson John (1995) Strategic planning for public and nonprofit organizations – A guide to strengthening and sustaining organizational achievement, Jossey Bass.

atmosphere or environment in the organization. This environment is encompassed by two concepts, namely culture and climate.[8] Culture evolves and can usually be traced back to the organization's founder. The founder gathers around people of like mind and values and these become the role models for new entrants. Culture has a strong influence on people's behaviour but is not easily changed. It is an invisible force that consists of deeply held beliefs, values and assumptions that are so ingrained in the fabric of the organization that many people might not be conscious that they hold them. People who are oblivious to the rites, symbols, customs, norms, language etc. may not advance and will become de-motivated. There is however, no evidence to suggest a right or wrong culture. What is important is that the culture actually helps an organization to achieve its goals – that it is pervasive and a positive force for good.

Climate is allied to culture and although people experience both, climate tends to be something of which there is more awareness. Culture provides a code of conduct that defines acceptable behaviour whereas climate tends to result in a set of conditions to which people react. Culture is more permanent whereas climate is temporary and is thought of as a phase the organization passes through. In this context therefore, the work environment will be affected by a change in the organizational climate. Several external forces cause changes in the climate such as economic factors, political factors and market factors. These can result in feelings of optimism or pessimism, security or insecurity, complacency or anxiety.

Values

Cultural characteristics include values and beliefs. Values are confirmed by top management expressing what they believe are the fundamental principles that guide the organization in accomplishing its goals. One discovers values rather than determines them. You can't impose a set of values on an organization but by using appropriate measures, shared values will emerge. The values may be discovered by asking 'What do we stand for?' or 'What principles will guide us on our journey?' or 'What do we believe characterises our culture?'

The value is usually expressed by a single word then translated into a belief expressed in a sentence or two. Two examples of this approach are given below.

Mars Confectionery's Five Principles

Quality   The consumer is our boss, quality is our work and value for money is our goal.

Responsibility As individuals, we demand total responsibility from ourselves; as associates, we support the responsibilities of others.

Mutuality   A mutual benefit is a shared benefit; a shared benefit will endure.

Efficiency   We use resources to the full, waste nothing and do only what we can do best.

Freedom   We need freedom to shape our future; we need profit to remain free.

---

[8] Rollinson, Broadfield and Edwards, (1998). *Organizational behaviour and analysis*, Addison Wesley Longmans.

Mars has operated by a set of values since its inception. The values are displayed and demonstrated on their web site (Appendix A). They call their values "our Five Principles" which they strongly believe are the real reason for their success.

ENRON also had a set of values but unlike Mars it failed to live up to them.

ENRON's Values

Integrity	We work with customers and prospects openly, honestly and sincerely. When we say we will do something we do it. When we say we cannot or will not do something then we won't do it.
Respect	We treat others as we would like to be treated ourselves. We do not tolerate abusive or disrespectful treatment. Ruthlessness, callousness and arrogance don't belong here.
Excellence	We are satisfied with nothing less than the very best in everything we do. We will continue to raise the bar for everyone. The great fun here will be for all of us to discover just how good we can really be.
Communication	We have an obligation to communicate. Here, we take the time to talk with one another ... and to listen. We believe that information is meant to move and that information moves people.

There is an interaction between values, beliefs and behaviours as shown in Fig. 9-2. The values translate into beliefs that produce behaviours that reflect the values. So, if integrity is a value and this is translated into the belief that we work with customers and prospects openly, honestly and sincerely, you would expect the companies customer service personnel to be open and honest when explaining a stock evaluation to customers. However, if there is a person or group of people inside or outside the organization who hold a position of influence and they exert pressure on others and persuade them that integrity now means protecting the company from collapse at any cost, or that loyalty is more important than integrity, the value is compromised by beliefs that are no longer a faithful translation.

Values are easily tested by examining actions and decisions and passing them through the set of values and establishing what they mean in practice. For example, if our value is "We treat others as we would like to be treated ourselves; how come we pay part timers below the minimum hourly rate?" "If we value quality, how come we allowed the installation team to commission a system that had not competed acceptance tests?"

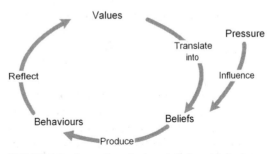

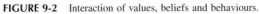

FIGURE 9-2 Interaction of values, beliefs and behaviours.

If there's a misalignment between what you say is important and how people behave it needs to be fixed immediately. Use examples and role models to get the message across rather than a series of rules, as invariably people learn better this way.

Ethics

Ethics concern a person's behaviour towards others and therefore within a particular work environment there will be some accepted norms of right and wrong. These values or standards vary from group to group and culture to culture. For example, bribery is an accepted norm in some countries but in others it is illegal.

Unfortunately it is often not until a situation arises that challenges the ethical standards of the individual or the group that the conflict becomes apparent. People may be content to abide by the unwritten code of ethics under normal conditions but when an important prize is within reach, the temptation to put the principles to one side is too great and some will succumb to the pressure and put self interest or profits first causing harm to the interests of others.

Employees may be easily led by other less ethical employees in a desire to follow the pack and those that do challenge their peers and their managers get accused of 'rocking the boat' and being 'troublemakers'. Management may turn a blind eye to unethical practices if in doing so they deliver the goods and no one appears to be harmed but such action weakens the values. Sometimes managers are simply unaware of the impact of their decisions. A one-off instruction to let a slightly defective product to be despatched because it was needed urgently, gets interpreted by the employees as permission to deviate from requirements. Employees naturally take the lead from the leader and can easily misread the signals. They can also be led by a manager who does not share the same ethical values and when under threat of dismissal; an otherwise law-abiding citizen can be forced into falsifying evidence.

Managing

Priorities

There will be many things that need to happen, not all are of equal importance. The effective manager prioritises decisions and the work required by his/her team but not everyone will necessarily share these priorities. It is easy to meet either quality, delivery or cost objectives but it requires particular commitment and certain competences to meet them all at the same time, but that is what is required. In order to do this the manager needs to know the relative importance of actions and decisions.

Stephen Covey's third habit is 'Put first things first – the principle of managing time and priorities around roles and goals'. This means that having understood the goals and knowing the role you are undertaking, your first priority is to that goal and role. Table 9-1 adapts Stephen Covey's Time Management Matrix, to show in one table the activities and the result of dealing with those activities. You deal first with the important matters that progress your plan towards its goal. There will be matters that are important and urgent to others but you have to be diplomatic and refuse unless of course you have time to spare. If you are drawn to solving problems and reacting to situations this will be because you have not spent enough time dealing with prevention activities.

Table 9-1 Time Management Matrix (Adapted from 7 Habits of Highly Effective People by Stephen Covey)

	Urgent		Not urgent	
	Quadrant I		Quadrant II	
	Activities	Results	Activities	Results
Important (Contribute to the goals)	Crisis, Pressing problems, Deadlines	Stress, Burnout, Crisis management, Always putting out fires	Prevention, Production capability activities, Relationship building, New opportunities, Planning, Recreation	Vision perspective, Balance, Discipline, Control, Few crisis
	Shrink time spent here by spending more time in Quadrant II		Invest more time in this Quadrant, these are the things you must put first	
	Quadrant III		Quadrant IV	
	Activities	Results	Activities	Results
Not Important (Doesn't contribute to the goals)	Interruptions, Some mail, Some calls, Some reports, Some meetings, Pressing matters, Popular activities	Short term focus, Crisis management, Reputation – chameleon character, See goals and plans as worthless, Feel victimized out of control, Shallow or broken relationships	Trivia, Some mail, Some calls, Time wasters, Pleasant activities	Total irresponsibility, Fired from jobs, Dependent on others or institutions for basics
	Try to avoid these even if they are popular		Resist these – say NO	

Management Style

Various styles of management can be employed depending on the culture, the skills and personality of the manager, the nature of the workforce, the task it needs to carry out and the economic climate at the time. It is often said that tough times call for tough measures; therefore, the style may need to change when the circumstances change. Certain styles can therefore be appropriate for certain circumstances. Knowing which style to use is the hallmark of an effective manager. A management style could either make or break an organization.

Robert Tannenbaum and Warren Schmidt[9] identified seven leadership patterns across a range from Boss centred leadership to Subordinate centred leadership which was

---

[9] Robert Tannenbaum and Warren Schmidt (1958) How to choose a leadership pattern. Harvard Business Review March–April.

further developed by Rensis Likert[10] and his associates into four styles of management. These are coupled in Table 9.2 with Douglas McGregor's Theory X and Theory Y[11] and William Ouchi's Theory Z[12] which represents Japanese management style although the match should not be taken as definitive.

Learning

One of the great challenges in our age is to impart understanding in the minds of those who have the ability and opportunity to make decisions that affect our lives. There is no shortage of information – in fact there is too much now we can search a world of information from the comfort of our armchair. We are bombarded with information but it is not knowledge – it does not necessarily lead to understanding. With so many conflicting messages from so many people, it is difficult to determine the right thing to do. There are those whose only need is a set of principles from which they are able to determine the right things to do. There are countless others who need a set of rules derived from principles that they can apply to what they do and indeed others, who need a detailed prescription derived from the rules for a particular task. At each level there is learning. Even the person who does the same boring job every day, working to the detailed prescription has probably learnt how to do it well without it driving him crazy.

Learning Styles

Managers need to accommodate different learning styles.

- Accommodators learn from concrete experience and active experimentation.
- Convergers learn through abstract conceptualization and active experimentation.
- Divergers learn through concrete experience and reflective observation.
- Assimilators learn through abstract conceptualization and reflective observation.

In the translation from principles to prescription, inconsistencies arise. Those translating the principles into rules or requirements are often not the same as those translating the rules into a detailed prescription. Rules are often an imperfect translation of principles and yet they are enforced without regard to or even an understanding of the principles they were intended to implement. This is no more prevalent than in local government where officials enforce rules without regard to what the rules were intended to achieve.

There is nothing intrinsically wrong with wanting a prescription. It saves time, it is repeatable, it is economic and it is the fastest way to get things done but it has to be right. The receivers of prescriptions need enough understanding to know whether what they are being asked to do is appropriate to the circumstances they are facing.

Outside the principles and prescription is the approach a person takes to the job. Only robots continue to make mistakes hour after hour until someone notices or an automated sensor detects them. Organizations need people to not only see but also to observe, not only hear but also listen, use their initiative and act accordingly. They need people to

---

[10] Likert Rensis (1967) The Human Organization. McGraw-Hill.
[11] McGregor Douglas (1960) The Human Side of Enterprise. McGraw-Hill.
[12] Ouchi, William G. (1981). Theory Z: How American business can meet the Japanese challenge, Addison-Wesley.

Table 9-2 Management Styles, Leadership and Motivation

Leadership patterns	Management styles	Motivational theory
Manager makes decision and announces it Manager sells decision	**Exploitive – Authoritative** • Manager has little confidence in subordinate • Manager issues orders/directives • Uses coercion to get things done • Atmosphere of distrust between superiors and subordinates • Generally there is opposition to goals of the formal organization	**Theory X assumptions** • People inherently dislike work and when possible will avoid it • They have little ambition, shun responsibility and prefer direction • They want security • It is necessary to use coercion to get them to achieve goals, control them and threaten punishment • There is always someone to blame
Manager presents ideas and invites questions Managers present tentative decision subject to change	**Benevolent – Authoritative** • Manager acts in a condescending manner • Manager issues orders/directive with opportunity to comment • Some decision making at lower levels within a prescribed framework • Carrot and stick approach • Staff appear cautious and fearful • Does not oppose all goals of the formal organization	**Theory Y assumptions** • If conditions are favourable people will accept responsibility and even seek it • If people are committed to organization objectives they will exercise self-direction and self-control • Commitment is a function of the rewards associated with goal attainment • The capacity of creativity in solving organizational problems is widely distributed in the population and the intellectual potential in average people is only partially utilized
Manager presents problem, gets suggestions, makes decision Manager defines limits, asks group to make decision	**Consultative – Democratic** • Manager has confidence in subordinates • Manager issues orders only after discussion with subordinates • Important decisions made at top with tactical decisions at lower levels • Two-way communication • Some trust between superiors and subordinates • Slight resistance to goals of the formal organization	**Theory Z characteristics** • Long-term employment and job security • Implicit, informal control with explicit, formalized measures • Slow evaluation and promotion • Moderately specialized careers • Concern for a total person, including their family
Manager permits subordinates to function within limits defined by superior	**Participative – Democratic** • Manager has complete confidence and trust in subordinates • Goals normally set by group participation • Decision making is highly decentralized • Communication up and down and sideways • Mutual trust • Formal and informal organizations are one and the same	

learn from mistakes, react to the unexpected, reflect on performance recognizing their weaknesses and undertake to do something about them.

Measuring

People naturally concentrate on what they are measured. It is therefore vital that leaders measure the right things. Deming advocated in his 14 points that we should *"Eliminate numerical goals and quotas for production"* as an obsession with numbers tends to drive managers into setting targets for things that the individual is powerless to control. A manager may count the number of designs that an engineer completes over a period. The number is a fact, but to make a decision about that person's performance on the basis of this fact is foolish, the engineer has no control over the number of designs completed and even if she did, what does it tell us nothing about the quality of the designs. If the measures used encourage people to cheat, the results you get might not be what they appear to be.

There is interaction between measures, behaviours and standards as illustrated in Fig. 9-3. This shows that measures produce behaviours that reflect the standards the group is actually following. If these standards are not the ones that should be followed it is likely that the measures being used are incorrect.

Selecting the wrong measure can have undesirable effects. Somewhere there will be a measure that encourages people to take a short cut or to deceive in order to get the job done or get the reward. With the wrong measures you can change good apples into bad apples. The person is either forced, coerced or encouraged to go down the wrong route by trying to achieve the measures upon which they are judged and often get their pay rises.

In the case of hospital waiting lists in the UK NHS, hospital administrators started to cheat in an attempt to meet the target. Patients were held in a queue waiting to get onto the waiting list thus making it appear that the waiting lists were getting shorter. As Kenniston observed in the 1970s,[13] *"We measure the success of schools not by the kinds of human beings they promote but by whatever increases in reading scores they chalk up. We have allowed quantitative standards, so central to the adult economic system, to become the principal yardstick for our definition of our children's worth."* Seems like the more we change the more we stay the same!

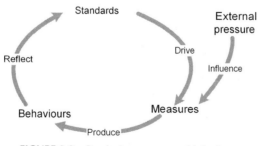

FIGURE 9-3   Standards, measures and behaviours.

---

[13] *Kenniston Kenneth. (1988).Originally quoted in "The 11-Year Olds of Today Are the Computer Terminals of Tomorrow," New York Times (February 19, 1976). Quoted in The Hurried Child, by David Elkind, ch. 3*

This tells us that the quality of the output is not only dependent upon there being relevant standards in place and a process for achieving them, but that the measures have to be in complete alignment with the standards, otherwise the wrong behaviours and hence the wrong results will be produced. All the measures should be derived from the stakeholder success measures or KPIs otherwise they will influence people in the wrong direction.

Motivating

People may not possess all the competences or share the values they need to do an excellent job in a particular culture but in general they do not go to work deliberately to do a bad job. However, with good leadership they will be motivated to acquire the necessary competences, discover and exhibit the necessary values.

Everything achieved in or by an organization ultimately depends on the activities of its workforce. It is therefore imperative that the organization is staffed by people who are motivated to achieve its goals. Everyone is motivated but not all are motivated to achieve their organization's goals. Many may be more interested in achieving their personal goals. Motivation is key to performance. The performance of a task is almost always a function of three factors: Environment, Ability and Motivation. To maximize performance of a task, personnel have not only to have the necessary ability or competence to perform it but also need to be in the right surroundings and have the motivation to perform it.[14] Motivation comes from within. A manager cannot alter employees at will despite what they may believe is possible.

Motivation has been defined as an inner mental state that prompts a direction, intensity and persistence in behaviour.[15] It is therefore a driving force within an individual that prompts him or her to achieve some goal.

The continual improvement principle means that everyone in the organization should be continually questioning its performance and seeking ways to reduce variation, continually questioning their methods and seeking better ways of doing things, continually questioning their targets and seeking new targets that enhance the organization's capability. However, this won't happen if the leaders have created an environment in which people are afraid to challenge the status quo, afraid to report problems, afraid to blow the whistle on unethical practices. Deming advocated in the eighth of his 14 points that leaders need to *"Drive out fear"* and this is done by changing the style of management, maybe also changing the people if they cannot change their behaviour.

In his two-factor theory, Hertzberg identified two quite separate groups of factors for worker motivation. He discovered that the factors that gave job satisfaction were different and unconnected to the factors that gave job dissatisfaction. He found that the key determinants of job satisfaction were achievement, recognition, work itself, responsibility and advancement. Whereas he found that company policy and administration, supervision, salary, interpersonal relationships and working conditions were prime causes of job dissatisfaction. He observed that when a company resolved the dissatisfiers it did not create satisfaction. Hertzberg concluded that when basic needs are not met, we become dissatisfied and meeting these needs does not make us satisfied, it

---

[14] Vroom V H, (1964). Work and Motivation, New York: John Wiley.
[15] Rollinson, Broadfield and Edwards, (1998). *Organizational behaviour and analysis*, Addison Wesley Longmans.

merely prevents dissatisfaction. He called these the *hygiene needs*. The needs that make us satisfied he called *motivator needs*. However, Hertzberg's theory is not universally accepted.

The social factors if disregarded cause unpredictable effects and some of these are the subject of legislation such as discrimination on the basis of religion, gender, race and disability. The issue is not whether product will be affected directly, but whether performance will be affected. It requires no more than common sense (rather than scientific evidence) to deduce that intimidation, sexual harassment, invasion of privacy and similarly unfair treatment by employees and employers, will adversely affect the performance of people and consequently the quality of their output. Social factors can have a psychological effect on employees causing de-motivation and mental stress. This is not to say that employees have to be mollycoddled, but it is necessary to remove the negative forces in the work environment if productivity is to be maximized and business continuity maintained.

Identifying the Barriers

The role of the manager in enabling a person to be motivated is that of removing barriers to worker motivation. There are two types of barriers that cause the motivation process to break down. The first barrier is job-related, i.e., there is something about the job itself that prevents the person performing it from being motivated. For example, boring and monotonous work in mass production assembly lines. The second barrier is goal-related, i.e., attainment of the goals is thwarted in some way. For example, unrealistic goals, insufficient resources and insufficient time for preparation. When targets are set without any regard for the capability of processes this often results in frustration and a decline in motivation (see also Chapter 16 in *Expressing quality objectives*).
Common barriers are:

- fear of failure, of reprisals, of rejection, of losing, of conflict, of humiliation and of exploitation;
- distrust of management, favouritism and discrimination;
- work is not challenging or interesting;
- little recognition, respect and reward for a job well done;
- no authority and responsibility.

Measuring Employee Satisfaction

The very idea that employees should be satisfied at work is a comparatively recent notion but clearly employee dissatisfaction leads to lower productivity. The measurement of employee satisfaction together with the achievement of the organization's objectives would therefore provide an indication of the quality of the work environment, i.e., whether the environment fulfils its purpose.

Many companies carry out employee surveys in an attempt to establish their needs and expectations and whether they are being satisfied. It is a fact that unsatisfied employees may not perform at the optimum level and consequently product quality may deteriorate. As with customer satisfaction surveys, employee satisfaction surveys are prone to bias. If the survey hits the employee's desk following a reprimand from a manager, the result is likely to be negatively biased. The results of employee satisfaction surveys are also often disbelieved by management. Management believe their decisions are always in their employees' best interests whereas the employees may not believe what management says if its track record has not been all that great. Employee satisfaction has less to do with product quality and more to do with relationships. However, employee relationships can begin to adversely affect product quality if no action is taken.

Design the employee satisfaction survey with great care and treat the results with an open mind because they cannot be calibrated. A common method for measuring satisfaction is to ask questions that require respondents to check the appropriate box on a scale from 'strongly agree' to 'strongly disagree'. An alternative is for an outsider appointed by management to conduct a series of interviews. In this way you will obtain a more candid impression of employee satisfaction. The interviewer needs some knowledge of the management style, the efforts management has actually made to motivate their workforce and not the rhetoric they have displayed through newsletters, briefings etc. On hearing what management has actually done, the employees may react differently. They also have short memories and are often reacting to immediate circumstances forgetting the changes that were made some time ago. The interviewer is also able to discover whether the employee has done anything about the feelings of dissatisfaction. It could be that a supervisor or middle manager is blocking communication. Whatever the method, management needs unbiased information of the level of employee satisfaction to do the job.

Empowering

Empowerment is said to motivate employees because it offers a way of obtaining higher level of performance without the use of strict supervision. However, it is more theory and rhetoric than a reality. To *empower* employees, managers not only have to delegate authority but to release resources for employees to use as they see fit and to trust their employees to use the resources wisely. If you are going to empower your employees, remember that you must be willing to cede some of your authority but also as you remain responsible for their performance, you must ensure your employees are able to handle their new authority. Employees have to be trained not only to perform tasks but also need a certain degree of experience in order to make the right judgements and therefore need to be competent. Some employees may acknowledge that they are willing to accept responsibility for certain decisions but beware, they may not be ready to be held accountable for the results when they go sour. It is also important that any changes arising from the empowering of employees to improve the process be undertaken under controlled conditions. However, empowerment does not mean that you should give these individuals the right to change policies or practices that affect others without due process. Empowerment would be an outcome of applying *Motivation Theory Y* or *Z*.

Rewarding

You can't motivate personnel solely by extrinsic rewards such as financial incentives; it requires a good understanding of an individual's pattern of needs. People desire psychological rewards from the work experience or like to feel a part of an organization or team. People can be motivated by having their efforts recognized and appreciated or included in discussions. However, this will only occur if the conditions they experience allow them to feel this way. Managers cannot motivate their staff; all they can do is to provide conditions in which staff are motivated. Targets don't motivate people to achieve them. You can raise the bar as high as an Elephant's eye, but if the conditions aren't conducive to get the best out of people, the staff won't be motivated to jump over the bar.

Motivation comes from satisfying personal needs and expectations of work, therefore the motivation to achieve quality objectives must be

FIGURE 9-4 Maslow's hierarchy of needs.

triggered by the expectation that achievement of objectives will lead to a reward that satisfies a need of some sort. The hierarchical nature of these needs was identified by Abraham Maslow (see Fig. 9-4). The physiological needs are at the base, those needs essential to sustain life like food, water, clothing and shelter. A person who lacks the basic necessities of life will be motivated by money; not money itself but what money can buy. Next is a concern for safety and security such as protective equipment, life insurance, job security and health care. At the third level the individual wants friendship and a social interaction at work rather than isolation. Fourth is the need for esteem satisfied by recognition of contribution, status relative to the individual's peers and influence through power, authority or respect. Deming advocated in the 12th of his 14 points that we should *"Remove barriers that rob hourly workers as well as management of their right to pride of workmanship"* which expressed a need for esteem. Lastly when all other needs are satisfied, the individual needs to realize their full potential and achieve their life goal. The order of needs should not be seen as being rigid. Some individuals may place esteem more basic than safety and security. For the manager, it is important that they know which need requires satisfaction.

Pleasing the Bosses

Peter Senge, author of the Fifth Discipline The Art and Practice of the Learning Organization received comment for the jacket of the first edition from Dr Edwards Deming which he has now incorporated into the Preface of the second edition by adding a middle section. The words in quotes are by Dr Deming.

"The relationship between a boss and a subordinate is the same as the relationship between a teacher and a student". The teacher sets the aims and the student responds to those aims. The teacher has the answer, the student works to get the answer. Students know when they have succeeded because the teacher tells them. By the time all children are 10 they know what it takes to get ahead in school and please the teacher – a lesson they carry forward through their careers of "pleasing the bosses and failing to improve the system that serves the customers".

Communicating

Stephen Covey advocated that we should "Seek first to understand before being understood" what he calls the principle of empathetic communication and so it is with many human interactions that we commence a dialogue by wanting to be understood and forcing our views upon people which leads to resentment. If we want others to do things we must first understand their view point, listen to what they have to say. Covey says that *"if you want to influence me you first need to understand me and you can't do that with technique alone".*

Many disputes arise through a lack of trust, a distrust of the other person's motives and manipulative techniques. The other person has to feel safe in our company, safe enough for them to open up, to tell us what we need to know if our plans are going to be carried thorough successfully. The ancient Greeks had a philosophy for communication.

- Ethos – first you project your personal credibility, your integrity and competency;
- Pathos – second you empathize with the other person, aligning yourself with their emotion;
- Logos – thirdly you present your reasoning, showing the logic of what you want done which has been modified as a result of your empathy with the other person.

What the authoritative manager (*Theory X*) does is project their anger, arrogance or superiority, ignore the other person's feelings and tell the other person what they want done. The democratic manager would try to convince other people on the validity of their logic after first taking ethos and pathos into account.

Key Messages from Part 2

1. There are several approaches that can be used to achieve, sustain and improve quality but not one of these approaches is sufficient on its own. For an organization to sustain success it needs to adopt approaches that incorporate the best of each into its culture.
2. The task or functional approach of Frederick Taylor still prevails in many organizations. Workers are grouped by speciality and do the work managers have assigned to them. They may be well trained to do the job but, its just a job. They may have no vision of what the end product will be, they may not be involved in planning the job so take little interest in it.
3. The risk-based approach gives each worker an objective, that of identifying, eliminating, reducing and controlling risks. This approach tends to create specialists such as quality, safety, environment, reliability specialists etc. On its own a risk-based approach to quality will not produce quality products because it depends on there being work processes to analyse for the risks to be identified.
4. There is only one management system in an organization and thinking of an organization in terms of distinct systems each having distinct objectives leads to sub-optimization.
5. The systems approach views the organization as a system of managed processes and it is the interaction of these processes that is critical to the achievement of quality. We now know that these processes include behaviours, therefore we need to adopt all these approaches to achieve, sustain and improve quality.
6. When outputs are based on speciality or function there is a tendency for there to be sub-optimization, i.e., each function works to grow, to improve and to achieve its objectives regardless of its impact on the other functions.
7. The primary difference between the task or functional approach and the process approach is that the functional approach results in outputs based on speciality whereas the process approach results in outputs derived from the needs of stakeholders.
8. All work is a process and results in outputs and those producing these outputs should be placed in a state of self-control so that they can be held accountable for the results.
9. People don't need written instructions to make them function; a management system is not a computer program.
10. Whether you want to manage, the organization, a business process, a work process or a task; in adopting a process approach:
    a. You need objectives that define what you want to achieve;
    b. You need realistic measures to determine how you are performing relative to these objectives;

    c. You need soundly based targets to set the standard of performance to be achieved;

    d. You need soundly based methods of measurement to measure performance;

    e. You need to ensure the integrity of measurement so you can rely on the results;

    f. You need a series of activities that generate the required outputs and add value;

    g. You need capable physical resources and competent human resources to carry out these activities as planned;

    h. You need to periodically measure performance and act on any unacceptable variation from target;

    i. You need to periodically evaluate the process and continually find better ways of doing things;

    j. You need to periodically reassess whether the objectives, measures and targets remain relevant to the needs of the stakeholders.

11. For processes to be effective they have to embody the behavioural approach. Human behaviour cannot be ignored in process design and where it should be evident is in the competence requirements.

12. A manager of people needs to understand that all people are different and the performance of anyone is largely governed by the system in which they work.

13. Quality is influenced by personal relationships and how well the interaction between customers and suppliers, employers and employees is managed; a purely technical approach to quality will not lead to sustained success.

14. The way people behave affects the decisions they make and these decisions are what drives the processes.

15. If auditors apply the process approach they would firstly look at what results were being achieved and whether they were consistent with the intent of ISO 9001, then they would discover what processes were delivering these results and only after doing this, they would establish whether these processes complied with stated policies, procedures and standards.

16. The assertive manager would ask, 'Why would I want to do that?' and if the auditor or consultant could not give a sound business case for doing it, the manager would rightly take no action.

17. A manager will decide what objectives to pursue, what to measure, what to prioritize, in what sequence the jobs are to be carried out, how the resources will be utilized and these decisions shape the process outcomes.

18. Pressure from others can influence people's beliefs and change outcomes and can influence what is measured and will change standards, beliefs and values for the worse unless resisted.

19. Somewhere there will be a measure that encourages people to cheat, to take a short cut or to deceive in order to get the job done or get the reward – when you find it, get it changed for a measure that is derived from what stakeholders look for as evidence that their needs are being met.

20. Neither the systems approach nor process approach will be effective unless the interaction between the people in the system is managed effectively.

21. Creating anything but a harmonious relationship with its employees, customers, suppliers and the community can only be detrimental to an organization's success.

22. Everything achieved by an organization ultimately depends on activities of its workforce. It is therefore imperative that the organization is staffed by people who are motivated to achieve its goals.

Complying with ISO 9001 Section 4 Requirements on Quality Management System Development

INTRODUCTION TO PART 3

Structure of ISO 9001 Section 4

Section 4 of ISO 9001 contains the basic requirements for establishing a management system rather than any particular component of the system. In some instances they are duplicated in other clauses of the standard but this is no bad thing because it emphasizes the principle actions necessary to develop and manage such a system. Unlike previous versions, the focus has moved away from documentation towards processes and therefore these general requirements capture some of the key activities that are required to develop an effective system.

Linking Requirements

Although the clauses in Section 4 of ISO 9001 are not intended as a sequence there is a relationship that can be represented as a cycle, but first we have to lift some clauses from Section 5 to commence the cycle. The words in bold indicate the topics covered by the clauses within Sections 4 and 5 of the standard. The cycle commences with the **Organization's purpose** (Clause 5.3 requires the quality policy to be appropriate to the organization's purpose or mission) through which are passed **customer requirements** (Clause 5.2 requires customer requirements to be determined) from which are developed **objectives** (Clause 5.4.1 requires objectives to be consistent with the quality policy). In planning to meet these objectives the **processes** and their **sequence and interaction** are determined. Once the relationship between processes is known, the **criteria and methods** for effective operation and control can be developed and **documented**. The processes are described in terms that enable their effective communication and a suitable way of doing this would be to compile the process descriptions into a **quality manual** that not only references the associated **procedures** and **records** but also shows how the processes interact. Before implementation the processes need to be **resourced** and the **information** necessary to operate and control them deployed and brought under **document control**. Once operational the processes need to be **monitored** to ensure that they are functioning as planned. **Measurements** taken to verify that the processes are delivering the required output and actions taken to **achieve the planned results**. The data obtained from monitoring and measurement that is captured on **controlled records** needs to be **analysed** and opportunities for **continual improvement** identified and the agreed actions **implemented**. Here we have the elements of the process development process that would normally be part of mission management but that process is largely addressed in the standard through Management Responsibility.

If every quality management system reflected the above linkages the organization's products and services would consistently satisfy customer requirements.

Chapter 10

Establishing a Quality Management System

CHAPTER PREVIEW

This chapter is aimed primarily at those with responsibility for developing the organization's management system or assisting with its development such as quality managers and consultants. However, it will also be of interest to those senior managers who decided to adopt ISO 9001 as the basis for assessing the capability of their management system so as to provide an insight into the work involved.

As stated in ISO 9001:2008, Clause 0.1, the adoption of a quality management system should be a strategic decision of an organization, thus implying that organizations have a choice. If you have entered this chapter after reading previous chapters you will be aware that all organizations have a management system whether or not they adopt one. This is because the system is the way the organization functions. A strategic decision that does need to be made is whether to adopt a particular approach for the management of quality and whether to use ISO 9001:2008 as a means to demonstrate its commitment to quality to its stakeholders. If it has been decided to use ISO 9001:2008 in this way, it is important that these decision makers understand what is involved.

In this chapter we examine the requirements in Clause 4.1 of ISO 9001:2008 for establishing a quality management system in terms of what they mean, why they are necessary and how compliance can be demonstrated and in particular:

- The determination of processes;
- The sequence and interaction of processes;
- The criteria and methods for effective operation and control;
- Information and resource availability (see also *Determining and providing resources* in Chapter 19);
- Measuring, monitoring and analysing processes (see also *Measuring and monitoring of products and processes* in Chapter 31);
- The management of processes;
- System implementation;
- System maintenance;
- Continual improvement (see also *Continual improvement* in Chapter 35);
- Outsourcing.

These requirements are repeated in other sections in more detail as denoted in parenthesis. As the documentation requirements are split between sections 4.1 and 4.2 we have consolidated these into the next chapter for convenience.

MANAGEMENT SYSTEM DEVELOPMENT (4.1)

The standard requires the organization *to establish a quality management system in accordance with the requirements of ISO 9001.*

What Does this Mean?

To *establish* means to set up on a permanent basis, install, or create and therefore in establishing a management system, it has to be developed, resourced, installed and integrated into the organization.

Establishing a system in accordance with the requirements of ISO 9001 means that the characteristics of the system have to meet the requirements of ISO 9001. However, the requirements of ISO 9001 are not expressed as system requirements of the form 'The system shall …' but are expressed as organization requirements of the form 'The organization shall …'. It would appear, therefore, that the system has to cause the organization to comply with the requirements and this will only happen if the system is the organization. Some organizations regard the management system as the way they do things but merely documenting what you do does not equate with establishing a system for the reasons given in *Approach to system development* in Chapter 6.

Why is this Necessary?

This requirement responds to the System Approach Principle.

ISO 9001 contains a series of requirements which if met will provide the management system the capability of supplying products and services that satisfy the organization's customers. All organizations have a management system – a way of working, but in some it is not formalized – in others it is partially formalized and in a few organizations the management system really enables its objectives to be achieved year after year. In some organizations, a management system has been established rather than allowed to evolve and if an organization desires year after year success, it needs a formal mechanism to accomplish this – it won't happen by chance. This requires management to think of their organization as a system as a set of interconnected processes that include tasks, resources and behaviours as explained under *Systems approach vs process approach* in Chapter 7.

How is this Demonstrated?

The terms 'establish', 'document', 'implement', 'maintain' and 'improve' are used in the standard as though this is a sequence of activities. In reality, in order to establish a system it has to be put in place and putting a system in place requires a number of interrelated activities. These can be grouped into four stages of a System Development Project as outlined below. These are consistent with the requirements of Clause 4.1

Addressing Requirements

In addressing the requirements of ISO 9001 it is not necessary for you to use the same wording as in the standard. It is the intent of a requirement that should be addressed in one way or another. An Exposition or Compliance table can provide a translation between the standard and your system (see Chapter 30).

FIGURE 10-1   Cycle of sustained success.

and of Clause 2.3 of ISO 9000:2005 that addresses the Systems Approach. What you have to remember is that you are more than likely not starting from scratch. A system already exists, if it didn't, you and many others would not be working in the organization. Therefore, your task is to reveal this system and in doing so, identify opportunities for improvement. When developing a management system you should have in your mind the cycle of sustained success as shown previously in Chapter 7 and again in Fig. 10-1. Clearly the input to the system of managed processes is the mission and vision with an output of delivering results that produce satisfied stakeholders.

You can demonstrate that you have established an effective system by answering the following questions. The order of the questions is not indicative of a sequence in which an organization is established even though you may be viewing the organization as a system. It is also an iterative process therefore as answers to questions emerge it may cause you to rethink answers already given.

Establish the Goals

By goals we mean the direction in which the organization is going and how it is planning to get there. This is expressed in a number of ways.

1. Ask what is the organization trying to do and in response present the **Mission** statement (see *Ensuring policy is appropriate* in Chapter 15).
2. Ask who are the organization's beneficiaries and in response present a list of **Stakeholders** (see *Identifying the stakeholders* in Chapter 3).
3. Ask what the stakeholder's needs and expectations are relative to the mission and in response present the **Business Outcomes** identified in the Stakeholder Analysis (see *Determining processes* below).
4. Ask what will the stakeholders look for to assess if their needs and expectations have been met and in response present the **Stakeholder success measures or KPIs** as part of the stakeholder analysis (see *Determining processes* and also *Expressing quality objectives* in Chapter 16).
5. Ask what outputs will deliver successful outcomes and in response present the **Business Outputs** (see *Determining processes* below).
6. Ask what limits our ability now or in the future to deliver these outputs and in response present the **Strategic issues** (see under *Quality objectives* in Chapter 16).
7. Ask in what area are we going to concentrate our resources, which segment of the market, with what products, what services and with what values are we going to be

the leader and in response identify the **strategy** (see *Strategic planning* in Chapter 16).

8. Ask what should the organization look like as it successfully implements its strategies and fulfils its mission and in response establish its **Vision**? (see also *Establishing quality policy* in Chapter 15).

Model the Processes

9. Ask what factors affect our ability to deliver the business outputs and in response present the critical activities and results, i.e., the **Critical Success Factors** (see *Determining processes* and also Chapter 37).

10. Ask which processes deal with these factors and in response present the **enabling processes** (see *Determining processes*).

11. To identify the generic business processes ask 'what contribution do this enabling process makes to the business' and in response present the **Business Processes** and their associated sub-processes (see *Determining processes*).

12. Ask what measures will reveal whether the business process objectives have been met and in response present the vision, translated into **process measures**.

13. Ask what provisions have been built into the process design to prevent these processes failing to deliver the required outputs and in response present the **preventive action plans** (see *Determining processes*).

Measure and Analyse Performance

14. Ask to what extent do the organization's products and services satisfy the needs and expectations of customers and in response present the results of **customer satisfaction surveys** (see Chapter 29).

15. Ask what results these processes are delivering and in response present an **analysis of performance** showing actual performance against the KPIs (see Chapters 29 and 34).

16. Ask what methods of measuring customer satisfaction and process capability are used and how often measurements are taken and in response present the **measurement methods** (see *Measuring and monitoring* in Chapter 31).

17. Ask how the integrity of these measurements is assured and in response present **calibration and/or verification records** (see *Control of measurement and monitoring equipment* in Chapter 32).

18. Ask what checks are carried out to verify that processes are operating as planned and in response present the results of **process audits** (see *Internal audit* in Chapter 30).

19. Ask what analysis is performed to determine whether the objectives are being achieved in the most effective way and in response present the results of **process studies** (see *Analysis of data* in Chapter 34).

20. Ask what analysis is performed to determine whether the objectives being achieved remain relevant to stakeholder needs and in response present the results of a recent **Stakeholder analysis** (see *Quality objectives* in Chapter 16 and *Customer satisfaction* in Chapter 29).

Review and Improve Capability

21. Ask what changes have been made as a result of reviewing the data analysis to bring about improvement by better control and in response present the result of **corrective and preventive actions** (see Chapters 36 and 37, respectively).

22. Ask what changes have been made as a result of reviewing the data analysis to improve process efficiency and in response present the results of **continual improvement** (see Chapter 35).

23. Ask what changes have been made as a result of reviewing the data analysis to improve system effectiveness and in response present the results of **management reviews** (see Chapter 18).

Answers to these 23 questions will demonstrate the extent to which you have established an effective quality management system.

DETERMINING PROCESSES (4.1a)

The standard requires the organization *to determine the processes needed for the quality management system and their application throughout the organization.*

What Does this Mean?

Processes produce results of added value. Processes are not procedures (see *Processes versus procedures* in Chapter 8). The results needed are those that serve the organization's objectives. Processes needed for the management system might be all the processes needed to achieve the organization's objectives and will therefore form a chain of processes from corporate goals to their accomplishment. The chain of processes is a *value chain* and therefore should extend from the needs of the stakeholders to the satisfaction of these needs. This is illustrated by the system model of Fig. 7-15.

The change in the 2008 edition from 'identity processes' to 'determine processes' is a change in intent. With the term 'identify' in this context, it could be interpreted that all one had to do was to give a series of existing processes an identity, a label, whereas the intent is to decide, lay down, declare or conclude from reasoning what processes are needed to deliver certain outputs.

The phrase 'needed for the quality management system' in this requirement implies that there are processes that are needed for other purposes. It also implies that the quality management system is something that needs processes rather than something that describes processes. The phrase is ambiguous and needs rewording. A note in Clause 4.1 does not shed any light on the matter as it simply suggests that these processes include processes for management activities, provision of resources, product realization and measurement, analysis and improvement which covers most processes but omits the processes needed to create a demand. Process needed for management activities could be anything from strategic planning to redundancy planning.

A better way of expressing the intent would be to say 'determine the processes needed to enable the organization to achieve its objectives and satisfy its customers'. One might go further and suggest it should be 'all stakeholders', but ISO 9001 limits requirements to customers as one of the stakeholders whereas ISO 9004 addresses all stakeholders.

FIGURE 10-2   End-to-end processes.

Published interpretation RFI 029 (see published interpretations in Chapter 4) states that the expression 'needed for the QMS' in Clause 4.1a does not limit the processes to product realization only. Traditionally work has been organized into functions of specialists, each performing tasks that serve functional objectives. By thinking of the organization as a collection of interacting processes rather than a series of interconnected functions, each focused on the needs of stakeholders, the chain of processes cuts across the functions. This is not to say that functions are unnecessary but we must recognize that work does not only flow through functions, it flows through and across the functions by means of a process. The processes needed for the management system are those processes with a purpose that is aligned to the organization's objectives (Fig. 10-2).

As illustrated in Table 8-5, the processes identified in the Generic System Model clearly show the same stakeholder at each end.

Why is this Necessary?

This requirement responds to the Process Approach Principle.

The management system consists of a series of interacting processes and therefore these processes need to be determined.

How is this Demonstrated?

As the standard is not specific to the types of processes, we must assume it is all processes and this would therefore include both business processes and work processes. This distinction is important for a full appreciation of this requirement. The relationship between these two types of processes is addressed further in *Types of processes* in Chapter 8.

In many cases, organizations have focused on improving the work processes believing that as a result there would be an improvement in business outputs but often such efforts barely have any effect. It is not until you stand back that the system comes into view. A focus on work processes and not business processes is the primary reason why ISO 9001, TQM and other quality initiatives fail. They resulted in sub-optimization – not optimization of the organizational performance. If the business objectives are functionally oriented, they tend to drive a function-oriented organization rather than a process-oriented organization. Establish process-oriented objectives, measures and targets, focused on the needs and expectations of external stakeholders, the functions will come into line and you will be able to optimize organizational performance. When Texas Instruments re-engineered its processes in 1992, their process map showed only six processes.[1] Hammer remarks that hardly any company contains more than 10 or so principle processes.

---

[1] Hammer, Michael and Champy, James (1993). *Reengineering the corporation*, Harper Business.

There are several ways of determining processes but the three that we will discuss here are the stakeholder method, objectives driven method and activity driven method.

The Stakeholder Driven Method

An approach for aligning the mission, vision and values with the needs of the stakeholders and for defining appropriate performance indicators is a Stakeholder Analysis which goes further than the Balanced Scorecard[①] and addresses all stake-holders and links stakeholder needs with the processes that deliver outputs that satisfy them. With this method we determine the processes needed from key performance indicators[①]. The questions we ask to reveal the information that will lead us to determining the processes are taken from those detailed above. By way of an example we have analysed a fictional fast food business and presented the results in Table 10-1. The questions asked are included in the heading to each column. This shows a direct link between the stakeholder's needs and the processes that will deliver outputs that satisfy those needs. Although the analysis is quite detailed, it is presented to demonstrate the technique and should not be assumed to be representative of any particular fast food outlet.

This is only part of a full stakeholder analysis as you would also gather information on their judgment about your organization's performance. John Bryson provides a practical approach to stakeholder analysis in *Creating and implementing your strategic plan*[2] but it does not address processes to any depth. The business processes were identified in the *System models* of Chapter 7 and under *Types of processes* in Chapter 8.

Objective Driven Method

The stakeholder driven method provides a close alignment between stakeholder needs and process activities but can be time consuming and difficult to do but well worth while. If the quality management system is to be limited in scope to satisfying customers, you could simply take the business outputs that are derived from customer needs in the stakeholder analysis shown in Table 10-1. Another method is to derive the processes from the strategic objectives, business objectives or project objectives but these will include objectives that serve more than customers. The processes identified in the system or organization model can be regarded as Level 0 implying there are further levels in a hierarchy as shown above. It is important to remember that the purpose of any process is to achieve an objective and therefore whether the objective is strategic such as a vision or mission, or is related to the completion of task such as serving a customer, there is still a process to achieve it. However, if the decomposition reaches a level where to go any further in the hierarchy you would be in danger of noting arm movements, you have gone a level too far.

To identify the processes, sub-process or activities you need to know what objectives need to be achieved or what outputs are required and for this you will need access to the business plans, project plans etc. Objectives are simply outputs expressed differently. For example, if the output is growth in the number of enquires the process objective is to

---

[2] Bryson John M. (2004) Creating and implementing your strategic plan – A workbook for public and non-profit. Jossey Bass.

TABLE 10-1 Example of Determining Processes from an Analysis of Stakeholder

Stakeholders	Stakeholder needs (business outcomes)	Indicators of stakeholder satisfaction	Outputs that will deliver successful outcomes	Critical success factors	Enabling processes	Related business process
Who are the beneficiaries?	What are their needs relative to the mission?	What will they look for as evidence of satisfaction?	What outputs will deliver successful outcomes?	What factors affect our ability to deliver these outputs?	Which process deals with this factor?	What contribution does this enabling process make to the business?
Investors	A financial return that meets target	Profitability	Positive cash flow	Maintaining demand	Maintain cash flow	Resource management
			Effective business strategy	Commitment to the planning process	Develop strategy	Mission management
			Growth in demand	Attracting customers	Product promotion	Demand creation
	Growth	Sales	Growth in demand		Product promotion	Demand creation
			Keeping ahead of the competition	Market intelligence	Competitor analysis	Demand creation
				Service innovation	Service design	
		Market share	Consumer change in preference	Reputation	Performance review	Mission management
			Food and service quality and price	Value for money	Develop pricing strategy	Demand creation
				Customer satisfaction	Service design	Demand creation
					Service delivery	Demand fulfilment

Stakeholder	Requirement	Measure		Leadership	Create internal environment	Mission management
Employees	Competitive pay and conditions	Staff grievances	Motivated workforce	Leadership	Create internal environment	Mission management
		Accident levels	High safety record	Ergonomics	Serving area design	Resource management
		Absenteeism	Low sickness record	Management style	Create internal environment	Mission management
		Entry to exit time	Short queue length	Management style	Design food outlet	Resource management
			Adequate dining area capacity	Serving area design	Staff development	Resource management
				Staff competence	Design food outlet	Resource management
				Dining area design		
Customers	Fast, safe and nutritious food	After effects	Safe food	Living by our values	Store food	Resource management
			Food handled hygienically		Prepare food	
			Cleanliness of equipment		Serve food	Demand fulfilment
					Staff development	
		Food taste and texture	Food cooked at the right temperature	Competent staff	Staff development	Resource management
			Food stored in right environment		Staff development	Resource management

Continued

TABLE 10-1 Example of Determining Processes from an Analysis of Stakeholder—cont'd

Stakeholders	Stakeholder needs (business outcomes)	Indicators of stakeholder satisfaction	Outputs that will deliver successful outcomes	Critical success factors	Enabling processes	Related business process
	Safe, clean and hygienic environment	Hazard-free public areas	Efficient design of preparation, serving and dining areas	Competent design agency	Service design	Demand creation
			Clean and dry surfaces	Competent staff	Staff development	Resource management
		Appearance and practices in all areas	Competent staff	Training budget	Staff development	Resource management
	Value for money	Competitiveness	Competitively priced menu	Market intelligence	Determine pricing strategy	Demand creation
			Attractive presentation of fresh food in consistent portions	Container design	Serving area design	Demand creation
				Delivery flow		
				Portion control	Serve food	Demand fulfilment
	Efficient counter service	Waiting time	Well designed serving area	Competent design agency	Serving area design	Demand creation
			Competent staff	Training budget	Staff development	Resource management
		Staff courtesy	Selection criteria	Training budget	Staff development	Resource management

Stakeholder	Mutually beneficial relationship	Business continuity	Repeat purchases	Profit margins	Financial planning	Resource management
Suppliers	Prompt Payment	Payment on or before due date	On-time payment on invoices	Cash flow	Accounts	Resource management
	Compliance with statutory regulations and instruments	Compliance	Food handling meets hygiene regulations		Prepare food	Resource management
			Handling and storage of cleaning fluids meet COSHH regulations	Shared values	Serve food	Demand fulfilment
			Exit doors meet fire regulations		Facility maintenance	Resource management
			Sanitation meets building and public health regulations			
Society	Corporate and social responsibility	Ethical policies and practices	Decisions implement soundly based ethics policy	Shared values	Create internal environment	Mission management
	A local environmental responsibility	Score on environmental survey	Decisions implement soundly based environmental policy	Management commitment	Waste collection	Demand fulfilment
					Waste disposal	Resource management
	Jobs for those who live in the community	Staff ratios	Locals given preference in selection	Conditions of employment	Recruitment	Resource management
	Support for other traders in the community	Source ratios	Produce sourced from local suppliers	Maintaining profit margin	Purchasing	Resource management

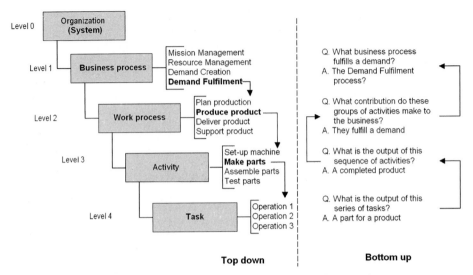

FIGURE 10-3   Process identification (top-down and bottom-up).

grow the number of enquiries. You can then ask a number of questions to determine the processes, sub-process or activities:

1. Ask what processes deliver these outputs or achieve these objectives;
2. Ask what activities produce these outputs or;
3. Ask what affects our ability to deliver these outputs.

The second question may generate different answers to the third question but asking both validates the answers to the second question. If you think you need to do X to achieve Y but when you pose question 3, there is no X in the list, you may have deduced that X is not critical to achieve Y and therefore is not a real process but part of another process.

Starting at Level 1 (see Fig. 10-3) answering question 1 of the business outputs will identify the business processes in the call-out text, e.g., an output of the business is a fulfilled demand therefore a Demand Fulfilment process is needed. Asking question 2 relative to the Demand Fulfilment process will identify the level 2 work processes, i.e., the activities that produce the outputs such as Plan production, produce product. Taking one of these work processes and repeating question 1 we identify level 3 processes such as set-up machine, make parts etc. Answering question 2 of the Make Parts process we identify the individual tasks at level 4.

Let us now suppose that when answering question 2 at level 3 you identified 'inspect' as an activity but when answering question 3 you deduced that making conforming parts was the key factor not inspection, thus inspection was an operation at level 4 and not a process at level 3 (Fig. 10-3).

Activity Driven Approach (Bottom-Up)

Instead of coming at it from objectives you can do it the other way around by identifying a sequence of activities and then:

- Ask "what is the output or objective of this activity" and thereby identify a Stage Output;
- Ask "where does this output go to" and thereby identify the next Stage in the process;

- Follow the trail until you reach the end of the chain of stages with a number of outputs;
- Collect the answers from group to group, department to department and then;
- Ask "what do these groups of outputs have in common" and thereby identify a series of Activity Groups and then;
- Ask "what contribution do these activity groups make to the business" and thereby identify the Business Processes? For example, advertising creates a demand therefore it is part of the Demand Creation Process.

The bottom-up approach involves everyone but has some disadvantages. As the teams involved are focused on tasks and are grouping tasks according to what they perceive are the objectives and outputs, the result might not align with the organizational goals; these groups may not even consider the organization goals and how the objectives they have identified relate to these goals. It is similar to opening a box of components and stringing them together in order to discover what can be made from them. It is not very effective if ones objective is to satisfy the external customer – therefore the top-down approach has a better chance of linking the tasks with the processes that will deliver customer satisfaction.

Case Study – Defining Processes

Some time ago we rewrote all our departmental procedures in the form of flow charts but we are unsure whether these are now documented processes or remain documented procedures as views on this matter differ. When auditing our system to ISO 9001:2008, what will the auditors be looking for to establish that we have actually defined and documented our processes?

When third party auditors ask you to show evidence that you have determined the processes that are needed by the QMS it is not uncommon for them to be satisfied with flow charts that relate to the activities the organization carries out. But from what you say, you took your procedures and rewrote them as flow charts – so in effect they are still procedures; it just so happens that they are now presented in flow chart form rather than text. The mistake you and many auditors make is that you perceive a flow chart to be a process but it appears that you are now prepared to question this logic.

If the charts simply depict activities in some sort of sequence, this would not be a complete process description. One of the problems exacerbated by the surfeit of flow charting tools following the launch of ISO 9001:2000 is that they have put in the hands of system developers a tool that makes every process or procedure look like a flow chart.

Possibly the question should be expressed differently. It is not a question of whether you have documented your processes but whether you have described how particular objectives are achieved or outputs produced. So return to the flow charts you produced and ask:

- Where on this chart does it define the objective that this series of activities is intended to accomplish?
- Where does it show how this objective relates to business objectives?
- Where on this chart does it define how performance of this process is measured, what the measures of success are and who reports the results?
- From which stakeholder does the input come from and the output go to?
- Where on this chart are the human and physical resources required to accomplish this objective?
- Where on the chart does it show where the resources are coming from?
- What changes were made the last time this process was reviewed that improved its performance, efficiency and effectiveness?

If the answer to any of these is 'I don't know?' You have not got an adequate process description.

In every organization there are sets of activities but each set or sequence is not necessarily *a process*. If the result of a sequence of activities adds no value, continue the
sequence until value is added for the benefit of customers – then you have defined
a *business* or *work process*.

Another technique is IDEF.[3] (Integrated Definition) developed by the US Air Force.
The objective of the model is to provide a means for completely and consistently
modelling the functions (activities, actions, processes and operations) required by
a system or enterprise, and the functional relationships and data (information or objects)
that support the integration of those functions. IDEF is a systematic method of
modelling that can reveal all there is to know about a function, activity, process or
system. Considering its pedigree it is more suited to very complex technical systems but
can result in *paralysis from too much analysis!* Many management systems do not
require such rigorous techniques. There can be a tendency to drill down through too
many layers such that at the lowest level you are charting movements of a person
performing an activity or identifying pens and pencils in a list of required resources. For
describing the management system processes it is rarely necessary to go beyond a task
performed by a single individual. As a rough guide you can cease the decomposition
when the charts stop being multifunctional.

SEQUENCE AND INTERACTION OF PROCESSES (4.1b)

The standard requires the organization *to determine the sequence and interaction of the
identified processes.*

What Does this Mean?

Sequence refers to the order in which the processes are connected to achieve a given
output. Interaction refers to the way the process outputs impact the other processes. A
common practice is to represent processes as boxes in a diagram with lines connecting them together. This shows the interconnections in a similar way to a circuit
diagram in a TV. If the lines have arrows it indicates the direction of flow for
information or product and to some extent the dependencies, the source of inputs and
the destination of outputs but not the interactions. Interaction between processes does
not have to be shown diagrammatically but Figs 7-13, 9-2 and 9-3 are examples
showing interactions.

Why is this Necessary?

This requirement responds to the System Approach Principle.

Objectives are achieved through processes, each delivering an output that serves as an
input to other processes along a chain that ultimately results in the objective being
achieved. It is therefore necessary to determine the sequence of processes. Some will
work in parallel; others in a direct line but all feeding results that are needed to
accomplish the objective. There will therefore be interactions between processes that
need to be understood.

---

[3] US Department of Defense (2001). www.idef.com. All IDEF models can be downloaded from this
web site.

How is this Demonstrated?

A practical way to show the sequence of processes is to produce a series of flow charts. However, charting every activity can make the charts appear very complex but by layering the charts in a hierarchy, the complexity is reduced into more digestible proportions.

You would start with a System Model similar to that in Fig. 10-4. This shows the relationship between the business processes and the stakeholders. It is a systems view of the organization. You could use the model in ISO 9001 but this is flawed as it identifies the sections of the standard rather than organizational processes. To some extent it also shows the interaction between the business processes but the nature of these interactions is omitted.

If we examine each of these business processes and ask "What affects our ability to deliver the process outputs?" we would identify the key stages, sub-processes or activities. By then placing these in a sequence in which they are executed you create the work processes at level 2. Do the same again for each work process and you create level 3 but this time, there may not be a sequence. Some activities may be activated by time or an event rather than by an input so these can be presented as shown in Figs 10-5 to 10-8. These process models have been derived using the stakeholder driven method in Table 10-1 for the fast food outlet.

All the processes require trained people, equipment and perhaps certain environmental conditions and therefore rely on the resource management process to deliver these outputs. The interface between these processes and resource management process creates an interaction when the processes are active. The reliance on resource management to

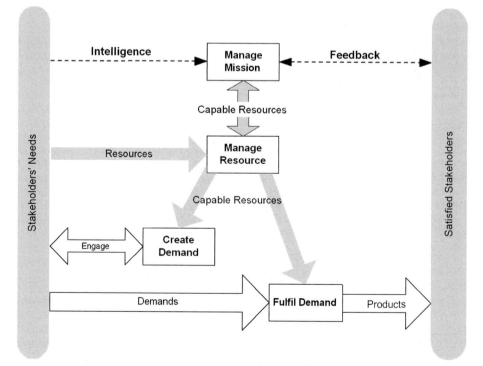

FIGURE 10-4   The generic system model.

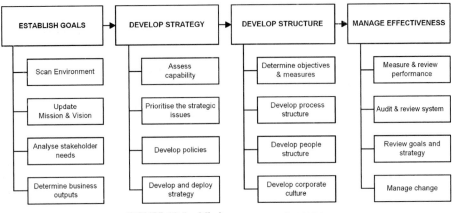

FIGURE 10-5   Mission management process.

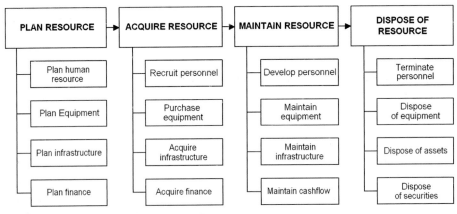

FIGURE 10-6   Resource management process.

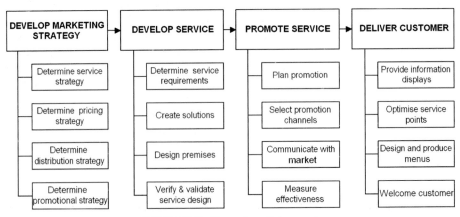

FIGURE 10-7   Demand creation process.

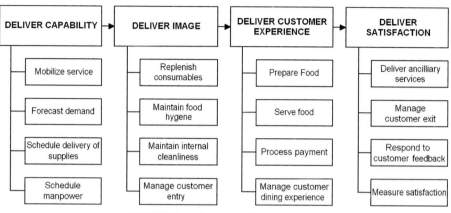

FIGURE 10-8 Demand fulfilment process.

provide inputs creates a dependency. Some of the interactions occur on demand and are therefore dynamic, others are passive and are often taken for granted but without which the process cannot deliver its required output. Analysis is needed to determine the delays in the system that create bottlenecks, the factors that create shortages, the effect of budget cuts on the quality of parts, materials and labour and so on.

CRITERIA AND METHODS FOR EFFECTIVE OPERATION AND CONTROL (4.1c)

The standard requires the organization *to determine criteria and methods required to ensure the effective operation and control of the identified processes.*

What Does this Mean?

The criteria that ensure effective operation are the standard operating conditions, the requirements, targets or success criteria that need to be met for the process to fulfil its objectives.

The methods that ensure effective operation are those regular and systematic actions that deliver the required results. In some cases the results are dependent on the method used and in other cases, any method might achieve the desired results. Use of the word 'method' in this context is interesting. It implies something different than had the standard simply used the word 'procedure'. Procedures may cover both criteria and methods but have often been limited to a description of methods. Methods are also ways of accomplishing a task that are not procedural. For example, information may be conveyed to staff in many ways – one such method might be an electronic display that indicates information on calls waiting; calls completed and call response time. The method of display is not a procedure although there may be an automated procedure for collecting and processing the data.

Why is this Necessary?

This requirement responds to the Process Approach Principle.

A process that is operating effectively delivers the required outputs of the required quality, on time and economically, while meeting the constraints that apply to the process. A process that delivers the required quantity of outputs that does not possess the required characteristics, are delivered late, waste resources and breach policy and safety, environmental or other constraints is not effective. It is therefore necessary to determine the criteria for the acceptability of the process inputs and process outputs and the criteria for acceptable operating conditions. Thus, it is necessary to ascertain the characteristics and conditions that have to exist for the inputs, operations and outputs to be acceptable.

How is this Demonstrated?

In order to determining the criteria for effective operation and control you need to identify the factors that affect success. Just ask yourself the question: *What are the factors that affect our ability to achieve the required objectives or deliver these outputs?* In a metal machining process, material type and condition, skill, depth of cut, feed and speed affect success. In a design process input requirement adequacy, designer competency, resource availability and data access affect success. In an auditing process, objectives, method, timing, auditor competence, site access, data access and staff availability affect success. There are starting conditions, running conditions and shutdown conditions for each process that need to be specified. If any one of these goes wrong, and whatever the sequence of activities, the desired result will not be achieved.

Determining the methods can mean determining the series of actions to deliver the results or simply identifying a means to do something. For example, there are various methods of control:

- Supervisors control the performance of their work groups by being on the firing line to correct errors.
- Automatic machines control their output by in-built regulation.
- Manual machines control their output by people sensing performance and taking action on the spot to regulate performance.
- Managers control their performance by using information.

The method is described by the words following the word 'by' as in the above list. A method of preventing failure is by performing an FMEA. FMEA[⊙] was developed by the aerospace industry in the mid 1960s. You don't have to detail how such an analysis is performed to have determined a method. However, in order to apply the method effectively, a procedure or guide may well be needed. The method is therefore the way the process is carried out which together with the criteria contributes to the description of the process.

ENSURING INFORMATION AVAILABILITY (4.1d)

The standard requires the organization *to ensure the availability of information necessary to support the operation and monitoring of the identified processes.*

What Does this Mean?

Information to support the operation of processes would include that related to:

- Process inputs,
- Planning activities,

- Preparatory activities,
- Result-producing activities,
- Routing activities,
- Process outputs.

Information to support the monitoring of processes would include that related to:

- Past and current performance, throughput, response time, downtime, etc;
- Operating conditions;
- Verification activities;
- Diagnostic activities.

Such information would include plans, specifications, standards, records and any other information required to be used in operating and monitoring the process.

Why is this Necessary?

This requirement responds to the Leadership Principle.

All processes require information whether they are automated or manually operated.

How is this Demonstrated?

Ensuring the availability of information is part of process management and is also addressed above. One would expect personnel to have information available in order to execute their work from preparation to completion.

To ensure availability of information you need to provide access at point of use and this is addressed under the heading *Control of documents*.

ENSURING THE AVAILABILITY OF RESOURCES (4.1d)

The standard requires the organization *to ensure the availability of resources necessary to support the operation and monitoring of the identified processes.*

What Does this Mean?

The resources necessary to support the operation of processes would include:

- Raw materials and consumables;
- Personnel;
- Utilities such as heat, light, power and water;
- Time;
- Equipment, plant machinery, facilities and work space;
- Money to fund the needs of the process.

The resources necessary to support the monitoring of processes would include:

- Instrumentation and equipment;
- Verification and certification services to ensure measurement integrity;
- Personnel to perform monitoring;
- Computers and other tools to analyse results;
- Utilities to energize the monitoring facilities;
- Money to fund the needs of monitoring.

Why is this Necessary?

This requirement responds to the Leadership Principle.

Without the necessary resources, processes cannot function as intended. All processes consume resources. If there are insufficient resources to monitor processes, it is hardly worthwhile operating them because you will not know how they are performing.

How is this Demonstrated?

The process owner or manager is responsible for ensuring the availability of resources. This commences with identifying resource needs, securing an available and qualified supply, providing for their deployment into the process when required and monitoring their utilization. Further guidance is given in *Resource management* in Chapter 6.

MONITORING, MEASURING, AND ANALYSING PROCESSES (4.1e AND 4.1f)

The standard requires the organization *to monitor, measure (where applicable) and analyse the processes and implement actions necessary to achieve planned results.*

What Does this Mean?

Measuring is concerned with determination of the quantities of an entity such as time, speed, and capability indices whereas monitoring is concerned with continual observation aside from periodic measurement. Measuring processes are rather different from measuring the output of processes – this is commonly referred to as inspection or output verification. Figure 10-9 illustrates this difference. Measurement of processes is only required where applicable which means, where the process characteristics are measurable. In reality there will always be a way of measuring performance but it might not always be by using conventional instruments.

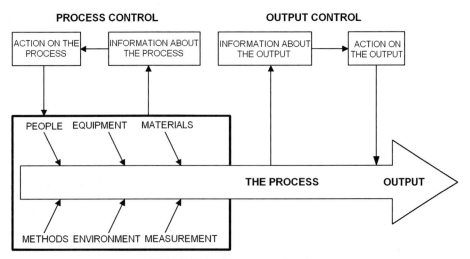

FIGURE 10-9   Process control model.

Analysing processes is concerned with understanding the nature and behaviour of processes for the purpose of their design, development and improvement. Measuring and monitoring take place following installation of the process whereas process analysis can be used as a design tool.

Action necessary to achieve planned results is addressed under *Taking action on process variation* in Chapter 31.

Why is this Necessary?

This requirement responds to the Factual Approach Principle.

You can't manage a process unless equipped with facts about its performance. Observations from monitoring provide this useful information. You cannot claim success, failure or make improvements unless you know the current performance of your processes. It is therefore necessary to install sensors to gather this data. The facts may tell you where you are, but further analysis is needed to establish whether it is an isolated occurrence, an upward or a downward trend and whether improvement is feasible. Process analysis is performed to enable the decision makers to make decisions based on fact.

How is this Demonstrated?

Process Measurement

In order to measure the process you need[4]:

1. Process objectives (what the process is designed to achieve). For example, a sales process may be designed to convert customer enquiries into sales orders and accurately convey customer requirements into the product or service generation process.
2. Indicators of performance (the units of measure). For example, for a sales process measures may be:
   - The ratio of confirmed orders to enquiries;
   - The ratio of customer complaints relative or order accuracy to total orders completed;
   - The ratio of orders lost due to price relative to total order won. Similar indicators could be applied for quality and delivery.

   There may be other indicators related to the behaviour of sales staff in dealing with customers and internal functions such as whether the sales promise matched the true capability of the organization etc.
3. Defined performance standard (the level above or below which performance is deemed to be sub-standard or inferior).
4. Sensors to detect variance before, during or after operations. There may be human or physical sensors, each of which has an element of measurement uncertainty.
5. Calibrated sensors so that you can be assured the results are accurate and precise. There are two types of measurements to be made.
   - Measurements that tell us whether the process is operating as intended;
   - Measurements that tell us whether the process is effective.

---

[4] Adapted from Juran, J. M. (1995). *Managerial Breakthrough, Second Edition*, McGraw-Hill.

The former measurements are taken using the process indicators and the later measurements are taken using process analysis. Most characteristics will be measurable, if not by variables in terms of mass, length or time, they will be measurable by attributes. (For more on measurement see Chapter 31.)

Process Monitoring

For process monitoring to be effective, the staff involved need to understand the process objectives and how they are measured. They need to be vigilant to potential and actual variations from the norm. A typical type of process monitoring takes place in a ship's engine room where there are lots of dials, gauges and data logging on a continuous basis. The watch engineers scan the log at the start of the watch for unusual occurrences that might account for variation in engine temperature. In monitoring a staff deployment process it may be noticed that staff are trained but there follows prolonged periods before the new skills are deployed. In the invoicing process it may be observed that a number of invoices go missing and have to be resent thus delaying receipt of revenue. In observing the design change process, it may be noticed that there is a burst in activity immediately prior to a holiday period without any additional resources being provided. Monitoring is looking for unusual occurrences or indicators of a potential change in performance.

Process Analysis

Process analysis can be used to implement Clauses 4.1a, 4.1b and 4.1c of ISO 9001 as well as Clause 4.1e and 4.1f.

Analysis in Process Design

Process analysis is performed to design a process and understand its behaviour. In this regard there are a number of activities that may be undertaken and there follows a sequence in which they could be implemented. This can be applied to level 3 activities as shown in Fig. 8-14.

- Define the key performance indicators[i];
- Define the method of measurement;
- Establish current performance against the indicators;
- Produce a process flow chart;
- Perform a task analysis to determine who does what, when, where, how and why;
- Identify constraints on the process and test their validity;
- Perform a control analysis to determine or verify the controls to be/being applied;
- Deploy known customer requirements (e.g., using QFD) to establish that the process will deliver the right output;
- Deploy the constraints (policies, regulations etc.) and identify any gaps;
- Identify failure modes and effects to establish the issues that could jeopardize success;
- Install failure prevention features to reduce, contain or eliminate potential failure modes;
- Conduct relationship analysis to establish conflicts of responsibility and authority and thus potential constraints;
- Perform productivity assessment to identify the number of transactions and their validity;

- Identify resources required to establish any deficiencies;
- Perform information needs analysis to identify or validate all the documentation needed;
- Perform a cultural analysis to establish the behavioural factors that will/are causing success or failure.

This analysis enables decisions to be made on the design or modification of processes and the conditions for their successful operation.

Analysis in Process Operation

In constructing the process, the identified measuring and monitoring stations should be installed. Process analysis is performed on the data generated by these sensors and includes several activities as follows.[5]

- Collect the data from monitoring activities;
- Sort, classify, summarize, calculate, correlate, present, chart and otherwise simplify the original data;
- Transmit the assimilated data to the decision makers;
- Verify the validity of the variation;
- Evaluate the economical and statistical significance of the variation;
- Discover the root cause of the variation;
- Evaluate the alternative solutions that will restore the status quo.

This analysis enables decisions to be made on the continued operation of the processes and whether to modify the conditions under which they operate. (See also under the heading *Taking action on process variation* in Chapter 31.)

MANAGING PROCESSES (4.1)

The standard requires the organization *to manage the identified processes in accordance with the requirements of ISO 9001.*

What Does this Mean?

Managing these processes in accordance with the requirements of ISO 9001 basically means that the way processes are managed should not conflict with the requirements in other words, managing processes means managing activities, resources and behaviours to achieve prescribed objectives.

The notes to Clause 4.1 of ISO 9001 need some explanation. It is stated that the processes needed for the management system include *management activities, provision of resources, product realization and measurement.* This note could cause confusion because it suggests that these are the processes that are needed for the management system. It would be unwise to use this as the model and far better to determine the processes from observing how the business operates. The term *provision of resources* should be *Resource Management* as addressed previously. *Product realization* is also a collection of processes such as design, production, service delivery, etc. *Measurement* is not a single process but a work process, activity or even a task within each process.

---

[5] Adapted from Juran, J. M. (1995). *Managerial Breakthrough, Second Edition,* McGraw-Hill.

Grouping all the measurement processes together serves no useful purpose except it matches the standard – a purpose of little value in managing the organization.

Why is this Necessary?

This requirement responds to the Process Approach Principle.

Desired results will not be achieved by chance – their achievement needs to be managed and as the processes are the means by which the results are achieved, this means managing the processes.

How is this Demonstrated?

The first stage in managing a process is to establish what it is you are trying to achieve, what requirements you need to satisfy, what goals you are aiming at; then establish how you will measure your achievements, what success will look like. If the objective were customer satisfaction, what would the customer look for as evidence that the requirements have been met? Would it be product fulfilling the need, delivered on time and not early, at the agreed price with no hidden costs?

Managing processes is primarily about ensuring that:

- Those involved in the process understand the objectives and how performance will be measured.
- Responsibility for actions and decision within the process is properly assigned and delegated.
- The processes providing the inputs are capable of meeting the demand required by the process.
- The required inputs are delivered when they need to be and are of the correct quality and quantity.
- The resources needed to perform the activities have been defined and conveyed to those who will deliver them.
- The forces that prevent, restrict limit or regulate some aspect of process performance are known and their effect minimized.
- The conditions affecting the behaviour of personnel and equipment are under control.
- The activities deemed necessary to deliver the required outputs and achieve the process objectives are carried out in the prescribed manner.
- Sensors are installed to detect variance in performance.
- The measurements taken provide factual data on which to judge performance.
- Provisions have been made for communicating unusual changes in the inputs, operating conditions and process behaviour and that these provisions are working effectively.
- Reviews are performed to verify outputs meet requirements.
- The causes of variation are determined and actions taken to restore the status quo and prevent recurrence of unacceptable variation.
- Reviews are performed to discover better ways of achieving the process objectives and actions are taken to improve the efficiency of operations.
- Objectives, targets and measures are reviewed and changed if necessary to enable the process to deliver outputs that continually serve the organization's objectives.

IMPLEMENTING A QUALITY MANAGEMENT SYSTEM (4.1)

The standard requires *the organization to implement a quality management system in accordance with the requirements of ISO 9001*.

What Does this Mean?

The notion of implementing a management system seems to imply that the management system is a set of rules, a procedure or a plan. One implements procedures but the management system is far more than a collection of procedures. Also the standard requires a management system to be established and as stated previously, to establish a management system you need to design and construct it and integrate it into the organization. If you were to write a book and put it on a bookshelf, you would not refer to the book being established. Implementation therefore applies to the use and operation of the management system following its construction and integration and is therefore concerned with the routine operation of an already established, documented and resourced system. Effective implementation means adhering to the policies and practices, doing what you say you will do.

Why is this Necessary?

This requirement responds to the Leadership Principle.

It goes without saying that it is necessary to use the management system that has been established because the benefits will only arise from using the system.

How is this Demonstrated?

There is no magic in meeting this requirement. You simply need to do what you said you would do, you have to keep your promises, honour your commitments, adhere to the policies, meet the objectives, improve the processes etc. – in other words manage your processes effectively, simply said but extremely difficult for organizations to do. Even if you documented what you do, your practices are constantly changing so little time would pass before the documents were out of date. A common failing with the implementation of documented practices is that they are not sold to the workforce before they become mandatory. Also, after spending much effort in their development, documented practices are often issued without any thought given to training or to verifying that practices have not in fact been changed. As a result, development is often discontinued after document release. It then comes as a shock to managers to find that all their hard work has been wasted. An effectively managed programme of introducing new or revised practices is a way of overcoming these shortfalls.

MAINTAINING A QUALITY MANAGEMENT SYSTEM (4.1)

The standard requires the organization *to maintain a quality management system in accordance with the requirements of ISO 9001*.

What Does this Mean?

For many working with previous versions of the standard this was interpreted as maintaining documents, but as the management system is the means by which the organization's objectives are achieved, it clearly means much more than this. Maintenance is concerned with both retaining something in and restoring something to a state in which it can perform its required function. In the context of a management system this entails maintaining processes and their capability and maintaining the organization to deliver that capability.

Why is this Necessary?

This requirement responds to the Process Approach Principle.

Without maintenance any system will deteriorate and management systems are no exception. A lack of attention to each of the factors mentioned above will certainly result in a loss of capability and therefore poor quality performance, financial performance and lost customers. Even to maintain performance a certain degree of improvement is necessary – in fact even raising standards can be perceived as a means of maintaining performance in a dynamic environment in which you adapt or die.

How is this Demonstrated?

In maintaining processes you need to keep:

- reducing variation,
- physical resources operational,
- human resources competent,
- financial resources available for replenishment of consumables, replace worn out or obsolete equipment,
- the process documentation up-to-date as changes in the organization, technology, resources occur,
- space available to accommodate input and output,
- buildings, land and office areas clean and tidy – remove the waste,
- benchmarking processes against best in the field.

In maintaining capability you need to keep:

- replenishing human resources as staff retire, leave the business or are promoted;
- renewing technologies to retain market position and performance;
- surplus resources available for unforeseen circumstances;
- up to date with the latest industry practices;
- refreshing awareness of the vision, values and mission.

Another set of actions that can be used is the Japanese 5-S technique.[6]

1. Seiri (straighten up): Differentiate between the necessary and unnecessary and discard the unnecessary.
2. Seiton (put things in order): Put things in order.
3. Seido (clean up): Keep the workplace clean.
4. Seiketsu (personal cleanliness): Make it a habit to be tidy.
5. Shitsuke (discipline): Follow the procedures.

---

[6] Imai, Masaaki, (1986). *KAIZEN, The key to Japanese Competitive Success,* McGraw-Hill

CONTINUAL IMPROVEMENT IN THE QUALITY MANAGEMENT SYSTEM AND ITS PROCESSES (4.1 AND 4.1f)

The standard requires the organization *to continually improve the effectiveness of the quality management system in accordance with the requirements of ISO 9001.*

What Does this Mean?

ISO 9000:2005 defines continual improvement as a recurring activity to increase the ability to fulfil requirements. As the organization's objectives are its requirements, continually improving the effectiveness of the management system means continually increasing the ability of the organization to fulfil its objectives.

Why is this Necessary?

This requirement responds to the Continual Improvement Principle.

If the management system is enabling the organization to accomplish its objectives when that is its purpose, why improve? The need for improvement arises out of a need to become more effective at what you do, more efficient in the utilization of resources so that the organization becomes best in its class. The purpose of measuring process performance is to establish whether or not the objectives are being achieved and if not to take action on the difference. If the performance targets are being achieved, opportunities may well exist to raise standards and increase efficiency and effectiveness.

How is this Demonstrated?

If the performance of a process parameter is currently meeting the standard that has been established, there are several improvement actions you can take:

- Raise the standard. For example, if the norm for the sales ratio of orders won to all orders bid is 60%, an improvement programme could be developed for raising the standard to 75% or higher.
- Increase efficiency. For example, if the time to process an order is within limits, identify and eliminate wasted resources.
- Increase effectiveness. For example, if you bid against all customer requests, by only bidding for those you know you can win, you improve your hit rate.

You can call all these actions *improvement actions* because they clearly improve performance. However, we need to distinguish between being better at what we do now and doing new things. Some may argue that improving efficiency is being better at what we do now, and so it is – but if in order to improve efficiency we have to be innovative we are truly reaching new standards. Forty years ago, supervisors in industry would cut an eraser in half in the name of efficiency rather than hand out two erasers. Clearly this was a lack of trust disguised as efficiency improvement and it had quite the opposite effect. In fact they were not only increasing waste but also creating a hostile environment.

Each of the improvement actions is dealt with later in the book and is the subject of continual improvement addressed again in Chapter 35.

There are several steps to undertaking continual improvement[7]

1. Determine current performance.
2. Establish the need for change.
3. Obtain commitment and define the improvement objectives.
4. Organize diagnostic resources.
5. Carry out research and analysis to discover the cause of current performance.
6. Define and test solutions that will accomplish the improvement objectives.
7. Product improvement plans which specify how and by whom the changes will be implemented.
8. Identify and overcome any resistance to change.
9. Implement the change.
10. Put in place controls to hold new levels of performance and repeat step one.

OUTSOURCING (4.1)

The standard requires the organization *to ensure control of any outsourced processes that affects product conformity to requirements and to identify such control within the quality management system.*

What Does it Mean?

In purchasing products and services to the supplier's own specification, the organization is not outsourcing processes or subcontracting. It is simply buying products and services. An outsourced process is one that is managed by another organization on behalf of the parent organization. The most common outsourced processes are manufacturing processes such as fabrication, assembly and finishing processes. But organizations also outsource information technology, human resources, cleaning, maintenance and accounting services. In the extreme, they unwisely outsource the internal auditing process and with it all responsibility. Often it is the activities that are outsourced not the process in its entirety. For a process to be outsourced, the supplier should be given an objective and given freedom to determine how that objective will be met. If the supplier merely performs activities dictated by the organization to the organization's specification it is not a process that has been outsourced but tasks.

The standard requires the type and extent of control to be applied to these outsourced processes to be defined within the quality management system. What this means is that in addition to identifying the outsourced processes in the system description, you need to describe how you manage these processes which will differ from the way ordinary purchases are managed.

Why is this Necessary?

At one time, an organization would develop all the processes it required and keep them in-house because it was believed it had better control over them. As trade became more competitive, organizations found that their none-core processes were absorbing a heavy overhead and required significant investment just to keep pace with advances in

---

[7] Adapted from Juran, J. M. (1995). *Managerial Breakthrough, Second Edition,* McGraw-Hill

technology. They realized that if they were to make this investment they would diminish the resources given to their core business and not make the advances they needed to either maintain or grow the business. A more cost-effective solution was to put the management of these non-core processes in the hands of organizations for which they were their core processes. However, the organization needs to have control over all the processes required for it to achieve its objectives otherwise it can't be confident that it will satisfy its customers.

How is this Demonstrated?

When managing an outsourced process, the organization is not simply placing orders for products and checking that the products received comply with the requirements but establishing process objectives and verifying that the supplier has developed a process that is capable of achieving those objectives. The processes should have a capability that enables the organization to avoid checking the outputs. All the rigor applied to the internal processes should be applied to the outsourced process. Data on the performance of these processes, their efficiency and effectiveness should be analysed by the organization and measures put in place to cause improvement action should the outcomes of the process not satisfy the organization's objectives.

To demonstrate that these outsourced processes are under control it will be necessary to:

- Define the contractual arrangements with the supplier of the outsourced services including:
  - Key performance indicators,
  - Performance targets for these indicators,
  - Reporting provisions,
  - Access rights to verify operations are proceeding in line with the agreed contractual arrangements,
  - Responsibilities of both parties.
- Show in the organization's budgets that you have provided resources for managing the suppliers as if they were part of your own organization;
- Show that you have assessed the provisions the supplier has made for planning, operating and controlling the process and indicated your approval of these plans;
- Show that the supplier shares the same values as you do with respect to customers' property and quality;
- Show that you are monitoring supplier performance and taking action on deviations;
- Show that you are discharging your obligations to the supplier as defined in the contract.

It is important to remember that when you outsource services, the people doing the work are not paid by you but by your contractor. These workers will do what they are paid to do, not what you expect them to do, as you lost the right to control their work when you outsourced the service.

Chapter 11

Documenting a Quality Management System

CHAPTER PREVIEW

This chapter is aimed primarily at those with responsibility for documenting the organization's processes and associated activities or involved with the preparation of documentation. It will also be of interest to those senior managers who decided to adopt ISO 9001 as the basis for assessing the capability of their management system by providing some insight into the work involved.

The first requirement for documentation is encountered in Clause 4.1 of the standard where it requires the organization to document a quality management system but the main requirements for documentation are in Clause 4.2. A fundamental principle of quality assurance (see *The assurance principles* in Chapter 1) is that the organization be prepared to substantiate by objective evidence that they have maintained control over activities affecting the quality of products supplied to customers. This objective evidence has three components:

1. A declaration of the intentions and planned arrangements for meeting customer requirements. These are conveyed through the documented policies, plans, processes, standards and procedures.
2. Records that these intentions and planned arrangements have been effectively implemented.
3. Records that the products supplied meet the specified requirements.

A common criticism of the approach adopted by ISO 9001 is the emphasis it places on documentation. The early editions of ISO 9001 gave the impression that documentation was the key to the pursuit of quality when in fact there were more important factors such as management commitment and human interaction. Although requirements for documentation still feature in the 2008 version, a more balanced approach is now taken. There is a reduction in the requirements for documentation, leaving it to management discretion and increasing the requirements on management commitment. However, there is still a long way to go for the standard to fully recognize the impact of human interaction on the pursuit of quality.

The old document what you do and do what you document approach (see *Approach to documentation* in Chapter 6) resulting in everything being documented should now be replaced with a more balanced approach. The basic premise of the three components above remains but the problem lies in determining the minimum level of documentation necessary to manage the processes effectively.

In this chapter we examine the requirements in Clause 4.2 of ISO 9001:2008 and in particular:

- What should be documented?
- The quality manual;
- Documented procedures;
- Planning and operational documents needed for effective control of processes including:
  - Policies and practices,
  - Process descriptions,
  - Procedures,
  - Standards,
  - Guides,
  - Derived documents.

The provisions needed for controlling the documents referred to in this chapter are dealt with in Chapter 12.

WHAT SHOULD BE DOCUMENTED?

The standard explains that the extent of quality management system documentation can differ due to the size and type of organization, complexity of the processes, and the competency of personnel.

The factors mentioned in the standard apply equally to documentation generated by a process and documentation supporting the process. For example, a person may need documented policies and practices to execute the processes reliably and also may produce documents that are required inputs for other processes. In both cases the factors of size, complexity etc. apply to the extent of the supporting documentation as well as to the extent of the output documentation – therefore documentation producers need to be aware of the documentation needs of the interacting processes.

Size of Organization

If we think about it, what has size of the organization got to do with the amount of information you document? A large organization could be large because of the quantity of assets – 2000 offices with two people in each. Or it could be large because it employs 6000 people, 5500 of whom do the same job. Or we could find that of the 6000, there are 200 departments, each providing a different contribution and each staffed with people of different disciplines. Therefore, size in itself is not a factor and size without some units of measure is meaningless.

Complexity of Processes

Complexity is a function of the number of processes and their interconnections in an organization. The more processes there are, the greater the number of documents. The more the interconnections, the greater the detail within documents. Complexity is also a function of the relationships. The more relationships, there are, the greater the complexity and channels of communication. Reducing the number of relationships can reduce complexity. Assigning work to fewer people reduces the number of transactions. Many documents exist simply to communicate information reliably and act as a point of reference, should our memory fail us, which introduces another factor – that of man's limited ability to handle unaided large amounts of data.

In the simplest of processes, all the influencing facts can be remembered accurately. As complexity increases, it becomes more difficult to remember all the facts and recall them accurately. A few extraordinary people have brilliant memories, some have learnt memory skills but the person of average ability cannot always remember a person's name or telephone number. The word 'password' is the most common password for Internet transaction because many of us would forget the password if it was something else. It would therefore be unreasonable to expect people to perform their work without the use of recorded information of some kind. What you should record and what you remember is often a matter of personal choice but in some cases you cannot rely on people remembering facts by chance. You therefore need to identify the dependencies in each process and perform a risk assessment to establish what must be documented.

Type of Organization

The type of organization will affect what you document and what documents you use but again not the amount of information you document. An organization that deals primarily with people may have little documentation. One that moves product may also have little documentation but one that processes information may have lots of documentation. A software house is different from a gas installation service, a bank is different from a textile manufacturer and therefore the content of the documentation will differ but they may use the same types of documents.

Factors Affecting the Amount of Information you Document

There has to be a limit on what you document. At school we are taught reading, writing and arithmetic so documents should not attempt to define how these activities are performed. But it depends on what you are trying to do. The documents in regular use need only detail what would not be covered by education and training. A balance should be attained between training and procedures. If you rely on training rather than employing documented procedures, you will need to show that you have control over the quality of training to a level that will ensure its effectiveness. We expect staff to know how to do the various tasks that comprise their trade or profession, how to write, how to design, how to type, answer the telephone, how to paint, lay bricks etc. You may feel it necessary to provide handbooks with useful tips on how to do these tasks more economically and effectively and you may also use such books to bridge gaps in education and training but these are not your procedures. If you need something to be done in a particular way because it is important to the outcome, the method will need to be documented so that others may learn the method.

You can combine several procedures in one document, the size of which depends on the complexity of your business. The more complex the business, the greater is the

Case Study – Begin with the End in Mind

In the early 1990s a team of consultants was leading an ISO 9000 Project in a large organization in the public sector. The challenge was to describe how a multi-site and very complex organization functioned in a way that met the requirements of ISO 9001. It was believed it would be too difficult to engage the whole organization in one giant quality management system and hence one certification, so it was decided to tackle it in stages; one covering the headquarters functions, another the operations centres and a third the field units. The net result was several certifications, each awarded at different times. After some of the systems had been documented and certificated there was a change in top management. The new Managing Director took one look at the shelves of manuals and instructed the Quality Director to cut it by half. A team was created to strike through statements they felt were unnecessary. They left policies and removed practices. They used red pens so they called it the Red Team. After the slash and burn the guys in the field units created their own Departmental Manuals from all the documented practices taken out of the previous manuals. So the iceberg was still there, only this time less of it appeared in the Managing Director's bookcase.

This was in the days before intranets so the volume of paper was very visible. What can now be buried in a linked file for the reader to click in and out of in a moment, had to be contained within the document which sometimes led to long documents. With an intranet the volume is not as visible but may still appear very complex with all the levels and linkages.

Because of the nature of the work they placed a lot of emphasis on prescription primarily because of their safety responsibility. If anything went wrong and they had followed the approved prescription they had exercised due diligence.

We don't know whether the Managing Director was disturbed by the physical volume of information that his troops had to know or that by being in his bookcase, whether he believed he had to understand it all to discharge his responsibilities. He had not authorized the work that resulted in the large volume of documentation so was not concerned about the cost but it changed the attitude of managers towards documentation.

There are several lessons from this story;

1. Don't distribute manuals of documentation to impress the recipients by the amount of work that has been undertaken.
2. Recognize the needs of individual managers and present them with information in a form that will give them confidence that they have the information they need and no more to discharge their responsibilities.
3. Discover what the real problem is with documentation before taking action and if necessary restructure it. If someone thought it is necessary to describe a function, process or activity in a particular way find out what they thought they were doing and why they thought they were doing it before you trash it. Someone's life might depend on those instructions being followed to the letter.
4. When describing how an organization functions, it is easy to lose sight of your objective. Follow Stephen Covey's advice and *"Always begin with the end in mind"*.

number of documents. The more variation in the way that work is executed, the larger the system description will be. If you have a small business and only one way of carrying out work, your system description will tend to be small. Your management system may be described in one document of no more than 30 pages. On the other hand a larger business may require several volumes and dozens of documents of over 10 pages each to adequately describe the system.

Process descriptions need to be user-friendly and so should be limited in size. Remember you can use other documents such as guides, standards, and operating procedures to extend what you have written. The documents should not, however, be so short as to be worthless as a means of controlling activities. They need to provide an adequate degree of direction so that the results of using them are predictable. Staff should be trained for routine activities, making procedures unnecessary. However, when dealing with activities that are not routine, if you neglect to adequately define what needs to be done and how to do it, don't be surprised that staff don't know what to do or constantly make mistakes. It is also important to resist the desire to produce manuals that are impressive rather than practical. Printing the documents on expensive paper with coloured logo does not improve their effectiveness and if they are not written simply and understood by a person of average intelligence, they will not be used.

Reasons for Not Documenting Information

There are several reasons for not documenting information.

- If the course of action or sequence of steps cannot be predicted, a procedure or plan cannot be written for unforeseen events.
- If there is no effect on performance by allowing freedom of action or decision, there is no mandate to prescribe the methods to be employed.
- If it cannot be foreseen that any person might need to take action or make a decision using information from a process, there is no mandate to require the results to be recorded. (However you need to look beyond your own organization for such reasons if demonstrating due diligence in a product liability suit requires access to evidence.)
- If the action or decision is intuitive or spontaneous, no manner of documentation will ensure a better performance.
- If the action or decision needs to be habitual, documentation will be beneficial only in enabling the individual reach a level of competence.

Apart from those aspects where there is a legal requirement for documentation, the rest is entirely at the discretion of management but not all managers will see things the same way. Some will want their staff to follow rules and others will want their staff to use their initiative (see Chapter 9).

THE QUALITY MANUAL (4.2)

Preparing the Quality Manual (4.2.1b and 4.2.2)

The standard requires *a quality manual to be established and maintained and included in the quality system documentation.*

What Does this Mean?

ISO 9000:2005 defines a quality manual as *a document specifying the quality management system of an organization.* It is therefore not intended that the manual be a response to the requirements of ISO 9001. As the top-level document describing the management system it is a system description describing how the organization ensures it consistently delivers product that satisfies customer and applicable statutory and regulatory requirements.

Quality manuals that paraphrase the requirements of the standard add no value. They are of no use to managers, staff or auditors. Often thought to be useful to customers, customers would gain no more confidence from this than would be obtained from the registration certificate.

Why is this Necessary?

This requirement responds to the System Approach Principle.

A description of the management system is necessary as a means of showing how all the processes are interconnected and how they interact to collectively deliver the business outputs. It has several uses as:

- a means to communicate the vision, values, mission, policies and objectives of the organization;
- a means of showing how the system has been designed;
- a means of showing linkages between processes;
- a means of showing who does what;
- an aid to training new people;
- a tool in the analysis of potential improvements;
- a means of demonstrating compliance with external standards and regulations.

How is this Demonstrated?

When formulating the policies, objectives and determining the processes to achieve them, the manual provides a convenient vehicle for containing such information. If left as separate pieces of information, it may be more difficult to see the linkages.

The requirement provides the framework for the manual. If the manual is a printed document its content may include the following:

- Introduction,
- Purpose (of the manual),
- Scope (of the manual),
- Applicability (of the manual),
- Definitions (of terms used in the manual),
- Business overview,
  - Nature of the business/organization – its scope of activity, its products and services,
  - The organization's stakeholders (customers, employees, regulators, investors, suppliers, owners etc.),
  - A context diagram showing the organization relative to its external environment,
  - Vision and values,
  - Mission,
- Organization,
  - Function descriptions,
  - Organization chart,
  - Locations with scope of activity,

- Business processes,
  - ○ The system model showing the key business processes and how they are interconnected,
  - ○ System performance indicators and method of measurement,
  - ○ Mission management process description,
  - ○ Resource management process description,
  - ○ Demand creation process description,
  - ○ Demand fulfilment process description,
- Function matrix (relationship of functions to processes),
- Location matrix (relationship of locations to processes),
- Requirement deployment matrices,
- ISO 9001 compliance matrix,
- ISO 14001 compliance matrix,
- Regulation compliance matrices (Environment, Health, Safety, Information Security etc.),
- Approvals (list of current product, process and system approvals).

The process descriptions can be contained in separate documents (see *Documents that ensure effective planning, operation and control of processes*).

As the manual contains a description of the management system, a more apt title would be a Management System Manual (MSM) or maybe a title reflecting its purpose might be Management System Description (MSD) or Business Management System Description (BMSD).

In addition, a much smaller document could be produced that does respond to the requirements of ISO 9001, ISO 14001, and the regulations of regulatory authorities. A model of such relationships is illustrated in Fig. 11-1. Here we show a number of management system standards hitting the organization and the aspects of the existing

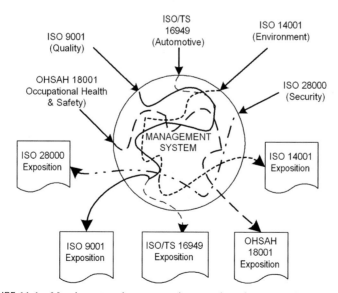

FIGURE 11-1 Mapping external system requirements through your management system.

management system that comply are referenced in an Exposition (see also Chapter 39 in *Preparing for certification*).

The process descriptions that emerge from the Management System Manual describe the core business processes and are addressed in Chapter 4 under *Documents that ensure effective operation and control of processes.*

If you choose to document your management system and make it accessible through a web browser, the 'quality manual' will not be a definable entity unless for the purposes of the external audit you refer to the complete Management System Description as the Quality Manual. The items in the list above could be accessed through hyperlinks in the navigation structure.

Case Study – Quality Manuals

One expects well-known companies to have a better level of understanding of quality issues but we have found that the quality manuals they submit as part of the tender are almost a duplicate of the standard with very minor changes. However, there are others that submit manuals that reflect their business, showing core and support processes, their inter-relationships and their key performance indicators but experience has shown that well-prepared documents do not necessarily reflect that a well-established and effective QMS exists. Why is there so much variation when all these companies are certified to ISO 9001:2000?

ISO 9000:2005 defines a quality manual as a document specifying the quality management system of an organization. Therefore, it could easily be a document that specifies the elements of the system or specifies how the organization satisfies the requirements of the standard. In both cases a document with the same headings as the standard might be the result.

ISO 9001 simply requires the manual to include the scope, exclusions, procedures and a description of the interaction between QMS processes, which is probably not very useful for qualifying suppliers.

It all comes down to purpose and how purpose is expressed.

The key factor in the background to your question is that the manuals are being submitted with a tender presumably for the purpose of evaluating the companies system for compliance with ISO 9001. They are given the title quality manual because that is what you probably requested, so the problem lies in the invitation to tender and that can easily be corrected. These companies might have several manuals, some for promotional purposes and others for internal use. Some organizations won't disclose proprietary information unless a confidentiality agreement is in force and there is nothing in ISO 9001 that compels them to supply customers with the same manual they show to the external auditors.

If you were to request a description of the processes that will be used to manage the work required to execute the contract you might receive a totally different type of document to the one you refer to as a quality manual.

However, you are right to observe that well-prepared documents do not necessarily reflect that a well-established and effective QMS exists.

Don't ask for a quality manual and you won't be disappointed.

Scope of the Quality Management System (4.2.2a)

The standard requires the quality manual to *include the scope of the quality management system including details of justification for any exclusion.*

What Does this Mean?

The system may not cover all activities of the organization and therefore, those that are addressed by the quality management system or excluded from it need to be identified. As there is no function in an organization that does not directly or indirectly serve the satisfaction of stakeholders, it is unlikely that any function or process will be excluded from the management system. However, don't mistake the scope of the quality management system with the scope of registration (see Chapter 39). The standard may also address activities that may not be relevant or applicable to an organization. The requirements for which exclusion is permitted are limited to those in Section 7 of the standard. Exclusions are also addressed in Chapter 39.

Why is this Necessary?

This requirement responds to the System Approach Principle.

It is sensible to describe the scope of the management system so as to ensure effective communication. The scope of the management system is one area that generates a lot of misunderstanding particularly when dealing with auditors, consultants and customers. When you claim you have a management system that meets ISO 9001, it could imply that you design, develop, install and service the products you supply, when in fact you may only be a distributor. Why you need to justify specific exclusions is uncertain because it is more practical to justify inclusions.

How is this Demonstrated?

The quality manual might define a scope that is less than that of the organization but the problem with this is that it sends out a signal that the manual exists only to address a requirement in ISO 9001 instead of having an important role in managing the organization.

If the manual does cover the whole organization, the scope of the system can be expressed as follows. 'This manual describes the processes that enable the organization to satisfy its customers and other stakeholders'. If you want to limit the scope to customers simply omit 'and other stakeholders' You would probably still include all departments in the organization because all are engaged in one way or another in satisfying customers. Marketing attracts customers, Sales secure orders from customers, Finance receives money from customers and pay suppliers, without which you would have no customers, HR develops the people who produce the products that satisfy customers, Reception greets customer when they visit, Maintenance keeps the lights on and the place looking nice so that customers go away with a good impression – therefore you can't omit anyone from the system.

Referencing Procedures in the Quality Manual (4.2.2b)

The standard requires the quality manual to *include the documented procedures established for the quality management system or reference to them.*

What Does this Mean?

As the standard now only requires six documented procedures, it is unclear whether it is these procedures that should be included or referenced or all procedures. A practical interpretation is to include or reference the Process Descriptions which themselves reference all the other documents used to manage processes. The procedures can be placed in a separate manual provided they are referenced in the quality manual (confirmed by published interpretation RFI 037).

Why is this Necessary?

This requirement responds to the Process Approach Principle.

By including or referencing the documentation that describes the management system, you are providing a road map that will help people navigate through the system. Take the road map away and people won't know which documents to use. All documentation in the management system should be related and serve a defined purpose. By expressing the documentation in a hierarchy you provide a baseline and thus a means of configuration control. This will prevent new documents being created or existing documents withdrawn without reference to the baseline.

How is this Demonstrated?

The retention of a requirement for the manual to include or refer to documented procedures indicates that management system documentation is still perceived by the standard makers as primarily comprising procedures. It is as though every requirement constitutes an activity that requires a documented procedure rather than it forming part of a process. However, the requirement for the interaction of processes to be described in the manual provides a means for correcting this inconsistency. If the manual is structured as suggested previously, each business process will be described and in doing so the relevant procedures and other documents needed for performing tasks within the process can be listed or included. With electronic documentation it is often not practical to duplicate lists of documents that would appear in a directory of a computer. Duplication creates a need for the synchronization of two or more lists to be maintained – thus causing additional effort and the possibility of error. Adding a new document to a file structure on a computer is the same as adding a new document to a list. A problem with file structures is that the configuration is changed when a document is added or deleted and therefore the status at any time in the past cannot be established unless a record is kept. Auditors and investigators certainly need to be able to establish the status of the system documentation at intervals so that they can determine the documentation that was current when a particular event took place. Ideally, the database containing the documents needs the capability to reconstruct the file structure that existed at a particular date in the past.

Describing the Interaction Between Processes (4.2.2c)

The standard requires the quality manual to *include a description of the interaction between the processes of the quality management system.*

What Does this Mean?

Each of these processes within the management system interacts with the others to produce the required outputs. A description of this interaction means describing the effects of the outputs from one process on the other processes in the system and how this generates the system outcomes. This is more like a system structure diagram than a flow chart. For example, in describing an air conditioning system, the system drawing would show all the components and how they linked together to form the system. The circuit diagrams would show the values of the components so that an engineer could find out how the system functioned and what signals to expect at any junction. Any component not linked into the system would have no function in the system and therefore would not be essential to its performance. Similarly with a management system, any process not connected to the system cannot perform a useful purpose within the system and can therefore be ignored. Any component in an air conditioning system indicating an incorrect output may cause system failure just as a process in the management system delivering the wrong output would cause system failure.

Interactions are more than interconnections or relationships. Interconnections are the channels along which data or product flow, like the wires between the components in an electronic circuit or the roads between the towns on the map. The relationships are the relative positions of the processes in a sequence as the data or product flows through the system. The interactions explain the dynamics of the system, so it is possible to see the effects that varying outputs on one process has on another process. Delays create bottlenecks, increased demand creates shortages, budget cuts result in cost cutting which impact the quality of parts, materials and labour and so on. These are the interactions.

Why is this Necessary?

This requirement responds to the System Approach Principle.

The management system comprises the processes required to achieve the organization's objectives and therefore they are not only linked together, they depend on each other to function. It is necessary to describe the behaviour of the processes so that it can be demonstrated that a coherent system exists and how it operates.

How is this Demonstrated?

The sequence and interaction of processes were addressed previously. In this case the interaction is required to be described and one way of doing this is through a Process Failure Modes and Effects Analysis (PFMEA), see Chapter 37. The process descriptions would define the standard operating conditions, i.e., the conditions that have to exist for the outputs to be acceptable. The PFMEA[1] would identify the effect on the outputs of variation in the inputs by anticipating how the inputs might vary. for example, what happens if the inputs are delayed, if process is supplied with inaccurate information, if the people are deployed to the process before being fully competent? Processes can be designed to detect certain variation in the inputs and feedback to the supplying process. This feedback is one of the interactions referred to.

Another approach is to carry out system modelling to discover what might happen when certain scenarios prevail. For instance, if the purchasing process is geared to replenish stocks when they reach a certain level, what might happen if there was a fall in customer demand, would the purchasing process react quickly enough? The lead time for purchase orders might be six months behind customer orders therefore when the stock arrives six months later there are no customers. A good example of this is described in The Fifth Discipline by Peter Senge.[1]

Figure 10-4 illustrates a high level description of the organization when viewed as a system of interacting processes. Expanding each of the four processes in this model as we have shown in Figs 10-5 to 10-8, would create a series of diagrams that adequately responds to this requirement and avoids following the undesirable practice of showing clauses of the standard in boxes with the boxes linked together to form some type of interrelationship. There are some examples of this on the Internet but most organizations have taken a hybrid approach. Figure 11-2 depicts an example of an approach to describing processes that combines processes, functions and clauses. The central flow is intended to symbolize the main production flow with the support groups (functions/departments) shown providing inputs or receiving outputs at various stages, thus implying that the people are somehow outside the process and remain in their silos! There are some support groups that appear to have no interface with the identified processes. The clauses of the standard are overlaid to indicate where they apply. There are a number of clauses that are not shown because they either apply to all processes and support groups or none of them.

Having scoured the Internet, this approach is not uncommon but it is not the process approach. Stringing the main result producing activities of the organization together in a chain is not the process approach. It looks like the central flow is the order to cash process or demand fulfilment therefore the boxes identified as processes are more like steps within a process than separate processes. Figures 10-4 to 10-8 provide a representation of the process approach that drives stakeholder needs and expectations through a number of processes that deliver stakeholder satisfaction. This is done without showing which department executes the activities or which requirements of the standard are satisfied. Such aspects are revealed at a much lower level in the decomposition – a level where one can be very specific and avoid the generalizations and ambiguities that arise from over descriptive high-level representations.

DOCUMENTED PROCEDURES AND RECORDS (4.2.1c)

The standard requires the management system documentation *to include documented procedures and records required by ISO 9001.*

What Does this Mean?

There are very few procedures actually required by the standard but this does not imply you don't need to produce any others. It is also often believed that the standard requires just six documented procedures but this is untrue. There are six specific requirements for documented procedures but Note 1 following Clause 4.2.1 of the 2008 version states that "a single document may address the requirements for one or more procedures and a requirement for a documented procedure may be covered by more than one document". Therefore, where the standard requires a documented procedure it may be satisfied by a single document, multiple documents or indeed part of a document.

---

[1] Senge Peter M. (2006) The Fifth Discipline. Random House Business Books London.

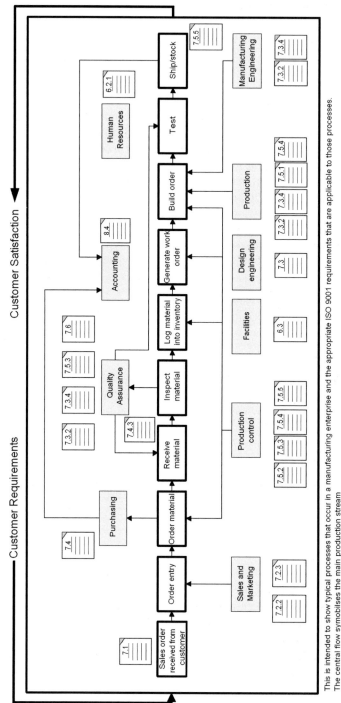

This is intended to show typical processes that occur in a manufacturing enterprise and the appropriate ISO 9001 requirements that are applicable to those processes. The central flow symoblises the main production stream

Adapted from California Manufacturing Technology Center graphic Implementing ISO 9001:2000

FIGURE 11-2 A hybrid interpretation of the process approach.

The standard requires documented procedures that address the following topics:

1. Document control,
2. Control of records,
3. Conducting audits,
4. Nonconformity control,
5. Corrective action,
6. Preventive action.

These areas all have something in common. They are what the authors of the early drafts of ISO 9001:2000 referred to as system procedures – they apply to the whole system and are not product, process or customer specific although it is not uncommon for customers to specify requirements that would impact these areas. Why procedures for these aspects are required and not for other aspects of the management system is unclear but it seems that the authors of ISO 9001, felt these were not processes or activities but tasks – a conclusion difficult to justify. They are certainly not business processes but could be work processes or activities within a work process.

The references to Clause 4.2.4 give some guidance to the types of records required but should not be interpreted as definitive. There are 19 references to records as indicated below:

1. Management review records (Clause 5.6.3);
2. Records of education, experience, training and qualifications (Clause 6.2.2);
3. Records needed to provide evidence that realization processes and resultant product meet requirement (Clause 7.1);
4. Customer requirement review records (Clause 7.2.2);
5. Design and development inputs (Clause 7.3.2) – this does not fit the criteria for a record (see *Design and development inputs* in Chapter 25);
6. Design and development review records (Clause 7.3.4);
7. Design verification records (Clause 7.3.5);
8. Design validation records (Clause 7.3.6);
9. Design and development change review records (Clause 7.3.7);
10. Supplier evaluation records (Clause 7.4.1);
11. Validation arrangements for processes to include requirements for records (Clause 7.5.2);
12. Product identification records (Clause 7.5.3);
13. Records of unsuitable customer property (Clause 7.5.4);
14. Calibration records (Clause 7.6);
15. Internal audit results are to be recorded (Clause 8.2.2);
16. Product verification records (Clause 8.2.4);
17. Nonconformity records (Clause 8.3);
18. Results of corrective actions taken are to be recorded (Clause 8.5.2);
19. Results of preventive actions taken are to be recorded (Clause 8.5.3).

Again it does not mean that you need to produce no more records than those listed above.

Why is this Necessary?

This requirement responds to the Process Approach Principle.

It is uncertain why the authors of ISO 9001:2008 deemed it necessary to require any specific procedures or records when there is a general requirement for the system to be documented. This should have been sufficient. Clearly procedures are required so that people can execute tasks with consistency, economy, repeatability and uniformity but there is no logical reason why procedures are required for only six subjects. One of the dangers in requiring specific records is that it imposes an artificial limit.

How is this Demonstrated?

Documenting the Six Procedures

One solution is to produce a number of procedures that collectively address the six subjects as required. The question is, why would you want to do this when in all other aspects, you may have documented your processes? All these aspects are inherent in every process to some degree and will go under different names therefore rather than to use the labels designated in ISO 9001, look for situations where an activity has the same objective and reference this activity and the provisions devised for controlling it to one or more of the six required procedures (see also Chapters 12, 31, 33, 36 and 37).

Documented Records

Regarding the inclusion of records, the issue is a different one from that of procedures dealt with above. With procedures there is an implication that the auditors might be looking for six procedures bearing the title of the six topics, whereas with records it is fairly clear that what are being referred to are types of records and not records with these titles. One way of demonstrating compliance with this requirement is to match your records with those in the list, taking care to ensure your records fulfil the same purpose as those in the list. Where you identify gaps the related activity may or may not be applicable in your organization but if it is, your practices will need to change for you to be fully compliant.

DOCUMENTS THAT ENSURE EFFECTIVE PLANNING, OPERATION AND CONTROL OF PROCESSES (4.2.1d)

The standard requires management system documentation *to include documents (including records) determined by the organization to be necessary to ensure the effective planning, operation and control of its processes.*

What Does this Mean?

The documents required for effective planning, operation and control of the processes would include several different types of documents. Some will be product and process specific and others will be common to all processes. Rather than stipulate the documents that are needed, other than what is required elsewhere in the standard, ISO 9001 provides for the organization to determine what it needs for *the effective operation and control of its processes.* This phrase is the key to determining the documents that are needed.

There are three types of controlled documents:

- Policies and practices including process descriptions, guides, operating procedures and internal standards.

- Documents derived from these policies and practices, such as drawings, specifications, plans, work instructions, technical procedures, records and reports.
- External documents referenced in either of the above.

There will always be exceptions but in general the majority of documents used in a management system can be classified in this way.

Derived documents are those that are derived by executing processes, e.g., audit reports result from using the audit process, drawings result from using the design process, procurement specifications result from using the procurement process. There are, however, two types of derived documents: prescriptive and descriptive documents. *Prescriptive documents* are those that prescribe requirements, instructions, guidance etc. and may be subject to change. They have issue status and approval status, and are implemented in doing work. *Descriptive documents* result from doing work and are not implemented. They may have issue and approval status. Specifications, plans, purchase orders, drawings are all prescriptive whereas audit reports, test reports, inspection records are all descriptive. This distinction is only necessary because the controls required will be different for each class of documents.

Why is this Necessary?

This requirement responds to the Process Approach Principle.

The degree of documentation varies from a simple statement of fact to details of how a specific activity is to be carried out. To document *everything* you do would be impractical and of little value. Several good reasons for documenting information are listed under *What should be documented* at the beginning of this Chapter.

How is this Demonstrated?

The identification of documentation needs was addressed under the heading *Documenting a quality management system*. In this section, the specific types of documents are described in more detail.

Policies and Practices

Policies

Any statement made by management at any level that is designed to constrain the actions and decisions of those it affects is a *policy*. There does not have to be a separate document dedicated to policies. Policies can take many forms; the purpose and mission of the organization become a policy when expressed by the management, so do the principles or values guiding people's behaviour – what is or is not permitted by employees, whether or not they are managers. Policies are therefore essential in ensuring the effective planning of processes because they lay down the rules to be followed to ensure that actions and decisions taken in the design and operation of processes serve the business objectives and the needs and expectation of the stakeholders.

The policies can be integrated within any of the documents.

There are different types of policy that may impact the business processes:

- Government policy, which when translated into statutes applies to any commercial enterprise;

- Corporate policy, which applies to the business as a whole and expresses its intentions on particular strategic issues such as the environment, quality, financial matters, marketing, competitors etc.;
- Investment policy – how the organization will secure the future;
- Expansion policy – the way in which the organization will grow, both nationally and internationally;
- Personnel policy – how the organization will treat its employees and the labour unions;
- Safety policy – the organization's intentions with respect to hazards in the work place and to users of its products or services;
- Social policy – how the organization will interface with society;
- Operational policy, which applies to the operations of the business, such as design, procurement, manufacture, servicing and quality assurance. This may cover, e.g.:
  - Pricing policy – how the pricing of products is to be determined,
  - Procurement policy – how the organization will obtain the components and services needed,
  - Product policy – what range of products the business is to produce,
  - Inventory policy – how the organization will maintain economic order quantities to meet its production schedules,
  - Production policy – how the organization will determine what it makes or buys and how the production resources are to be organized,
  - Servicing policy – how the organization will service the products its customers have purchased,
- Department policy, which applies solely to one department, such as the particular rules a department manager may impose to allocate work, review output, monitor progress etc.;
- Industry policy, which applies to a particular industry, such as the codes of practice set by trade associations for a certain trade.

All policies set boundary conditions so that actions and decisions are channelled along a particular path to fulfil a purpose or achieve an objective. Many see policies as requirements to be met – they are requirements but only in so far as constraining an action or decision.

In organizations that have a strong value-based culture, policies are often undocumented. Rules are appropriate to a command and control culture. In all cases you need to ask, what would be the effect on our performance as an organization if this were not documented? If the answer is nothing or a response such as – well somebody might do xyz – forget it! Is it likely? If you cannot predict with certainty that something will happen that should be prevented, leave people free to choose their own path unless it's a legal requirement.

However, even in a command and control culture one does not need to write *everything* down, as policies are needed only for important matters where the question of right or wrong does not depend on circumstances at the time, or when the relevance of circumstances only rarely comes into the picture.

A common practice is to paraphrase the requirements of ISO 9001 as operational policy statements. Whilst this approach does provide direct correlation with ISO 9001 it renders the exercise futile because users can read the same things by referring to ISO 9001. Operational policies should respond to the needs of the organization, not paraphrase the standard.

Process Descriptions

Process descriptions are necessary in ensuring effective operation and control of processes because they contain or reference everything that needs to be known about a process. Applying the theory in Chapter 8, the essential elements would be as follows:

Process purpose – why the process exists expressed as what the process has been set up to do.
Process objectives – what the process aims to achieve as expressed in Table 11-1.

TABLE 11-1 Level of Detail on Process Objectives

Objective	Measure	Method	Target	Frequency of measurement	Responsibility
The measurable results the process is intended to deliver	The characteristics by which performance is judged (not the level of performance; this is the target value)	The method of measurement. For example, inspection, test, analysis, demonstration, simulation validation of records	The level of performance to be achieved. For example, standard, specification, requirement, budget, quota, plan	How often the measurements are taken	The person or role responsible for measuring performance

---

Process risk – what could go wrong in producing the required outputs within the designated constraints as expressed in Table 11-2.

TABLE 11-2 Level of Detail on Process Risk

Risk	Effect	Cause	Probability	Controls
The potential failure mode or hazard How might this part or process fail to meet the requirements? What could happen which would adversely affect performance? What would an stakeholder consider to be unacceptable?	The anticipated effect of this failure mode/hazard on the process outcome	The probable causes of the failure/hazard	The probability that this failure/ hazard could occur	The controls in place to prevent this failure/ hazard from occurring or the actions needed to eliminate, reduce or control this mode of failure

Process activators – what event, time or input triggers or energizes the process.

Process operation – what the process does expressed in a diagram or flow chart that identifies:

- **Inputs** expressed in terms of their nature and origin,
- **Activities** expressed as a sequence or series of actions and decisions necessary to plan, produce, check and rectify the process outputs together with assigned responsibilities. The criteria and methods are defined in work instructions when appropriate. Where there are exceptions, options or alternatives courses of action for particular cases these should be included but don't try to address all possible exceptions – apply the 80/20 rule;
- **Constraints** expressed in terms of the policies, regulations, codes of practice, codes of conduct and other conditions that govern the manner in which the activities are carried out;
- **Outputs** expressed in terms of their nature and destination.

Human resources – what the process needs to do what it is supposed to do, when expressed as the capacity and competency needed to achieve the process objectives (Table 11-3). See also Chapter 20.

TABLE 11-3 Level of Detail on Human Resources

Process stage	Results	Units of competence	Assessment method	Evidence required	Performance criteria	Number required
What action or decision must be must taken?	What must be achieved?	What must the person be able to do to produce this output?	How should assessment be conducted?	What evidence should be collected?	How well must this be achieved?	Number of people required with this competence at this stage

Physical resources expressed as the capacity and capability needed to achieve the process objectives in terms of the plant, facilities, machinery, floor space, instrumentation, monitoring equipment, calibration etc.

Process constraints expressed as the statutory and legal regulations, corporate policies and other conditions resulting from the analysis of critical success factors that constrain the manner in which the process operates, can be contained in tables with cross-references to other process elements.

Process reviews – how the process is controlled, expressed as:

- **Performance reviews** performed to an established schedule in accordance with defined methods to determine whether the process objectives are being achieved;
- **Improvement reviews** performed to an established schedule in accordance with defined methods to establish whether there are better ways of achieving the process objectives than those that are currently prescribed;
- **Effectiveness reviews** performed to an established schedule in accordance with defined methods to establish whether the process objectives, measures and targets remain aligned to the business objectives and stakeholder needs.

Clearly this description goes well beyond the content of a procedure. It also goes well beyond flow charts. Flow charts depict the steps in a process but do not fully describe a process. However, there should only be an arrow between the boxes if the activity is triggered by an input from the previous box. This is not always the case.

Don't try to include everything; the process description is a model of reality not a complete specification that has to explain everything. There may be aspects that are nice to include but make maintenance of the description very difficult. There may be many unknowns that can stay unknowns but going through the process of asking the questions will by itself reveal much about the process that you need to know to manage it effectively.

Procedures

Procedures are necessary in ensuring the effective operation and control of processes because they layout the steps to be taken in setting up, operating and shutting down a process.

A procedure is a sequence of steps to execute a routine task. It informs a user *how* a task should be performed. Procedures prescribe how one should proceed in certain circumstances in order to produce a desired output. Procedures are documented when it is important for the routines to be undertaken in the same way each time. Sometimes the word procedure can imply formality and a document of several pages but this is not necessarily so. A procedure can be of five lines, where each line represents a step in the execution of a task. (See also under the heading *Plans* below.)

Procedures can only work where judgment is no longer required or necessary. A procedure can specify the criteria to be met by a decision but will not make the decision; that is down to the individual weighing up the facts and making a judgement within their competence. Once you need to make a judgement, you cannot prescribe what you might or might not do with the information in front of you. A form of judgement-based procedure is a decision tree that flows down a chain of questions to which either a yes or a no will route you down a different branch. The chart does not answer your questions but is a guide to decision-making.

There remains confusion between processes and procedures and this is addressed in Chapter 8.

The relationship between processes and the defining documentations such as procedures, instructions, standards and guides is illustrated in Fig. 11-3.

Standards

Standards are essential in ensuring the effective operation and control of processes because they define the criteria required to judge the acceptability of the process capability and product quality.

Standards define the acceptance criteria for judging the quality of an activity, a document, a product or a service. There are national standards, international standards, standards for a particular industry and company standards. Standards may be in diagrammatic form or narrative form or a mixture of the two. Standards need to be referenced in process descriptions or operating procedures. These standards are in fact *your* quality standards. Product standards describe features and characteristics that *your* products and

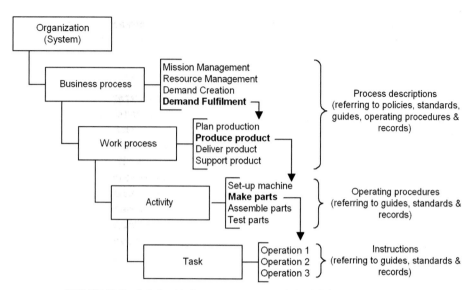

FIGURE 11-3 Relationship between processes and the defining documentation.

services must possess. Some may be type-specific, others may apply to a range of products or types of products and some may apply to all products whatever their type.

Process standards are essential for defining the acceptance criteria for a capable process.

Guides

Guides are necessary in ensuring the effective operation and control of processes because they provide information of use during the preparation, operation, shut down and troubleshooting of processes.

Guides are aids to decision-making and to the conduct of activities. They are useful as a means of documenting your experience and should contain examples, illustrations, hints and tips to help staff perform their work as well as possible. Without such guides, organizations become vulnerable after laying off staff in a recession. Unless you capture best practice, staff take their experience with them and the organization looses its capability.

Derived Documents

There are several types of derived documents.

Specifications

These are crucial in ensuring the effective planning and control of processes because they govern the characteristics of the inputs and outputs and are thus used in process and product design and measurement.

Plans

These are crucial in ensuring the effective operation and control of processes because they layout the work that is to be carried out to meet the specification. Unfortunately the word 'plan' can be used to describe any intent, will or future action so that specifications, procedures and process descriptions could be called plans when they are part of what you intend to do. A plan is therefore a statement of the provisions that have been made to achieve a certain objective. It describes the work to be done and the means by which the work will be done and managed.

Reports

These are useful in ensuring the effective operation and control of processes because they contain information about the process or the product being processed. They may be used to guide decision-making both in the design and operation of processes and in product realization.

Records

These are essential in ensuring the effective operation and control of processes because they capture factual performance from which decisions on process performance can be made. The reason for establishing records is to provide information necessary for managing processes, meeting objectives and demonstrating compliance with requirements – both customer requirements and legal requirements.

Records are defined in ISO 9001 as *documents stating the results achieved or providing evidence of activities performed.* Records are therefore produced during an event or immediately afterwards. Records do not arise from contemplation. They contain facts, the raw data as obtained from observation or measurement and produced manually or automatically.

There is no requirement to produce records solely to satisfy an auditor. The records concerned are those for the effective operation of the organization's processes. If a record has no useful purpose within the management system, there is no requirement that it be established or maintained.

Previously we have shown that there are 19 specific types of record required by the standard. Such an analysis is only useful to illustrate that not all records required for effective quality management are identified in ISO 9001. It is unnecessary to produce a record testifying that each requirement of ISO 9001 has been met. For example there is no obvious benefit from *maintaining* records that a document is legible (Clause 5.5.6e), that a customer enquired about some information that was later provided (Clause 7.2.3) or that a particular filing system was improved (Clause 8.1). At the time, the facts may be recorded temporarily but disposed of as the event passes into history. Such information may not be needed for future use. However, there are obvious benefits from requiring records to be established for:

- Customer complaints,
- Warranty claims,
- Failure in analysis reports,
- Process capability studies,
- Service reports,
- Concessions,
- Change requests,
- Subcontractor assessments,
- Performance analysis,

- Deviations and waivers,
- Contract change records,
- Quality cost data,
- External quality audit records.

Instructions

These are crucial in ensuring the effective operation and control of processes because they cause processes to be initiated and define variables that are specific to the date and time, location, product or customer concerned. Work instructions define the work required in terms of who is to perform it, when it is to commence and when it is to be completed. They also include what standard the work has to meet and any other instructions that constrain the quality, quantity, delivery and cost of the work required. Work instructions are the product of implementing an operating procedure or a document standard (see further explanation below).

In simple terms, *instructions* command work to be done, *procedures* define the sequence of steps to execute the work to be done. Instructions may or may not refer to procedures that *define how an activity is performed.* In some cases an instruction might be a single command such as 'Pack the goods'.

For example, you may issue an instruction for certain goods to be packed in a particular way on a specified date and the package to be marked with the contents and the address to which it is to be delivered. So that the task is carried out properly you may also specify the methods of packing in a *procedure*. The procedure would not contain specific details of a particular package – this is the purpose of the instruction. The procedure is dormant until the instruction to use it is initiated or until personnel are motivated to refer to it.

Internal References

These documents are useful in ensuring the effective operation and control of processes because they will contain data relevant to the equipment, people, facilities or other factors on which set-up or operation of the process depends. Reference documents differ from other types of documents in that they should be neither prescriptive nor instructional. They should not be descriptive like reports, proposals or records but should contain data that is useful in carrying out a task.

External Reference Documents

These documents are those not produced by the organization but used by the organization as a source of information, consequently the categories of external reference documents are identified by their source. There are several types including National and international standards

- Public data,
- Customer data,
- Supplier data,
- Industry data.

Competence and Documentation

Competence may depend on the availability of documentation. For example, a designer will refer to data sheets to assist in selecting components not because of a lack of competency but because of a person's limited memory and a desire for accuracy. The designer can remember where to look for the relevant data sheets, but not the details. If the document containing the relevant data cannot be found, the designer is unable to do the job and therefore cannot demonstrate competence.

When personnel are new to a job, they need education and training. Documentation is needed to assist in this process for two reasons. Firstly, to make the process repeatable and predictable and secondly to provide a memory bank that is more reliable than the human memory. As people learn the job they begin to rely less and less on documentation to the extent that in some cases, no supporting documentation may be used at all to produce the required output. This does not mean that once the people are competent you can throw away your documentation. It may not be used on a daily basis, but you will inevitably have new staff to train and improvements to make to your existing processes. You will then need the documentation as a source of information to do both.

Chapter 12

Document Control

CHAPTER PREVIEW

This chapter is aimed primarily at those with responsibility for controlling the organization's documentation or involved with the use, review, approval or disposal of documentation.

In the days when documents were only available in paper form, document control was an aspect that attracted a great deal of attention from internal and external auditors. Ensuring everyone had access to and was working with the correct version of a document was not easy. In the modern organization these issues have been overcome by documents being held and maintained on an electronic database that enables users to access the authorized version at their workstation or from other locations. Nonetheless, the technology may have eliminated some problems but inevitably creates new ones such as, how do you know that what you are looking at is the full version, what happens when the system crashes or if the document doesn't display on your terminal and of course what happens when you need a document in a place where there is no computer access?

In this chapter we show you how to overcome these and other issues and examine the requirements in Clause 4.2.3 of ISO 9001:2008 for controlling the documentation that was the subject of Chapter 11. In particular we address:

- Controlling documents required for the management system;
- Document control procedures;
- Document review, approval and re-approval after change;
- Revision of documents;
- Identifying changes;
- Identifying the current revision of documents;
- Ensuring the availability of controlled documents;
- Ensuring documents are legible and identifiable;
- Control of external documents;
- Preventing unintended use of obsolete documents.

CONTROL OF DOCUMENTS (OTHER THAN RECORDS) (4.2.3)

Controlling Documents Required for the Management System (4.2.3)

The standard requires *documents required by the quality management system to be controlled*.

249

What Does this Mean?

Documents required by the management system are those documents used by or generated by a process that forms part of the management system. The documentation used and generated by a process could include any of the documents identified in Chapter 11. However, records are subject to different controls from the other types of documents because they are time-related and once produced they must not be changed unless they contain errors (see *Control of records*).

The term document should be taken to include data or any information that is recorded and stored either on paper or magnetic media in a database or on a disk. It may be both an audio and visual record although the controls that will be applied will vary depending on the media. There is often confusion between quality system documents and quality documents and also between technical documents and quality documents. Such fine distinctions are unnecessary. Whether the document has the word *quality* in its title is irrelevant. The only question of interest is 'Is the document used or generated by this process?' If the answer is 'yes' the document should be controlled in some respects. Notes that you make when performing a task and then discard are not documents generated by a process – they are merely an aide-memoir.

Controlling documents means regulating the development, approval, issue, change, distribution, maintenance, use, storage, security, obsolescence or disposal of documents. We will call these the 'document functions'. A controlled document is a document for which one or more of the document functions are controlled. Therefore, if document security is controlled but not changed, the subject document can still be classed as being controlled.

Why is this Necessary?

This requirement responds to the Process Approach Principle.

It is necessary to regulate the documentation to ensure that:

- documents fulfil a useful purpose in the organization;
- resources are not wasted in the distribution of non-essential information;
- only valid information is used in the organization's processes;
- people have access to appropriate information for them to perform their work;
- information is kept up-to-date;
- information is in a form that can be used by all relevant people;
- classified information is restricted to only those with a need to know;
- information is important for the investigation of problems; improvement opportunities or potential litigation is retained.

How is this Demonstrated?

For those using document control software with the documents located on a secure server, many of the recommendations in this part of the book may seem unnecessary. However, even proprietary software claiming to meet the requirements of ISO 9001 for document control may not contain all the features you need. There are also many organizations that still use paper for good reasons. Paper does not crash without warning. Paper can be read more easily at a meeting or on a train. Comments can be added more easily to paper.

In the world of documents there are two categories: those that are controlled and those that are not controlled. You do not need to exercise control over each of the document functions for a document to be designated as a controlled document. Controlling documents may be limited to controlling their revision. On the other hand, you cannot control the revision of national standards but you can control their use, storage, obsolescence, etc. Even memoranda can become *controlled documents* if you impose a security classification on them.

The standard acknowledges that records are indeed documents but require different controls to those that apply to other documents.

In order to control documents, a document control process needs to be established that provides an adequate degree of control over all types of documents generated and used in the management system.

The process stages are common to all documents but the mechanisms for controlling different types of documents may differ. There are many software packages available that can be used to develop documents and control their issue, access, storage, maintenance and disposal. Unfortunately, few can handle all types. You may not wish to trust all your documentation to one package. The software has automated many of the procedural issues such that it is no longer the 'Achilles Heel' of a management system.

Document Control Procedures (4.2.3)

The standard requires that *a documented procedure be established to define the controls needed.*

What Does this Mean?

This requirement means that the methods for performing the various activities required to control different types of documents should be defined and documented in one or more procedures.

Why is this Necessary?

This requirement responds to the Process Approach Principle.

Documents are recorded information and the purpose of the document control process is to firstly ensure that the appropriate information is available wherever needed and secondly to prevent the inadvertent use of invalid information. At each stage of the process are activities to be performed that may require documented procedures in order to ensure consistency and predictability. Procedures may not be necessary for each stage in the process.

How is this Demonstrated?

Document Development and Maintenance

Every process is likely to require the use of documents or generate documents and it is in the process descriptions that you define the documents that need to be controlled. Any document not referred to in your process descriptions is therefore, by definition, not essential to the achievement of quality and not required to be under control. It is not necessary to identify uncontrolled documents in such cases. If you have no way of tracing documents to a governing process, a means of separating controlled from uncontrolled may well be necessary.

The procedures that require the use or preparation of documents should also specify or invoke the procedures for their control. If the controls are unique to the document, they should be specified in the procedure that requires the document. You can produce one or more common procedures that deal with the controls that apply to all documents. The stages in the process may differ depending on the type of document and organizations involved in its preparation, approval, publication and use. One procedure may cater for all the processes but several may be needed.

The aspects you should cover in your document control procedures (some of which are addressed further in this chapter) are as follows:

- planning new documents, funding, prior authorization, establishing need etc.;
- preparation of documents, who prepares them, how they are drafted, conventions for text, diagrams, forms, security etc.;
- standards for the format and content of documents, forms and diagrams;
- document identification conventions;
- issue notation, draft issues and post-approval issues;
- dating conventions, date of issue and date of approval or date of distribution;
- document review, who reviews them and what evidence is retained;
- document approval, who approves them and how approval is denoted;
- document proving prior to use;
- printing and/or publication, who does it and who checks it;
- distribution of documents, who decides, who does it and who checks it;
- use of documents, limitations, unauthorized copying and marking;
- revision of issued documents, requests for revision, who approves the request and who implements the change;
- denoting changes, revision marks, reissues, sidelining and underlining;
- amending copies of issued documents, amendment instructions, and amendment status;
- indexing documents, listing documents by issue status;
- document maintenance, keeping them current and periodic review;
- document accessibility inside and outside normal working hours;
- document filing, masters, copies, drafts, and custom binders;
- document storage, libraries and archive, who controls location and loan/access arrangements;
- document retention and obsolescence;
- document disposal/deletion, who does it and what records are retained.

With electronically stored documentation, the document database may provide many of the above features and may not need to be separately prescribed in your procedures. Only the tasks carried out by personnel need to be defined in your procedures but you need to know what the software does and how it works. A help file associated with a document database is as much a documented procedure as a conventional paper based procedure.

Document Security

While many of the controls are associated with developing new documents and managing change they address controls for causing the right things to happen. Controls are also needed to prevent, unauthorized changes, copying and disposal as well as computer viruses, fire and theft.

One solution used by many organizations is to publish their documents in a portable document format (PDF) as this provides built-in security measures. Users access the documents through a web browser or directly from a server with controlled access. The users can't change the document but may be permitted to print it and naturally, printed versions would be uncontrolled. Some organizations state this on the document but it is really unnecessary. Such a practice suggests the system has been documented for the auditors rather than the employees. However, you obviously need to make sure that everyone understands what the management system is and how to access and use the associated documents.

If original documents are available for users, inadvertent change can be a real problem. A document that has been approved might easily be changed simply because the 'current date' has been used in the approval date field. This would result in a change in the 'approval date' every time a user accesses the document.

Whatever the controls, they need rigorous testing to ensure that the documents are secure from unauthorized change.

> **You Never Know Who Might Access Your Documents**
>
> In auditing a particular management system the auditor was left alone while his host attended to a phone call. Using a remote terminal the auditor started to browse the file structure that was visible and opened a published FMEA report. He found that the data could be changed and the document saved to the server without any permissions being requested.

Document Approval (4.2.3a)

The standard requires that *documents be approved for adequacy prior to issue*.

What Does this Mean?

Approval prior to issue means that designated authorities have agreed the document before being made available for use. Whilst the term adequacy is a little vague, it should be taken as meaning that the document is judged as fit for the intended purpose. In a paper-based system, this means approval before the document is distributed. With an electronic system, it means that the documents should be approved before they are published or made available to the user community.

Published interpretation RFI 030 states that documents do not have to be reviewed as well as approved prior to issue because some degree of checking, examination or assessment by the person or persons approving is inherent in 'approval for adequacy'. It should also be noted that 'a review' is another look at something so in the context of 'prior to issue' the notion of a review is only relevant after a document is issued. Another published interpretation RFI 004 confirms that documented inspection and test procedures have to be *approved for adequacy prior to issue*.

Why is this Necessary?

This requirement responds to the Factual Approach Principle.

By subjecting documentation to an approval process prior to its use you can ensure that the documents in use have been judged by authorized personnel and found fit for purpose. Such a practice will also ensure that no unapproved documents are in circulation, thereby preventing errors from the use of invalid documents.

How is this Demonstrated?

Adequacy of Documents

The document control process needs to define the process by which documents are approved. In some cases it may not be necessary for anyone other than the approval authority to examine the documents. In others it may be necessary to set up a panel of reviewers to solicit their comments before approval is given. It all depends on whether the approval authority has all the information needed to make the decision and is therefore 'competent'.

Users should be the prime participants in the approval process so that the resultant documents reflect their needs and are fit for the intended purpose. If the objective is stated in the document, does it fulfil that objective? If it is stated that the document applies to certain equipment, area or activity, does it cover that equipment, area or activity to the depth expected of such a document? One of the difficulties in soliciting comments to documents is that you will gather comment on what you have written but not on what you have omitted. A useful method is to ensure that the procedures requiring the document specify the acceptance criteria[0] so that the reviewers and approvers can check the document against an agreed standard.

To demonstrate that documents have been deemed as adequate prior to issue, you will need to show that the document has been processed through the prescribed document approval process. Where there is a review panel, a simple method is to employ a standard comment sheet on which reviewers can indicate their comments or signify that they have no comment. During the drafting process you may undertake several revisions. You may feel it necessary to retain these in case of dispute later, but you are not required to do so. You also need to show that the current issue has been reviewed so your comment sheets need to indicate document issue status.

Approval Authorities

The standard no longer contains a specific requirement for documents to be approved by authorized personnel. The person approving a document derives his or her authority from the process. The process descriptions or procedures should identify who the approval authorities are, by their role or function, preferably not their job title and certainly not by their name because both can change. The procedure need only state that the document be approved, e.g., by the Chief Designer prior to issue. Another method is to assign each document to an owner. The owner is a person who takes responsibility for its contents and to whom all change requests need to be submitted. A separate list of document owners can be maintained and the procedure need only state that the Owner approves the document. It is not necessary for all approval authorities to be different from the author. You only need separate approval authorities where there is added value by having an extra pair of eyes (see text box in Chapter 6 on *Organizational freedom*).

Denoting Approval

The standard doesn't require that documents visibly display approval but it is necessary to be able to demonstrate through the process controls that the only documents that can be released into the user domain are approved. A process that is robust in this aspect will remove the need for approval signatures or other indications of approval because any document not bearing a certain designation will be judged invalid for use.

With paper-based systems, approval can be denoted directly on the document, on a change or issue record, in a register or on a separate approval record. All you need is a means of checking that the person who signed the document was authorized to do so. If below the signature you indicate the position of the person and require his or her name to be printed alongside his or her signature, you have exercised due diligence.

With electronic systems, indication of approval is accomplished by electronic signature captured by the software as a function of the security provisions. These can be set up to permit only certain personnel to enter data in the approval field of the document. The software is often not as flexible as paper-based systems and therefore provisions need to be made for dealing with situations where the designated approval authority is unavailable. If you let competency determine authority rather than position, other personnel will be able to approve documents because their electronic signature will provide traceability.

With most electronic file formats you can access the document properties from the Toolbar. Document properties can tell you when the document was created, modified, accessed and printed. But it can also tell you who the author was and who approved it provided the author entered this information before publication. Open the 'properties' of any document and you may get some surprises. The information is often inserted automatically and is not erased when a file is moved from one computer to another. If you use an old document as the basis for creating a new document you will carry over all the old document's properties so don't be surprised if you get a call one day questioning why you are using a competitor's information!

Issuing Documents

The term *issue* in the context of documents means that copies of the document are distributed. You will of course wish to *issue* draft documents for comment but obviously they cannot be reviewed and approved beforehand. The sole purpose of issuing draft documents is to solicit comments. The ISO 9001 requirement should have been that the documents are reviewed and approved prior to *use*. Some organizations insist that even drafts are approved for issue. Others go further and insist that copies cannot be taken from unapproved documents. This is nonsense and not what is intended by the standard. Your draft documents need to look different from the approved versions either by using letter issue notation (a common convention) or by printing the final versions with a watermark. If the approved document would carry signatures, the absence of any signature indicates that the document is not approved. With electronic systems, the draft documents should be held on a different server or in a different directory and provisions made to prohibit draft documents being published into the user domain.

Approving External Documents

The requirements for document approval do not distinguish between internal and external documents. However, there is clearly a need to review and approve external documents prior to their internal release in order to establish their impact on the organization, the product, the process or the management system. The external document control procedure should make provision for new documents and amendments to such documents to be reviewed and approved for use prior to their issue into the organization.

Approving Data

ISO 9000:2005 defines a document *as information and its support medium.* This means that databases containing contacts, problems, sales, complaints, inventory etc. are documents and yet we don't call them 'document bases'. We prefer the term database. The term data is not defined in ISO 9000:2005 but is commonly understood to be information organized in a form suitable for manual or computer analysis. When data is recorded it becomes information and should therefore be controlled. All data should be examined before use otherwise you may inadvertently introduce errors into your work. The standard does not require common controls for all information so you are at liberty to pitch the degree of control appropriate to the consequences of failure.

Regarding approval of data, you will need to define which data needs approval before issue as some data may well be used as an input to a document which itself is subject to approval. It all depends on how we interpret 'approved prior to issue'. This should be taken to mean 'issue to someone else'. Therefore, if you use data that you have generated yourself it does not need approval prior to use. If you issue data to someone else, it should be approved before distributing in a network database. If your job is to run a computer program in order to test a product, you might use the data resulting from the test run to adjust the computer or the program. You should be authorized to conduct the test and therefore your approval of the data is not required because the data has not in fact been issued to anyone else. The danger hiding in this requirement is that an eagle-eyed auditor may spot data being used without any evidence that it has been approved. As a precaution, ensure you have identified in your procedures those types of data that require formal control and that you know the origin of the data you are using.

Document Review (4.2.3b)

The standard requires that *documents be reviewed.*

What Does this Mean?

A review is another look at something. Therefore, document review is a task that is carried out at any time following the issue of a document.

Why is this Necessary?

This requirement responds to the Continual Improvement Principle.

Reviews may be necessary when:

- taking remedial action (i.e., correcting an error);
- taking corrective action (i.e., preventing an error recurring);

- taking preventive action (i.e., preventing the occurrence of an error);
- taking maintenance action (i.e., keeping information current);
- validating a document for use (i.e., when selecting documents for use in connection with a project, product, contract or other application);
- taking improvement action (i.e., making beneficial change to the information).

How is this Demonstrated?

Reviews may be random or periodic. Random reviews are reactive and arise from an error or a change that is either planned or unplanned. Periodic reviews are proactive and could be scheduled annually to review the policies, processes, products, procedures, specification etc. for continued suitability. In this way obsolete documents are culled from the system. However, if the system is being properly maintained there should be no outdated information available in the user domain. Whenever a new process or a modified process is installed, the redundant elements including documentation and equipment should be disposed.

Revision of Documents (4.2.3b)

The standard requires that *documents be updated as necessary and re-approved following their review.*

What Does this Mean?

Following a document review, action may or may not be necessary. If the document is found satisfactory, it will remain in use until the next review. If the document is found unsatisfactory there are two outcomes:

- The document is no longer necessary and should be withdrawn from use – this is addressed by the requirement dealing with obsolescence.
- The document is necessary but requires a change – this is addressed by this requirement.

The standard implies that updating should follow a review. The term *update* also implies that documents are reviewed only to establish whether they are current when in fact document reviews may be performed for many different reasons. A more appropriate term to *update* would be *revise*. A revision process is executed before a document is subject to re-approval.

Why is this Necessary?

This requirement responds to the Continual Improvement Principle.

It is inevitable that during use a need will arise for changing documents and therefore provision needs to be made to control not only the original generation of documents but also their revisions.

How is this Demonstrated?

The Change Process

The document change process consists of a number of key stages, some of which are not addressed in ISO 9001:

- identification of need (addressed by document review);
- request for change (not addressed in the standard);
- permission to change (not addressed in the standard);
- revision of document (addressed by document updates);
- recording the change (addressed by identifying the change);
- review of the change (addressed under the heading *Quality management system planning*);
- approval of the change (addressed by document re-approval);
- issue of change instructions (not addressed in the standard);
- issue of revised document (addressed by document availability).

As stated previously, to control documents it is necessary to control their development, approval, issue, change, distribution, maintenance, use, storage, security, obsolescence or disposal and we will now address those aspects not specifically covered by the standard.

What is a Change?

In controlling changes it is necessary to define what constitutes a change to a document. Should you allow any markings on paper copies of documents, you should specify those that have to be supported by change notes and those that do not. Markings that add comment or correct typographical errors are not changes but annotations. Alterations that modify instructions are changes and need prior approval. The approval may be in the form of a change note that details the changes that have been approved.

Depending on the security features of the PDF files it may be possible to annotate documents and save them locally. These are not changes, but the software may also allow text and object changes which are more serious and should be disabled in an approved document.

Request for Change

Anyone can review a document but approved documents should only be changed/ revised/amended under controlled conditions. The document review will conclude that a change is necessary; a request for change should be made to the issuing authorities. Even when the person proposing the change is the same as would approve the change, other parties may be affected and should therefore be permitted to comment. The most common method is to employ Document Change Requests. By using a formal change request it allows anyone to request a change to the appropriate authorities.

Change requests need to specify:

- the document title, issue and date;
- the originator of the change request (who is proposing the change, his or her location or department);
- the reason for change (why the change is necessary);
- what needs to be changed (which paragraph, section, etc. is affected and what text should be deleted);
- the changes in content required where known (the text or image which is to be changed, inserted or deleted).

effect than 20 minor changes. With small documents, say between three and six pages, it is often easier to reissue the whole document for each change.

Identifying Changes (4.2.3c)

The standard requires *that changes to documents be identified.*

What Does this Mean?

The requirement means that it should be possible to establish what has been changed in a document following its revision.

Why is this Necessary?

This requirement responds to the Factual Approach Principle.
   There are several benefits in identifying changes:

- Approval authorities are able to identify what has changed and so speed up the approval process.
- Users are able to identify what has changed and so speed up the implementation process.
- Auditors are able to identify what has changed and so focus on the new provisions more easily.
- Change initiators are able to identify what has changed and so verify whether their proposed changes were implemented as intended.

How is this Demonstrated?

There are several ways in which you can identify changes to documents:

- by sidelining, underlining, emboldening or similar technique;
- by a change record within the document (front or back) denoting the nature of change;
- by a separate change note that details what has changed and why;
- by appending the change details to the initiating change request.

If you operate a computerized documentation system, your problems can be eased by the versatility of the computer. Using a database you can provide users with all kinds of information regarding the nature of the change, but be careful. The more you provide, the greater the chance of error and the harder and more costly it is to maintain. Staff should be told the reason for change and you should employ some means of ensuring that, where changes to documents require a change in practice, adequate instruction is provided. A system that promulgates change without concern for the consequences is out of control. The changes are not complete until everyone whose work is affected by them both understands them and are equipped to implement them when necessary. Although not addressed under document control, the requirement for the integrity of the management system to be maintained during change in Clause 5.4.2 implies that changes to documents should be reviewed before approval to ensure that the compatibility between documents is maintained. When evaluating the change you should assess the impact of the requested change on other areas and initiate the corresponding changes in the other documents.

Identifying the Current Revision of Documents (4.2.3c)

The standard requires *the current revision status of documents to be identified.*

What Does this Mean?

When a document is revised its status changes to signify that it is no longer identical to the original version. This status may be indicated by date, by letter or by number or may be a combination of issue and revision. Every change to a document should revise the revision index. Issue 1 may denote the original version. On changing the document an incremental change to the revision index is made so that the new version is now Issue 2 or issue 1.1 depending on the convention adopted.

Why is this Necessary?

This requirement responds to the Factual Approach Principle.

It is necessary to denote the revision status of documents so that firstly, planners can indicate the version that is to be used and secondly, that users are able to clearly establish which version they are using or which version they require so as to avoid inadvertent use of incorrect versions.

How is this Demonstrated?

There are two aspects to this requirement. One is the identity denoted on the document itself and the other is the identity of documents referred to in other documents.

Revision Conventions

Changes may be *major* causing the document to be reissued or re-released, or they may be *minor* causing only the affected pages to be revised. You will need to decide on the revision conventions to use. Software documents often use a different convention to other documents such as release 1.1, or version 2.3. Non-software documents use conventions such as Issue 1, Issue 2 Revision 3, and Issue 4 Amendment 2. A convention often used with draft documents is letter revision status whereby the first draft is Draft A, second draft is Draft B and so on. When the document is approved, the status changes to Issue 1. During revision of an approved document, drafts may be denoted as Issue 1A, 1B etc. and when approved the status changes to Issue 2. Whatever the convention adopted, it is safer to be consistent so as to prevent mistakes and ambiguities.

Revision letters or numbers indicate maturity but not age. Dates can also be used as an indication of revision status but dates do not indicate whether the document is new or old and how many changes there have been. In some cases this is not important, but in others there are advantages in providing both date and revision status therefore denoting date and revision status is often the simplest solution.

Document Referencing

Staff should have a means of being able to determine the correct revision status of the documents they should use. You can do this through the work instructions, specification or planning documents, or by controlling the distribution, if the practice is to work to the latest issue. However, both these means have weaknesses. Documents can get lost, errors can creep into specifications and the cost of changing documents sometimes prohibits keeping them up-to-date. The issuing authority for each range of documents should maintain a register of documents showing the progression of changes that have been made since the initial issue. With configuration documents (documents which prescribe the features and characteristics of products and services), the relationship between documents of various issue states may be important. For example, a Design Specification at issue 4 may equate with a Test Specification at issue 3 but not with the Test Specification at issue 2. This record is sometimes referred to as a Master Record Index (MRI) but there is a distinct difference between a list of documents denoting issue state and a list of documents denoting issue compatibility state. The former is a Document Record Index and the later a Configuration Record Index. If there is no relationship between the document issues care should be taken to not imply a relationship by the title of the index.

The index may be issued to designated personnel or, so as to preclude the use of obsolete indices, it may be prudent not to keep hard copies. With organizations that operate on several sites using common documentation, it may well be sensible to issue the index so that users have a means of determining the current version of documents.

It is not necessary to maintain one index. You can have as many as you like. In fact if you have several ranges of documents it may be prudent to create an index for each range.

Regarding electronically controlled documents, arranging them so that only the current versions are accessible is one solution. In such cases and for certain type of documents, document numbers, issues and dates may be of no concern to the user. If you have configured the security provisions so that only current documents can be accessed, providing issue status, approval status, dates etc. adds no value for the user, but is necessary for those maintaining the database. It may be necessary to provide access to previous versions of documents. Personnel in a product-support function may need to use documentation for various models of a product as they devise repair schemes and perform maintenance. Often documentation for products no longer in production carries a different identity but common components may still be utilized in current models.

Re-approving Documents after Change (4.2.3b)

The standard requires that documents *be re-approved after revision.*

What Does this Mean?

Following a change the revised document needs to be subject to approval as verification of its fitness for purpose.

Why is this Necessary?

This requirement responds to the Factual Approach Principle.

As the original document was subject to approval prior to issue, it follows that any changes should also be subject to approval prior to issue of the revised version. The approval does not have to be by the same people or functions that approved the original although this may be the case in many situations. The criteria are not whether the people or functions are the same, but whether the approvers are authorized. Organizations change and therefore people and functions may take on different responsibilities.

How is this Demonstrated?

Depending on the nature of the change, it may be necessary to provide the approval authorities with factual information on which a decision can be made. The change request and the change record should provide this information. The change request provides the reason for change and the change note provides details of what has changed.

The change should be processed in the same way as the original document and be submitted to the appropriate authorities for approval. If approval is denoted on the front sheet of your documents, you will need to reissue the front sheet with every change. This is another good reason to use separate approval sheets. They save time and paper. With electronically controlled documents, archived versions provide a record of approvals provided they are protected from revision.

Ensuring the Availability of Controlled Documents (4.2.3d)

The standard requires *that relevant versions of applicable documents are available at points of use.*

What Does this Mean?

The relevant version of a document is the version that should be used for a task. It may not be the latest version because you may have reason to use a different version of a document such as when building or repairing different versions of the same product. Applicable documents are those that are needed to carry out work. *Availability at points of use* means the users have access to the documents they need at the location where the work is to be performed. It does not mean that users should possess copies of the documents they need, in fact this is undesirable because the copies may become outdated and not withdrawn from use.

Why is this Necessary?

This requirement responds to the Process Approach Principle.

This requirement exists to ensure that access to documents is afforded when required. Information essential for the performance of work needs to be accessible to those performing it otherwise they may resort to other means of obtaining what they need that may result in errors, inefficiencies and hazards.

How is this Demonstrated?

Availability

In order to make sure that documents are available you should not keep them under lock and key (or password protected) except for those where restricted access is necessary for security purposes. You need to establish who wants which documents and when they need them. The work instructions should specify the documents that *are* required for the task so that those documents not specified are not essential. It should not be left to the individual to determine which documents are essential and which are not. If there is a need for access out of normal working hours, access has to be provided. The more the copies the greater the chance of documents not being maintained so minimize the number of copies. Distribute the documents by location, not by named individuals. Distribute to libraries, or document control centres so that access is provided to everyone and so that someone has responsibility of keeping them up-to-date. If using an *Intranet*, the problems of distribution are less difficult but there will always be some groups of people who need access to hard copy.[1]

The document availability requirement applies to both internal and external documents alike. Customer documents such as contracts, drawings, specifications and standards need to be available to those who need them to execute their responsibilities. Often these documents are only held in paper form and therefore distribution lists will be needed to control their location. If documents in the public domain are required, they only need be available when required for use and need not be available from the moment they are specified in a specification or procedure. You should only have to produce such documents when they are needed for the work being undertaken at the time of the audit. However, you would need to demonstrate that you could obtain timely access when needed. If you provide a lending service to users of copyrighted documents, you would need a register indicating to whom they were loaned so that you can retrieve them when needed by others.

A document that is not ready for use or is not used often may be archived. But it needs to be accessible otherwise when it is called for it won't be there. It is therefore necessary to ensure that storage areas, or storage mediums provide secure storage from which documents can be retrieved when needed. Storing documents off-site under the management of another organization may give rise to problems if they cannot be contacted when you need the documents. Archiving documents on magnetic tape can also present problems when the tape cannot be found or read by the new technology that has been installed! Electronic storage presents very different problems to conventional storage and gives rise to the retention of 'insurance copies' in paper for use when the retrieval mechanism fails.

Relevant Versions of Internal Documents

A question often asked by Auditors is 'How do you know you have the correct issue of that document?' The question should not arise with an electronically controlled documentation system that prohibits access to invalid versions. If your system is not that sophisticated, one way of ensuring the latest issue is to control the distribution of documents so that each time a document changes, the amendments are issued to the same staff who received the original versions. If you identify authentic copies issued by the issuing authority in some way, by coloured header, red stamp or other means, it will be immediately apparent which copies are authentic and under control and which are uncontrolled. Another way is to stamp uncontrolled documents with an 'Uncontrolled Document' stamp. All paper documents should carry some identification as to the issuing authority so that you can check the current issue if you are in doubt. The onus

---

[1] Hoyle, David, (1996). ISO 9000 *Quality System Development Handbook,* Butterworth Heinemann.

should always rest with the user. It is the users' responsibility to check that they have the correct issue of a document before work commences. One way of signifying authenticity is to give document copy numbers in red ink as a practical way of retaining control over their distribution. If documents are filed in binders by part or volume, the binder can be given a copy number, but you will need a cross-reference list of who holds which copy.

Where different versions of the same document are needed, you will need a means of indicating which issue of which document is to be used. One method is to specify the pertinent issues of documents in the specifications, drawings, work instructions or planning documents. This should be avoided if at all possible because it can cause maintenance problems when documents change. It is sometimes better to declare that staff should use the latest version unless otherwise stated and provide staff with a means of determining what the latest version is.

Relevant Versions of External Documents

In some cases the issues of public and customer specific documents are stated in the contract and therefore it is important to ensure that you possess the correct version before you commence work. Where the customer specifies the issue status of public domain documents that apply you need a means of preventing their withdrawal from use in the event that they are revised during the term of the contract.

Where the issue status of public domain documents is not specified in a contract, you may either have a free choice as to the issue you use or, as more likely, you may need to use the latest issue in force. Where this is the case you will need a means of being informed when such documents are revised to ensure you can obtain the latest version. The ISO 9000 series, for instance is reviewed every five years so could well be revised at five-year intervals. With national and international legislations the situation is rather different because this can change at *any* time. You need some means of alerting yourself to changes that affect you and there are several methods from which to choose:

- subscribing to the issuing agency of a standards updating service;
- subscribing to a general publication which provides news of changes in standards and legislation;
- subscribing to a trade association which provides bulletins to its members on changes in the law and relevant standards;
- subscribing to the publications of the appropriate standards body or agency;
- subscribing to a society or professional institution that updates its members with news of changes in laws and standards;
- joining a business club which keeps its members informed of such matters;
- as a registered company you will receive all kinds of complementary information from government agencies advising you of changes in legislation.

As an ISO 9001 registered company you will receive bulletins from your certification body on matters affecting registration and you can subscribe to ISO Management Systems to obtain worldwide news of events and changes in the management systems arena.

The method you choose will depend on the number and diversity of external documents you need to maintain and the frequency of usage.

Issuing Change Instructions

If you require an urgent change in a document, a legitimate means of issuing change instructions is to generate a Document Change Note. The Change Note should detail the changes to be made and be authorized by the appropriate authorities. On receipt of the Change Note the recipients make the changes in the manuscript or by page replacement, and annotate the changes with the serial number of the Change Note. This practice primarily applies to paper systems but with electronically controlled documents, changes can be made to documents in a database without any one knowing and therefore it is necessary to provide an alert so that users are informed when a change has been made that may affect them. If the information is of a type that users invariably access rather than rely on memory, change instructions may be unnecessary.

Ensuring Documents are Legible and Identifiable (4.2.3e)

The standard requires documents *to remain legible and readily identifiable.*

What Does this Mean?

Legibility refers to the ease with which the information in a document can be read or viewed. A document is readily identifiable if it carries some indication that will quickly distinguish it from similar documents. Any document that requires a reader to browse through it looking for clues is clearly not readily identifiable.

Why is this Necessary?

This requirement responds to the Process Approach Principle.

The means of transmission and use of documents may cause degradation such that they fail to convey the information originally intended. Confusion with document identity could result in a document being misplaced, destroyed or otherwise being unobtainable. It can also result in incorrect documents being located and used.

How is this Demonstrated?

Legibility

This requirement is so obvious that it hardly needs to be specified. As a general rule, any document that is printed or photocopied should be checked for legibility before distribution. Legibility is not often a problem with electronically controlled documents. However, there are cases where diagrams cannot be magnified on screen so it would be prudent to verify the capability of the technology before releasing documents. Not every user will have 20:20 vision! Documents transmitted by fax present legibility problems due to the quality of transmission and the medium on which the information is printed. Heat-sensitive paper is being replaced with plain paper but many organizations still use the old technology. You simply have to decide your approach. For any communication required for reference, it would be prudent to use photocopy or scan the fax electronically and dispose of the original. Documents used in a workshop environment may require protection from oil and grease. Signatures are not always legible so it is prudent to have a policy of printing the name under the signature. Documents subject to frequent photocopying can degrade and result in illegible areas.

Identification

It is unusual to find documents in use that carry no identification at all. Three primary means are used for document identification – classification, titles and identification numbers. Classification divides documents into groups based on their purpose – policies, procedures, records, plans etc. are classes of documents. Titles are acceptable provided there are no two documents with the same title in the same class. If you have hundreds of documents it may prove difficult to sustain uniqueness. Identification can be made unique in one organization but outside it may not be unique. However, the title as well as the number is usually sufficient. Electronically controlled documents do not require a visible identity other than the title in its classification. Classifying documents with codes enables their sorting by class.[2] Some documentation platforms automatically assign a unique reference to each new document.

Control of External Documents (4.2.3f)

The standard requires *documents of external origin determined by the organization to be necessary for the planning and operation of the quality management system are identified and their distribution controlled.*

What Does this Mean?

An external document is one produced externally to the organization's management system. There are two types of external documents, those in the public domain and those produced by specific customers and/or suppliers. Controlling distribution means designating those who need external documents and ensuring that any change of ownership is known about and approved. The requirement limits the control to those external documents which the organization has determined are necessary rather than those that second or third parties might consider necessary.

Why is this Necessary?

External documents used by the organization are as much part of the management system as any other document and hence require control although the control that can be exercised over external documents is somewhat limited. You cannot for instance control the revision, approval or identification of external documents therefore all the requirements concerning document changes will not apply. You can, however, control the use and amendment of external documents by specifying which versions of external documents are to be used and you can remove invalid or obsolete external documents from use or identify them in a way that users recognize as invalid or obsolete. You can control the amendment of external documents by controlled distribution of amendment instructions sent to you by the issuing agency.

How is this Demonstrated?

Those external documents that are necessary for the planning and operation of the quality management system need to be identified within the documentation of the management system and this is not limited to policies and procedures but includes any

---

[2] Hoyle, David, (1996). ISO 9000 *Quality System Development Handbook,* Butterworth Heinemann.

product related documents such as drawings and specifications. External documents are likely to carry their own identification that is unique to the issuing authority. If they do not carry reference numbers the issuing authority is normally indicated which serves to distinguish them from internal documents. Where no identification is present other than a title, the document may be invalid. This sometimes happens with external data and forms. If the source cannot be confirmed and the information is essential, it would be sensible to incorporate the information into an appropriate internal document.

In order to control the distribution of external documents you need to designate the custodian in the appropriate process descriptions or procedures and introduce a mechanism for being notified of any change in ownership. If the external documents are classified, prior approval should be granted before ownership changes. This is particularly important with military contracts because all such documents have to be accounted for. Unlike the internal documents, many external documents may only be available in paper form so that registers will be needed to keep track of them. If electronic versions are provided, you will need to make them 'read only' and put in place safeguards against inadvertent deletion from the server.

Preventing Unintended use of Obsolete Documents (4.2.3g)

The standard requires *the unintended use of obsolete documents to be prevented and a suitable identification to be applied to obsolete documents retained for any purpose.*

What Does this Mean?

In simple terms an obsolete document is one that is no longer required for operational purposes. If an obsolete document has not been removed from the workplace there remains a possibility that it could be used. A suitable identification is one that readily distinguishes a current version of a document from an obsolete version.

Regrettably, the standard no longer refers to invalid documents as well as obsolete documents. Invalid documents may not be obsolete and may take several forms. They may be:

- documents of the wrong issue status for a particular task;
- draft documents which have not been destroyed;
- documents which have not been maintained up-to-date with amendments;
- documents which have been altered or changed without authorization;
- copies of documents which have not been authenticated;
- unauthorized documents or documents not traceable through the management system;
- illegal documents such as copies taken of copyright or classified documents without permission from the appropriate authority.

Why is this Necessary?

This requirement responds to the Process Approach Principle.

A means of distinguishing current documents from obsolete documents is needed to prevent their unintended use. Use of obsolete information may lead to errors, failures and hazards.

However, there are several reasons why you may wish to retain documents that are replaced by later versions:

- As a record of events (What did we do last time?);
- For verifying that the correct changes were made (What did it say in the last version or the first version?);
- For justifying rejection of a discarded solution (We've had this one before – what did we decide on that occasion?);
- For investigating problems that did not occur previously (What was it that we did differently?);
- For preparing a defence in a product liability case (What controls were in place when we made this product?);
- For learning from previous designs (How did we solve that problem?);
- For restoring or refurbishing a product (They don't make them like they used to – do they?);
- For reference to explanations, descriptions, translations etc. in order to preserve consistency in subsequent revisions or new documents (What wording did we use last time we encountered this topic or term?).

How is this Demonstrated?

With an electronic documentation system, access to obsolete documents can be barred to all except the chosen few. All it needs is for operational versions to be held in an operational directory and archived versions to be transferred into an archive directory automatically when a new version is released. On being transferred the revision status should be changed automatically to 'obsolete' indicating that later versions have been released.

With paper documents, it is more difficult. There are two options. You either remove the obsolete documents or mark them in some way that they are readily recognizable as obsolete.

It is unnecessary to remove invalid or obsolete documents if you provide staff with the means of determining the pertinent issues of documents to use. If you do not have a means of readily distinguishing the correct version of a document, amendment instructions should require that the version being replaced is destroyed or returned to the document controller. If you allow uncontrolled copies to be taken, you will need to provide a means of distinguishing controlled and uncontrolled documents.

One way of identifying obsolete documents is to write SUPERSEDED or OBSOLETE on the front cover, but doing this requires that the custodian be informed. When a version of a document is replaced with a new version, the withdrawal of the obsolete version can be accomplished in the amendment instructions that accompany the revision. When documents become obsolete by total replacement, their withdrawal can also be accomplished with the amendment instruction. However, where a document becomes obsolete and is not replaced, there needs to be a Document Withdrawal Notice which informs the custodian of the action to be taken and the reason for withdrawal.

There is no simple way of identifying invalid documents because the reasons that they are invalid will vary. The onus must rest with the user who if properly trained and motivated will instinctively refrain from using invalid documents.

With electronically controlled documents the invalid documents can be held in a database with access limited to those managing the documents. In such cases, an approved document becomes invalid when its status is denoted as draft or withdrawn.

CONTROL OF RECORDS (4.2.1e and 4.2.4)

Controlling Records (4.2.4)

The standard requires those records *established to provide evidence of conformity to requirements and the effective operation of the quality management system to be controlled.*

What Does this Mean?

A record is defined in ISO 9000:2005 as *a document stating results achieved or providing evidence of activities performed.* However, this clause does not require any records to be established, only that those that have been established for a particular purpose are controlled. The requirement for records is in Clause 4.2.1d.

Although a record is a document, the document control requirements of Clause 4.2.3 do not apply to *records* primarily because records are not issued, neither do they exhibit revision status simply because they are results that are factual when recorded. If the facts change, a new record is usually created rather than the previous record revised. Even where a record is revised and new facts added, the old facts remain identified as to their date. The only reason for revising facts contained in a record without changing the identity of the record or the date when they were collected is where the facts were incorrectly recorded. This subtle difference demands different treatment for documents that are classed as records to those that are classed as informative. As with other types of documents, records result from processes and may be used as inputs to other processes.

It is the purpose of the records that is important here. Records that have not been established to '*provide evidence of conformity to requirements*' or evidence of '*the effective operation of the quality management system*' would not need to be controlled. This means that in addition to all the records you might need to manage the organization and its operations, only certain ones are required to be controlled. (see Chapter 11 under *Documented procedures and records*).

Why is this Necessary?

This requirement responds to the Factual Approach Principle.

The reason for controlling records is to secure their access, integrity, durability and disposal. If information of this nature cannot be located, or is illegible, or its integrity is suspected, or is obsolete, the decisions that require this information cannot be made on the basis of fact and will therefore be unsound or not as robust as intended. Lost, altered or illegible records can result in an inability to prove or disprove liability.

How is this Demonstrated?

To demonstrate that your records are under control you would need to show that:

- Any record can be retrieved or show when it was destroyed;
- Records are up-to-date;

- The information in the records is up-to-date where applicable as some records may be designed to collect data as they pass through the process and need to be promptly updated with current information;
- Records in good condition (see under the heading *Protection and storage*);
- Records cannot be altered without authorization following their completion;
- Any alterations to records have been authorized.

The filing provisions should enable your records to be readily retrievable, however, you need to maintain your files if the stored information is to be of any use. In practice, records will collect at the place they are created and unless promptly transferred to secure files they may be mislaid, lost or inadvertently destroyed. Once complete, records should not be changed. If they are subsequently found to be inaccurate, new records should be created. Alterations to records should be prohibited because they bring into doubt the validity of any certification or authentication because no one will know whether the alteration was made before or after the records were authenticated. In the event that alterations are unavoidable due to time or economic reasons, errors should be struck through in order that the original wording can still be read, and the new data added and endorsed by the certifying authority.

Records held electronically present a different problem and is the reason why a requirement for the protection of records is introduced (see later).

Records that are established in order to:

a) *provide evidence of conformity to requirements and*
b) *provide evidence of the effective operation of the quality management system*

should be classified as *controlled records* so as to distinguish them from other records. If information needs to be recorded in order to manage the organization effectively, these records should be identified in the governing procedures as being required. This will then avoid arguments on what is or is not a controlled record, because once you have chosen to identify a record as a controlled record you have invoked all the requirements that are addressed in Clause 4.2.4.

Regarding the effectiveness of the management system, the very existence of a document is not evidence of its effectiveness but it could be regarded as a record. To be a *record*, the document would need to contain results of an examination into the effectiveness of the system.

One can demonstrate the effective operation of the management system in several ways:

- by examination of audited results against the organization's objectives;
- by examination of customer feedback;
- by examination of system, process and product audit results;
- by examination of the management review records;
- by examination of quality cost data.

Showing records that every requirement of the standard has been met will not demonstrate that the system is effective. You may have met the requirement but not carried out the right tasks or made the right decisions. The effectiveness of the management system should be judged by how well it fulfils its purpose. Although there is no specific requirement for you to do this, you can deduce this meaning from the requirement in Clause 5.6.1 for the system to be reviewed for its continuing suitability.

Some auditors may quote this requirement when finding that you have not recorded a particular activity that is addressed in the standard. They are not only mistaken but also attempting to impose an unnecessary burden on companies that will be perceived as bureaucratic nonsense. One can demonstrate the effectiveness of the system simply by producing and examining one or more of the above records.

The subcontractor records that are delivered to you should form part of your records. However, the controls you can exercise over your subcontractor's records are somewhat limited. You have a right to the records you have paid for but no more unless you invoke the requirements of this clause of the standard in your subcontract. Your rights will probably only extend to your subcontractor's records being made available for your inspection on their premises therefore you will not be able to take away copies. It is also likely that any subcontractor records you do receive are *copies* and not originals. Before placing the contract you will need to assess what records you require delivered and what records the contractor should produce and retain.

Establishing a Records Procedure (4.2.4)

The standard requires *a documented procedure to be established to define the controls needed for the identification, storage, protection, retrieval, retention and disposition of records.*

What Does this Mean?

Records have a life cycle. They are generated during which time they acquire an identity and are then assigned for storage for a prescribed period. During use and storage they need to be protected from inadvertent or malicious destruction and as they may be required to support current activities or investigations, they need to be brought out of storage quickly. When their usefulness has lapsed, a decision is made as to whether to retain them further or to destroy them.

Although the requirement implies a single procedure, several may be necessary because there are several unconnected tasks to perform. A procedure cannot ensure a result. It may prescribe a course of action which if followed may lead to the correct result, but it is the process that ensures the result not the procedure.

Customers may specify retention times for certain records as might regulations applicable to the industry, process or region in which the organization operates. It may be a criminal offence to dispose of certain records before the limit specified in law.

Why is this Necessary?

This requirement responds to the Process Approach Principle.

The requirement for a procedure is not as important as the topics the procedure is required to address. Records are important in the management of processes and building customer confidence; therefore it is important that the provision made for controlling them be defined in a way that ensures consistent results.

How is this Demonstrated?

Records Procedures

You may only need one procedure which covers all the requirements but this is not always practical. The provisions you make for specific records should be included in the documentation for controlling the activity being recorded. For example, provisions for inspection records should be included in the inspection procedures; provisions for design review records should be included in the design review procedure. Within such procedures you should provide the forms (or content requirement for the records), the identification, collection or submission provisions, the indexing and filing provisions. It may be more practical to cover the storage, disposal and retention provisions in separate procedures because they may not be type-dependent. Where each department retains their own records, these provisions may vary and therefore warrant separate procedures.

Storage of Records

Records soon grow into a mass of paper and occupy valuable floor space. To overcome this problem you may choose to scan the records into PDF format and archive them on a remote server. In both cases you need to control the process and the conditions of storage. With paper archives you will need to maintain records of what is where and if the server is under the control of another group inside or outside the organization, you will need adequate controls to prevent loss of identity and inadvertent destruction.

A booking in/out system should be used for completed records when they are in storage in order to prevent unauthorized removal.

You will need a means of ensuring that you have all the records that have been produced and that none are missing or if they are, you know the reason. One solution is to index records and create and maintain registers listing the records in numerical order as reference or serial numbers are allocated. The records could be filed in sequence so that you can easily detect if any are missing, or you can file the records elsewhere providing your registers or your procedures identify the location.

Records should also be stored in a logical order (filed numerically or by date) to aid retrieval. With electronically held records their storage should be secured from inadvertent deletion. If archived on CD ROM or a controlled server, protection methods should be employed (see below).

Protection of Records

The protection of records applies when records are in use and in storage and covers such conditions as destruction, deletion, corruption, change, loss and deterioration arising from wilful or inadvertent action.

On the subject of loss, you will need to consider loss by fire, theft, and unauthorized removal. If using computers, you will also need to consider loss through computer viruses and unauthorized access, deletion or the corruption of files.

It is always risky to keep only one copy of a document. If computer-generated, you can easily take another copy provided you always save it, but if manually generated, its loss can be very costly. It is therefore prudent to produce additional copies of critical records as an insurance against inadvertent loss. These 'insurance copies' should be stored in a remote location under the control of the same authority that controls the original records. Insurance copies of computer disks should also be kept in case of problems with the hard disk or the file server. Data back-up at defined periods should be conducted and the backed-up data stored securely at a different location than the original data.

Records, especially those used in workshop environments can become soiled and therefore provisions should be made to protect them against attack by lubricants, dust, oil and other materials which may render them unusable. Plastic wallets can provide adequate protection whilst records remain in use.

Retention of Records

It is important that records are not destroyed before their useful life is over. There are several factors to consider when determining the retention time for records.

- *The duration of the contract* – some records are only of value whilst the contract is in force.
- *The life of the product* – access to the records will probably not be needed for some considerable time, possibly long after the contract has closed. In some cases the organization is required to keep records for up to 20 years and for product liability purposes, in the worst-case situation (taking account of appeals) you could be asked to produce records up to 17 years after you made the product.
- *The period between management system assessments* – auditors may wish to see evidence that corrective actions from the last assessment were taken.

You will also need to take account of the subcontractor records and ensure that adequate retention times are invoked in the contract.

The document in which the retention time is specified can present a problem. If you specify it in a general procedure you are likely to want to prescribe a single figure, say five years for all records. However, this may cause storage problems – it may be more appropriate therefore to specify the retention times in the procedures that describe the records. In this way you can be selective.

You will also need a means of determining when the retention time has expired so that if necessary you can dispose of the records. The retention time doesn't mean that you must dispose of them when the time expires – only that you must retain the records for at least that stated period. Not only will the records need to be dated but also the files that contain the records need to be dated and if stored in an archive, the shelves or drawers also dated. It is for this reason that all documents should carry a date of origin and this requirement needs to be specified in the procedures that describe the records. If you can rely on the selection process a simple method is to store the records in bins or computer disks that carry the date of disposal.

While the ISO 9001 requirement applies only to records, you may also need to retain tools, jigs, fixtures, and test software – in fact anything that is needed to repair or reproduce equipment in order to honour your long-term commitments.

Should the customer specify a retention period greater than what you prescribe in your procedures, special provisions will need to be made and this is a potential area of risk. Customers may choose not to specify a particular time and require you to seek approval before destruction. Any contract that requires you to do something different creates a problem in conveying the requirements to those who are to implement them. The simple solution is to persuade your customer to accept your policy. You may not want to change your procedures for one contract. If you can't change the contract, the only alternative is to issue special instructions. You may be better off storing the records in a special contract store away from the normal store or alternatively attach special labels to the files to alert the people looking after the archives.

Disposition of Records

Disposition in this context means the disposal of records once their useful life has ended. The requirement should not be confused with that on the retention of records. Retention times are one thing and disposal procedures are quite another.

As stated previously, records are the property of the organization and not personal property so their destruction should be controlled. Controls should ensure that records are not destroyed without prior authorization and, depending on the medium on which data are recorded and the security classification of the data, you may also need to specify the method of disposal. The management would not be pleased to read details in the national press of the organization's performance, collected from a waste disposal site by a zealous newspaper reporter – a problem often reported as encountered by government departments and civic authorities!

Validation of Records

The standard does not specifically require records to be authenticated, certified or validated other than product verification records in Clause 8.2.4. A set of results without being endorsed with the signature of the person who captured them or other authentication lacks credibility. Facts that have been obtained by whatever means should be certified for four reasons:

- They provide a means of tracing the result to the originator in the event of problems.
- They indicate that the provider believes them to be correct.
- They enable you to verify whether the originator was appropriately qualified.
- They give the results credibility.

If the records are generated by computer and retained in computerized form, a means needs to be provided for the results to be authenticated. This can be accomplished through appropriate process controls by installing provisions for automated data recording or preventing unauthorized access.

Accessibility of Records (4.2.4)

The standard requires records *to remain legible, readily identifiable and retrievable.*

What Does this Mean?

Readily retrievable means that records can be obtained on demand within a reasonable period (hours not days or weeks). Readily identifiable means that the identity can be discerned at a glance.

Why is this Necessary?

Records are important in the management of processes and building customer confidence; therefore, it is important that they be accessible.

How is this Demonstrated?

Legibility of Records

Unlike prescriptive documents, records may contain handwritten elements and therefore it is important that the handwriting is legible. If this becomes a problem, you either improve discipline or consider electronic data capture. Records also become soiled in a workshop environment so may need to be protected to remain legible. With electronically captured data, legibility is often not a problem. However, photographs and other scanned images may not transfer as well as the original and lose detail so care has to be taken in selecting appropriate equipment for this task. Many organizations now scan records to store in portable document format (PDF) and make available over a network but there are variations in the quality of scanned images which can render the records illegible. Checks on image quality should therefore be performed before destroying the original.

Identification of Records

Whatever the records, they should carry some identification in order that you can determine what they are, what kind of information they record and what they relate to. A simple way of doing this is to give each record a reference number and a name or a title in a prominent location on the record.

Records can take various forms – reports containing narrative, computer data, and forms containing data in boxes, graphs, tables, lists and many others. Where forms are used to collect data, they should carry a form number and name as their identification. When completed they should carry a serial number to give each a separate identity. Records should also be traceable to the product or service they represent and this can be achieved either within the reference number or separately, provided that the chance of mistaken identity is eliminated. The standard does not require records to be identifiable to the product involved but unless you do make such provision you will not be able to access the pertinent records or demonstrate conformance to specified requirements.

Retrieving Records

You need to ensure that the records are accessible to those who will need to use them. This applies not only to current records but also to those in the archive and any 'insurance copies' you may have stored away. A balance has to be attained between security of the records and their accessibility. You may need to consider those who work outside normal working hours and those rare occasions when the trouble-shooters are working late, perhaps away from base with their only contact via a computer link. In providing record retrieval you need to consider two aspects. You need to enable authorized retrieval of records and prohibit unauthorized retrieval. If records are held in a locked room or filing cabinet, you need to nominate certain persons as key holders and ensure that these people can be contacted in an emergency. Your procedures should define how you provide and prohibit access to the records. With electronically held records, password protection will accomplish this objective provided you control the enabling and disabling of passwords in the records database. For this reason it is advisable to install a personnel termination or movement process that ensures that passwords are disabled or keys returned on departure of staff from their current post.

Remember these records are not personal property or the property of a particular department. They belong to the organization and are a record of the organization's performance. Such records should not be stored in personal files. The filing system you create should therefore be integrated with the organization's main filing system and the file location should either be specified in the procedure that defines the record or in a general filing procedure.

If you operate a computerized record system, filing will be somewhat different although the principles are the same as for paper records. Computerized records need to be located in named directories for ease of retrieval and the locations identified in the procedures.

Key Messages from Part 3

1. The system should be established to enable the organization to achieve goals that have been derived from the needs of stakeholders; meeting the requirements of ISO 9001 is secondary.

2. All organizations have a management system – a way of working, but in some it is not formalized – in others it is partially formalized and in a few organizations the management system really enables its objectives to be achieved year after year.

3. If you want to limit the scope of the system to customers simply omit the other stakeholders but you would probably still include all departments in the organization because all are engaged in one way or another in satisfying customers. Marketing attract customers, Sales secure orders from customers, Finance receive money from customers and pay suppliers, without which you would have no customers, HR develop the people who produce the products that satisfy customers, Reception greet customer when they visit, Maintenance keep the lights on and the place looking nice so that customers go away with a good impression – in fact you can't omit anyone from the system.

4. Processes exist in a functioning organization whether you deliberately set out to design them or whether you simply allow people to interact as they go about their work.

5. Delays create bottlenecks, increased demand creates shortages, budget cuts result in cost cutting which impact the quality of parts, materials and labour and so on. These are the interactions to which clause 4.1 of ISO 9001 refers.

6. If you establish process-oriented objectives, measures and targets, focused on the needs and expectations of external stakeholders, the functions will come into line and you will be able to optimize organizational performance.

7. An effective way to determine the processes is to derive the business outputs from an analysis of stakeholder needs then identify the processes that deliver these outputs. If you do it by working from the bottom upwards you may not find alignment between your processes and your stakeholder's needs.

8. If the result of a sequence of activities adds no value, continue the sequence until value is added for the benefit of customers – then you have defined a *business* or *work process*.

9. It is not a question of whether you have documented your processes but whether you have described how particular objectives are achieved or outputs produced.

10. If we think about it, what has size of the organization got to do with the amount of information you document?

11. When describing how an organization functions, it is easy to lose sight of your objective. Follow Stephen Covey's advice and *"Always begin with the end in mind"*.

12. Apart from those aspects where there is a legal requirement for documentation, the rest is entirely at the discretion of management but not all managers will see things the same way. Some will want their staff to follow rules and others will want their staff to use their initiative.

13. Ignore the requirements for specific procedures and records in ISO 9001. Keep the end in mind, don't produce more documentation that is needed to manage the system, its processes and the outputs.

14. The system you document will be a model of the system you perceive as delivering the organization's outcomes. It will not be the system itself, for that is the organization and all the interactions.

15. Describe your system in a way that makes it easy for the users of the documentation to learn, control and improve. Don't put in anything just to satisfy external auditors – always satisfy your needs first.

16. As information can now be available at the workplace at the click of a button the old problems with document control have been replaced with new ones. Computer viruses and security, corrupted storage media that can't be accessed so don't let the new technology make you complacent.

17. Implementing the management system is not about following procedures. It is about operating the processes as they were designed to operate.

18. Managing processes means managing activities, resources and behaviours to achieve prescribed objectives.

19. Monitoring, measuring, and analysing processes means being aware of what processes are doing, how they are behaving, what the interactions are and whether prescribed objectives and targets are being achieved.

20. Maintaining the management system is not maintaining the documentation. It is about maintaining the organization's capability to satisfy its customers.

21. Outsourced processes should not be out of sight – they require just as much supervision, if not more, than when they were in-house particularly ensuring the suppliers share the same values and give the same priority as you would to issues affecting your customers.

Complying with ISO 9001 Section 5 Requirements on Management Responsibility

INTRODUCTION TO PART 4

Structure of ISO 9001 Section 5

While conformity to *all* requirements in ISO 9001 is strictly management's responsibility, those in Section 5 of the standard are indeed the responsibility of top management. All clauses in this section commence with the phrase "Top management shall …". The first four clauses clearly apply to the strategic planning processes of the organization rather than to specific products. However, it is the board of directors that should take note of these requirements when establishing their vision, values, mission and objectives. These requirements are amongst the most important in the standard. There is a clear linkage between customer's needs, policy, objectives and processes. One leads to the other in a continuous cycle as addressed previously under *Establish the goals* in Chapter 10. Although the clauses in Section 5 are not intended as a sequence, each represents a part of a process that establishes direction and keeps the organization on course.

Linking Requirements

If we link the requirements together in a cycle (indicating the headings or topics from ISO 9001 in bold italics type) the cycle commences with a **Mission** – an expression of purpose, what the organizations exists to do, and then a **Focus on customers** for it is the customer that will decide whether or not the organization survives. It is only when you know what your market is, who your customers will be and where they will be that you can define the **Vision** of the organization (what success will look like) and from the vision come the **Policies or Values** that will guide you on your journey. These policies help frame the **Objectives**, the milestones en route towards your destination. The policies won't work unless there is **Commitment** so that everyone pulls in the same direction. **Plans** have to be made to achieve the objectives and these plans need to identify and layout the **Processes** that will be employed to deliver the results – for all work is a process and without work nothing will be achieved. The plans also need to identify the **Responsibilities and Authority** of those who will be engaged in the endeavour. As a consequence it is essential that effective channels of **Internal Communication** be established to ensure that everyone understands what they are required to achieve and how they are performing. No journey should be undertaken without a means of knowing where you are, how far you have to go, what obstacles are likely to lie in the path ahead or what forces will influence your success. It is therefore necessary to collate the facts on current performance and predictions of what lies ahead so that a **Management Review** can take place to determine what action is required to keep the organization on course or whether any changes are necessary to the course or the capability of the organization for it to fulfil its purpose and mission – and so we come full circle. What the requirements of Section 5 therefore address is the mission management process with the exception of process development, which happens to be addressed in Section 4 of the standard.

Chapter 13

Management Commitment

CHAPTER PREVIEW

This Chapter is aimed primarily at top management whose commitment is necessary for the management system to enable the organization to achieve sustained success. It is also applicable to any manager responsible for delivering business outputs or for results that impact business outputs.

As explained in the *Preview* to Chapter 10, a strategic decision does need to be made to adopt a particular approach to the management of quality and whether to use ISO 9001:2008 as a means to demonstrate its commitment to quality to its stakeholders. If it has been decided to use ISO 9001:2008 in this way, it is important that these decision makers understand what such a commitment involves. It is not simply a matter of delegating the task of developing a quality management system to a project manager or quality manager. What you would be doing by such an action would be delegating authority for how the organization should be managed and that is probably not what you intended.

In Part 1 we explained what the ISO 9000 family of standards was all about and by now you should be aware that it is primarily about achieving sustained success, not simply producing a series of documents that describes what you do or getting a certificate of conformity to ISO 9001. In Part 2 we described a number of approaches for achieving sustained success all of which dwelt on particular aspects that collectively would, if supported, lead to sustained success. These approaches go beyond what is required to meet the requirements of ISO 9001 simply because ISO 9001 assessments only determine whether your organization has the capability to satisfy the requirements of your customers. In order to achieve sustained success top management has to demonstrate its commitment to maintaining a capability of meeting the needs and expectations of all stakeholders not just its customers.

In this chapter we examine the requirements in Clause 5.1 of ISO 9001:2008 and in particular we explain:

- What top management means in the context of ISO 9001?
- What commitment means in general and in the context of a management system?
- Why commitment from top management is essential?
- What commitment involves on a day to day basis?

Several of the requirements in Clause 5.1 are duplicated elsewhere in the standard and for a more detailed explanation on these topics refer to the chapters denoted in brackets.

- Both Clause 5.2 and 5.3 require top management to establish a quality policy (see Chapter 15).
- Both Clause 5.2 and 5.4.1 require top management to establish quality objectives (see Chapter 16).
- Both Clause 5.2 and 5.6 require top management to conduct management reviews (see Chapter 18).

COMMITMENT TO THE QMS (5.1)

The standard requires that *top management provides evidence of its commitment to the development and implementation of the quality management system and continually improving its effectiveness.*

What Does this Mean?

Top management

Throughout this section of the standard the term *top management* is used which brings the actions and decisions of top management into the development and operation of the quality management system and makes them full partners in its success. Top management is defined in ISO 9000:2005 as *the person or group of people who direct and control an organization* and therefore sit at the top of the organization tree.

Commitment

A commitment is an obligation that a person (or a company) takes on in order to do something. It is very easily tested by examination of the results. It is also easy for you to make promises to resolve immediate problems hoping that the problem may go away in due course. This is dishonest and although you may mean well, the problem will return to haunt you if you can't deliver on your promise.

A commitment exists if a person agrees to do something and informs others of their intentions. A commitment that is not communicated is merely a personal commitment with no obligation except to one's own conscience.

Commitment therefore means:

- Doing what you need to do to meet the organization's objectives;
- Doing what you say you will do;
- Not accepting work below standard;
- Not shipping product below standard;
- Not walking by problems, not overlooking mistakes;
- Improving processes;
- Honouring plans, procedures, policies and promises;
- Listening to the workforce;
- Listening to the stakeholders.

Commitment to a Management System

Although the standard does indicate how commitment to a management system should be demonstrated, it presupposes that top management understands what it is committing itself to. There is a presumption that a management that is committed to the development of a management system will be committed to quality because it believes that the management system is the means by which quality will be achieved. This is by no means obvious because it depends on top management's perception of quality and the role of the management system in its achievement. (This argument was also addressed with respect to Clause 0.1 of the standard that we discussed in Chapter 5 under the heading *Levels of Attention*). Many managers have given a commitment to quality without knowing what impact it would have on their business. Many have given a commitment without being aware that their own behaviour may need to change let alone the behaviour of their managers and staff. The phrase 'continually improving its effectiveness' clearly implies that top management has to ensure that the management system continually fulfils its purpose and hence need to understand what its purpose is. A study of Clauses 4.1 and 5.1 together with the definitions in ISO 9000:2005 should leave one in no doubt that the purpose of the management system is to enable the organization to satisfy its customers. Expressing a commitment to the development and improvement of the management system therefore means a belief that the management system is the means by which quality is achieved and that resources will be provided for its development, implementation, maintenance and improvement.

The perceptions of quality may well be customer focused but the managers may perceive the management system as only applying to the quality department and consequently may believe that a commitment to the development of a management system implies that they have to commit to maintaining a quality department. This is not what is intended. A quality department is the result of the way work is structured to meet the organization's objectives. There are other solutions that would not result in the formation of a department dedicated to quality and there is certainly no requirement in ISO 9001 for a department that is dedicated to quality.

Why is this Necessary?

This requirement responds to the Leadership Principle.

Management must be seen to do what it says it will do so as to demonstrate its understanding. There is no doubt that actions speak louder than words for it is only when the words are tested that it is revealed whether the writers were serious. It is not about whether management can be trusted but whether they understand the implications of what they have committed themselves to. The message is that the management system is supposed to make the right things happen and not simply be regarded as a set of procedures.

How is this Demonstrated?

Defining Top Management

In order to clarify who in the organization is a member of top management, it will be advantageous to specify this in the Quality Manual. It is then necessary to ensure that the positions of the personnel performing the specified actions are from top management.

A search of Section 5 of the standard will reveal that Top management are required to:

- provide evidence of commitment to the development and improvement of the management system;
- communicate the importance of meeting customer requirements;

- establish the quality policy and ensure it meets prescribed criteria;
- ensure that customer needs and expectations are determined, converted into requirements and fulfilled to achieve customer satisfaction;
- establish quality objectives and ensure they are established at each relevant function
  and level;
- ensure that resources needed to achieve the objectives are identified, planned and are
  available;
- appoint members of management who shall have responsibility for ensuring the
  management system is established etc.;
- review the management system to ensure its continuing suitability and effectiveness.

Merging requirements in Sections 5.1–5.6 of the standard produced this list. It is
interesting that in some of the above references top management is required to *ensure*
something rather than *do* something. To ensure means to make certain and top
management can't make certain that something will happen unless it is in control.
However, it does mean that it can delegate to others the writing of the policy, the
objectives and provision of resources but it can't escape responsibility for them and the
way they are implemented and the results they achieve.

In some organizations, there are two roles, one of Management Representative and
another of Quality Manager with the former only being in top management. It is clear
that in such cases, the Quality Manager should not be the person carrying out these
actions but the person who is a member of top management.

Commitment

Once communicated, a commitment can be tested by establishing:

- if resources have been budgeted for discharging the commitment;
- that resources are allocated when needed;
- that performance of the tasks to which the person has given his or her commitment
  are progressed, monitored and controlled;
- that deviations from commitment are not easily granted.

In managing a management system, such tests will need to be periodically carried out
even though it will be tedious for both the person doing the test and the person being
subjected to it. It is less tedious if such tests are a feature of the programme that the
management has agreed to, thereby making it impersonal and by mutual consent.

The management has to be committed to quality, in other words it must not knowingly ship defective product, give inferior service or in any other way knowingly
dissatisfy its stakeholders. It must do what it says it will do and what it says it will do
must meet the needs and expectations of the stakeholders. A manager who signs off
waivers without customer agreement is not committed to quality whatever the reasons. It
is not always easy, however, for managers to honour all their commitments when the
customer is screaming down the phone for supplies that have been ordered or employees
are calling for promised pay rises. The standard now requires proof that managers are
committed to quality through their actions and decisions. When they start spending time
and money on quality, diverting people to resolve problems, motivating their staff to
achieve performance standards, listening to their staff and to customers, there is
commitment. It will also be evident from customer feedback, internal and external audits

and sustained business growth. Increased profits do not necessarily show that the company is committed to quality. Profits can rise for many reasons not necessarily because of an improvement in quality. Managers should not just look at profit results to measure the success of the improvement programme. Profits may go down initially as investment is made in management system development. If managers abandon the programme because of short-term results, it shows not only a lack of commitment but also a lack of understanding. Every parent knows that a child's education does not bear fruit until he or she is an adult. It is therefore much better to tailor the programme to available resources than abandon it completely.

Commitment to a Management System

Although the standard defines how top management should demonstrate its commitment, there are some obvious omissions. Commitment to the development and improvement of a management system could be demonstrated by evidence that top management:

- understands the role of the management system in relation to achieving business objectives and customer satisfaction;
- is steering management system development and improvement effort and monitoring its performance;
- is promoting core values, adherence to policy, best practice and continual improvement;
- undertakes the top management actions in a timely manner;
- does not defer decisions that are impeding progress in management system development and improvement;
- is implementing the processes that have been developed to achieve business objectives;
- is actively stimulating system improvements.

Chapter 14

Customer Focus

CHAPTER PREVIEW

This chapter is aimed primarily at top management and those engaged in market research, sales, order processing or in fact anyone who faces the customer and whose understanding of customer needs and expectations is critical to the performance of the organization.

Customer focus is one of the eight quality management principles (see Chapter 1) and means putting your energy into satisfying customers and understanding that profitability or avoidance of loss comes from satisfying customers. If people were to ask themselves before making a decision, what does the customer need or how will this decision affect our customers? – the organization would begin to move its focus firmly in the direction of its customers.

In this chapter we examine the requirements in Clause 5.2 of ISO 9001:2008 on customer focus and in particular the meaning, necessity and evidence needed for:

- Determining customer requirements;
- Communicating the importance of requirements;
- Meeting customer requirements;

The requirements for customer focus feature in all stages in the managed process of Fig. 8-11 because all activities should be customer focused.

DETERMINING CUSTOMER REQUIREMENTS (5.2)

The standard requires top management *to ensure that customer requirements are determined.*

What does this Mean?

We defined who an organization's customer is, and where the customers are in Chapter 3 and it is worth repeating that customers are not simply the person or organization that pay the invoice but include users as well as makers.

Determining customer needs and expectations to many is very different from determining customer requirements. The former implies that the organization should be pro-active and seek to establish customer needs and expectations before commencing the design of products and services and offering them for sale. The latter implies that the organization should react to the receipt of an order by determining what the customer wants. However, ISO 9000:2005 defines a requirement as *a need or expectation that is stated, customarily implied or obligatory*. We can therefore use the term *requirement* or the terms *needs, expectations and obligations* as synonymous. As there is no indication in the statement that such requirements are limited to those in an order or contract, it can be interpreted as requiring both proactive and reactive action.

The requirement is specific in that it requires top management *to ensure* customer requirements are determined – meaning that it has to ensure an effective process is in place for determining customer requirements.

Why is this Necessary?

This requirement responds to the Customer Focus Principle.

All organizations have customers. Organizations exist to create and retain satisfied customers – those that do not do so, fail to survive. Not-for-profit organizations have customers even though they may not refer to them as customers but whatever they call them, the people they serve may not purchase anything, but they give and they take and if the organization fails to fulfil their needs, it ceases to exist. Governments are a prime example. If they fail to satisfy the voters they fail to be re-elected. It is therefore essential for the survival of an organization that it determines customer requirements.

How is this Demonstrated?

The requirement is implemented through the Demand Creation Process. This requirement extends the management system beyond the processes required to satisfy current customers and clearly brings the marketing process as well as the sales process into the management system.

The marketing process is primarily concerned with finding out what customers want and attracting them to the organization so that these wants are satisfied. In this process it is important to keep the organization's purpose and mission in focus because all too easily, the organization may become entangled in pursuing opportunities that others may be far better equipped to satisfy – "stick to the knitting" as Tom Peters would say.[1] There are millions of opportunities out there. The key is to discover those that your organization can exploit better than any other and generate a profit by doing so.

The sales process is primarily concerned with making contact with customers for existing products and services and converting enquiries into firm orders. In this process, it is important that the customer requirements are determined so as to match the benefits of existing products and services with the needs and expectations of customers. The tender or contract review process is therefore important in ensuring that needs are understood before a commitment to supply is accepted.

Understanding Customer Needs

In order to determine customer needs and expectations you need to answer some important questions.

---

[1] Peters, Tom and Waterman, Robert, (1995). *In search of excellence*, Harper Collins.

- Who are our customers?
- Where are our customers?
- What do customers buy?
- What is value to the customer?
- Which of the customer's wants are not adequately satisfied?

The answers to the above questions will enable marketing objectives to be established for:

- existing products and services in present markets;
- abandonment of obsolete products, services and markets;
- new products and services for existing markets;
- new markets;
- service standards and service performance;
- product standards and product performance.

The results of market research will be a mix of things. It will identify:

- new potential customers for existing products and services;
- new potential markets;
- opportunities for which no technology exists;
- opportunities for which no product or service solution exists;
- enhancements to existing products and services.

The organization needs to decide which of these to pursue and this requires a process that involves all the stakeholders. A process for developing the marketing strategy that only involves the marketeers will not exploit the organization's full potential. The contributions from design, production, service delivery, legal and regulation experts are vital to formulating a robust set of customer requirements from which to develop new markets, new products and new services. The research may identify a need for improvement in specific products or a range of products, but the breakthroughs will come from studying customer behaviour. For example, research into telecommunications brought about the mobile phone and technology has reduced it in size and weight so that the phone now fits into a shirt pocket. Further research on mobile phones has identified enhancements such as access to e-mail and the Internet, even TV through the mobile phone but whether these are essential improvements is debatable. The fear of radiation and driving laws means that a breakthrough will arise by eliminating manual interaction so that the communicator is worn like a hat, glove or a pair of spectacles, being voice activated and providing total hands free operation.

Determining customer needs and expectations should not be limited to your present customer-base. Customers may want your products but may be unable to obtain them. If your products and services are limited to the home market either due to import regulations or distribution policies you could satisfy a new sector of the market with your existing products and extend their life. Austin Morris did this in the 1960s with their Morris Cowley and Oxford by exporting the technology to India.

From analysing the results of the research a design brief or requirement can be developed that translates customer needs and expectations into performance, physical and functional characteristics for a product or service. This forms the basis of the input into the product and service design processes. Often this requirement is no more than a couple of lines on a memorandum. In the 1970s, the Managing Director of Ferranti

Computer Systems issued an instruction to the design staff that a 16-bit digital computer was required. No more information was provided but in subsequent discussions the target customers (Royal Navy) and functional purpose (Command and Control Systems) were established and by further research the detail performance, physical and functional characteristics and regulatory standards were identified. The process of converting needs into requirements can therefore be quite protracted and iterative. With some projects, establishing requirements is a distinct phase that is put out for tender and where the winner may not necessarily win a subsequent contract to develop the product.

If you misunderstand customer needs and expectations you will produce an inadequate set of requirements, often not knowing they are inadequate until you launch the product into the market. It is therefore the most important stage in the demand creation process where ideas and beliefs are tested and retested to ensure that they really do reflect customer needs and expectations.

Teruyuki Minoura, who in 2003 was managing director of global purchasing for Toyota Motor Corporation warned that suppliers needed to shift their focus to the car user instead of the carmaker. "You are going to have to start analysing the needs and wants of the end user. You're going to be finding out what end users want and working to develop suitable components. Then you're going to be offering what you've developed to carmakers like us, who are going to incorporate these components into our designs."

Gathering the Data

Decisions affecting the future direction of the organization and its products and services are made from information gleaned through market research. Should this information be grossly inaccurate, over optimistic or pessimistic the result may well be the loss of many customers to the competition. It is therefore vital that objective data is used to make these decisions. The data can be primary data (data collected for the first time during a market research study) or secondary data (previously collected data). However, you need to be cautious with secondary data because it could be obsolete or have been collected on a different basis than needed for the present study.

The marketing information primarily identifies either *problems* or *opportunities*. *Problems* will relate to your existing products and services and should indicate why there has been a decline in sales or an increase in returns. In order to solve these problems a search for possible causes should be conducted and one valid method for doing this is to use the Fishbone Diagram or Cause and Effect Diagram an example of which was shown in Fig. 2-2. *Opportunities* will relate to future products and services and should indicate unsatisfied wants. There are three ways of collecting such data: by observation, survey and by experiment.

Observation studies are conducted by actually viewing the overt actions of the respondent. In the automotive industry this can either be carried out in the field or in the factories where subcontractors can observe their customer using their materials or components.

Using surveys is the most widely used method for obtaining primary data. Asking questions that reveal their priorities, their preferences, their desires, their unsatisfied wants etc. will provide the necessary information. Information on the profile of the ultimate customers with respect to location, occupation, life style, spending power, leisure pursuits etc. will enable the size of market to be established. Asking questions about their supplier preferences and establishing what these suppliers provide that you don't provide is also necessary. Knowing what the customer will pay more for is also necessary, because many may expect features that were once options, to be provided as standard.

A method used to test the potential of new products is the *controlled experiment* – using prototypes, alpha models etc. distributed to a sample of known users. Over a limited period these users try out the product and compile a report that is returned to the company for analysis.

A source of secondary data can be trade press reports and independent reviews. Reading the comments about other products can give you some insight into the needs and expectations of potential customers.

Case Study ~Defining Customer Needs

We design and manufacture products for the computer industry but sell only proprietary products. When a customer places an order with us, we establish that we understand the requirements and have the capability to meet them before accepting the order. However, in ISO 9001 we note that we may have to determine customer needs and expectation as well. Does this mean that we should bring our marketing function within the QMS or could we be compliant merely by extending our contract review procedure to gather data on customer needs and expectations as well as customer requirements?

You can't be compliant simply by extending your contract review procedure.

The answer depends upon what the Marketing function does, what you define as the scope of the QMS and what you want to be assessed.

Understanding future customer needs and expectations is obviously essential for you to determine what opportunities there are in the market for your type of business. Understanding current customer needs and expectations is necessary for you to satisfy current customers so it's not a question of whether or not you should do either of these. Extending the contract review procedure will only meet part of the requirement.

Your QMS does need to include the processes necessary to determine current and future customer needs and expectations to be compliant with clause 5.1 of the standard.

Your Marketing function may have responsibility for establishing the needs and expectations of future customers and therefore its activities in this respect would need to be brought within the QMS. But your Marketing function may have other responsibilities such as sales promotion, public relations and market research broader than simply customer preferences. It may for example conduct research into new technology, the economic climate, legislation, trading conditions and social changes. If the scope of the QMS is the scope of the organization, your QMS would include all activities of the Marketing function.

However, while the ISO 9000 family addresses all stakeholders, the scope of ISO 9001 is limited to customers and although they are most important stakeholder, they are only one. You therefore have a choice:

Extend the QMS so that it includes all functions and submit all functions to the external audit

Extend the QMS so that it includes all functions but limit the scope of registration to those activities directly concerned with satisfying customer requirements

Limit the QMS to those activities directly concerned with satisfying customer requirements but why would you want to do this unless of course you were of the opinion that the only purpose of a QMS is to gain and retain ISO 9001 certification

The external auditors will want you to identify the process in which current and future customer needs and expectations are determined. They will also want to know who does this and if it is split between Marketing and Sales function they will audit them both for this and any other activities that fall within the scope of ISO 9001.

COMMUNICATING THE IMPORTANCE OF REQUIREMENTS (5.1a)

The standard requires that *top management communicate to the organization the importance of meeting customer as well as statutory and regulatory requirements.*

What Does this Mean?

Communication

This requirement is the primary application of the *customer focus principle*. Effective communication consists of four steps: attention, understanding, acceptance and action. It is not just the sending of messages from one source to another. It is therefore not enough for top management to publish a quality policy that declares the importance of meeting customer, regulatory and legal requirements – this is only the first step and depending on how this is done it may not even accomplish the first step.

Statutory and Regulatory Requirements

Although the ISO 9000 family addresses all stakeholders, ISO 9001 focuses on customers and therefore the statutory and regulatory requirements referred to in this requirement are those pertaining directly or indirectly to the product or service supplied. However, there are few regulations that in some way would not impact the customer if not complied with. Customers not only purchase products and services but also often desire a sustaining relationship so that they can rely on consistent quality, delivery and cost. If the operations are suspended through non-compliance with environmental, health and safety legislation, existing customers will be affected and potential customers will be lost. The absence of key personnel due to occupational health reasons may well impact deliveries and relationships. Some customers require their suppliers to operate ethically and conserve the environment. If the financial laws are breached causing the organization to cut costs to pay the fines, credit worthiness may also be affected. Such breaches also distract management from focusing on customers; therefore it is not only the product related regulations that can affect customer satisfaction.

There is also a secondary meaning. Some customers may require the organization to provide products and services that do not comply with statutory and regulatory requirements.

Why is this Necessary?

This requirement responds to the Customer Focus Principle.

Top management need to attract the attention of the workforce and this requires much more than pinning-up framed quality policy statements all over the place. In organizations where the producers are remote from the customer, it is necessary to convey customer expectations down the line so that decisions are made primarily on the basis of satisfying these expectations and not on the secondary internal requirements. It is also necessary for management to signal its intentions with respect to customer requirements so that personnel know of the factors affecting the work priorities.

How is this Demonstrated?

Before embarking on a communication programme, top management should examine its culture relative to profit. If asked why the organization exists, many would say it is to make a profit – clearly indicating they are a profit-focused organization – not a customer-focused organization. Profits are important but are the *result* of what we do – not the *reason* for doing it. Profits are needed to pay off loans, for investment in new technology, new plant, and new skills and to award higher salaries but they come from satisfying customers and operating efficiently and effectively. Focusing on profit creates a situation where people constantly observe the bottom line rather than the means to achieve it so that most decisions are based on cost rather than quality. Managers in such organizations rule by fear and treat a lack of capability as an excuse rather than a valid reason why targets are not met. It is right to question the value of any activity but with the focus on customer satisfaction not profit. It is vital therefore that top management consider the culture they have created before communicating the importance of customer require-ments. Where profit is the first priority, stressing the importance of customer require-ments will create a conflicting message. Top management needs to make employees understand that the priorities have changed so that the aim is to satisfy customers profitably.

If the customer's requirements cannot be satisfied profitably, business should not be transacted unless there are opportunities for recovering the loss on another transaction. It is also important that before accepting a commitment to supply, a clear understanding of customer needs and expectations as well as regulatory and legal requirements is established – hence sales personnel need to understand that they should not make promises which the organization cannot honour.

Meeting this requirement may therefore require a realignment of priorities and careful consideration as to the way management will attract the attention, understanding, acceptance and correct action of the workforce.

MEETING REQUIREMENTS (5.2)

The standard requires customer requirements *to be met with the aim of enhancing customer satisfaction.*

What Does this Mean?

This requirement changes the focus from *doing what you say you do* to *doing what you need to do to satisfy your customers*. It also means that if your interpretation of customer requirements is incorrect in some way, you have an obligation to go beyond the requirements and aim for customer satisfaction. It does not mean however, that you must satisfy customers regardless of their demands. Some customers are unreasonable and expect something for nothing; other may want you to turn a blind eye to statutory regulations so it is your choice not to supply them if that is the case. Often when dealing with a customer you are dealing with a particular individual who you believe shares the values of their organization. If you believe the individual is not espousing such values you would be justified in contacting other customer personnel before declining to supply.

Why is this Important?

This requirement responds to the Customer Focus Principle.

Organizations only stay in business by satisfying their customers. ISO 9001 focuses on customers but organizations can fail to survive if they do not also satisfy other stakeholders such as their employees, regulators and society. People will only continue to work for organizations that treat them fairly. Regulators can close down businesses if they breach the rules and if all else fails, society can influence government and so change the law to stop organizations behaving in a way that harms the population not only of the host country but even the earth itself.

How is this Demonstrated?

The whole standard addresses the elements of a management system that aims to achieve customer satisfaction and therefore by constructing and operating a system that meets the intent of ISO 9001, this goal will be achieved.

Chapter 15

Quality Policy

CHAPTER PREVIEW

This chapter is aimed primarily at top management for it is they who must determine the quality policy and no one else although others will be involved such as those who are assisting top management develop their strategy for sustained success.

All organizations have policies. In fact any guide to action or decision can be classed as a policy and so policies exist at all levels in an organization but they might not be referred to as policies, going under names such as guides, codes of practice, beliefs, traditions etc. At the corporate level policies generally act as constraints on how the mission and vision are to be accomplished. There are no requirements in ISO 9001:2008 for a mission, vision or strategy so that the quality policy is placed in context except that Clause 5.3 refers to the policy being appropriate to the organization's purpose which can be understood to be its mission.

In this chapter we examine the requirements in Clauses 5.1 and 5.3 of ISO 9001:2008 on quality policy and in particular:

- Explain what a quality policy is and its purpose;
- How the quality policy relates to mission and vision statements and statements of values;
- How to ensure the policy is appropriate to the organization;
- How to express a commitment to meeting requirements and continual improvement;
- The problems with using the policy as a framework for setting quality objectives;
- Issues with communicating the policy;
- The process for reviewing the policy.

The position where the requirements for quality policy feature in the managed process is shown in Fig. 15-1. In this case they form part of the mission management process and represent the constraints as translated from stakeholder needs and expectations that condition the strategy represented by the 'Plan' box.

ESTABLISHING QUALITY POLICY (5.1b)

The standard requires that top management *establish the quality policy.*

What Does this Mean?

ISO 9000:2005 defines a quality policy as the overall intentions and direction of an organization related to quality as formally expressed by top management. It also

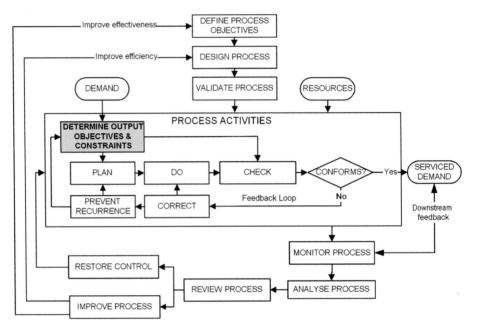

FIGURE 15-1   Where the requirements of Clause 5.3 apply in a managed process.

suggests that the policy be consistent with the overall policy of the organization and provides a framework for setting quality objectives. Furthermore, ISO 9000:2005 advises that the eight quality management principles be used as a basis for forming the quality policy. The quality policy can therefore be considered as the values, beliefs and rules that guide actions, decisions and behaviours. Values and rules were addressed in Chapter 9. Values and rules guide actions, decisions and behaviours and hence may be termed *policies*. They are not objectives because they are not achieved – they are demonstrated by the manner in which actions and decisions are taken and the way your organization behaves towards others.

The relationship between policy and objectives is important because ISO 9001 implies one is derived from the other but in reality the relationship is quite different as illustrated in Fig. 15-2.

The detail of quality policy will be addressed later. What is important in this requirement is an understanding of why a quality policy is needed, what is required to *establish* a quality policy and where it fits in relation to other policies.

Why is this Necessary?

This requirement responds to the Leadership Principle.

Defining the purpose or mission of the business is one thing but without some guiding principles, the fulfilment of this mission may not happen unless effort is guided in a common direction. If every manager chooses his or her direction and policies, the full potential of the organization would not be realized.

The purpose of corporate policies is to influence the short- and long-term actions and decisions and to influence the direction in which the mission will be fulfilled. If

> ## Corporate Terminology
>
> There are so many similar terms that are used by upper management that it is not surprising that their use is inconsistent. We use the term *goal* when we mean purpose or perhaps we really did mean goal or should we use the term *mission* or *objective*, or perhaps a better term would be *target* ... and so on. The problem is that we often don't know the intension of the user. Did he or she carefully select the term to impart a specific meaning or would the alternatives have been equally appropriate? What follows are the meaning used in the book.
>
> | | |
> |---|---|
> | Purpose | Why we exist, why we do what we do |
> | Mission | As Purpose |
> | Vision | What the organization should look like as it successfully fulfils its mission |
> | Goals | The strategic objectives for the long term based on the mission and vision |
> | Values | The beliefs that will guide our behaviour |
> | Strategy | How we are going to get there |
> | Policy | Rules that guide our actions and decisions – the signposts en route |
> | Principles | A fundamental truth |
> | Objectives | What we want to achieve at key milestones along the journey |
> | Measures | What will indicate achievement |
> | Targets | What we aim at to achieve objectives |

there were policies related to the organization's customers, they could be fulfilled at the expense of employees, investors and society. If there were policies related to profit, without other policies being defined, profit is positioned as a boundary condition to all actions and decisions. Clearly this may not direct the organization towards its mission.

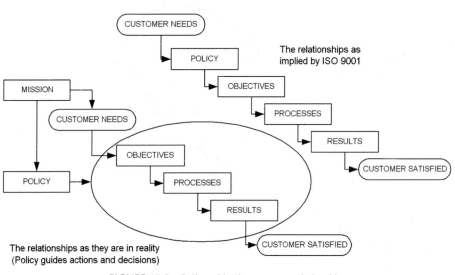

FIGURE 15-2 Policy–objectives–process relationship.

How is this Demonstrated?

To establish means to put in place permanently. A quality policy that is posted in the entrance hall is published, but not established. For a policy to become established, it has to reflect the beliefs of the organization and underpin every conscious thought and action. This will only arise if everyone shares those beliefs. For this to happen, managers need to become the role model so that by their actions and decisions they exemplify the policy. Belief in the policy is unlikely if the quality policy is merely perceived as something written only to satisfy ISO 9001.

The definition of quality in ISO 9000 goes beyond relations with the purchaser and embraces all stakeholders. The eight quality management principles also go far beyond the mechanical processes for achieving product quality and embrace the softer factors that influence the behaviour of people in an organization. Clearly, a quality policy that addresses all these issues comes close to reflecting all the policies of the organization and hence can be termed the corporate policy. While organizations may have a safety policy, an environmental policy, a personnel policy, a servicing policy etc. these are really topics within the corporate policy. Quality is not just another topic but a term that embraces all the topics. It is hard to think of any policy that could not be classed as a quality policy when quality is defined as *the degree to which a set of inherent characteristics fulfils requirements*. This definition does not limit *quality* to the fulfilment of customer requirements, but extends it to the fulfilment of all stakeholder's requirements.

The phrase 'a set of inherent characteristics' might be perceived as being limited to products and services, but an organization, a person, a process, a decision, a document or even an environment has a set of inherent characteristics.

Safety policies are quality policies because they respond to employees and customers as stakeholders. Environmental policies are quality policies because they respond to society as a stakeholder. There is therefore no advantage in issuing a separate quality policy – it is more effective if the organization formulates its corporate policies and within them addresses the topics covered by the eight quality management principles.

ENSURING POLICY IS APPROPRIATE (5.3a)

The standard requires the quality policy *to be appropriate to the purpose of the organization*.

What Does this Mean?

The purpose of an organization is quite simply the reason for its existence and this is referred to as the *mission*. It is what the organization has been formed to do. As Peter Drucker so eloquently puts it – "there is only one valid definition of business purpose and that is: to create a customer".[1] In ensuring that the quality policy is appropriate to the purpose of the organization, it must be appropriate to the customers the organization desires to create. It is therefore necessary to establish who the customers are, where

---

[1] Drucker, Peter F. (1977). *Management,* Harper's College Press.

the customers are, what they buy or wish to receive and what these customers regard as value.

Some people take the view that the purpose of an organization is to make money but they are confusing objectives with purpose. Money is important but as a resource for accomplishing objectives not the *reason* for existing. Even banks have a greater purpose than making money (this view is now questionable in view of the 2009 banking crisis. However, only the mint actually makes money) as it is the value brought to society that money brings that is important. Profit is needed in order to grow the organization so that it may create more wealth for its stakeholders.

Why is this Necessary?

This requirement responds to the Leadership Principle.

As stated above, the quality policy is the corporate policy and such policies exist to channel actions and decisions along a path that will fulfil the organization's mission. A goal of the organization may be the attainment of ISO 9001 certification and thus a quality policy of meeting the requirements of ISO 9001 would be consistent with such a goal, but in reality this is a constraint and not a goal as people work to fulfil the mission in a way that is consistent with the requirements of ISO 9001.

How is this Demonstrated?

Policies are not expressed as vague statements or emphatic statements using the words *may, should* or *shall*, but clear intentions by use of the words *'we will'* – thus expressing a commitment or by the words *'we are, we do, we don't, we have'* expressing shared beliefs. Very short statements tend to become slogans which people chant but rarely understand the impact on what they do. Their virtue is that they rarely become outdated. Long statements confuse people because they contain too much for them to remember. Their virtue is that they not only define what the company stands for but how it will keep its promises.

As we showed in Fig. 8-16, when the mission is passed through the stakeholder requirement it filters objectives and constraints. The constraints need to be studied and those that will impact the organization as a whole need to be translated into policies.

EXPRESSING A COMMITMENT (5.3b)

The standard requires that the quality policy *includes a commitment to comply with requirements and continually improve the effectiveness of the quality management system.*

What Does this Mean?

A commitment to comply with requirements means that the organization should undertake to meet the requirements of all stakeholders. This means meeting the requirements of customer, suppliers, employees, investors, owners and society. Customer requirements are those either specified or implied by customers or determined by the organization and these are dealt with in more detail under Clauses 5.2 and 7.2.1. The requirements of employees are those covered by legislation such as access, space, environmental conditions, equal opportunities and maternity leave but also the legislation appropriate to minority groups such as the disabled and any agreements made with unions or other representative bodies. Investors have rights also and these will be addressed in the investment agreements. The requirements of society are those obligations resulting from laws, statutes, regulations etc.

An organization accepts such obligations when it is incorporated as a legal entity, when it accepts orders from customers, when it recruits employees, when it chooses to trade in regulated markets and when it chooses to use or process materials that impact the environment.

The effectiveness of the management system is judged by the extent to which it fulfils its purpose. Therefore, improving effectiveness means improving the capability of the management system. Changes to the management system that improve its capability, i.e., its ability to deliver outputs that satisfy all the stakeholders, are a certain type of change and not all management system changes will accomplish this. This requirement therefore requires top management to pursue changes that bring about an improvement in performance.

Why is this Necessary?

This requirement responds to the Leadership Principle.

Policies guide action and decision. It is therefore necessary for top management to impress on their workforce that they have entered into certain obligations that commit everyone in the enterprise. Such commitments need to be communicated through policy statements in order to ensure that when taking actions and making decisions, staff give top priority to meeting the requirements of the stakeholders. This is not easy. There will be many difficult decisions where the short-term interests of the organization may need to be subordinated to the needs of customers. Internal pressures may tempt people to cut corners, break the rules and protect their own interests. Committing the organization to meet requirements may be an easy decision to take – but difficult to honour.

Over 1 million organizations worldwide have obtained ISO 9001 certification but many (perhaps the majority) have not improved their performance as a result. They remain mediocre and not top performers primarily because the management system is not effective. If it were effective the organization would meet its objectives year on year and grow the number of satisfied customers. Resources, technology, market conditions, economical conditions and customer needs and expectations continually change, thus impacting the organization's capability. The effectiveness of the management system in meeting this challenge therefore needs to be continually improved.

How is this Demonstrated?

A policy containing a commitment to meeting the needs and expectations of all stakeholders would meet the first part of this requirement. In making a commitment to meet all these requirements the organization is placed under an obligation to

- identify the relevant requirements;
- design and install processes that will ensure the requirements are met;
- verify compliance with the identified requirements;
- demonstrate to relevant authorities that the requirements have been met.

The second part can be dealt with by including a policy that commits the organization to improve the effectiveness of the system by which the organization's objectives are achieved. Policies are more easily understood when expressed in terms that are understood by the employees. The term *interested parties* is 'ISO speak' and may not be readily understood. Spell it out if necessary – in fact it is highly desirable where relevant to state exactly what you mean rather than use the specific words from the standard.

PROVIDING A FRAMEWORK FOR QUALITY OBJECTIVES (5.3c)

The standard requires the quality policy *to provide a framework for establishing and reviewing quality objectives.*

What Does this Mean?

There are two different interpretations of this requirement. One interpretation is that it implies that each statement within the quality policy should have an associated quality objective and each lower level quality objective should be traceable to higher-level quality objectives that have a clear relationship with a statement within the quality policy. Through this relationship, the objectives deploy the quality policy. In this respect the quality policy would appear to be the mission statement of the organization.

The other is that the quality policy represents a set of guiding principles, values or constraints and therefore when setting as well as reviewing quality objectives, these principles should be employed to ensure that the objectives are appropriate to the purpose of the organization. This means that the policies are used to measure or frame (put boundaries around) the objectives. It does not mean that the words used in the quality policy should somehow be translated into objectives.

Why is this Necessary?

This requirement responds to the Leadership Principle.

There needs to be a link between policy and objectives otherwise the processes designed to achieve them would be unlikely to implement the policy. A consequence of displaying a quality policy and impressing on everyone how important it is, is that it becomes disconnected from real work. It does not get used unless it is also linked to the work that people do and such a link is made by using the policy to frame or measure objectives.

Examples of Corporate Policies

On Customers

We will listen to our customers, understand their needs and expectations and endeavour to satisfy those needs and expectations in a way that meets the expectations of our other stakeholders.

On Leadership

We will establish and communicate our vision for the organization and through our leadership exemplify core values to guide the behaviour of all to achieve our vision.

On People

We will involve our people in the organization's development, utilize their knowledge and experience, recognize their contribution and provide an environment in which they are motivated to realize their full potential.

On Processes and Systems

We will take a process approach towards the management of work and manage our processes as a single system of interacting processes that produce outcomes which satisfy all our stakeholders.

On Continual Improvement

We will provide an environment in which every person is motivated to continually improve the efficiency and effectiveness of our products, processes and our management system.

On Decisions

We will base our decisions on the logical and intuitive analysis of data collected where possible from accurate measurements of product, process and system characteristics.

On supplier relationships

We will develop alliances with our suppliers and work with them to jointly improve performance.

On profits

We will satisfy our stakeholders in a manner that will yield a surplus that we will use to develop our capabilities and our employees, reward our investors and contribute to improvement in our society

On the environment, health and safety

We will operate in a manner that safeguards the environment and the health and safety of those who could be affected by our operations.

How is this Demonstrated?

In the ISO 9000:2005 definition of quality policy it is suggested that the eight quality management principles be used as a basis for establishing the policy and an example of such policies is presented in the text box.

If we take just anyone of these policy statements they will look good in the lobby – visitors will be impressed but the bottom line is whether actual performance meets the expectations set. If no one thinks through the process for ensuring the links between the policy, the objectives and the outputs the policy won't be met with any consistency.

Without being linked to the business processes, these policies remain dreams. There has to be a means to make these policies a reality and it is by using the policies as measures of success for objectives that this is accomplished. By deriving objectives from customer needs and expectations and the policies from the constraints governing how these objectives will be achieved you will produce a series of objectives and measures for the enabling processes.

Purely as an exercise and not as a real policy we could fabricate a policy that does provide a framework for setting objectives but it is not recommended.

Let us assume the policy is as follows:

We are committed to providing products that are delivered on time and meet customer requirements while yielding a profit and increasing sales. We accomplish this through product and process innovation, cost reduction activities and compliance with ISO 9001 requirements.

If we use the policy as a framework it might look like that in Table 15-1.

TABLE 15-1 Fabricating Quality Objectives from a Quality Policy

Quality policy	Quality objectives
We are committed to providing products that are delivered on time	98% on time delivery as measured by the customer
and meet customer requirements	99.9% of monthly output to be defect free as measured by customer returns
while yielding a profit	5% profit on annual sales
and increasing sales	25% increase in annual sales volume
We accomplish this through product and process innovation,	20% of our product range will contain new products and 50 improvement teams will be set up to seek process improvement
cost reduction activities,	15% reduction in cost of poor quality as a percentage of sales
compliance with ISO 9001 requirements	Certified to ISO 9001 by July 2005

This seems to be derived from what top management wants. It passes the scrutiny of the external auditors simply because it meets the requirement. One cannot deny that, but does it meet the intent? Isn't the intent of the organization is to satisfy its stakeholders and therefore wouldn't one expect the objectives to be derived from the needs and expectations of these stakeholders?

ENSURING POLICY IS COMMUNICATED AND UNDERSTOOD (5.3d)

The standard requires that the quality policy *is communicated and understood within the organization.*

What Does this Mean?

For a policy to be communicated it has to be brought to the attention of personnel. Personnel have to be made aware of how the policy relates to what they do so that they understand what it means before action is taken. Without action there is no demonstration that communication has been effective. If you are already doing it, publishing the policy merely confirms that this is your policy. If the organization does not exhibit the right characteristics, there will be a need to change the culture to make the policy a reality.

Why is this Necessary?

This requirement responds to the Leadership Principle.

As has been stated previously, a policy in a nice frame positioned in the lobby of an organization may impress the visitors but unless it is understood and adhered to, it will have no effect on the performance of the organization.

How is this Demonstrated?

It is difficult to imagine how a policy could be understood if it wasn't communicated but it signifies that the understanding has to come about by top management communicating the policy rather than the policy being deployed via the grape vine.

Whilst it is important that management shows commitment towards quality, policy statements can be one of two things – *worthless* or *obvious*. They are worthless if they do not reflect what the organization already believes in and is currently implementing. They are obvious if they do reflect the current beliefs and practices of the organization. It is therefore foolish to declare in your policy what you would like the organization to become.

This is perhaps the most difficult requirement to achieve. Any amount of documentation, presentations by management, and staff briefings will not necessarily ensure that the policy is understood. Communication of policy is about gaining understanding but you should not be fooled into believing that messages delivered by management are effective communication. Effective communication consists of four steps: attention, understanding, acceptance and action. It is not just the sending of messages from one source to another. So how do you *ensure* that the policy is understood?

Within your management system you should prescribe the method you will employ to ensure that all the policies are understood at all appropriate levels in the organization. One way of doing this is for the policies to be used as measures of success referred to in the process descriptions (see Chapter 11). There will be other levels in the organization where a clear understanding of the corporate policy is necessary for the making of sound decisions. At other levels, staff may work to instructions, having little discretion in what they can and cannot do. At these levels relevant aspects of the policy may be translated by the local manager into words that the staff understand. These can be conveyed through local procedures or notices or again process measures.

One method to ensure understanding is for top management to do the following:

- Debate the policy together and thrash out all the issues. Don't announce anything until there is a uniform understanding among the members of the management team. Get the managers to face the question, "Are we all agreed about why we need this policy?", "Do we intend to adhere to this policy?" and remove any doubt before going ahead.

- Ensure the policy is presented in a user-friendly way.
- Announce to the workforce that you now have a policy that affects everyone from the top down.
- Publish the policy to the employees (including other managers).
- Display the policy in key places to attract peoples' attention.
- Arrange and implement training or instruction for those affected.
- Test understanding at every opportunity, e.g., at meetings, when issuing instructions or procedures, when delays occur, when failures arise and when costs escalate.
- Audit the decisions taken that affect quality and go back to those who made them if they do not comply with the stated policy.
- Take action every time there is misunderstanding. Don't let it go unattended and don't admonish those who may have misunderstood the policy. It may not be their fault!
- Every time there is a change in policy, go through the same process. Never announce a change and walk away from it. The change may never be implemented!
- Give time for the understanding to be absorbed. Use case studies and current problems to get the message across.

The audit programme is another method of testing understanding and is a way of verifying whether the chosen method of ensuring understanding is being effective.

In determining whether the policy is understood, auditors should not simply ask, "What is the quality policy?" All this will prove is whether the auditee remembers it! The standard does not require that everyone knows the policy, only that it be communicated and understood. To test understanding therefore, you need to ask, e.g.,

- How does the quality policy affect what you do?
- What happens if you can't accomplish all the tasks in the allotted time?
- What would you do if you discovered nonconformity immediately prior to delivery?
- How would you treat a customer who continually complains about your products and services?
- What action would you take if someone asked you to undertake a task for which you were not trained?
- What are your objectives and how do they relate to the quality policy?
- What action would you take if you noticed that someone was consuming food and drink in a prohibited area?
- What action would you take if you noticed that a product for which you were not responsible was in danger of being damaged?

ENSURING THAT THE POLICY IS REVIEWED (5.3e)

The standard requires the quality policy *to be reviewed for continuing suitability.*

What Does this Mean?

This requirement means that the policy should be examined in light of planned changes in the organization or stakeholder needs and expectations to establish whether it will remain suitable for guiding the organization towards its mission.

Why is this Necessary?

This requirement responds to the Leadership Principle.

Nothing remains static for very long. As the organization grows and seeks new opportunities, its size and character will need change as it responds to the markets and economic climate in which it operates. A policy established under different circumstances may therefore not be appropriate for what the organization needs to become to meet these challenges – it may not be suitable for guiding the future organization towards its mission. The policy is required to be appropriate for the organization's purpose (mission) and while the mission may not change, the environment in which the organization operates does change. These changes will impact the corporate policy. The policy will need to be reviewed in light of changes in the economic, social and technological environment for its suitability to enable the organization to fulfil its purpose.

How is this Demonstrated?

The quality policy should be reviewed whenever a change in the market, the economic climate, statutory and legal requirements, technology or a major change in the organizational structure is contemplated. This review may be conducted through process reviews and system reviews.

The review may conclude that no change is needed to the actual words but the way they are being conveyed might need to change. If the environment or the organization has changed, the policy might be acceptable but needs to be interpreted differently, conveyed to different people using different examples than were used previously.

Changes in policy have wide impact and therefore should not be taken lightly. They should be reviewed by top management with the full participation of the management team and therefore should be debated during strategic planning reviews, business reviews or process reviews. We are not talking about tinkering with the wording but a real change in direction. Changes in technology might mean that the workforce ceases to be predominantly on site as it becomes more effective to promote home or remote working. This change will impact the policy regarding leadership and people. Changes in the economic climate might mean that the workforce ceases to consist primarily of employees as it becomes more effective to outsource work to subcontractors and consultants. This change will impact the policy regarding leadership, suppliers and people.

Once the decision is made to change the policy it has to be communicated and the process for educating the workforce initiated.

Chapter 16

Quality Objectives and Planning

CHAPTER PREVIEW

This chapter is aimed primarily at managers at all levels for it is they who must determine the quality objectives although others will be involved such as those who are assisting managers and develop plans and processes for implementing the corporate strategy.

There are a number of terms used to define the ends or outcomes an organization seeks to achieve. These terms include mission, vision, goals, aims, objectives, targets and milestones and all tend to create confusion. The confusion is compounded when qualifying words are added to these terms such as quality objectives, environmental targets, project goals etc. In a generic sense they could all be classed as objectives but what generally sets them apart are timescales and levels.

In this chapter we examine the requirements in Clause 5.4.1 of ISO 9001:2008 and in particular:

- What quality objectives are and how they differ from other objectives?
- The various levels at which quality objectives should be established;
- The process for establishing quality objectives;
- How objectives should be expressed?
- Defining measures that will indicate whether objectives have been achieved;
- Planning to meet quality objectives;
- Planning for change.

The position where the requirements for quality objectives and planning feature in the managed process is shown in Fig. 16-1. In this case they form part of the mission management process and perform two functions: one where the product objectives and plans are the business objectives and plans and the other where the process objectives are the objectives of all the business processes and the planning is the process design as Clause 5.4.2 invokes Clause 4.1 and thus addresses process design characteristics.

QUALITY OBJECTIVES (5.4.1)

Establishing Objectives (5.1c and 5.4.1)

The standard requires that top management *ensure that quality objectives, including those needed to meet requirements for product, are established at relevant functions and levels within the organization.*

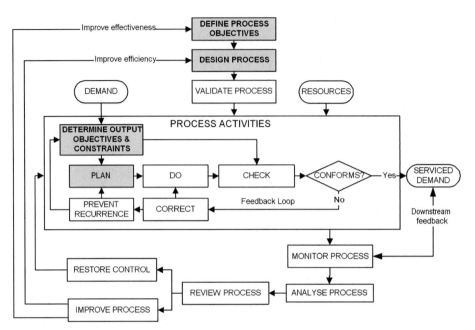

FIGURE 16-1   Where the requirements of Clause 5.4 apply in a managed process.

What Does this Mean?

An objective is a result that is aimed for and is expressed as a result that is to be achieved. ISO 9000:2005 defines quality objectives as *results sought or aimed for related to quality*. Bryson[1] defines objectives as "a measurable target that must be met on the way to attaining a goal" and this suggests that a goal is attained by achieving a number of objectives, one group of which would be quality objectives.

Objectives remain effective until achieved, whereas policies remain effective until changed. Objectives are therefore not the same as policies (see Chapter 15). Objectives are required at levels within the organization not levels within the organization *structure*. This is clarified by the requirement for objectives to include those needed to meet requirements for product. There are five levels at which control and improvement objectives need to be established:

- Corporate level where the objectives are for the whole enterprise to enable it to fulfil its mission/vision;
- Process level where the objectives are for specific processes to enable them to fulfil corporate goals;
- Product or service level where the objectives are for specific products or services or ranges of products or services to enable them to fulfil *or create* customer needs and expectations;
- Departmental or function level where the objectives are for an organizational component to enable it to fulfil corporate goals;
- Personal level where the objectives are for the development of individual competency.

[1] Bryson John M. (2004) Creating and implementing your strategic plan – A workbook for public and non-profit. Jossey Bass.

Why is this Necessary?

This requirement responds to the Leadership Principle.

Management needs to ensure that the objectives are established as a basis for action. All work serves an objective and it is the objective that stimulates action. The reason for top management ensuring that objectives are established is to ensure that everyone channels their energies in a positive direction that serves the organization's mission.

The requirement for defining objectives is one of the most important requirements. Without quality objectives there can be no improvement and no means of measuring how well you are doing. Without objectives any level of performance will do. If you don't know where you are going, any destination will do! Objectives are therefore necessary as a basis for measuring performance, to give people something to aim for, to maintain the status quo in order to prevent decline and to advance beyond the status quo for the enterprise to grow.

How is this Demonstrated?

Process for Establishing Objectives

Achievable objectives do not necessarily arise from a single thought. There is a process for establishing objectives. At the strategic level, the subjects that are the focus for setting objectives are the outputs that will produce successful outcomes for the organization's stakeholders (see Table 10-1). Customer needs, regulations, competition and other external influences shape these objectives and cause them to change frequently. The measures arise from an analysis of current performance, the competition and the constraints of customers and other stakeholders and there will emerge the need for either improvement or control. The steps in the objective setting process are as follows[2]

- Identifying the need from an analysis of stakeholders or the strategic issues;
- Drafting preliminary objectives;
- Proving the need to the appropriate level of management in terms of:
  - whether the climate for change is favourable,
  - the urgency of the improvement or controls,
  - the size of the losses or potential losses,
  - the priorities;
- Identifying or setting up the forum where the question of change or control is discussed;
- Conducting a feasibility study to establish whether the objective can be achieved with the resources that can be applied;
- Defining achievable objectives for control and improvement;
- Communicating the objectives.

The standard does not require that objectives be achieved but it does require that their achievement be planned and resourced. It is therefore prudent to avoid publishing objectives for meeting an unproven need which has not been rigorously reviewed and assessed for feasibility. It is wasteful to plan for meeting objectives that are unachievable and it diverts resources away from more legitimate uses.

---

[2] Juran, J. M. (1995). Managerial Breakthrough Second Edition. McGraw-Hill

Objectives are not established until they are understood and therefore communication of objectives must be part of this process. Communication is incomplete unless the receiver understands the message but a simple yes or no is not an adequate means of measuring understanding. Measuring employee understanding of appropriate quality objectives is a subjective process. Through the data analysis carried out to meet the requirements of Clause 8.4 you will have produced metrics that indicate whether your quality objectives are being achieved. If they are being achieved you could either assume your employees understand the quality objectives or you could conclude that it doesn't matter. Results alone are insufficient evidence. The results may have been achieved by pure chance and in six months time your performance may have declined significantly. The only way to test understanding is to check the decisions people make. This can be done with a questionnaire but is more effective if one checks decisions made in the work place. Is their judgement in line with your objectives or do you have to repeatedly adjust their behaviour?

For each objective you should have a plan that defines the processes that will achieve them. Assess these processes and determine where critical decisions are made and who is assigned to make them. Audit the decisions and ascertain whether they were consistent with the objectives and the associated measures. A simple example is where you have an objective of producing conforming product in a way that decreases dependence on inspection (the measure). By examining corrective actions taken to prevent recurrence of nonconformities you can detect whether a person decided to increase the level of inspection in order to catch the nonconformities or considered alternatives. Any person found increasing the amount of inspection has clearly not understood the objective.

For an objective to be established it has to be communicated, translated into action and become the focus of all achievement. Objectives are not wish lists. The starting point is the mission statement and although ISO 9001 suggests that the quality objectives should be based on the quality policy, is it more likely to be the strategic issues arising from an analysis of current performance, competition, opportunities and new technology that drive the objectives.

Objectives for Control and Improvement

A management system is not a static system but a dynamic one and if properly designed and implemented can drive the organization forward towards world-class quality. All managerial activity is concerned either with maintaining performance or with making change. Change can retard or advance performance. That which advances performance is beneficial. In this regard, there are two classes of quality objectives, those serving the control of quality (maintaining performance) and those serving the improvement of quality (making beneficial change).

The objectives for quality control should relate to the standards you wish to maintain or to prevent from deteriorating. To maintain your performance and your position in the market you will have to continually seek improvement. Remaining static at whatever level is not an option if your organization is to survive. Although you will be striving for improvement it is important to avoid slipping backwards with every step forwards. The effort needed to prevent regression may indeed require innovative solutions. While to the people working on such problems, it may appear that the purpose is to change the status quo, the result of their effort will be to maintain their present position not raise it to higher levels of performance. Control and improvement can therefore be perceived as one and the same thing depending on the standards being aimed for and the difficulties in meeting them.

Strategic Objectives

The identification of strategic issues is the heart of the strategic planning process. Many issues will emerge from a PEST[①] and SWOT[②] Analysis, Stakeholder needs Analysis and Competition Analysis. Bryson covers strategic issues[3] in depth and suggests that these issues fall into three main categories:

* Current issues that probably require immediate action;
* Issues that are likely to require action in the near future but can be handled as part of the organization's regular planning cycle;
* Issues that require no action at present but need to be continuously monitored.

From the first step in establishing the goals (see Chapter 10) a number of issues are likely to emerge that are more tactical than strategic. It is important to capture the tactical issues as people will think these have the most impact on their work and dealing with these often engages people in the strategic planning process. Also addressing the tactical issues often removes barriers to confronting the strategic issues. The issues identified may be grouped into one or more categories. The following list of categories was identified by Peter Drucker in the 1970s but it is timeless[4].

Marketing

* Existing products in current markets,
* Abandoning products,
* New products in current markets,
* New markets,
* New products in new markets,
* Standards and performance.

Innovation

* Reaching market goals in near future,
* Taking advantage of technological advances in the distant future.

Human resources

* Supply of managers and their development,
* Supply of staff and their development,
* Relationships with representative bodies,
* Relationship with suppliers,
* Employee attitudes and competencies.

Physical resources

* Supply of raw materials and components,
* Supply of capital equipment,
* Supply of buildings and facilities.

---

[3] Bryson John M. (2004) Creating and implementing your strategic plan – A workbook for public and non-profit. Jossey Bass.
[4] Drucker, Peter F. (1977). *Management,* Harper's College Press

Financial resources

- Investment and attracting capital,
- Obtaining financial resources.

Productivity

- Utilization of knowledge,
- Utilization of physical resources,
- Utilization of time,
- Utilization of financial resources,
- Making workers productive,
- Utilizing experience and ability,
- Utilizing reputation.

Social responsibility

- Disadvantaged people,
- Protection of the environment,
- Education of potential employees,
- Contribution to professions,
- Health and safety of employees at work,
- Minimizing impact on society, economy, community and individual.

Profit requirement

- Producing the minimum profit needed to accomplish the other objectives.

As part of the strategic planning you might analyse competitor products and benchmark inside and outside the industry. There are many books[5] and organizations you can turn to for advice on benchmarking[①]. With benchmarking you analyse your current position, find an organization that is performing measurably better and learn from them what they are doing that gives them the competitive edge. You then set objectives for change as a result of what you learn.

Strategic objectives need to be deployed through the processes otherwise they will remain pipedreams and therefore the strategic plan should show the linkage between processes and objectives in a way as shown in Table 10-1 for business outputs.

The statements of objectives may be embodied within business plans, product development plans, improvement plans and process descriptions.

Constraints as Objectives

Many constraints are expressed as objectives; for example, Motorola's Safety Objectives include the following:

Motorola seeks to provide a workplace that is free of occupational injuries and illnesses. Motorola reports in 2004 that its global injury and illness rate has been significantly below both the U.S. manufacturing average and the U.S. electronics industry average. However, we know that Motorola did not stop producing product in order to achieve this objective. What is more likely is that managers looked for ways of

---

[5] Codling, Sylvia. (1998). *Benchmarking*, Gower.

satisfying as many if not more customers whilst at the same time imposing constraints on those activities that impact employees' health.

At Legal and General the environmental objectives are to achieve continual improvement in the environmental performance of operational properties. Their targets include ensuring that less than 35% of waste produced at selected sites is disposed of directly to landfill. For Legal and General staff, this objective constrains what they do with waste.

In the case of Motorola, although the constraints are called objectives, the true objective is to create and satisfy customers under conditions that are *free of occupational injuries and illnesses* and in the case of Legal and General, it is to create and satisfy customers *while continually improving environmental performance*. In both cases the constraint has modified the description of the objective.

When organizations include among their objectives, external certification, cost reduction, energy conservation, safety and increase in profit, they face the difficult task of trying to balance these competing constraints. They then run the risk of dissatisfying their customers because they traded one off against the other. Whatever the demands are called by those who make them, they need to be placed in the right category so that they are treated appropriately. If increase in profit was an objective, this could be achieved easily by increasing the margin between cost and selling price. However, customers might not be willing to pay the increased price and go elsewhere thus reducing sales and consequently decreasing profit. Treating profit as a constraint rather than an objective causes the designers and producers to look for ways of reducing costs.

Process Objectives

There are two types of processes, business processes and work processes. Business processes deliver business outputs and work processes deliver outputs required by business processes. At the process level the objectives are concerned with process performance – addressing process capability, efficiency and effectiveness, use of resources, and controllability. As a result, objectives for control may focus on reducing errors and reducing waste, increasing controllability but may require innovative solution to achieve such objectives. Objectives for improvement might include increasing throughput, turnaround times, response times, resource utilization, environmental impact, process capability and use of new technologies, etc.

Product Objectives

At the product level, objectives are concerned with product or service performance addressing customer needs and competition. Again these can be objectives for control or improvement. Objectives for control might include removing nonconformities in existing products (improving control) whereas objectives for improvement might include the development of new products with features that more effectively satisfy customer needs (improving performance), use of new technologies, and innovations. A product or service that meets its specification is only of good quality if it satisfies customer needs and requirements. Eliminating all errors is not enough to survive – you need the right products and services to put on the market. Objectives for satisfying the identified needs and expectations of customers with new product features and new service features will be quality objectives.

Departmental Objectives

At the departmental level objectives are concerned with organizational performance – addressing the capability, efficiency and effectiveness of the organization, its responsiveness to change, the environment in which people work etc. Control objectives might be to maintain expenditure within the budget, to keep staff levels below a certain level, to maintain moral, motivation or simply to maintain control of the department's operations. Objectives for improvement might be to improve efficiency by doing more with less resources, improving internal communication, interdepartmental relationships, information systems etc.

Personal Objectives

At the personal level, objectives will be concerned with worker performance addressing the skills, knowledge, ability, competency, motivation and development of people. Objectives for control might include maintaining time-keeping, work output and objectivity. Objectives for improvement might include improvement in work quality, housekeeping, interpersonal relationships, decision-making, computer skills etc.

Measuring Quality Objectives (5.4.1)

The standard requires quality objectives *to be measurable and consistent with the quality policy.*

What Does this Mean?

There should be a tangible result from meeting an objective and a defined time period should be specified when appropriate. The objective should therefore be expressed in the form – *what is to be achieved* and *what will success look like, i.e., what will be happening or have happened as a result of achieving this objective?* The success measures indicate when the objective has or has not been achieved – namely the passing of a date, a level of performance, or the absence of a problem, a situation or a condition that currently exists or the presence of a condition that did not previously exist.

All of the organization's objectives should in some way serve to fulfil requirements of customers and other stakeholders. Objectives at lower levels should therefore be derived from those at higher levels and not merely produced to satisfy the whim of an individual.

Why is this Necessary?

This requirement responds to the Leadership Principle.

Where objectives are not measurable there is often some difficulty in establishing whether they have been achieved. Achievement becomes a matter of opinion and therefore variable. Measures provide consistency and predictability and produce facts on which decisions can be made.

How is this Demonstrated?

Consistency with Policy

The relationship between policy and objective is addressed in Chapter 15. But let us say you have a policy that addresses customer focus. Your objectives might include marketing objectives that were customer focused thereby linking the policy with the objectives. You may have a human resource objective for improving employee motivation. However, in this instance the process designed to achieve this objective would need to demonstrate adherence to a policy for the involvement of people. Here the process and not the objective links with the policy.

Testing Objectives

Objectives need to be consistent with the quality policy so that there is no conflict. For example, if the policy is "*We will listen to our customers, understand their needs and expectations and endeavour to satisfy those needs and expect in a way that meets the expectations of our other stakeholders.*" an objective which penalizes suppliers for poor performance would be inconsistent with the policy.

A technique has evolved to test the robustness of objectives and is identified by the letters SMART[1] meaning that objectives should be Specific, Measurable, Achievable, Realistic and Timely.

Although the SMART technique for objective setting is used widely, there is some variance in the words used. If you search the Internet on the key words "SMART

Case Study – Establishing Quality Objectives

When our management team read ISO 9001, we had long debates over what constitutes a quality objective. We have a business plan full of objectives but we do not label them with respect to quality, productivity, safety, financial etc. Is there a simple definition that will enable us to distinguish quality objectives from other types of objectives or is it unimportant provided we deploy these objectives to lower levels and can demonstrate that we have processes for achieving them?

It is not unusual for third party auditors to want to know what quality objectives have been defined but they are less inclined to challenge you now if you show them the business objectives within the Business Plan. However, you would be expected to indicate what it was that you were showing the auditor. Simply presenting a sheet a paper or a computer screen containing a series of statements should evoke a number of questions. "What are these? Which are the quality objectives?" or "I don't see anything here related to quality."

Auditors may well hold the view that quality objectives are business objectives, but may not hold that the opposite is true – i.e. all business objectives are quality objectives. If none of the business objectives related to the satisfying of customers, an auditor might feel justified in again asking to see the quality objectives primarily because the standard your organization is being assessed against is ISO 9001 and this requires a system that enhances customer satisfaction.

It is unimportant what you call these objectives and you are correct in assuming that it is more important that you deploy these objectives to lower levels and can demonstrate that you have processes for achieving them. However, you should be asked to explain how the objectives were defined – what process had these objectives as its output. While listening to your tale (and as it is an ISO 9001 audit) the auditor would be looking for evidence that some of the objectives were derived from the needs and expectations of customers. If it were an ISO 14001 audit the auditor would be looking for evidence that some of the objectives or the measures were derived from the needs and expectations of both customers and society or the local community and hence took into account the appropriate laws and regulations protecting the environment.

Objectives" you will discover several variations on this theme. The S of SMART has been used to denote Small meaning not too big to be unachievable – one small step at a time. The A of SMART has been used to denote Attainable, Accountable and Action oriented and the R of SMART has been used to denote Resource-consuming action and Relevant. These differences could arise out of different uses of the technique.

Measures (See also Chapter 8)

It is important to determine the measures that will be used to verify achievement of the objectives. If you have an objective for being World Class, what measures will you use that indicate when you are World Class? You may have an objective for improved delivery performance. What measures will you use that indicate delivery performance has improved? You may choose to use percent delivered on time. You will also need to set a target relative to current performance. Let us say that currently you achieve 74% on-time delivery so you propose a target of 85%. However, targets are not simply figures better than you currently achieve. The target has to be feasible and therefore it is necessary to take the steps in the process described previously for setting objectives.

In the last 30 years or so there has emerged an approach to management that focuses on objectives. Management by objectives or management by results has dominated boardrooms and management reports. In theory management by objectives or results is a sensible way to manage an organization but in practice this has led to internal competition, sub-optimization and punitive measures being exacted on staff that fail to perform. Deming advocated in the 11th of his 14 points[6] "Eliminate management by objectives" for the simple reason that management derives the goals from invalid data. They observe that a goal was achieved once and therefore assume it can be achieved every time. If they understood the process they would realize that the highs and lows are a characteristic of natural variation. They observe what the competition achieves and raise the target for the organization without any analysis of capability or any plan for its achievement. Management sets goals and targets for results that are beyond the capability of staff to control. Targets for the number of invoices processed, the number of orders won, the hours taken to fix a problem. Such targets not only ignore the natural variation in the system but are also set without any knowledge about the processes that deliver the results. If a process is unstable, no amount of goal setting will change its performance. If you have a stable process, there is no point in setting a goal beyond the capability of the process – you will get what the process delivers.

The published interpretation, RFI 035, states that objectives having Yes/No criteria are deemed measurable. An example would be "Develop a new product to meet the requirements of the 'YYYYY' market by March 2003".

Atsushi Niimi, CEO of Toyota Motor Manufacturing North America is reported to have said in October 2003: "We have some concerns about sustaining high quality because North American parts suppliers average 500 defects per million parts versus 15 per million in Japan. But if it works, and Lexuses made here are equivalent to those from Japan, Toyota will have exported a major upgrade of its already respected production system." So from this we can clearly express at least one quality objective.

---

[6] Deming, W. Edwards, (1982). *Out of the crisis*, MITC.

QUALITY MANAGEMENT SYSTEM PLANNING (5.4.2)

Planning to Meet Quality Objectives (5.4.2a)

The standard requires top management *to ensure that the planning of the quality management system is performed to meet the quality objectives and the requirement in Clause 4.1.*

What Does this Mean?

Planning is performed to achieve objectives and for no other purpose and therefore the requirement clearly indicates that the purpose of the management system is to enable the organization to meet its quality objectives. This is reinforced by the definition of quality planning in ISO 9000:2005 which states that *it is part of quality management focused on setting objectives and specifying necessary operational processes and related resources to fulfil the quality objectives.*

We have deduced that quality objectives are corporate objectives and therefore quality planning will be synonymous with *strategic planning*. (Remember that Clause 0.1 advises us that the adoption of a QMS was a strategic decision!)

The additional requirement for management system planning to meet the requirements of Clause 4.1 means that in planning the processes of the management system, you need to put in place provisions to measure, monitor and analyse processes, determine their sequence and interaction and determine criteria and methods to ensure effective operation and control. In addition you will need to provide resources and information necessary to support the operation and monitoring of these processes.

Why is this Necessary?

This requirement responds to the Process Approach Principle.

For objectives to be achieved the processes for achievement need to be planned. This means that the management system should be result-oriented with the objectives employed to drive performance.

How is this Demonstrated?

The process of defining objectives was outlined above indicating that planning proceeds only after the feasibility of achieving an objective has been established. One plans only to achieve an objective and remembering that planning consumes resources, an effective management system would need to ensure that dreams, wish lists and ambitions do not become the subject of any formal planning.

As objectives are required to be defined at relevant functions and levels, it follows that planning is also required at relevant functions and levels thereby requiring planning at corporate, divisional and department levels, product, process and system levels.

The planning referred to in this clause is focused on that needed to meet the organization's objectives. It is not focused on that needed to meet specific contracts or orders or for specific products and services – this type of planning is addressed under the heading *Product realization*. The organizational and resource planning needed for developing a new range of products or services would be considered to be part of corporate planning.

Objectives are achieved through processes and therefore in planning to meet an objective, the planner should identify the process or processes involved. At a high level this may be no more than an outline strategy for achieving the objectives, minimizing risks and measuring success. Responsibilities may identify no more than a function or department although in some organizations ownership is deemed paramount and individual managers are named. At the lower level, the plan may extend to detail activities with a bar chart showing timescales and responsibilities.

Processes for Planning

In Taylor's Scientific Management, there was a planning function that did all the planning. In the complex organizations created since Taylor's time, it has become too unwieldy for one function to do all the planning. Planning has been deployed to the function that will achieve the objective. At the corporate level there will be a strategic planning process that runs on a cycle of one, three and five years. Every function will be involved in providing their inputs. As each cycle ends a new one begins. The planning process includes the objective setting process and it is quite common for the ideas to come from below and float to the top. Here selections are made and passed down again for feasibility studies which go to the Board for approval and return to the source for detailed budgets and justification. Depending on the resources involved and the urgency the process may take months or even years to gain approval for the plans. The sanction to spend is often based on approval levels requiring budgets and detail plans before approval can be given. Even after such a lengthy process, there is likely to be another process for acquiring the resources that also requires approval, indicating that approved plans do not necessarily signify that permission to spend has been granted. This is often because of timing. When the time comes to acquire the resources, the priorities may have changed and plans once approved may be put on ice. It is interesting that ISO 9001 does not require plans to be implemented – it requires the *system* to be implemented and within this system may be provisions to abort plans when circumstances dictate that necessity for the survival of the organization.

It is therefore necessary to define the planning processes so that there is a clear linkage between objectives and plans to meet them.

Strategic Planning

The steps in the strategic planning process were identified in the Mission Management process of Fig. 10-6 as:

- Assess capability,
- Prioritize the strategic issues,
- Develop strategies and budgets,
- Develop and deploy strategy.

In Bryson's 10-step process the above steps match steps 4–8 inclusive. Steps 2 and 3 are addressed in the "Establish the gaols" stage of the Mission Management process, step 9 is stage 3 and step 10 is addressed by stage 4 of the same process.

The strategy is usually defined in corporate or business plans and it is quite common to produce separate business plans for the year ahead and three to five years ahead. Such plans may exist for each profit centre and consolidate the plans of all functions within that profit centre. These plans typically contain the budgets and other provisions such as head count and inventory required to meet the declared objectives. Corporate planning is not usually referred to as corporate quality planning, although in the larger enterprises, corporate quality planning may be one part of the corporate plan. In such cases, the corporate quality plan may address objectives related to improvement by better control, leaving the objectives related to improvement by innovation to be defined in product development or process development plans. The labels are not important. The scope of the set of plans should address all the defined objectives, regardless of what they are called.

> **Strategic Planning Cycle**
>
> 1. Initiate and agree on a strategic planning process;
> 2. Clarify organizational mandates;
> 3. Identify and understand stakeholders and develop and refine mission and values;
> 4. Assess the environment to identify strengths, weaknesses, opportunities, and challenges;
> 5. Identify and frame strategic issues;
> 6. Formulate strategies to manage the issues;
> 7. Review and adopt the strategic plan;
> 8. Establish an effective organizational vision for the future;
> 9. Develop an effective implementation process;
> 10. Reassess strategies and the strategic planning process.
>
> (John M Bryson)

As the quality management system is a series of managed processes, planning of the quality management system to meet the quality objectives and the requirements of Clause 4.1 equates with establishing a quality management system and this is addressed in Chapter 10.

Planning for Change (5.4.2b)

The standard requires *the integrity of the quality management system to be maintained when changes to the quality management system are planned and implemented.*

What Does this Mean?

This requirement refers to *change* in general not simply changes to the management system documentation. As the management system is the means by which the organization's objectives are achieved (not just a set of documents), it follows that any change in the enabling mechanism should be planned and performed without adversely affecting its capability. Changes needed to accomplish these objectives should be managed and the processes required to execute the changes should be part of the management system.

It means that the linkages and the compatibility between interfaces should be maintained during a change. By being placed under *planning*, there is recognition that the plans made to meet the defined objectives may well involve changing the organization, the technology, the plant, machinery, the processes, the competency levels of staff and perhaps the culture.

Why is this Necessary?

This requirement responds to the Process Approach Principle.

If changes in the management system are permitted to take place without consideration of their impact on other elements of the management system, there is likely to be deterioration in performance. In the past it may have been common for changes to be made and some months later the organization charts and procedures to be updated indicating that these documents are perceived as historical records – certainly not documents used in executing the change. The updating activities were often a response to the results of the change process creating in people's minds that it was a housekeeping or administrative chore. To meet this requirement, change management processes need to be designed and put in place. The integrity of the management system will be maintained only if these processes are made part of the system so that in planning the changes, due consideration is given to the impact of the change on the organization, its resources, processes and products, and any documentation resulting from or associated with these processes.

How is this Demonstrated?

To control any change there are some basic steps to follow and these are outlined under *The Improvement process* in Chapter 35. To maintain the integrity of the management system you need to do several things:

- Use the change processes to plan and execute the change. If they don't exist, the management system does not reflect how the organization operates. These processes should be part of the mission management process under "Manage Effectiveness" (see Fig. 10-5).
- Determine the impact of the change on the existing system and identify what else needs to change to maintain system effectiveness.
- Plan and execute the change concurrently with associated changes to documentation.
- Don't remove the old processes until the new processes have been proven effective.
- Measure performance before, during and after the change.
- Don't revert to routine management until the changes have been integrated into the culture – people perform the new tasks without having to be told.

The management system should not be perceived as a set of discrete processes. All processes should be connected therefore a change in one process is likely to have an effect on others. For example, if new technology is to be introduced, it may not only affect the process in which it is to be used but also the HR development process, the resource management process and the demand creation process because the new technology will improve the organization's capability and thus enabling it to create new markets, attract different customers etc. On a more mundane level, if a new form is introduced, it is not only the process in which the form is used that may be affected but also the interfacing processes that receive the form when complete. It is the information management processes that make the form available and secure its contents.

Chapter 17

Responsibility, Authority and Communication

CHAPTER PREVIEW

This chapter is aimed primarily at managers at all levels for it is they who must determine the responsibilities and authority of those who will develop and implement the strategy and manage and operate the organization's processes.

Having a well-defined quality policy and well-defined quality objectives will change nothing. The organization needs to empower its people to implement the policy and achieve the objectives through effectively managed processes. Only then will it be able to accomplish the mission, realize the vision and achieve the goals. This starts by assigning responsibility and delegating authority for the work to be done and from this determining the competences of those concerned. This will be a cascading process so that as the processes are developed and the activities identified, the responsibilities and authority for carrying them out can be determined.

In this chapter we examine the requirements in Clause 5.5 of ISO 9001:2008 and in particular:

- What responsibility and authority mean and how they are defined
- The principles that apply to the assignment of responsibility and delegation of authority;
- How responsibility and authority is communicated through organization charts, function descriptions, Job descriptions Procedures and flow charts

The position where the requirements for responsibility and authority feature in the managed process is shown in Fig. 17-1. In this case they represent a part of the planning effort within the mission management process.

RESPONSIBILITY AND AUTHORITY (5.5.1)

Defining Responsibilities and Authority (5.5.1)

The standard requires that *the responsibilities and authority be defined.*

What Does this Mean?

The term responsibility is commonly used informally to imply an obligation a person has to others. However, the term authority has increasingly become associated with

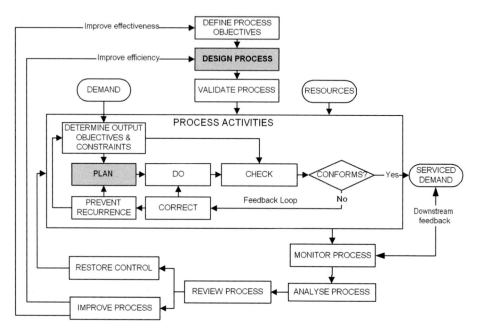

FIGURE 17-1   Where the requirements of Clause 5.5 apply in a managed process.

power and public bodies but in principle one cannot have responsibility without authority and vice versa. Problems arise when these two are not matched, where one is greater or less than the other.

Responsibility is in simple terms an area in which one is entitled to act on one's own accord. It is the obligation of staff to their managers for performing the duties of their jobs. It is thus the obligation of a person to achieve the desired conditions for which they are accountable to their managers. If you cause something to happen, you must be responsible for the result just as you would if you cause an accident – so to determine a person's responsibility ask, 'What can you cause to happen?'

Authority is, in simple terms, the right to take actions and make decisions. In the management context it constitutes a form of influence and a right to take action, to direct and co-ordinate the actions of others and to use discretion in the position occupied by an individual, rather than in the individuals themselves. The *delegation of authority* permits decisions to be made more rapidly by those who are in more direct contact with the problem.

The requirement applies to all personnel responsible for carrying out activities defined by the processes included in the quality management system. This will include the top management, heads of department and operational staff whose work directly or indirectly affects the quality of the organization's outputs to its customers (confirmed by published interpretation RFI 051).

Why is this Necessary?

This requirement responds to the Leadership Principle.

It is necessary for management to define who should do what in order that the designated work is assigned to specific individuals for it to be carried out. In the absence of the delegation of authority and assignment of responsibilities, individuals assume duties that may duplicate those duties assumed by others. Thus, jobs that are necessary but unattractive will be left undone. It also encourages decisions to be made only by top management which can result in an increased management workload but also engender a feeling of mistrust by the workforce.

How is this Demonstrated?

Principles of Responsibility and Authority

A person's job can be divided into two components: actions and decisions. Responsibilities and authority should therefore be described in terms of the actions assigned to an individual to perform and discretion delegated to an individual, i.e., the decisions they are permitted to take together with the freedom they are permitted to exercise. Each job should therefore have core responsibilities that provide a degree of predictability and innovative responsibilities that in turn provide the individual with scope for development.

In defining responsibilities and authority there are some simple rules that you should follow:

- Through the process of delegation, authority is passed downwards within the organization and divided among subordinate personnel whereas responsibility passes upwards.
- A manager may assign responsibilities to a subordinate and delegate authority; however, he or she remains responsible for the subordinate's use of that authority.
- When managers delegate responsibility for something, they remain responsible for it. When managers delegate authority, they lose the right to make the decisions they have delegated but remain responsible and accountable for the way such authority is used. Accountability is one's control over the authority one has delegated to one's staff.
- It is also considered unreasonable to hold a person responsible for events caused by factors over which they are powerless to control.
- Before a person can be in a state of control they must be provided with three things:
  ○ Knowledge of what they are supposed to do, i.e., the requirements of the job, the objectives they are required to achieve;
  ○ Knowledge of what they are doing, provided either from their own senses or from an instrument or another person authorized to provide such data;
  ○ Means of regulating what they are doing in the event of failing to meet the prescribed objectives. These means must always include the authority to regulate and the ability to regulate both by varying the person's own conduct and varying the process under the person's authority. It is in this area that freedom of action and decision should be provided.
- The person given responsibility for achieving certain results must have the right (i.e., the authority) to decide how those results will be achieved, otherwise, the responsibility for the results rests with those who stipulate the course of action.
- Individuals can rightfully exercise only that authority which is delegated to them and that authority should be equal to that persons' responsibility (not more or less than it). If people have authority for action without responsibility, it enables them to walk by problems without doing anything about them. Authority is not power itself. It is quite possible to have one without the other! A person can exert influence without the right to exert it.

Process Responsibilities

It is important to recognize that ISO 9001 does not dictate the assignment of responsibility or the delegation of authority, this is left for the top management to decide but the principles above will help to guide managers into making effective decisions. Among these will be process responsibilities. As processes cross functional/departmental boundaries, it is often difficult to assign the responsibility for a business process to one individual. However, if the decision is made on the basis of process outputs, the task is made simpler. In this way there is no single process owner but as many as there are process outputs. In the event that an output emanates from several sites, you would have a manager responsible for each output from a site. Although the process will be the same on each site, the implementation responsibility can be assigned to specific managers and when it comes to change the process design, all these managers come together in an Improvement Team to improve the process for all sites.

Communicating Responsibilities and Authority (5.5.1)

The standard requires that the responsibilities and authority *be communicated within the organization*.

What Does this Mean?

Communication of responsibility and authority means that those concerned need to be informed and to understand their obligations so that there is no doubt on either side about what they will be held accountable for.

Why is this Necessary?

This requirement responds to the Leadership Principle.

There are several reasons for why is it necessary to communicate this information:

- to convey consistency and avoid conflict;
- to show which functions make which contributions and thus serve to motivate staff;
- to establish channels of communication so that work proceeds smoothly without unplanned interruption;
- to indicate from whom staff will receive their instructions, to whom they are accountable and to whom they should go to seek information to resolve difficulties.

How is this Demonstrated?

There are several ways in which responsibilities and authority can be communicated:

- in an organization structure diagram or *organigram,*
- in function descriptions,
- in job descriptions,
- in terms of reference,
- in procedures,
- in process descriptions and flow charts.

The standard does not stipulate which method should be used. In very small companies a lack of such documents defining responsibility and authority may not prove detrimental to quality provided people are made aware of their responsibilities and adequately trained. However, if you are going to rely on training, then there has to be some written material that is used so that training is carried out in a consistent manner.

Organigrams are a useful way of showing interrelationships but imprecise as a means of defining responsibility and authority. They do illustrate the lines of authority and accountability but only in the chain of command. Although it can define the area in which one has authority to act, it does not preclude others having responsibilities within the same area, e.g., the title Design Manager – Computer Products, implies that the person could be responsible for all aspects of computer product design when in fact they may not have any software, mechanical engineering or reliability engineering responsibilities. Titles have to be kept brief because they are labels for communication purposes and are not usually intended for precision on the subject of responsibilities and authority. One disadvantage of organigrams is that they do not necessarily show the true relationships between people within the company. There are also no customers or suppliers on the charts thereby omitting external relationships. Horizontal relationships can be difficult to depict with clarity in a diagram. They should therefore not be used as a substitute for policy.

Function descriptions are useful in describing the role and purpose of a function, its objective and its primary responsibilities and authority. Function in this context refers to business functions rather than product functions and is a collection of activities that make a common and unique contribution to the purpose and mission of a business (see Chapter 8). Function is determined by the contribution made rather than the skill that the contributors possess. A Function/Department/Group description is needed to define the role the function executes in each process to which it contributes. These become useful in staff induction as a means of making new staff aware of who is who and who does what without getting into too much detail. They are also useful to analysts and auditors because they enable a picture of who does what to be quickly assimilated.

Job descriptions or job profiles are useful in describing what a person is responsible for; however, it rather depends on the reason for having them as to whether they will be of any use in managing quality. Those produced for job evaluation, recruitment, salary grading, etc. may be of use in the management system if they specify the objectives people are responsible for achieving and the decisions they are authorized to take.

Terms of reference are not job descriptions but descriptions of the boundary conditions. They act as statements that can be referred to in deciding the direction in which one should be going and the constraints on how to get there. They are more like rules than a job description and more suited to a committee than an individual. They rarely cover responsibilities and authority except by default.

Procedures are a common way of defining peoples' responsibilities and authority because it is at the level of procedures that one can be specific as to what someone is required to do and what results they are responsible for. Procedures specify individual actions and decisions. By assigning actions or decisions to a particular person or role a person carries out you have assigned to them a responsibility or given them certain authority. They do present problems however. It may be difficult for a person to see clearly what his or her job is by scanning the various procedures because procedures often describe tasks rather than objectives. This is the advantage of the process description. When writing procedures never use names of individuals because they will inevitably change. The solution is to use position or role titles and have a description for a particular position or role that covers all the responsibilities assigned through the process descriptions and procedures. Individuals only need to know what positions they occupy or roles they perform.

Process descriptions can be used to allocate responsibilities within a process either at function, role or job level, e.g., in the demand creation process the responsibility for the product design process could be assigned to the design department (the process owner) and then at activity level, the responsibilities would be denoted in flow charts including those personnel located in other departments.

Flow charts make it clear that anyone with the indicated title carries responsibility for performing action or decision described within the shape on the chart. Colour coding could be used to make global changes less tedious when job titles change. It does have the advantage of being concise but this could be a disadvantage as it does not allow qualification or detail to be added.

In organizations that undertake projects rather than operate continuous processes or production lines, there is a need to define and document project related responsibilities and authority. These appointments are often temporary, being only for the duration of the project. Staff are assigned from the line departments to fulfil a role for a limited period. To meet the requirement for defined responsibility, authority and interrelationships for project organizations you will need Project Organization Charts and Project Job Descriptions for each role, such as Project Manager, Project Design Engineer, Project Systems Engineer, Project Quality Engineer etc.

As project structures are temporary, there needs to be processes in place that control the interfaces between the line functions and project teams.

Some organizations have assigned responsibility for each element of the standard to a person, but such managers are not thinking clearly. There are 51 clauses and many are interrelated. Few can be taken in isolation therefore such a practice is questionable. When auditors ask 'Who is responsible for purchasing?' ask them to specify the particular activity they are interested in. Remember you have a system which delegates authority to those competent to do the job.

MANAGEMENT REPRESENTATIVE (5.5.2)

Appointing Management Representatives (5.5.2)

The standard requires the top management *to appoint a member of the organization's management who has certain defined responsibility and authority for ensuring the quality management system is established, implemented and maintained.*

What Does this Mean?

This implies that one manager in particular is delegated the authority and responsibility for managing the system but not for ensuring the system produces the desired results. In principle the Management Representative's role is similar to the roles of the Financial Director, Security Director, Safety Director etc. It is a role that exists to set standards and monitor and influence performance thus giving an assurance to management, customers and regulators that specified objectives are being achieved. The management system comprises all the processes required to create and retain satisfied customers, therefore it does not mean that this manager must also manage each of the processes but should act as a coordinator, a facilitator and a change agent, and induce change through others who in all probability are responsible to other managers.

There is a note in Clause 5.5.3 of ISO 9001, which states: The responsibility of a management representative may also include liaison with external parties on matters relating to the supplier's quality management system. Logically a representative carries the wishes of the people they represent to a place where decisions are taken that affect them – Members of Parliament, Union Representatives, Committee Members etc. The 'note' would appear to address the need for representation outside the business. Inside the business, the person represents management to the workforce but not in the same sense. The person carries the wishes of management (i.e., the policies) to the workforce so that the workforce makes decisions that take into account the wishes of management. What the organization needs is not so much a representative but a director who can represent management when necessary and influence other managers to implement and maintain the system. Such a person is unlikely to be under the direct authority of anyone other than the CEO.

In the standard the term 'management representative' appears only in the title of the requirement. The emphasis has been put on management appointing a member of its own management, indicating that the person should have a managerial appointment in the organization. This implies that a contractor or external consultant cannot fill the role. It also implies that the person should already hold a managerial position and be on the payroll. However, it is doubtful that the intention is to exclude a person from being promoted into a managerial position as a result of no current manager being available for the appointment. It is also doubtful that it would preclude the authority of the management representative being delegated to a contractor, provided that responsibility for the tasks is retained within the company. In fact TC176 interpretation RFI 027 states that it is allowable for a person who works full-time on a contract basis to act as the organization's management representative.

Why is this Necessary?

This requirement responds to the Systems Approach Principle.

As everyone in some way contributes to the quality of the products and services provided, everyone shares the responsibility for the quality of these products and services. The achievement of quality, however, is everyone's job but only in so far as each person is responsible for the quality of what they do. You cannot hold each person accountable for matters over which they have no control. It is a trait of human nature that there has to be a leader for an organization to meet an objective. It does not do it by itself or by collective responsibility – someone has to lead.

If quality is vital to survival it makes sense to appoint someone to direct the programme for establishing quality policy and putting in a system that will ensure quality is not compromised. As with finance, security and personnel these directors do not implement the policies (except those that apply to them), they regulate compliance through their influence. The other functional managers are appointed to deal with other factors critical to the company's survival and each is bound by the others' policies. This way of delegating authority works because it establishes a champion for each key factor who can devote resources for achieving specified objectives. Each manager is responsible for some aspect of security, finance, quality, personnel etc. Their responsibilities extend to implementing policy and achieving objectives. This means that the Production Director, for example, is responsible for implementing the quality policy and achieving quality objectives within a system that is under the control of the Quality Director. Likewise, the Production Director is responsible for implementing the design solution that is under the control of the Design Director.

If you were to make every manager responsible for setting policy, setting up systems and ensuring compliance then you would have as many management systems as there were managers. This is not an effective way to run a business. In such a structure, you would not have one company but as many companies as there were managers. If each manager is to serve common objectives, then we have to divide the objectives between them and either permit one manager to impose requirements on other managers (known as functional authority) or have one manager propose requirements that are agreed by the others with every manager responsible for the results (consensus management) or with the other managers contributing their ideas/views leaving the responsibility for results with the proposing manager (participative management) (see also Chapter 9).

How is this Demonstrated?

There are two schools of thought: one is that the management representative is a figurehead rather than a practitioner and has a role established solely to meet the ISO 9001 requirement. It is doubtful that any organization not registered to ISO 9001 will have made such an appointment. Those organizations not registered would not perceive there was a system to be managed. The CEO would either take on the role or would appoint one of the executive directors as the management representative in addition to his or her regular job – the role being to ascertain that a quality management system is being established, implemented and maintained. Such a person may not necessarily employ the resources to do this. These resources would be dispersed throughout the organization. While the system is being developed, a project manager is assigned to co-ordinate resources and direct the project towards its completion. After the system is fully operational, a management system manager takes over to maintain and improve the system who, with a small staff, manages the audit and improvement programmes.

The other school of thought views the management representative as a practitioner and not a figurehead. Here you would appoint a senior manager as a quality director and assign him or her the role of management representative. This director takes on the role of project manager during the development phase and management system manager during the maintenance and improvement phase. He or she acts as the management representative with the customer and registrar and in effect is the eyes of the customer inside the organization. Depending on the size and complexity of the organization, there may be one person doing all of these jobs. In some cases a fairly large team of engineers, auditors, analysts, statisticians etc. may be appropriate. Before ISO 9001, organizations appointed quality managers not management representatives. The difference is that being a quality manager was a job whereas being a management representative is a role.

To give this appointment due recognition, an appointment at executive level would be appropriate. The title chosen should reflect the position and as stated previously need not be a full-time job. Often companies appoint a member of the executive to take on the role in addition to other responsibilities. It could be the Marketing, Sales, Engineering,

Production or any other position. The notion that there has to be independence is one that is now dated and a reflection of an age when delivery was believed to be more important than quality. A person with responsibility for delivery of product or service also carries a responsibility for the quality of his or her actions and decisions. A person who therefore subordinates quality to delivery is unfit to hold the position and should be enlightened or replaced.

If you have one management system, the role of management representative and job of quality director becomes difficult to separate and can cause a conflict of interest unless the management representative is the CEO. In large organizations with multiple sites, each with separate ISO 9001 registrations, a more appropriate solution is to have a management representative for each site and one quality director for the whole organization.

As with all assignments of responsibility one has to:

- define the actions and decisions for which the person is to be responsible ensuring no conflict with others;
- define the competency needed;
- select a person with the necessary abilities;
- ensure that you give the person the necessary authority to control the results for which they are responsible;
- provide and environment in which the person is motivated to achieve the results for which they are responsible;
- evaluate and develop the person's competency to perform the role effectively.

Responsibility for Establishing and Maintaining Processes (5.5.2)

The standard requires the management representative *to ensure that processes needed for the quality management system are established, implemented and maintained.*

What Does this Mean?

The management system consists of interacting processes each of which needs to be established, implemented and maintained. This requirement means that top management being responsible for the system, delegates authority to one manager (the Quality Director/Manager/Management Representative) to orchestrate the design, development, construction, maintenance and improvement of these processes.

Why is this Necessary?

This requirement responds to the Leadership Principle.

If the CEO assigned responsibility for this task to each functional manager, it is likely that a fragmented system would emerge rather than a coherent one. Someone has to lead the effort required, to direct resources and priorities and judge the resultant effectiveness.

How is this Demonstrated?

Primarily, the designated person is the system designer for the management system appointed by top management. This person may not design all the processes and produce the documentation but may operate as a system designer. They lay down the requirements needed to implement the corporate quality policy and verify that they are being achieved. As system designer, the person would also define the requirements for processes so as to ensure consistency and lead a team of process owners who develop, implement and validate business processes. In this regard the person needs the authority to:

- manage the design, development, construction and evaluation of the processes of the management system including the necessary resources;
- determine whether the processes meet the requirements of the standard, are suitable for meeting the business needs, are being properly implemented and cause unacceptable variation to be reduced;
- manage the change processes for dealing with changes to the processes of the system.

Responsibility for Reporting on QMS Performance (5.5.2)

The standard requires the management representative *to report to top management on the performance of the quality management system and the need for improvement.*

What Does this Mean?

This requirement means that the management representative collects and analyses factual data across all the organization's operations to determine whether the quality objectives are being achieved and if not, to identify opportunities for improvement.

Why is this Necessary?

This requirement responds to the Factual Approach Principle.

Each manager cannot measure the performance of the company relative to quality although individually they carry responsibility for the utilization of resources within their own area. The performance of the company can only be measured by someone who has the ability and authority to collect and analyse the data across all company operations. All managers may contribute data, but this needs to be consolidated in order to assess performance against corporate objectives just as a Finance Director consolidates financial data.

How is this Demonstrated?

To report on management system performance and identify opportunities for improvement in the management system the management representative needs the right to:

- determine the effectiveness of the management system;
- report on the quality performance of the organization;
- identify opportunities for improvement in the management system;
- cause beneficial changes in quality performance.

By installing data collection and transmission nodes in each process, relevant data can be routed to the management representative for analysis, interpretation, synthesis and assessment. It can then be transformed into a language suitable for management action and presented at the management review. However, this requirement imposes no reporting time period, therefore, performance should also be reported when considered necessary or on request of top management.

Responsibility for Promoting Awareness of Customer Requirements (5.5.2)

The standard requires the management representative *to ensure the awareness of customer requirements throughout the organization.*

What Does this Mean?

This means that the management representative encourages and supports initiatives by others to make staff at all levels aware of customer requirements. This is not necessarily the detail requirements as would be contained in specifications, but their general needs and expectations, what is important to them, what the product being supplied will be used for and how important the customer is to sustain the business.

Why is this Necessary?

This requirement responds to the Customer Focus Principle.

Unless staff are aware of customer requirements it is unlikely that they will be achieved. Customer satisfaction is the aim of the management system and hence it is important that all staff at all levels do not lose sight of this. Clearly all managers are responsible for promoting awareness of customer requirements but this does not mean it will happen as internal pressures can cause distractions. Constant reminders are necessary when making decisions in which customer satisfaction may be directly or indirectly affected. Often staff at the coalface are remote from the end user and unaware of the function or role of their output relative to the final product or service. Heightened awareness of customer requirements and the role they play in achieving them can inject a sense of pride in what they do and lead to better performance.

How is this Demonstrated?

There are general awareness measures that can be taken and awareness for specific customers. General awareness can be accomplished through:

- the quality policy and objectives;
- induction and training sessions;
- instructions conveyed with product and process documentation;
- bulletins, notice boards and staff briefings;
- brochures of the end product in which the organization's product features;
- videos of customer products and services featuring the organization's products.

It is also a responsibility of designers to convey (through the product specifications) critical features and special customer or CTQ[①] characteristics. Also production planners or service delivery planners should denote special requirements in planning documents so that staff are alerted about the requirements that are critical to customers. (See also *Competency training and awareness* in Chapter 20.)

Responsibility for External Liaison

Although only a note in Clause 5.5.3 of the standard, it is necessary to have someone who can liase with customers or other stakeholders on quality issues, who can co-ordinate visits by stakeholders or their representatives and who can keep abreast of the state of the art in quality management through conferences, publications and exhibitions. However, it does not have to be the same person who deals with all external liaison.

INTERNAL COMMUNICATION (5.5.3)

Establishing Communication Processes (5.5.3)

The standard requires *appropriate communication processes to be established within the organization.*

What Does this Mean?

Communication processes are those processes that convey information and impart understanding upwards, downwards and laterally within the organization. They include the people transmitting information, the information itself, the receivers of the information and the environment in which it is received. They also include all auditory and visual communication, the media that convey the information and the infrastructure for enabling the communication to take place. The medium of communication such as telephone, e-mail and video is the means of conveying information and forms part of the process of communication.

Why is this Necessary?

This requirement responds to the Leadership Principle.

The operation of a management system is dependent on effective transmission and reception of information and it is the communication processes that are the enablers. Information needs to be communicated to people for them to perform their role as well as possible. These processes need to be effective otherwise

- the wrong information will be transmitted;
- the right information will fail to be transmitted;
- the right information will go to the wrong people;
- the right information will reach the right people before they have been prepared for it;
- the right information will reach the right people too late to be effective;
- the communication will not be understood;
- the communication will cause an undesirable result.

How is this Demonstrated?

As the requirement focuses on processes rather than the subject of communication it follows that whatever information needs to be communicated, the communicator needs to select an appropriate communication process. There needs therefore to be some standard processes in place that can be used for communicating the majority of information. There also needs to be a communication policy that facilitates downwards communication and encourages upwards and lateral communication.

A simple solution would be to identify the various types of information that need to be conveyed and the appropriate process to be used. In devising such a list you need to consider the audience and their location along with the urgency, sensitivity, impact and permanency of the message.

- Audience influences the language, style and approach to be used (Who are they?).
- Location influences the method (Where are they?).
- Urgency influences the method and timing when the information should be transmitted (When it is needed?).
- Sensitivity influences the distribution of the information (Who needs to know?).
- Impact influences the method of transmission and the competency of the sender (How should they be told and who should tell them?).
- Permanency influences the medium used (Is it for the moment or the long-term?).

Communication processes should be established for communicating:

- **The vision, mission and values of the organization** – While displaying this information acts as a reminder, this does not communicate. You need to establish a process for gaining understanding, getting commitment and sharing vision and values. A method for developing shared vision is addressed by Bryson.[1]
- **Operating policies** – These are often conveyed through manuals and procedures but a communication process is needed to ensure that they are understood at all levels.
- **The corporate objectives** – A process is needed for conveying these down the levels in the hierarchy with translation possibly at each level as they are divided into departmental, group, section and personal objectives.
- **Plans for entering new markets, for new products and processes and for improvement** – A process is needed for communicating plans following their approval so that all engaged in the project have a clear understanding of the strategy to be adopted.
- **Customer requirements, regulations and statutory requirements** – A process is needed for ensuring that these requirements reach the point at which they are implemented and are understood by those who will implement them.
- **Product and process objectives** – These are often conveyed through plans but a communication process is needed to ensure that they are understood by those who are to use them or come into contact with them.
- **Product and process information** – A process is needed to ensure that all product and process information gets to those who need it, when they need it and in a form that they can understand.
- **Problems** – A process is needed for communicating problems from their source of detection to those who are authorized to take action.
- **Progress** – Managers need to know how far we are progressing through the plan and therefore a communication process is needed to ensure the relevant managers receive the appropriate information.

---

[1] Bryson John M. (2004) Creating and implementing your strategic plan – A workbook for public and non-profit. Jossey Bass.

- **Change** – All change processes should incorporate communication processes in order to gain commitment to the change.
- **Results and measurements** – A process is needed for communicating financial results, quality and delivery performance, accomplishments, good news, bad news and customer feedback.

Clearly, not everything can be communicated to all levels because some information will be sensitive, confidential or simply not relevant to everyone. Managers therefore need to exercise a 'need to know' policy that provides information necessary for people to do their job as well as creating an environment in which people are motivated. Other than national and commercial security, too much secrecy is often counterproductive and creates an atmosphere of distrust and suspicion that affects worker performance. If you don't want somebody to have information say so, don't give them a reason that is untrue just to get them off your back for it will return to bite you!

In communication processes there needs to be a feedback loop to provide a means for conveying questions and queries as well as acceptance. These feedback loops should be short – to the next level only. An Executive who demands to be kept informed of progress will soon stop reading the reports and if the process continues without change, the reports will just pile up in his or her office.[2] This is not an uncommon phenomenon. A manager may demand reports following a crisis but fail to halt further submissions when the problem has been resolved. The opposite is also not uncommon where a local problem is not communicated outside the area and action is subsequently taken which adversely affects the integrity of the management system.

Communicating the Effectiveness of the QMS (5.5.3)

The standard requires communication *to take place regarding the effectiveness of the quality management system.*

What Does this Mean?

This means that there should be communication from top management and from those on the scene of action (two-way communication) as to whether the management system is enabling achievement of the organization's objectives. Information from above should initiate improvement action. Information from below should prompt investigation and analysis in order to identify improvement actions.

Why is this Necessary?

This requirement responds to the Leadership Principle.

It is important that staff are kept informed of how effective the management system is to encourage continuation of the status quo or to encourage improvement. Also staff should be encouraged to report system effectiveness or ineffectiveness whenever it is encountered.

---

[2] Juran, J. M. (1995). Managerial Breakthrough, Second Edition. McGraw-Hill.

How is this Demonstrated?

After each management review, the results can be communicated to staff but care should be taken with the format of the message. Charts against each major objective showing how performance has changed are the most effective. New improvement initiatives should also be communicated indicating the project name, the project leader or champion, the project objectives and timescales. However, the application of this requirement should not be restricted to annual communication briefings. Update the charts monthly and display on notice boards or on the Intranet. Provide means for staff to alert management of ineffectiveness in the management system by opening channels through to the management representative. It is not uncommon for a particular practice to be changed locally or ignored altogether and subsequently discovered on periodic audit. There should be free communication so no one takes such action without consultation and agreement.

Chapter 18

Management Review

CHAPTER PREVIEW

This chapter is aimed primarily at top management for it is they who must review the management system. However, reviews may be conducted at other levels and managers at these levels will be involved along with people collecting and analysing the data such as project managers, change managers, statisticians, improvement coordinators and others whose role is to seek opportunities for improvement and push through changes in performance.

Having established a system for managing the organization, it follows that this system should be periodically reviewed in order to establish that it is working properly. By this we mean that it continues to enable the organization to achieve its mission, vision and strategic objectives in an efficient and effective manner.

In this chapter we examine the requirements in Clauses 5.1 and 5.6 of ISO 9001:2008 on management review and in particular:

- What the management review means and why it is necessary
- Where it fits in context with other reviews
- What the review should be designed to achieve
- How reviews should be planned
- What the scope of the reviews should be
- What records should be generated and retained
- What the inputs and outputs of the review should be

The position where the requirements for management review feature in the managed process is shown in Fig. 18-1. In this case they form part of the mission management process and perform two functions: (a) a process review and improvement function that leads to improvements by improved control, improved efficiency and improved effectiveness and (b) a performance review checking whether the strategic plans are achieving their objectives and if not initiating action to correct deficiencies (correction) or restore the status quo (remedial or corrective action). The performance review function is present because Clause 5.6.3 refers to improvement in product related requirements.

CONDUCTING MANAGEMENT REVIEWS (5.1d)

The standard requires that top management *conduct management reviews.*

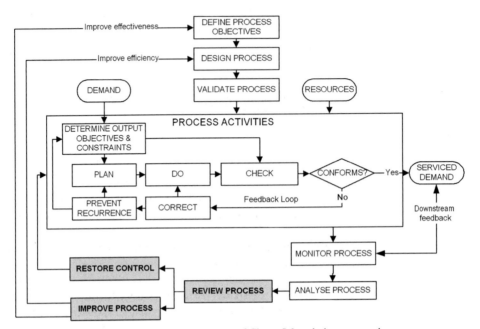

FIGURE 18-1   Where the requirements of Clause 5.6 apply in a managed process.

What Does this Mean?

The term review is defined in ISO 9000:2005 as an activity undertaken to ensure the suitability, adequacy, effectiveness and efficiency of the subject matter to achieve established objectives. The addition of the term management means that the management review can be perceived as a review of management rather than a review by management, although both the meanings are conveyed in the standard. The rationale for this is that the examples given in ISO 9000:2005 such as design review and nonconformity review clearly indicate it is design and nonconformity that are being reviewed. If the system was to be reviewed then the action should be called a system review. It is no doubt unintentional in the standard but, if the management system is the means the organization employs to achieve the ends; a review of results without review of the capability to achieve them (the means) would be ineffective. For these reasons the management review as referred to in ISO 9001 could well be the Strategic Review or Business Performance Review with separate committees or focus groups targeting specific aspects.

Why is this Necessary?

This requirement responds to the Factual Approach Principle.

Top management has set the policies and agreed the objectives and means for their achievement, i.e., the management system. It follows therefore that periodic checks are needed to establish that the management system continues to fulfil its purpose. Failure to do so will inevitably result in deterioration of standards and performance as inherent weaknesses in the system build up to eventually cause catastrophic failure, i.e., customer satisfaction will decline and orders, markets and business will be lost. It may be argued that this won't happen because people won't let it happen – they will take action. If these conditions persist, what will emerge is not a managed system but an unmanaged system that is unpredictable, unreliable with erratic performance. A return to the days before the management system (as defined in this book) was established.

How is this Demonstrated?

Top management will not regard the management review as important unless they believe it is essential for running the business. The way to do this is to treat it as a business performance review. This is simpler than it may appear. If the quality policy is now accepted as corporate policy and the quality objectives are accepted as corporate objectives, any review of the management system becomes a performance review and is no different from any other executive meeting. The problem with the former management reviews was that they allowed discussion on the means for achieving objectives to take place in other management meetings leaving the management review to a review of errors, mistakes and documentation that no one was interested in anyway. The management system *is* the means for achieving objectives therefore it makes sense to review the *means* when reviewing the *ends* so that actions are linked to results and commitment secured for all related changes in one transaction.

The requirement emphasizes that top management conduct the review – not the quality manager, not the operational manager – but *top management* – those who direct and control the organization at the highest level. In many ISO 9001 registered organizations, the management review is a chore, an event held once each year, on a Friday afternoon before a national holiday – perhaps a cynical view but nonetheless often true.

One of the reasons that the management review may not work is when it is considered something separate from management's job, separate from running the business, a chore to be carried out just to satisfy the standard. This is partially due to perceptions about quality. If managers perceive quality to be about the big issues like new product or service development, major investment programmes for improving plant, for installing computerization etc. then the management review will take on a new meaning. If on the other hand it is about reviewing nonconformities, customer complaints and internal audit records, it will not attract a great deal of attention unless of course the audits also include the big issues.

In order to provide evidence of its commitment to conducting management reviews, management would need to demonstrate that it planned for the reviews, prepared input material in the form of performance results, metrics and explanations, decided what to do about the results and accepted action to bring about improvement. It would also show that it placed the burden on all managers and did not single out the quality manager as the person responsible and accountable for the management review.

OBJECTIVES OF THE REVIEW (5.6)

The standard requires top management to *review the quality management system to ensure its continuing suitability, adequacy and effectiveness.*

TABLE 18-1 Elements of Management System Performance

Concept	ISO term	Other terms
Output meets requirements	Adequacy	Effectiveness
Results achieved in best way	Suitability	Efficiency
System fulfils needs	Effectiveness	Adaptability

What Does this Mean?

The three terms: adequacy, suitability and effectiveness are not included as three alternatives but as three different concepts. However, their meanings vary as illustrated in Table 18-1.

The adequacy of the management system is judged by its ability to deliver product or service that satisfies requirements, standards and regulations. It does what it was designed to do. In some cases this condition is referred to as effectiveness.

The suitability of the management system is judged by its ability to enable the organization to sustain current performance. If the management system is inefficient, the organization may not be able to continue to feed a resource-hungry system. In such cases we would be justified in claiming that the management system is not suitable for its purpose even though customers may be satisfied by the outputs, other stakeholders will soon express dissatisfaction. A better term would be efficiency.

The effectiveness of the management system is judged by how well it enables the organization to fulfil the needs of society. The system may deliver satisfied customers and minimize use of resources but if it is not responding to the changing needs of society, of customers, of regulators and of other stakeholders, it is not an effective system. In some cases this concept is referred to as adaptability. However, effectiveness is about doing the right things, not doing things right. Doing things right is about satisfying the customer. Doing the right things is choosing the right objectives. If the corporate objectives change or the environment in which the organization operates changes, will the system enable the organization to achieve these new objectives or operate successfully in the new environment? If the purpose of the system is merely to ensure customers are supplied with products and services which meet their requirements, then its effectiveness is judged by how well it does this and not how much it costs to do it. If the purpose of the management system is to enable the organization to fulfil its purpose, its effectiveness is judged by how well it does this. The measures of effectiveness are therefore different.

Why is this Necessary?

This requirement responds to the Systems Approach Principle.

There is a need for top management, as the sponsors of the system, to look again at the data the system generates and determine whether the system they established is actually doing the job they wanted it to do. If the organization is being managed as a system then this system is the management system, so a review of the system is a review of the organization. Financial performance is reviewed regularly and a statement of accounts is produced every year. There are significant benefits to be gained if quality performance is treated in the same way because it is quality performance that actually causes the financial performance. Under-performance in any area will be reflected in the financial results.

Treating the review as a chore, something that we do because we want to keep our ISO 9001 certificate, will send out the wrong signals. It will indicate that members of top management are not serious about quality or about the system they commissioned to achieve it – it will also indicate they don't understand and if they don't understand they clearly cannot be committed.

How is this Demonstrated?

In any organization, management will conduct reviews of performance so as to establish how well the organization is doing in meeting the defined objectives. As the objectives vary it is often more practical to plan reviews relative to the performance characteristic being measured. As a result, organizations may convene strategic reviews, divisional reviews, departmental reviews, product reviews, process reviews, project reviews etc. They all serve the same purpose, that of establishing

- whether performance is in line with objectives;
- whether there are better ways of achieving these objectives;
- whether the objectives are relevant to the needs of the stakeholders.

The review(s) are part of the improvement process and where it (they) fits in this process is illustrated in Fig. 18-2.

Such reviews should be part of the management system and will examine the capability of the system to deliver against objectives and fulfil the mission. The standard does not require only one review. In some organizations, it would not be practical to cover the complete system in one review. It is often necessary to consolidate results from lower levels and feed into intermediate reviews so that departmental reviews feed results into divisional reviews that feed results into corporate reviews. The fact that the lower level reviews are not performed by top management is immaterial provided the results of these reviews are submitted to top management as part of the system review. It is also not necessary to separate reviews on the basis of ends and means. A review of financial performance is often separated from technical performance and both of these are separated from management system reviews. This situation arises in cases where the management system is perceived as procedures and practices. In organizations that separate their performance reviews from their management system reviews, one has to question whether they are gaining any business benefit or in fact whether they have really understood the purpose of the management system.

In determining the effectiveness of the management system you should continually ask:

- Does the system fulfil its purpose?
- To what extent are our customers satisfied with our products and services?
- Are the corporate objectives being achieved as intended?
- Do measurements of process performance indicate the processes are effective?
- Do the results of the audits indicate that the system is effective?
- Are procedures being used properly?
- Are policies being adhered to?

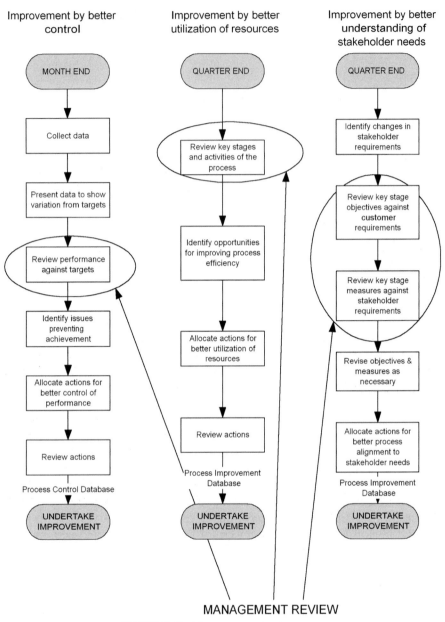

FIGURE 18-2   Management review in context.

If the answer is 'Yes' your system is operating effectively. If your answer is 'No' to any of these questions, your management system has not been effectively designed or is not being effectively used.

The review should be in three stages. Stage one is collecting and analysing the data, stage two is reviewing the data and stage three is a meeting to discuss the results and decide on a course of action. When you have a real understanding of the intentions of the review you will realize that its objectives cannot be accomplished entirely by a meeting.

PLANNING THE REVIEW (5.6.1)

The standard requires *management reviews at planned intervals.*

What Does this Mean?

Planned intervals mean that the time between the management reviews should be determined in advance, e.g., annual, quarterly or monthly reviews. The plan can be changed to reflect circumstances but should always be looking forward.

Why is this Necessary?

By requiring reviews at planned intervals it indicates that some forethought is needed so that performance is measured on a regular basis thus enabling comparisons to be made.

How is this Demonstrated?

This requirement responds to the Process Approach Principle.

A simple bar chart or table indicating the timing of management reviews over a given period will meet this requirement. The frequency of management reviews should be matched to the evidence that demonstrates the effectiveness of the system. Initially the reviews should be frequent say monthly, until it is established that the system is effective. Thereafter the frequency of reviews can be modified. If performance has already reached a satisfactory level and no deterioration appears within the next three months, extend the period between reviews to six months. If no deterioration appears in six months extend the period to 12 months. It is unwise to go beyond 12 months without a review because something is bound to change that will affect the system. Shortly after reorganization, the launch of a new product or service, breaking into a new market, securing new customers etc. a review should be held to establish if performance has changed. After new technology is planned, a review should be held before and afterwards to measure the effects of the change.

SCOPE OF REVIEW (5.6.1)

The standard requires the review *to include assessing opportunities for improvement and the need for changes to the quality management system including quality policy and quality objectives.*

What Does this Mean?

The review is of current performance and hence there will be some parameters where objectives or targets have not been accomplished thus providing opportunities for improvement. There will also be some areas where the status quo is not good enough for the growth of the organization or to meet new challenges.

Why is this Necessary?

This requirement responds to the Continuous Improvement Principle.

Top management should never be complacent about the organization's performance. Even maintaining the *status quo* requires improvement, just to maintain market position, keep customers and retain capability. If the management review restricts its agenda to examining audit results, customer complaints and nonconformities month after month without a commitment to improvement, the results will not get any better, in fact they will more than likely get worse.

There will be reports about new marketing opportunities, reports about new legislation, new standards, the competition and benchmarking[①] studies. All these may provide opportunities for improvement. In this context improvement means improvement by better control (doing things better) as well as improvement by innovation (doing new things). These changes may affect the quality policy (see also under the heading *Ensuring the policy is reviewed*) and will certainly affect the objectives. Objectives may need to change if they proved to be too ambitious or not far reaching enough to beat the competition.

How is this Demonstrated?

The implication of this requirement is that performance data on the implementation of quality policy and the achievement of quality objectives should be collected and reviewed in order to identify the need to change the system, the quality policy and quality objectives.

As the management system is the means by which the organization achieves its objectives, it follows that the management review should evaluate the need for changes in the objectives and the processes designed to achieve them. As no function, process or resource in the organization would exist outside the management system, the scope of the review is only limited by the boundary of the organization and the market and environment in which it operates.

The approach you take should be described in your quality manual – but take care! What you should describe is the process by which you determine the suitability, adequacy and effectiveness of the management system. In doing this you would describe all the performance reviews conducted by management and show how they serve this objective. Describing a single management review without reference to all the other ways in which performance is reviewed sends out the signal that there isn't an effective process in place.

RECORDS OF MANAGEMENT REVIEWS (5.6.1)

The standard requires records from management reviews *to be maintained.*

What Does this Mean?

A record from the review means the outcome of the review but the outcome won't be understood unless it is placed in the right context. The records therefore also need to include the criteria for the review and who made what decisions.

Why is this Necessary?

This requirement responds to the Factual Approach Principle.

Records from management reviews are necessary for several reasons to:

- convey the actions from the review to those who are to take them;
- convey the decisions and conclusions as a means of employee motivation;
- enable comparisons to be made at later reviews when determining progress;
- define the basis on which the decisions have been made;
- demonstrate system performance to stakeholders.

How is this Demonstrated?

The records from management reviews need to contain:

- Date of review (location might be necessary if the review is carried out at a meeting);
- Contributors to the review (the process owners, functional managers, management representative, auditors etc.);
- Criteria against which the management system is being judged for effectiveness (the organization's objectives, measures and targets);
- Criteria against which the management system is to be judged for continued suitability (future changes in the organization, legislation, standards, customer requirements, markets);
- The evidence submitted, testifying the current performance of the management system (charts, tables and other data against objectives);
- Identification of strengths, weaknesses, opportunities and threats (SWOT – the analysis of: What are we good at? What we are not good at? What can we change? What can't we change? What must we change?);
- Conclusions (is the management system effective or not and if not in what way?);
- Actions and decisions (what will stay the same and what will change?);
- Responsibilities and timescales for the actions (who will do it and by when will it be completed?).

REVIEW INPUTS (5.6.2)

The standard requires inputs to management review to include *information about various aspects of the system.*

What Does this Mean?

This means that data from audits, product measurements, process measurements, customers, end users, suppliers, regulators, etc. has to be analysed relative to defined objectives to establish current performance (How are we doing?) and identify improvement opportunities (Can we do better?). Data on planned changes in the organization, resources, the infrastructure, legislation and standards have to be examined for their impact.

Why is this Necessary?

This requirement responds to the Factual Approach Principle.

It is necessary to gather sufficient data relative to the objectives being measured to provide a sound basis for the review. Any review of the management system needs to be based on fact. The factors identified in this clause cover most of the parameters influencing the effectiveness of the management system.

How is this Demonstrated?

The key questions to be answered are: Is the system delivering the desired outputs? Is it suitable for continued operation without change? Is the system effective? And what information do we need to answer these questions in full? The standard identifies several inputs to the review, which are addressed below, but these should not be seen as limiting. The input data should be that which is needed to make a decision on the effectiveness of the system. Rather than use a generic list as is presented in the standard, a list tailored to business needs should be developed and continually reviewed and revised as necessary.

System Performance

The most important factor that has been omitted from the list of inputs in Clause 5.6.2 of ISO 9001 is system performance. System performance data should be used to establish whether the defined quality objectives are being met. It may also be used to establish whether there is conflict between the stated quality policy, the quality objectives and the organizational purpose and the expectations and needs of the stakeholders. There may be a small number of factors on which the performance of the organization depends and these above all others should be monitored. For example, in a telephone data centre, processing thousands of transactions each day, system availability is paramount. In a fire department or an ambulance service, response time is paramount. In an air traffic control centre the number of near misses is paramount because the centre exists to maintain aircraft separation in the air. Analysis of the data that the system generates should reveal whether the targets are being achieved. It is also important to establish whether the system provides useful data with which to manage the business. This can be done by providing evidence showing how business decisions have been made. Those made without using available data from the management system show either that poor data is being produced or management is unaware of its value. One of eight quality management principles is the *factual approach to decision making* and therefore implies decisions should be made using data generated from the management system.

Improvement opportunities may cover:

- the identification of major projects to improve overall performance;
- the setting of new objectives and targets;
- the revision of the quality policy;
- the adequacy of the linkages between processes.

Audit Results

Audit results should be used to establish whether the system is operating as planned and whether the commitments declared in the quality policy are being honoured. You can determine this by providing the results of all quality audits of the system, of processes and of products. An analysis of managerial decisions should reveal whether there is constancy of purpose or lip service being given to the policy. Audit results should also be used to establish whether the audit programme is being effective and you can determine this by providing the evidence of previous audit results and problems reported by other means. Current performance from audit results should compare the results with the quality objectives you have defined for the system as a whole and for the audit programme in particular.

Improvement opportunities relative to audit results may cover:

- the scope and depth of the audit programme;
- the suitability of the audit approach to detect problems worthy of management attention;
- the competency of the auditors to add value and discover opportunities that enhance the organization's capability;
- the relevance of audit results to the organization's objectives.

Customer Feedback

Customer feedback should be used to establish whether customer needs are being satisfied. You can determine this by providing the evidence of customer complaints, market share statistics, competitor statistics, warranty claims, customer satisfaction surveys etc. Current performance from customer feedback should compare the results with the quality objectives you have set for customer needs and expectations.

Improvement opportunities relative to customer feedback may cover:

- the extent to which products and services satisfy customer needs and expectations;
- programmes to eliminate the root cause of an increasing trend in customer rejects or returns;
- the adequacy of the means used to assess customer satisfaction and collect data;
- the need to develop new or enhanced products or services;
- the need to explore new markets or obtain more accurate data of current markets.

Process Performance

Process performance data should be used to establish whether process objectives are being achieved. Current performance of processes should compare process data with the quality objectives you have set for the processes. Improvement opportunities relative to process performance may cover:

- the efficiency of processes relative to the utilization of resources (physical, financial and human resources and the manner in which they are structured);
- the effectiveness of processes relative to the utilization of knowledge, experience in achieving process objectives;
- the need to change process design, methods and techniques including process measurement;
- the need to reduce variation;
- the need to meet or exceed new legal and regulatory requirements that apply to the process.

Product Performance

Product performance data should be used to establish whether products fulfil their intended purpose in both design and build quality. Current performance of products should compare product data with product specifications and product specifications with design intent (what the product was intended to accomplish). The data may be obtained from actual and potential field failures and studies undertaken to assess the impact on safety and the environment.

Improvement opportunities relative to product performance may cover:

- the need to change product design, technology and materials;
- the need to change product literature to match actual performance (reset expectations);
- the adequacy of the means used to measure product performance and collect data;
- the conditions of use and application.

Corrective Actions

Corrective action data should be used to establish whether the recurrence of problems is being prevented. A list of last year's problems by root cause against a list of this year's problems by root cause might indicate whether the solutions you implemented last year were effective. ("Today's problems come from yesterday's solutions" - Peter Senge) Current performance on the status of corrective actions should compare the results with the quality objectives you have set for dealing with corrective actions such as closure time and degree of recurrence. Improvement opportunities relative to corrective actions may cover:

- the adequacy of problem analysis and resolution techniques;
- the need for new training programmes;
- the capability of the system to maintain performance in line with objectives (its sensitivity to change).

Preventive Actions

Current performance on the status of preventive actions should compare the results with the quality objectives you have set for dealing with preventive actions such as closure time and degree of occurrence. Remember that preventive actions are supposed to prevent the occurrence of problems therefore a measure of status is the extent to which problems occur. However, this is an area that often causes confusion. Preventive action is often not an action identified as preventive but an action identified under guise of planning, training, research and analysis. Why else would you plan, but to achieve objectives and hence to prevent failure? Why else do you perform an FMEA[1], but to prevent failure? Therefore, don't just look for actions with the label *preventive*.

Improvement opportunities relative to preventive actions may cover:

- the adequacy of techniques to identify potential problems;
- the need for new tools and techniques, training programmes etc.;
- the re-organization of departments, resources etc.

Actions from Management Reviews

Current performance on follow-up actions from earlier management reviews should address not only whether they are open or closed but how effective they have been and how long they remain outstanding as a measure of planning effectiveness.

Improvement opportunities relative to actions from management reviews may cover:

- the prioritization of actions;
- the reclassification of problems relative to current business needs;
- the need to re-design the management review process.

Changes Affecting the Management System

It is difficult to foresee any change inside the organization that would not affect the management system in some way or other. However, the management system should be designed to cope with a degree of change without top management intervention. The change management processes to bring in new products, new processes, new people, new resources and new organization structures should be part of the management system as explained previously. Changes in products, processes, organization structures etc. will all affect the management system documentation but there should be processes in place to manage these changes under controlled conditions. In an environment in which perceptions of the management system have not been harmonized, it is likely that some change mechanisms will be outside the documented management system and in such circumstances, these changes need to be brought to the management review.

REVIEW OUTPUTS (5.6.3)

The standard requires the outputs from the management review to include *decisions and actions related to the improvement of the effectiveness of the quality management system and its processes, improvement of product and actions related to resource needs.*

What Does this Mean?

Improving the effectiveness of the management system is not about tinkering with documentation but enhancing the capability of the system so that it enables the organization to fulfil its objectives more effectively. The management system comprises processes therefore the effectiveness of these too must be improved. Improvement of product related to customer requirements means not only improving the degree of conformity of existing product but enhancing product features so that they meet changing customer needs and expectations.

Why is this Necessary?

This requirement responds to the Continual Improvement Principle.

The outcomes of the management review should cause beneficial change in performance. The performance of products is directly related to the effectiveness of the processes that produce them. The performance of these processes is directly related to the effectiveness of the system that connects them. Without resources no improvement would be possible.

How is this Demonstrated?

The implication of this requirement is that the review should result in decisions being made to improve products, processes and the system in terms of the actions required.

Actions related to improvement of the system should improve the capability of the system to achieve the organization's objectives by undertaking action to improve the interaction between the processes.

Actions related to improvement of processes should focus on making beneficial changes in methods, techniques, relationships, measurements, behaviours, capacity, competency etc. A quick fix to overcome a problem is neither a process change nor a process improvement because it only acts on a particular problem. If the fix not only acts on the present problem but will also prevent its recurrence, it can be claimed to be a system improvement. This may result in changes to documentation but this should not be the sole purpose behind the change – it is the performance that should be improved.

Actions related to improvement of products should improve:

- The quality of design. The extent to which the design reflects a product that satisfies customer needs.
- The quality of conformity. The extent to which the product conforms to the design.
- The quality of use. The extent to which the user is able to secure continuity of use and low cost of ownership from the product.

Such actions may result in providing different product features or better-designed product features as well as improved reliability, maintainability, durability and performance. Product improvements may also arise from better packaging, better user instructions, clearer labelling, warning notices, handling provisions etc.

Actions related to resource needs are associated with the resource planning process that should be part of the management system. If this process were operating effectively, no work would commence without adequate resources being available. If the resources cannot be provided, the work should not proceed. It is always a balance between time, effort and materials. If the effort cannot be provided the time has to expand accordingly.

Key Messages from Part 4

1. The management commitment required in ISO 9001 is a commitment by top management to produce products and deliver services that satisfy customer needs and expectations and a commitment to managing the organization as a system that enables this goal to be achieved consistently.

2. Such commitment means that top management cannot walk by problems that affect the quality of products and ignore them; they must act and set a good example of leadership by creating conditions that enable their employees and suppliers to achieve the objectives they have been set.

3. Customer requirements in ISO 9001 are not simply what the customer specifies in a written order but the customer's stated and implied needs and expectations and determining these requires the organization to be proactive i.e. discover what potential customers will be looking for before the product or service is designed.

4. There needs to be a link between policy and objectives otherwise the processes designed to achieve them would be unlikely to implement the policy.

5. The quality policy is a constraint on how you conduct your business. It is not an objective, the objective comes from the customer. The policy is therefore used to measure how well the objective is achieved.

6. There is nothing to be gained by labelling objectives 'Quality objectives'

7. Treating profit as a constraint rather than an objective causes the designers and producers to look for ways of reducing costs.

8. Having a well defined quality policy and well defined quality objectives will change nothing. The organization needs to empower its people to implement the policy and achieve the objectives through effectively managed processes and only then will it be able to accomplish the mission, the vision and achieve the goals

9. At the strategic level, the subjects that are the focus for setting objectives are the outputs that will produce successful outcomes for the organization's stakeholders

10. Deming advocated in the 11th of his 14 points "Eliminate management by objectives" for the simple reason that management derives the goals from invalid data. They observe that a goal was achieved once and therefore assume it can be achieved every time. If they understood the process they would realize that the highs and lows are a characteristic of natural variation

11. If people have authority for action without responsibility, it enables them to walk by problems without doing anything about them.

12. one cannot have responsibility without authority and vice versa. Problems arise when these two are not matched, where one is greater or less than the other

13. In the absence of the delegation of authority and assignment of responsibilities, individuals assume duties that may duplicate those duties assumed by others resulting in jobs that are necessary but unattractive left undone.

14. If quality is vital to survival it makes sense to appoint someone to direct the programme that will ensure quality is not compromised. As with finance, security and personnel these directors do not implement the policies (except those that apply to them), they regulate compliance. The other functional managers are appointed to deal with other factors critical to the company's survival and each is bound by the others' policies.

15. A person with responsibility for delivery of product or service also carries a responsibility for the quality of his or her actions and decisions. A person who therefore subordinates quality to delivery is unfit to hold the position and should be enlightened or replaced

16. Information from above should initiate improvement action. Information from below should prompt investigation and analysis in order to identify improvement actions.

17. The management system *is* the means for achieving objectives therefore it makes sense to review the *means* when reviewing the *ends* so that actions are linked to results and commitment secured for all related changes in one transaction.

18. If the management review restricts its agenda to examining audit results, customer complaints and nonconformities month after month without a commitment to improvement, the results will not get any better, in fact they will more than likely get worse.

19. The key questions to be answered are; Is the system delivering the desired outputs? Is it suitable for continued operation without change? Is the system effective? and What information do we need to answer these questions in full?

20. The outcomes of the management review should cause beneficial change in performance. The performance of products is directly related to the effectiveness of the processes that produce them. The performance of these processes is directly related to the effectiveness of the system that connects them. Without resources no improvement would be possible.

Complying with ISO 9001 Section 6 Requirements on Resource Management

INTRODUCTION TO PART 5

Structure of ISO 9001 Section 6

Section 6 of ISO 9001 draws together all the resource-related requirements that were somewhat scattered in previous versions. Resource management is a key business process in all organizations. In practice, resource management is a collection of related processes that are often departmentally oriented.

- Financial resources might be controlled by the Finance Department;
- Purchased materials, equipment and supplies might be controlled by the Purchasing Department;
- Measuring equipment maintenance might be controlled by the Calibration Department;
- Plant maintenance might be controlled by the Maintenance Department;
- Human resources might be recruited, developed and dismissed by HR Department;
- Building maintenance might be controlled by the Facilities Management Department.

These departments might control the resources in as much that they plan, acquire, maintain and dispose of them but do not manage them totally because they are not the sole users or customers of the resource. They might therefore only perform a few of the tasks necessary to manage resources. Collectively they may control the human, physical and financial resources of the organization.

Whatever the resource, firstly it has to be planned, then acquired, deployed, maintained and eventually disposed of. The detail of each process will differ depending on the type of resource being managed. Human resources are not 'disposed off' but their employment or contract terminated.

The standard does not address financial resources specifically (see ISO 9004) but clearly they are required to implement and maintain the management system and hence run the organization. Purchasing is not addressed under resource management but under product realization. However, the location of clauses should not be a barrier to the imagination because their location is not governed by the process approach but by user expectations.

TABLE P5 Clause Alignment with Process Model

| | Types of resource | | | | | |
| | | Physical resources | | | | |
Resource management processes	Human resource	Plant, equipment and materials	Buildings	Utilities and support services	Measuring devices	Financial resource
Resource planning	6.1 6.2.2a 7.1	6.1 6.3 7.1	6.1 6.3 7.1	6.1 6.3 7.1	7.1 7.6	6.1 7.1
Resource acquisition	6.1 7.4	6.1 6.3 7.4	6.1 6.3	6.1 6.3 7.4	7.4	6.1
Resource deployment	4.1d 6.1 6.2.1	4.1d 6.1 6.3	4.1d 6.1 6.3	4.1d 6.1 6.3	4.1d 7.5.1	4.1d 6.1
Resource maintenance	6.2.2b	6.3	6.3	6.3	7.6	NA
Resource disposal	Not addressed	Not addressed	Not addressed	Not addressed	Not addressed	Not addressed

Linking Requirements

The clauses of ISO 9001 are related to the processes identified in fig. 10-7 as indicated in Table P5.

Note that there are no clauses that address resource disposal. This is probably because the standard only focuses on intended product, whereas ISO 14001 would address resource disposal and unintended product. Resource disposal impacts the environment and other stakeholders and if a company discharges waste into the ground water, it could lead to prosecutions that displease the customer therefore it would be unwise to exclude it from your management system.

Chapter 19

Determining and Providing Resources

CHAPTER PREVIEW

This chapter is aimed at all levels of management for it is they who determine and provide resources but it is top management which holds the purse strings and distributes the funds to acquire other resources to lower levels of management.

In Chapter 18 we stated that the organization needs to empower its people to implement the policy and achieve the objectives through effectively managed processes and only then will it be able to accomplish the mission. However, it needs resources to acquire the people in the first place and once acquired the people become a resource. Resources are the fuel an organization needs to keep going and achieve its goals Therefore, it becomes critical for sustained success that organizations determine the resources they need to accomplish their goals and if they can't maintain an adequate supply, they might have to modify their goals. It is difficult maintaining a position of being world class if you are in a market where the competition can starve you of resources or you are a public body that depends on the political will to channel the necessary resources your way.

In this chapter we examine the requirements on resources in Clauses 5.1 and 6.1 of ISO 9001:2008 and in particular

- What top management does to ensure the availability of resources;
- The methods used to determined resources;
- The methods used to ensure that the identified resources are provided.

The chapter deals with resources in general, the next chapter deals with human resources in particular.

The position where the requirements for planning and providing resources feature in the managed process is shown in Fig. 19-1.

ENSURING THE AVAILABILITY OF RESOURCES (5.1e)

The standard requires that top management *ensure the availability of necessary resources.*

What Does this Mean?

A resource is something that can be called on when needed and therefore includes time, personnel, skill, machines, materials, money, plant, facilities, space, information,

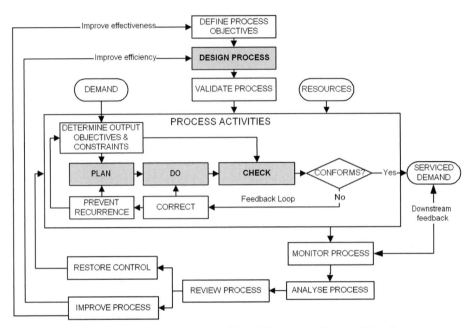

FIGURE 19-1   Where the requirements of Clause 6.1 apply in a managed process.(*Note the process we are examining is resource management therefore the Resource input shown in the figure relates to the resources needed in operating this process. The Demand arises from other processes.*)

knowledge etc. *To ensure* means to make certain. For top management *to ensure the availability of necessary resources*, it has to:

- Know what objectives it is committed to achieve,
- Know what resources are required to achieve these objectives,
- Know when the resources will be needed,
- Know of the availability of such resources,
- Know the cost of acquiring these resources.

The resources referred to here are not only those needed to document the management system, audit and update it and verify product conformity. They are the resources needed to run the organization for sustained success.

Why is this Necessary?

This requirement responds to the Leadership Principle.

No plans will be achieved unless the resources to make them happen are provided. It is therefore incumbent on top management to not only require a management system to be designed but also provide the means for it to function effectively. In many organizations, the quest for ISO 9001 certification has resulted in an existing manager being assigned additional responsibilities for establishing the management system without being provided with additional resources. Even if this person did have surplus time available, it is not a job for one person. Every manager should be involved and they too need additional resources. In the long term, the total resources required to maintain the organization will be less with an effective management system than without but to start with, additional time and skills are required and need to be made available.

How is this Demonstrated?

It is not uncommon for management to budget for certain resources and when the time comes to acquire them, the priorities have changed and the ambitious plans are abandoned. Top management will need to be more careful when agreeing to objectives and plans for their accomplishment. It will need to have confidence that, excluding events beyond its control, funds will be available to acquire the resources committed in the improvement plans. This does not mean that management will be forced to fund plans when it clearly has no funding available but such circumstances need to be monitored. Any lack of funding should be reviewed to establish whether it was poor estimation, forces outside the organization's control or a genuine lack of commitment. Opportunities often change and it would be foolish to miss a profitable business opportunity while pursuing an improvement programme that could be rescheduled. The risks need to be assessed and the objectives adjusted. It could also be that such business opportunities remove the need for the improvement programme because it removes the process or product line that is in need of improvement. However, managers at all levels need to be careful about approving plans which are over ambitious, impractical or not feasible with the anticipated resources that will be available.

DETERMINING RESOURCES (6.2.1)

The standard requires the organization *to determine the resources needed to implement and maintain the quality management system, continually improve its effectiveness and enhance customer satisfaction by meeting customer requirements.*

What Does this Mean?

Determination of resources means identifying resource needs. The resources needed to implement the management system include all human, physical and financial resources needed by the organization for it to function. The resources needed to maintain the management system are those needed to maintain a level of performance. The resources needed to continually improve the effectiveness of the management system are those needed to implement change in the organization's processes. Those resources needed to enhance customer satisfaction are no more than those needed to achieve the organization's objectives because of the linkage between customer requirements, policy, objectives and processes.

The phrase "*resources needed to implement and maintain the quality management system*" sends out the signal that the management system is a set of policies and procedures that are implemented and maintained. There is clearly inconsistency throughout the standard on this issue. What is missing from the requirement is the determination of resources needed to design and develop the management system.

If you put a boundary around the quality management system and perceive it only as a part of the overall management system, there will be those resources which serve the achievement of quality and those which serve other purposes, but such boundaries are not useful because they divert attention from the basic goal of satisfying the stake-holders. If you are faced with making a decision as to what to include and exclude, the questions you need to answer are 'Why would we want to exclude a particular resource from the management system? What business benefit is derived from doing so?' Hopefully, you will conclude that there are no benefits from their exclusion and many benefits from their inclusion. Whether or not you exclude a part of the management system from ISO 9001 certification is an entirely separate matter as is dealt with in Chapter 39 on *Dealing with exclusions*.

Why is this Necessary?

This requirement responds to the Process Approach Principle.

Without adequate resources, the management system will not function. As addressed previously, the management system is more than a set of documents. It is a dynamic system and requires human, physical and financial resources for it to deliver the required results. Starve it of resources and you starve the organization of resources; the planned results will not be achieved.

How is this Demonstrated?

Resource management is a common feature of all organizations and while it may be known by different titles, the determination and control of the resources to meet customer needs is a fundamental requirement and fundamental to the achievement of all other requirements.

There are two types of resource requirements: those needed to set up and develop the organization and those needed to execute particular contracts or sales. The former is addressed in Clause 6.1 and the latter in Clause 7.1.

The way many companies identify resource requirements is to solicit resource budgets from each department covering a 1–5-year period. However, before the managers can prepare budgets they need to know what requirements or objectives they will have to meet. They will need access to the corporate plans, sales forecasts, new regulations and statutes, new product development plans, marketing plans, production plans etc. as well as the policies, objectives, process descriptions and procedures.

In specifying resource needs for meeting product requirements there are three factors that need to be defined, namely, the quantity, delivery and quality expressed by questions such as: How many do you want? When do you want them? To what specification or standard do they need to be? These factors will affect cost directly and if not determined when establishing the budgets, you could have difficulty later when seeking approval to purchase. In specifying resource needs for meeting organizational requirements there are three factors that need to be defined. These are the objectives for maintaining the *status quo*, for improving efficiency and for improving effectiveness.

Although resources are normally configured around the organization's departments and not its processes, departmental resource budgets can divert attention away from process objectives. A more effective way to determine resources is by process and not by department. In this way the resources become focused on process objectives and overcome conflicts that can arise due to internal politics and the power structure. It then becomes less of a problem convincing top management of the need when it can be clearly demonstrated that the requested resource serves the organization's objectives.

A practical way of ensuring that you have adequate resources is to assign cost codes to each category of work and divide them into two categories: *maintenance* and *improvement*. Include all costs associated with maintaining the status quo under maintenance and all costs associated with change under improvement. You can then focus on reducing maintenance costs for the same level of sales without jeopardizing improvement. It is often difficult to obtain additional resources after the budget has been approved but provided they can be justified against the organization's objectives, it should not be a problem. Either the management is serious about achieving the objective or it isn't. It is lunacy to set goals then object to providing the resources to achieve the goals. Arguments are perhaps more about estimating accuracy than need but also might be about resource availability. Since the goals were set and the estimates produced, the environment might have changed with revenue falling and interest and exchange rates much higher.

Determining resources is not simply about quantities – the number of people, equipment, machines etc. It is also about capability and competence. It is of no benefit to possess the right number of people if they are not competent to deliver the outputs needed – no benefit to own the right number of machines if they can't produce product to the required accuracy.

There are four types of change that affect resources.

- The unplanned loss of capability (staff leave or die, equipment or software obsolescence, major breakdown, fuel shortage, man-made or natural disaster);
- An increase or reduction in turnover (doing more or less of what we do already);
- A change in the organization's objectives (aiming at new targets, new products and new processes);
- A change in the external standards, regulations, statutes, markets and customer expectations (we have to do this to survive).
- A change in the availability of certain types of materials, products, people, money, space etc. either locally, nationally or globally.

PROVIDING RESOURCES (6.1 AND 7.5.1)

The standard requires the organization *to provide the resources needed to implement and maintain the quality management system and continually improve its effectiveness.*

What Does this Mean?

Providing resources means acquiring and deploying the resources that have been identified as being needed. The acquisition process, should deliver the resource in the right quantity and quality when they are needed. The deployment process should transport and prepare the resource for use. Therefore, if there is an identified need for human resources, they have been provided to the process that needs them only when they are in position ready to assume their duties, i.e., deemed competent or ready to take up a position under supervision. Likewise with equipment, it has been provided to the process that needs it only when it is installed, commissioned and ready for use. The maintenance process should maintain stock levels, equipment, people, facilities etc. so that there is no shortage of supply of capable and suitable physical and human resources.

Why is this Necessary?

This requirement responds to the Process Approach Principle.

For plans to be implemented resources have to be provided but the process of identifying resources can be relatively none-committal. It establishes a need, whereas the provision of resources requires action on their acquisition which depending on prevailing circumstances may not lead to all needs being provided for when originally required.

How is this Demonstrated?

As indicated above, providing resources requires the implementation of the acquisition processes. The physical resources will be initiated through the purchasing process; the human resources through the recruitment process and the financial resources through the funding process. The purchasing process is dealt with further in Chapter 26, *Purchasing*. An example of a recruitment process is illustrated in Fig. 19-2. The acquisition of financial resources is beyond the scope of this book because such methods are so varied and specialized. For example, funds can be acquired by cutting costs, eliminating waste, downsizing, selling surplus equipment, stocks and shares or seeking a bank loan.

Inventory

Inventory is not addressed in ISO 9001 but clearly an adequate supply of materials and components is necessary for the organization to produce the required products and deliver the required services.

To enable you to achieve delivery requirements you may need adequate stocks of parts and materials to make the ordered products in the quantities required. In typical commercial situations, predicting the demand for your products is not easy – organizations tend to carry more inventory than needed to cope with unexpected demand. The possibility of an unexpected increase in demand leads to larger inventories as an out-of-stock situation may result in lost customer orders. Most companies have to rely on forecasts and estimates. Some customers may protect you to some extent from fluctuations in demand by giving you advanced notification of their production and service requirements in order that your production schedule can be 'order driven'. In the event that an increase in demand is necessary you should be given adequate warning in order that you can increase your inventory in advance of the need. If adequate warning cannot be given, you need suitable clauses in your contract to protect you against any unexpected fluctuations in demand that may cause you to fail to meet the delivery requirements.

Inventory management is concerned with maintaining economic order quantities in order that you order neither too much stock nor too little to meet your commitments. The stock level is dependent on what it costs both in capital and in space needed to maintain such levels. Even if you employ a 'ship-to-line' principle, you still need to determine the

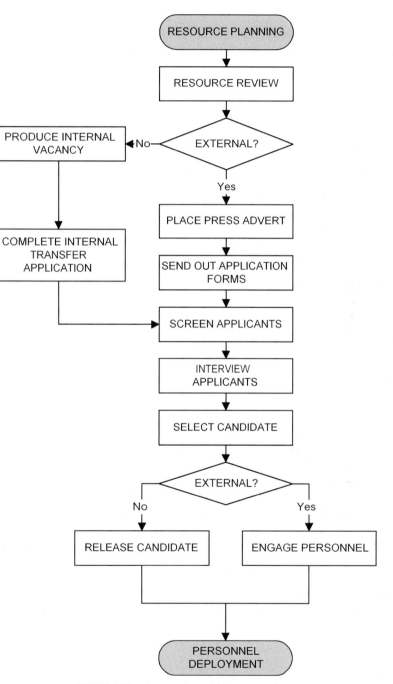

FIGURE 19-2　Personnel acquisition process flow.

economic order quantities. Some items have a higher value than others thereby requiring a higher degree of control. Use of the Pareto Principle will probably reveal that 20% of inventory requires a higher degree of control to enable you to control 80% of the inventory costs.

Whether or not 100% on-time delivery is a requirement of your customers, you won't retain customers for long if you continually fail to meet their delivery requirements regardless of the quality of the products you supply. It is only in a niche market that you can retain customers with a long waiting time for your products. In competitive markets you need to exceed delivery expectations as well as product quality expectations to retain your market position.

In addition to establishing an inventory management system, you should optimize inventory turns. This is a measure of how frequently the stock, raw material, work-in-progress and finished goods are turned over in relation to the sales revenue of a product. It is the ratio of sales turnover of product or value of stock. A high figure indicates better control over the manufacturing process. A low figure indicates excessive stock to accommodate process variations. To achieve optimum stock turns you will need metrics for receiving and storage times. You should also assure stock rotation, meaning that parts and materials are used on a first-in-first-out (FIFO) basis. The picking system will need to be date sensitive to operate FIFO[0].

Chapter 20

Human Resources

CHAPTER PREVIEW

This chapter is aimed at all levels of management for it is they who determine and provide resources but it is top management which holds the purse strings and distributes the funds to acquire other resources to lower levels of management.

It is often said by managers that people are our greatest asset but in reality it is not the people exactly but the contribution they can make that qualifies them as the assets. Unlike non-human resources, the human is adaptable, it is a flexible resource that can learn new skills, reach new levels of performance when given the right competences and when working in the right environment. But like non-human resources, the human is affected by the environment and even when equipped with the right competences may not function as well as expected if the conditions are not appropriate. The human resource is therefore the most difficult of resources for managers to manage effectively but is vital to enable organizations to achieve sustained success.

In this chapter we examine the requirements in Clause 6.2 of ISO 9001:2008 and in particular:

- What competence means and how it differs from qualifications and training?
- How competence is determined?
- Assessing competence;
- Providing training to fill the competence gap;
- Evaluating the effectiveness of personnel development measures;
- Increasing the sensitivity to the impact of activates;
- Maintaining training records.

The position where the requirements on human resources feature in the managed process is shown in Fig. 20-1. In this case the process is a work process within the resource management process. The product objective would be to provide human resources in the numbers and competence required to enable the organization to achieve its objectives. The input demand comes from the other processes and the output is competent personnel ready to carry out the work required in these processes.

COMPETENCE OF PERSONNEL (6.2.1)

The standard requires *personnel performing work affecting conformity to product requirements to be competent on the basis of appropriate education, training, skills and experience.*

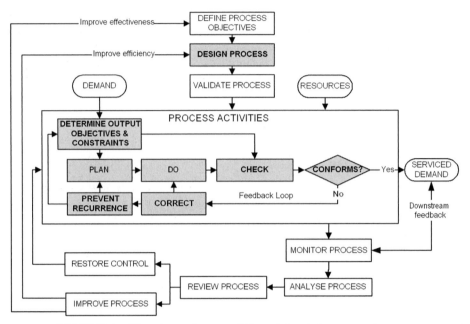

FIGURE 20-1   Where the requirements of Clause 6.2 apply in a managed process.

What Does this Mean?

The requirement makes a distinction between those personnel whose work affects *conformity to product requirements* and those personnel whose work does not affect *conformity to product requirements* otherwise known as product quality. Why would you not want all your personnel to be competent? It makes no sense! In principle, everyone's work affects the quality of the products and services supplied by the organization, some directly, others indirectly and this is now recognized in the 2008 edition with the added note.

If a person has the appropriate education, training and skills to perform a job, the person can be considered qualified. If a person demonstrates the ability to achieve the desired results, the person can be

Levels of Competence

- The beginner is an unconscious incompetent he doesn't know what he needs to know to do the job.
- The learner is a conscious incompetent – he knows what he needs to know but has not yet acquired the knowledge or the skill.
- The professional is a conscious competent – he has acquired the skills and knowledge and can do the job well but may have to be consciously aware of what he is doing from time to time.
- The master is an unconscious competent – he can do the job well without thinking about it.

considered competent. Qualified and competent are therefore not the same. Qualified personnel may not be able to deliver the desired results. While they may have the knowledge and skill, they may exhibit inappropriate behaviours (the expert who knows everything but whose interpersonal skills cause friction with staff to such an extent that it adversely affects productivity!).

A competent person may not be appropriately educated, trained or possess what are perceived to be the required skills (the pragmatist who gets the job done but can't explain how he does it). This requirement therefore presents a dichotomy because you can't be competent on the basis of appropriate education, training and skills unless you can also demonstrate you can achieve the required results. The standard fails to specify whether competence in this context is about what individuals *know* or what individuals *can do*. Taking a pragmatic approach, it would not serve the intent of the standard to simply focus on what people know therefore competence in the context of ISO 9001 must be about what people *can do*.

Competence is the ability to demonstrate *use* of knowledge, skills and behaviours to achieve the results required for the job. It is the ability to perform the whole work role not just specific tasks, the ability to meet standards that apply in the particular job not in a classroom or examination but an ability to perform in the real working environment with all the associated variations, pressures, relationships and conflicts. Competence is not a probability of success in the execution of one's job; it is a real and demonstrated capability. ISO 9000:2005 defines competence as the demonstrated ability to apply knowledge and skills. The subtle difference in this definition is that the frame of reference is not present by there being no reference to the results required for the job. Behaviours have been omitted but we must assume this is an oversight and not intentional.

Competence is concerned with outcomes rather than attributed abilities. Competence is more than a list of attributes. A person may claim to have certain ability but proof of competence is only demonstrated if the desired outcomes are achieved. While the opposite of competence is incompetence, this is perceived as a pejorative term so we tend to use the phrase 'not yet competent' to describe those who have not reached the required standard of competence. Competence does not mean excellence. Competence is meeting the established performance or behavioural standards. Excellence is exceeding the established standards. Competence is a quality of individuals, groups and organizations. Corporate competence empowers an organization to build a competitive advantage and is the result of having capable processes that deliver corporate goals. Core competences are the things the organization is good at and these are dependent on employing competent human resources.

Why is this Necessary?

This requirement responds to the Leadership and Involvement of People Principles.

Traditionally, personnel have been selected on the basis of certificated evidence of qualifications, training and experience rather than achievement of results. Here are some examples showing the inadequacy of this method:

- A person may have received training but not have had the opportunity to apply the knowledge and skills required for the job.
- A person may have practiced the acquired skills but not reached a level of proficiency to work unsupervised.
- A person may possess the knowledge and skills required for a job but may be temporarily or permanently incapacitated (a professional footballer with a broken

leg is not competent to play the game until his leg has healed and he is back on top form).

- A person may have qualified as a chemist 30 years ago but not applied the knowledge since.
- Airline pilots who spend years flying one type of aircraft will require some period in the flight simulator before flying another type because they are not deemed competent until they have demonstrated competency.
- A person may have been competent in maintaining particular air traffic control equipment but has not had occasion to apply the skills in the last 12 months. In this industry the engineers need to demonstrate competence before being assigned to maintain a particular piece of equipment because flight safety is at risk.

The above examples illustrate why qualifications, training certificates and years of experience are not necessarily adequate proof of competence.

How is this Demonstrated?

In any organization there are positions that people occupy, jobs they carry out and roles they perform in these positions. For each of these certain outcomes are required from the occupants. The starting point is therefore to define the outcomes required from a job and then define what makes those performing it successful and agree these with the role or jobholder. Having set these standards, they become the basis for competence assessment and competence development. The education and training provided should be consistent with enabling the individual concerned to achieve the agreed standards.

The outcomes required for a role or job has two dimensions. There are the hard results such as products and decisions and the soft results such as behaviours, influence and stamina. The outcomes are also dependent on the conditions or context in which the role or job is performed. Two examples might help to clarify this point:

- A manager of a large enterprise may produce the same outcomes as the manager of a small enterprise but under entirely different conditions. It follows therefore that a competent manager in one context may not be competent in another.
- A person who occupies the position of production manager in a glass factory performs the role of a manager but also performs several different jobs concerned with the production of glass and therefore may need to be competent to negotiate with suppliers, use computers, produce process specifications, blow glass, drive trucks, test chemicals, administer first aid etc. depending on the scale of the operations being managed.

Each job comprises a number of tasks that are required to deliver a particular result. Having a fork lift truck driving certificate is not a measure of competence. All this does is to prove that the person can drive a forklift truck. What the organization may need is someone who can move 4 tonnes of glass from point A to point B safely in 5 min using a fork lift truck. The management role may occupy some peoples' 100% of the time and therefore the number of competencies needed is less than those who perform many different types of jobs in addition to management. For this reason there is no standard set of competencies for any particular position because each will vary but there are national schemes for assessing competence relative to specific occupations. In the UK the National Vocational Qualification (NVQ) scheme has been operating since 1986. The central feature of NVQs is the National Occupational Standards (NOS) on which they are based. NOS are statements of performance standards that describe what competent people in a particular occupation are expected to be able to do. They cover all the main aspects of an occupation, including current best practice, the ability to adapt to future requirements and the knowledge and understanding that underpin competent performance. NVQs are work-related, competence-based qualifications. They reflect the skills and knowledge needed to do a job effectively and represent national standards recognized by employers throughout the country. Five levels of competence are defined covering knowledge, complexity, responsibility, autonomy and relationships.

While ISO 9001 does not require vocational qualifications such as the UK NVQ scheme, the implication of this requirement for competence is that managers will need to select personnel on the basis of their ability to deliver the outcomes required. Selecting a person simply because of the qualification they hold or that they are a member of the same club would not be appropriate unless of course they can also deliver the required outcomes!

The standard requires:

- personnel to be competent,
- the necessary competence to be determined,
- actions to be taken to satisfy these needs,
- the evaluation of the effectiveness of actions taken.

The standard is therefore implying that organizations should employ competence-based assessment techniques but as indicated previously, it is neither explicit nor clarified in ISO 9004 whether competence in this context is based on what people know or what they can do. The current method of selection may be on the basis of past performance but without performance standards in place and a sound basis for measurement, this method is not capable of delivering competent people to the workplace.

You will need to maintain documentary evidence that personnel are competent to perform the jobs assigned to them and to do this you need to identify the competence needed and demonstrate that a competence assessment process is employed to validate competence. This is addressed in the next section. Fletcher[1] provides useful guidance on designing competence-based assessment programmes. Competence-based assessment has a number of uses:

- Assessment for certification (not required for ISO 9001 but regulations or customers may require this. Airline pilots, welders and personnel performing non-destructive testing are some examples);
- Performance appraisal (indirectly required by ISO 9001 from the requirement for an evaluation of actions take to satisfy needed competence);
- Identification of training needs (required by ISO 9001);

---

[1] Fletcher, Shirley, (2000). *Competence-based assessment techniques*, Kogan Page.

- Skills audit (indirectly required by ISO 9001 from the requirement to ensure the organization has the ability to meet product requirements and from the requirement for selecting competent personnel);
- Accreditation of prior learning (not required by ISO 9001);
- Selection and recruitment (required by ISO 9001);
- Evaluating training (required by ISO 9001).

COMPETENCE TRAINING AND AWARENESS (6.2.2)

Determining Necessary Competence (6.2.2a)

The standard requires the organization to *determine the necessary competence for personnel performing work affecting conformity to product requirements.*

What Does this Mean?

Individual competence is concerned with the ability of a person to achieve a result whereas training is concerned with the acquisition of skills to perform a task and education is concerned with the acquisition of knowledge. It is therefore not a question of whether a person has the skills and knowledge to do a job but on whether a person is able to achieve the desired outcome. This is known as the competence-based approach. Academic qualifications tend to focus on theory or application of theory to work situations. In contrast, the competence-based approach focuses on the results the individuals are achieving. People are either competent or not yet competent. There are no grades, percentages or ratings. People are deemed competent when they have demonstrated performance that meets all the required standards. A person who is appropriately educated, trained and experienced is competent only if he has the ability to produce the desired results when required. If for some reason a competent person became incapacitated he would no longer be deemed competent to perform the job he was performing prior to the incapacity.

Why is this Necessary?

This requirement responds to the Leadership and Involvement of People Principles.

When assigning responsibility to people we often expect that they will determine what is needed to produce a good result and perform the job right first time. We are also often disappointed. Sometimes it is our fault because we did not adequately explain what we wanted or more likely, we failed to select a person that was competent to do the job. We naturally assumed that because the person had a college degree, had been trained in the job and had spent the last two years in the post, that they would be competent. But we would be mistaken, primarily because we had not determined the necessary competence for the job and assessed whether the person had reached that level of competence. In theory we should select only personnel who are competent to do a job but in practice we select the personnel we have available and compensate for their weaknesses either by close supervision or by providing the means to detect and correct their failures.

> ### Case Study – Demonstrating Competency
>
> *We have always taken the view that managers should select their staff on the basis of their suitability for the job rather than on having attended training courses or gained academic qualifications. However, the only evidence is a proven track record through the internal appraisal process which is of course confidential. In ISO 9001, it requires we identify competency needs and evaluate the effectiveness of training provided. How do we demonstrate competency?*
>
> Let us get one thing out of the way to begin with. Most of the information you provide to external auditors is confidential and you agree to provide this data when you contract the services of a certification body. Also auditors will not want to see staff records containing personal details. However, they are entitled to examine competence requirements and records of competence in the same way they might want to examine test specifications and equipment calibration certificates.
>
> The fact that your managers select their staff on the basis of their suitability for the job should not present a problem unless they are basing their judgment of suitability on length of service or a clean personnel record rather than current performance.
>
> Competence is the ability to achieve certain results therefore if staff cannot deliver the necessary results they should not be deemed competent. Regrettably, previous ISO 9001 certificates were granted on the basis of demonstrated training not demonstrated competence. When staff were found to be following the prescribed procedures, in the past, an auditor might simply ask about the training they had received and if it was relevant, there would be no nonconformity. Following the introduction of competency into the standard, auditors now need to seek evidence that staff can demonstrate an ability to achieve prescribed results. Therefore, in order to demonstrate competence, you will need to define the competence requirements for a task or activity and then produce evidence that those performing these tasks or activities satisfy these requirements.
>
> If a person needs to be able to operate a machine in such a way as to control the output between set limits, records showing the output is being controlled by this person between the set limits demonstrates competence. If in order for this person to attain this level of competence, he or she undertook a period of training, the same records demonstrate that the training was effective.
>
> With every task there will be a number of results that need to be achieved. The person will not only need to produce the output required but also satisfy the constraints such as doing the task safely, doing it with minimum use of resources, doing it politely and doing it without collateral damage.
>
> In essence this is no different from controlling product quality. You define the requirement (set the standard) then determine how you will establish whether the requirement has been achieved (verify conformity).

The competence-based movement developed in the 1960s out of a demand from businesses for greater accountability and more effective means of measuring and managing performance. This led to research into what makes people effective and what constitutes a competent worker. Two distinct competence-based systems have emerged. The British model focuses on standards of occupational performance and the American model focused on competency development. In the UK the standards reflect the outcomes of workplace performance. In the USA, the standards reflect the personal attributes of individuals who have been recognized as excellent performers[2] but what individuals achieved in the past is not necessarily an indication of what they achieve in the future; they age, they forget, their eyesight deteriorates and they may not be as agile both physically and mentally as they once were.

Competence is particularly important in the professions because the outputs result from an intellectual process rather than an industrial process. We put our trust in professionals and expect them to be competent but methods of setting standards of competence and their evaluation have only been developed over the last 15 years. It was believed that education, training and experience were enough, but the cases of malpractice particularly in the medical profession have caused the various health authorities to look again at clinical competence. The financial crisis of 2008 may trigger a similar review and outcome in the financial profession.

How is this Demonstrated?

Determining the competence necessary for performing a job is a matter of determining the outcomes required of a job, the performance criteria or standards to be achieved, the evidence required and the method of obtaining it. It is important that the individuals whose performance is to be assessed are involved in the setting of these standards.

The jobs that people perform must be related to the organization's objectives and as these objectives are achieved through processes, these jobs must contribute to achievement of the process objectives. In the decomposition from the system level where the business processes are identified through to work processes and sub-processes you will arrive at a level where the results are produced by a single person. The objectives for these processes or sub-processes describe outcomes 'what must be achieved?' If you then ask, 'What must be done for this to be achieved?' These are termed *units of competence*. Several units of competence will be necessary to achieve a given outcome. For example, a font line operator's primary output is conforming product. The operator needs to possess several competences for conforming product to be produced consistently.

An operator might need the ability to:

- understand and interpret technical specifications,
- set up equipment,
- operate the equipment so as to produce the required output,
- undertake accurate measurements,
- apply variation theory to the identification of problems,
- apply problem-solving methods to maintain control of the process.

Simply possessing the ability to operate a machine is not a mark of competence.

Table 20-1 shows the key questions to be asked, the terms used and one example. It should be noted that there may be several performance criteria and a range of methods used to collect the evidence. It should also be noted that the evidence should be against the unit of competence not against each performance criteria because it is competence to deliver the specified outcome that is required not an ability to produce discrete items of evidence.[3] Terminology in this area is not yet standardized and therefore there are some

---

[2] Fletcher, Shirley, (2000). *Competence-based assessment techniques,* Kogan Page.
[3] ibid.

TABLE 20-1 Competence-Based Assessment

Key question	What this is called?	Example
What must be achieved?	Outcome	Conforming product
What must be done for this to be achieved?	Unit of competence	The ability to apply variation theory to the identification of problems (this is one of several)
How well must this be achieved?	Performance criteria	Distinguishes special cause problems from common cause problems (this is one of several)
How should assessment be conducted?	Assessment method	Observation of performance
What evidence should be collected?	Evidence requirement	Run charts indicating upper and lower control limits with action taken only on special causes (this is one of several)

differences in the terms between the British and American competence-based systems. For instance, performance criteria seem to be clustered into elements of competence in the British system.

When considering the introduction of a competence-based assessment system Fletcher provides a useful check list

Is the proposed system:

- Based on the use of explicit statements of performance?
- Focused on the assessment of outputs or outcomes of performance?
- Independent of any specified learning programme?
- Based on a requirement in which evidence of performance is collected from observation and questioning of actual performance as the main assessment method?
- One which provides individualized assessment?
- One which contains clear guidance to auditors regarding the quality of evidence to be collected?
- One which contains clear guidelines and procedures for quality assurance?

The determination of competence requires that we have defined a standard for competence, measured performance and acquired evidence of attainment. We therefore need to ask:

- What are the key results or outcomes for which the person is responsible? (The units of competence.)
- What are the principal tasks the individual is expected to perform and the expected behaviours the individual is required to exhibit to achieve these outcomes? (The elements of competence.)
- What evidence is required to demonstrate competence?
- What method of measurement will be used to obtain the evidence?

The methods for setting competence standards are quite complex therefore the reader should consult the various references in Appendix B. A documented procedure for identifying competence or training needs would be written around the topics addressed above.

The purpose in determining competence is to identify the requirements for the job. Requirements for new competencies arise in several ways as a result of the following:

- Job specifications.
- Process specifications, maintenance specifications, operating instructions etc.
- Development plans for introducing new technologies.
- Project plans for introducing new equipment, services, operations etc.
- Marketing plans for launching into new markets, new countries, new products and services.
- Contracts where the customer will only permit trained personnel to operate customer-owned equipment.
- Corporate plans covering new legislation, sales, marketing, quality management etc.
- An analysis of nonconformities, customer complaints and other problems.
- Developing design skills, problem-solving skills or statistical skills.
- Introducing a quality management system thus requiring awareness of the topics covered by ISO 9001, the quality policies, quality objectives and training in the implementation of quality system procedures, standards, guides etc.

You have a choice as to the form in which the competences are defined. Some organizations produce a skills matrix, which shows the skills each person has in a department. This is not strictly a list of competences because it is not linked to outputs. It is little more than a training record but could have been developed from observation by supervisors. The problem is that it is people oriented not process oriented. A better way is for a list of competences to be generated for each process based on the required process outputs (see *Process descriptions* in Chapter 11). In identifying the units of competence, you are in effect identifying all the tools, techniques, methods and practices used by the organization such as design of experiments, CAD, SPC etc.

Providing for Training (6.2.2b)

The standard requires the organization *where applicable to provide training or take other actions to achieve the necessary competence.*

What Does this Mean?

Having identified the competence needs, this requirement addresses the competence gap but this gap is only established after assessing competence. Therefore, there are two types of action needed to satisfy these needs, namely, *competence assessment* and *competence development*.

Why is this Necessary?

This requirement responds to the Leadership and Involvement of People Principles.

Having identified the competence needed to achieve defined outcomes, it is necessary to determine the current level of competence and provide the means to develop the competence of personnel where it is found that staff are not yet competent in some areas of their job.

How is this Demonstrated?

Competence Assessment

To operate a competence-based approach to staff selection, development and assessment, it is necessary to:

- set criteria for the required performance,
- collect evidence of competence,
- match evidence to standards,
- plan development for areas in which a 'not yet competent' decision has been made.

A number of questions arise when considering the collection and assessment of evidence.

- What do we want to assess?
- Why do we want to assess it?
- Who will perform the assessment?
- How will we ensure the integrity of the assessment?
- What evidence should be collected?
- Where will the evidence come from?
- How much evidence will be needed?
- When should the assessment commence?
- Where should the assessment take place?
- How will we conduct the assessment?
- How will we record and report the findings?

Answers to these questions can be found in Fletcher's book on Competence-based assessment techniques.

Bridging the Competence Gap

Once the results of the competence assessment are known, the gap may be bridged by a number of related experiences:

- Training courses where an individual undertakes an internal or external course;
- Mentoring where a more senior person acts as a point of contact to give guidance and support;
- Coaching on-the-job where a more experienced person transfers knowledge and skill;
- Job rotation where a person is temporarily moved into a complementary job to gain experience or relieve boredom;
- Special assignments where a person is given a project that provides new experiences;
- Action learning where a group of individuals work on their own but share advice with others and assist in solving each other's problems;
- On the job learning where the individual explores new theories and matches these with organizational experience.

On-the-Job-Training

In many cases, formal assessment may not be required simply because the gap is glaringly obvious. A person is competent to produce a particular result and the result required or the method of achieving it is changed. A need therefore arises for additional training or instruction. After completing the training or instruction the records need to be updated indicating the person's current competence. This would apply equally to contractors or agency personnel.

External Training

Beware of training courses that are no more than talk and chalk sessions where the tutor lectures the students, runs through hundreds of slides and asks a few questions! There is little practical gain from these kinds of courses. A course that enables the participants to learn by doing, to learn by self-discovery and insight is a *training* course. The participants come away having had a skill changing experience. Just look back on your life, and count the lessons you have learnt by listening and watching and compare that number with those you have learnt by doing. The latter will undoubtedly outnumber the former.

Encourage your staff to make their mistakes in the classroom not on the job or if this is not practical, provide close supervision on the job. Don't reprimand staff under training – anyone can make mistakes. An environment in which staff are free to learn is far better than one in which they are frightened of doing something wrong.

Training Aids

If training is necessary to improve skills involving the operation or maintenance of tools or equipment, you need to ensure that any practical aids used during training:

- represent the equipment that is in use on the production line;
- adequately simulate the range of operations of the production equipment;
- are designated as training equipment and only used for that purpose;
- are recorded and maintained indicating their serviceability and their design standards including records of repairs and modifications.

Students undertaking training may inadvertently damage equipment. It may also be necessary to simulate or inject fault conditions so as to teach diagnostic skills. Training activities may degrade the performance, reliability and safety of training equipment and so it should be subject to inspection before and after training exercises. The degree of inspection required would depend on whether the equipment has been designated for use only as training equipment or whether it will be used either as test equipment or to provide operational spares. If it is to be used as test or operational equipment, it will need to be re-certified after each training session. During the training sessions, records will need to be maintained of any fault conditions injected, parts removed and any other act which may invalidate the verification status of the equipment. In some cases it may be necessary to refurbish the equipment and subject it to the full range of acceptance tests and inspections before its serviceability can be assured. Certification can only be maintained while the equipment remains under controlled conditions. As soon as it passes into a state where uncontrolled activities are being carried out, its certification is immediately invalidated. It is for such reasons that it is often more economical to allocate equipment solely for training purposes.

Evaluating the Effectiveness of Personnel Development Activities (6.2.2c)

The standard requires the organization *to evaluate the effectiveness of the actions taken.*

What Does this Mean?

All education, training, experience or behavioural development should be carried out to achieve a certain objective. The effectiveness of the means employed to improve competence is determined by the results the individual achieves from doing the job. Regardless of how well the education, training or behavioural development has been designed, the result is wholly dependent on the intellectual and physical capability of the individual. Some people learn quickly while others learn slowly and therefore personnel development is incomplete until the person has acquired the appropriate competence, i.e., delivering the desired outcomes.

Why is this Necessary?

This requirement responds to the Factual Approach and Involvement of People Principles.

The mere delivery of education or training is not proof that it has been effective. Many people attend school only to leave without actually gaining an education. Some may pass the examinations but are not educated because they are often unable to apply their knowledge in a practical way except to prescribed examples. The same applies with training. A person may attend a training course and pass the course examination but may not have acquired the necessary proficiency – hence the necessity to evaluate the effectiveness of the actions taken.

How is this Demonstrated?

Competence is assessed from observed performance and behaviours in the workplace not from an examination of education and training programmes far removed from the workplace.

Having established and agreed standards of competence for each role and job in the organization, those who can demonstrate attainment of these standards are competent. Thus, competent personnel have the opportunity to prove their competence.

Fletcher[4] identifies the following key features of a competence-based assessing system:

- Focus on outcomes,
- Individualized assessment,
- No percentage rating,
- No comparison with other individual's results,
- All standards must be met,
- Ongoing process,
- Only 'competent' or 'not yet competent' judgements made.

---

[4] Fletcher, Shirley, (2000). *Competence-based assessment techniques,* Kogan Page.

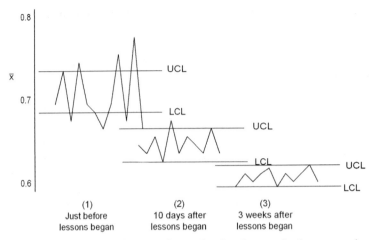

FIGURE 20-2 Average daily scores for a patient learning to walk after an operation.

There are three parts to the evaluation:

- An evaluation of the personnel performance activity before development;
- An evaluation of the personnel performance immediately on completion of development activity;
- An evaluation of the personnel development activity within weeks of its completion.

Deming illustrates this as a run chart[5] an example of which is shown in Fig. 20-2.

Development Activity Evaluation (the Initial Stage)

Activity evaluation by the students themselves can only indicate how much they felt motivated by the event. It is not effective in evaluating what has been learnt. This is more likely to be revealed by examination at the end of the event or periodically throughout the development period. However, the type of examination is important in measuring the effectiveness of the personnel development, e.g., a written examination for a practical course may test the theories behind the skills but not the personal mastery of the skills themselves. A person may fail an examination by not having read the question correctly, so examination by itself cannot be a valid measure of training effectiveness. You need to examine the course yourself before sending your staff on it. If you want information to be conveyed to your staff, a lecture with accompanying slide show may suffice. Slide shows are good for creating awareness but not for skill training. Skills cannot be acquired by any other means than by doing.

Development Activity Effectiveness – Short Term (the Intermediate Stage)

We often think of training as a course away from work. We go on training courses. But the most effective training is performed on the job. Training should be primarily about learning new skills not acquiring knowledge – that is education. On returning to work or normal duties after a course it is important that the skills and knowledge learnt are put to

---

[5] Deming, W. Edwards, (1982). *Out of the crisis,* MITC. Chapter 8, page 253.

good effect as soon as possible. A lapse of weeks or months before the skills are used will certainly reduce the effectiveness. Little or no knowledge of skill may have been retained. Training is not about doing something once and once only. It is about doing something several times at frequent intervals. One never forgets how to ride a bicycle or drive a car regardless of the time-lapse between each attempt, because the skill was embedded by frequency of opportunities to put the skill into practice in the early stages. Therefore, to ensure effectiveness of training you ideally need to provide opportunities to put into practice the newly acquired skills as soon as possible. The person's supervisor should then examine the students' performance through sampling work pieces, reading documents he or she produces, observing the person doing the job and reviewing the decision they make. If you have experts in the particular skills then in addition to appraisals by the supervisor, the expert should also be involved in appraising the person's performance. Pay particular attention to the person's understanding of customer requirements. Get this wrong and you could end up in trouble with your customer!

Development Activity Effectiveness – Long Term (the Final Stage)

After several months of doing a job and applying the new skills, a person will acquire techniques and habits. The techniques shown may not only demonstrate the skills learnt but also those being developed through self-training. The habits may indicate that some essential aspects of the training have not been understood and that some reorientation is necessary. It is also likely that the person may have regressed to the old way of doing things and this may be due to matters outside their control. The environment in which people work and the attitudes of the people they work with can have both a motivating and de-motivating effect on an individual. Again the supervisor should observe the person's performance and engage the expert to calibrate their judgement. Pay particular attention to customer requirements and whether the trainee really understands them. If there are significant signs of regression you will need to examine the cause and take corrective action.

Records of the Effectiveness of Personnel Development

There is no requirement for records of results of the effectiveness of personnel development (published interpretations RFI 044). However, it would be prudent to maintain records for the duration of any learning programme so that reviews may take place using factual data. Once the individual moves onto another programme of learning new records might be created and the old ones destroyed as they have served their purpose.

Increasing Sensitivity to the Impact of Activities (6.2.2d)

The standard requires the organization to ensure that *its personnel are aware of the relevance and importance of their activities and how they contribute to the achievement of the quality objectives.*

What Does this Mean?

Every activity an individual is required to perform should serve the organization's objectives either directly or indirectly. All activities impact the organization in some way and the quality of results depends on how they are perceived by the person performing them. In the absence of clear direction, personnel use intuition, instinct, knowledge and experience to the select activities they perform and how they should behave. Awareness of the relevance of an activity means that individuals are more able to select the right activities to perform in a given context. Awareness of the importance of an activity means that individuals are able to approach the activity with the appropriate behaviour. Some activities make a significant contribution to the achievement of objectives and others make less of a contribution but all make a contribution. Awareness of this contribution means that individuals are able to apportion their effort accordingly.

Why is this Necessary?

This requirement responds to the Customer Focus and Involvement of People Principles.

Other than those at the front end of the business, personnel often don't know why they do things, why they don't do other things, why they should behave in a certain way and why they should or should not put a lot of effort into a task. Some people may work very hard but on activities that are not important, not relevant or not valued by the organization. Working smart is much better and is more highly valued and why awareness of the relevance and importance of activities and their contribution to the organization's objectives is essential for enabling an organization to function effectively.

Awareness of contribution also puts a value on the activity to the organization and therefore awareness of the contribution that other people make puts a job in perspective, overcomes grievances and discontent. Personnel can sometimes get carried away with their sense of self-importance that may be based on a false premise. When managers make their personnel aware of the context in which activities should be performed, it helps redress the balance and explain why some jobs are paid more than others, or more highly valued than others.

There are perhaps thousands of activities that contribute to the development and supply of products and services some of which create features that are visible to the customer or are perceived by the customer as important. They may be associated with the appearance, odour, sound, taste, function or feel of a product where the activity that creates such features is focused on a small component within the product the customer purchases. They may also be associated with the actions, appearance or behaviour of service delivery personnel where the impact is immediate because the personnel come face to face with the customer. However, close to or remote from the customer and seemingly insignificant, the result of an activity will impact customer satisfaction. Explaining the relationship between what people do and its effect on customers can have a remarkable impact on how personnel approach the work they perform. Awareness creates pride, a correct sense of importance and serves to focus everyone on the organization's objectives.

How is this Demonstrated?

Designing an Awareness Process

As with any other process, the place to start is by defining what you want to achieve, i.e., what are the process objectives? Then, establish how you will measure whether these objectives have been achieved. If staff are aware of the relevance and importance of their activities and how they contribute to the achievement of the quality objective, they will be doing certain things and not doing other things, what you need to do is to define what these are. Then you need to define the activities that will be carried out to achieve these objectives as measured.

An obvious way of implementing this requirement is for managers to advise staff before an event of the type of actions and behaviours that are considered appropriate. Also managers should advise staff during or immediately after an event that their action or behaviour is inappropriate and explain the reasons for this. If done in a sympathetic and sensitive manner the effect can be productive and people will learn. If done insensitively, abruptly and in a condescending manner, the effect can be counterproductive and people will not learn. However, the success of this method depends on managers being on the scene of action to observe intended or actual behaviour.

There are other ways of building awareness.

- Induction to a new job.
- Training for a new or changed job.
- Product briefings.
- Chart displays and warning notices.
- Performance result briefings.
- Videos showing activities in context, where components are used, safety incidents etc.
- Coaching of personnel by demonstrating appropriate behaviour that they may follow.

A source of information is the Failure Modes and Effects Analysis carried out on the product and process as a means of identifying preventive action. In this analysis the possible modes of failure are anticipated and measures are taken to eliminate, reduce or control the effect. Such measures may include staff training and awareness as to the consequence of failure or nonconformity. However, it is sometimes not enough to explain the consequences of failure; you may need to enable them to see for themselves the effect using simulations, prototypes or case studies. Staff may have no idea of the function that the part they are producing performs where it fits and how important it is or where the role they are performing fit in the big picture. This education is vital to increasing sensitivity. In many organizations this sensitivity is low. The manager's task is to heighten sensitivity so that everyone is in no doubt what effect nonconformity has on the customer.

You can take a horse to water but you can't make it drink so the saying goes! It is the same with people! Making personnel aware of the quality issues and how important they are to the business and to themselves and the customer may not motivate certain individuals. The intention should be to build an understanding of the collective advantages of adopting a certain style of behaviour. From an awareness you are actually trying to modify behaviour so this is what you measure.

Measuring Effectiveness

Measuring employee understanding of appropriate quality objectives is a subjective process but measuring the outputs employees produce is not. Competence assessment would therefore be an effective means of measuring the effectiveness of the awareness process. In this way you don't have to measure it twice. Competence assessment serves to indicate whether staff have the ability to do the job and also serves to demonstrate the awareness process has or has not been effective.

If you don't employ competence assessment techniques there is an alternative. Through the data analysis carried out to meet the requirements of Clause 8.4 you will have produced metrics that indicate whether your quality objectives are being achieved. If they are being achieved you could either assume your employees understand the quality objectives or you could conclude that it doesn't matter. However, the standard requires a measurement. Results alone are insufficient evidence. You need to know how the results were produced. The results may have been achieved by pure chance and in six months time your performance may have declined significantly. The only way to test understanding is to check the decisions people make. This can be done with a questionnaire but is more effective if one checks decisions made in the work place. Is their judgement in line with your objectives or do you have to repeatedly adjust their behaviour? Take a walk around the plant and observe what people do, how they behave, what they are wearing, where they are walking and what they are not managing. You might conclude that:

- A person not wearing eye protection obviously does not understand the safety objectives.
- A person throwing rather than placing good product into a bin obviously does not understand the product handling policy.
- A person operating a machine equipped with unauthorized fittings obviously does not understand either the safety objectives, the control plan or the process instructions.
- An untidy yard with evidence of coolant running down the public drains indicates a lack of understanding of the environmental policy.

For each quality objective you should have a plan that defines the processes involved in its achievement. Assess these processes and determine where critical decisions are made and who is assigned to make them. Audit the decisions and ascertain whether they were contrary to the objectives (see also *Process for establishing objectives* in Chapter 16).

Maintaining Training Records (6.2.2e)

The standard requires the organization *to maintain records of education, training, skills and experience.*

What Does this Mean?

As the requirement references Clause 4.2.4, the records referred to are records that provide:

- Evidence of the extent to which a person's abilities fulfil certain competence requirements;
- Evidence of activities performed to specify, develop or verify the abilities of a person who is intended to fulfil certain competence requirements.

Such records will include:

- A personal development plan indicating the actions to be taken by the organization and the individual in meeting competence requirements;
- Records of the actions taken;
- Records of any measurement and verification of competence.

Therefore, the records required need to extend beyond lists of training courses, academic qualifications and periods of experience because these only record actions taken and not whether they were planned or whether they achieved the desired result.

Calling these records 'training records' becomes misleading because they will contain evidence of other activities as well as training. They are part of the personnel records but do not constitute all the personnel records because these will undoubtedly include confidential information. A more suitable label for the records that contain the results of competence assessment would be Personnel Competence Records.

Why is this Necessary?

This requirement responds to the Factual Approach Principle.

The justification for records is provided under *Documented procedures and records* in Chapter 11.

How is this Demonstrated?

The standard is somewhat inconsistent in this requirement. Previously it required personnel to be competent, next it required the necessary competence to be determined but instead of requiring records of competence it reverts back to requiring only records of education, skills, training and experience. It would be more useful to generate records of the competence assessments.

Comparing product records with personnel records can be a useful way to determine the information required in competence records (see Table 20-2).

TABLE 20-2 Contrasting Product Records and Personnel Records

Product record	Personnel record
Identity of product	Identity of person
Product specification reference	Job specification reference
Required characteristics	Required competences
Product verification stages	Competence assessment stages
Product verification method	Competence assessment method
Inspector or tester	Competence assessor
Product verification results	Assessment results
Nonconformity reports	Opportunities for improvement
Remedial action plan	Personal development plan
Re-verification results	Re-assessment results
Certification of conformity	Certification of competence
Certifying authority	Certifying authority

The Job Specification identifies the competence needed and the Personnel Development Plan (PDP) identifies the education, training and behavioural development required to bridge the gap in terms of courses of study, training and development together with dates. Re-verification therefore provides evidence of education, training and behavioural development undertaken together with dates completed. The Certification of Competence provides evidence that the actions were effective. However, certification of competence is not required unless it is necessary for regulatory purposes. With this method you will also need to maintain separately in the personnel records, historical records of education, training and experience to provide a database of capability that can be tapped when searching for potential candidates for new positions.

Whenever any personnel development is carried out you should record on the individual's personal file, details of the course taken, the dates, duration and examination results (if one was taken). Copies of the certificate should be retained on file as evidence of training but these are not necessarily evidence of competence. You may find it useful to issue each individual with a PDP that includes personal development log, but do not rely on this being maintained or retained by the person in question. Often personnel development records are held at some distance away from an individual's workplace and in certain cases, especially for certificated personnel performing special processes, individuals should carry some identification of their proficiency so as to avoid conflict if challenged.

The records should indicate whether the prescribed level of competence has been attained. In order to record competence, formal training needs to be followed by on-the-job assessment. The records should also indicate who has conducted the education, training or behavioural development and there should be evidence that this person or organization has been assessed as competent to deliver and evaluate such activities.

Competence records should contain evidence that the effectiveness of action taken has been evaluated and this may be accomplished by a signature and date from the auditor against the stages of evaluation.

Periodic reviews of Competence records should be undertaken to clearly identify personnel development needs.

You will need two types of competence records – those records relating to a particular individual and those relating to particular activities. The former is used to identify an individual's competence and the latter to select competent people for specific assignments.

Chapter 21

Infrastructure (6.3)

CHAPTER PREVIEW

This chapter is aimed at all levels of management responsible for determining and providing the infrastructure that enables the organization to function in its operating environment.

ISO 9000:2005 defines infrastructure as the system of permanent facilities and equipment of an organization. Infrastructure also includes basic facilities, equipment, services and installations needed for growth and functioning of the organization. Such basics would include the buildings and utilities such as electricity, gas, water and telecommunications. Within the buildings it would include the office accommodation, furniture, fixtures and fittings, computers, networks, dining areas, medical facilities, laboratories, plant, machinery and on the site it would include the access roads and transport. In the offices and workshops it would include the IT systems for planning and resourcing operations and the computer systems for processing, storing and displaying information. In fact everything an organization needs to operate other than the financial, human and consumable resources. In many organizations infrastructure is classified under the heading of *capital expenditure* because it is not order-driven, i.e., it does not change on receipt of an order but might change before a big contract is signed that was bid on the basis of changes in the infrastructure such as new IT systems, buildings for new assembly lines etc.

In this chapter we examine the requirements in Clause 6.3 of ISO 9001:2008 and in particular:

- Determining the components and systems that comprise the infrastructure;
- Providing the components and systems that comprise the infrastructure; and
- Maintaining the components and systems specifically plant and facilities and planned, preventive and corrective maintenance.

The position where the requirements for managing the infrastructure feature in the managed process is shown in Fig. 21-1. In this case the process is a work process within the resource management process. The product objective would be to provide an infrastructure that enabled the organization to achieve its objectives. The PDCA in the diagram is planning, installing, checking and improving this infrastructure. The demand comes from the mission management process and the output goes through every process back to the mission management process where it is assessed.

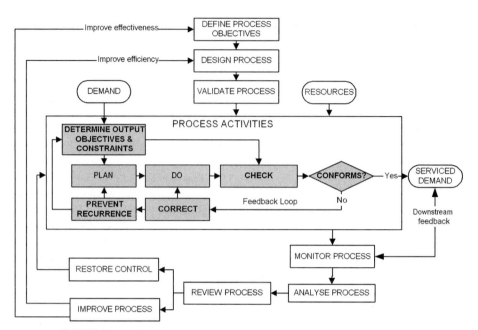

FIGURE 21-1   Where the requirements of Clause 6.3 apply in a managed process.

DETERMINING INFRASTRUCTURE NEEDS (6.3)

The standard requires the organization *to determine the infrastructure needed to achieve conformity to product requirements.*

What Does this Mean?

The emphasis in this requirement is on infrastructure needed to achieve conformity to product requirements. As the conforming product is the organization's output, it follows that most of the infrastructure exists for this purpose. However, there will be areas, buildings, facilities etc. that may not be dedicated to this purpose but to meeting requirements of stakeholders other that the customer of the organization's products. The requirement is not implying that these other facilities do not need to be identified, provided and maintained, but that such provision is not essential to meet ISO 9001. As with determining resources previously, ask: 'Why would we want to exclude a particular resource from the management system? What business benefit is derived from doing so?'

Why is this Necessary?

This requirement responds to the Process Approach Principle.
    The design, development and supply of products and services do not exist in a vacuum. There is always an infrastructure within which these processes are carried out and on which these processes depend for their results. Without an appropriate

infrastructure the desired results will not be achieved. A malfunction in the infrastructure can directly affect results.

How is this Demonstrated?

In determining the resources needed to implement the quality system, some are product, project, contract or order specific, others are needed for maintenance and growth of the organization. These are likely to be classified as capital assets. The management of the infrastructure is a combination of asset management (knowing what assets you have, where they are, how they are depreciating and what value they could realize) and of facilities management (identifying, acquiring, installing and maintaining the facilities).

As the infrastructure is a critical factor in the organization's capability to meet customer requirements, and ability to continually meet customer requirements, its management is vital to the organization's success. Within the resource management process there are therefore several work processes related to the management of the infrastructure. It would be impractical to put in place one process because the processes will differ depending on the services required. Based on the generic model for resource management (see Figure 10-7) several planning processes will be needed for identifying and planning the acquisition, deployment, maintenance and disposal of the various assets. In describing these processes, you need to cover the aspects addressed in Chapter 11 on *Process descriptions* and in doing so, identify the impact of failure on the organization's ability to achieve conforming product.

PROVIDING INFRASTRUCTURE

The standard requires the organization *to provide the infrastructure needed to achieve conformity to product requirements*.

What Does this Mean?

Providing infrastructure simply means acquiring and deploying the infrastructure that has been determined as being necessary to achieve conformity to product requirements.

Why is this Necessary?

This requirement responds to the Process Approach Principle.

Unless the required infrastructure is provided the desired results will not be achieved. However, it is often the case that resources are not available to fund infrastructure that has been planned and as a result contingency plans have to be brought into effect or else difficulties will be encountered. A failure to provide adequate infrastructure can adversely affect the organization's performance.

In many cases the plans may address the worst case not taking account of human ingenuity.

How is this Demonstrated?

Providing the infrastructure is associated with the acquisition and deployment of resources. It follows therefore that processes addressing the acquisition and deployment of buildings, utilities, computers, plant, transport etc. need to be put in place. Many will use the purchasing process but some require special versions of this process because provision will include installation and commissioning and all the attendant architectural and civil engineering services. Where the new facility is required to provide additional capability so that new processes or products can be developed, the time to market becomes dependent on the infrastructure being in place for production to commence. Careful planning is often required because orders for new products may well be taken on the basis of projected completion dates and any delays can adversely affect achievement of these goals and result in dissatisfied customers. Customers may have bought tickets for a flight to a new destination, a rail journey through a new tunnel, a football match at a new stadium relying on the airport, the tunnel or the stadium being open for business on time. A new microprocessor, television, automobile may be dependent on a new production plant and orders may be taken on the basis of projected completion dates. With major capital works, plans are made years in advance with predictions of completion dates based on current knowledge. A new Air Traffic Control Centre at Swanwick in the UK was planned to replace West Drayton ATC outside London because when the plans were made in the late 1980s it was predicted that capacity would be exceeded by the mid-1990s. The Operational Handover date was set as December 1996. After many delays, by the summer of 2000 it was still not operational and the problem that it was designed to solve was getting worse. Air traffic congestion continued to delay flights and dissatisfy customers until it was finally brought into service two years later in January 2002. It now controls 200,000 square miles of airspace among the busiest and most complex in the world.[1] These examples as do others illustrate the importance of infrastructure on performance and the link with customer satisfaction.

MAINTAINING INFRASTRUCTURE

The standard requires the organization *to maintain the infrastructure needed to achieve conformity to product requirements.*

What Does this Mean?

The identification and provision of the infrastructure needs little explanation but in maintaining it the implications go beyond the maintenance of what exists. Maintenance is more to do with maintaining the capability the infrastructure provides. Plant and facilities can be relatively easily maintained, but maintaining their capability means continually providing a capability even when the existing plant and facilities are no longer serviceable. Such situations can arise due to man-made and natural disasters. Maintaining the infrastructure means maintaining output when there is a power cut, a fire, a computer virus, a flood or a gas explosion. Maintaining the infrastructure therefore means making provision for disaster recovery and therefore maintaining business continuity.

---

[1] Source: http://www.nats.co.uk/services/swanwick.html.

Why is this Necessary?

This requirement responds to the Process Approach Principle.

Unless the infrastructure is maintained the desired results will not be achieved. A failure to maintain adequate infrastructure can adversely affect the organization's performance.

How is this Demonstrated?

Maintenance of Plant and Facilities

There are two aspects to maintenance as addressed previously. Maintaining the buildings, utilities and facilities in operational condition is the domain of planned preventive and corrective maintenance. Maintaining the capability is the domain of contingency plans, disaster recovery plans and business continuity provisions. In some industries there is no obligation to continue operations as a result of *force majeure*⊕ including natural disasters, war, riots, air crash labour stoppage, illness, disruption in utility supply by service providers etc. However, in other industries, provisions have to be made to continue operations albeit at a lower level of performance in spite of force majeure.

Although such events cannot be prevented, their effects can be reduced and in some cases eliminated. Contingency plans should therefore cover those events that can be anticipated where the means to minimize the effects are within your control. What may be a force majeure situation for your suppliers does not need to be the same for you.

Start by doing a risk assessment and identify those things on which continuity of business depends namely power, water, labour, materials, components, services etc. Determine what could cause a termination of supply and estimate the probability of occurrence. For those with a relatively high probability (1 in 100) find ways to reduce the probability. For those with lower probability (1 in 10,000 chance) determine the action needed to minimize the effect. The FMEA⊕ technique works for this as well as for products and processes (see Chapter 37).

If you are located near a river and it floods in the winter, can you claim it to be an event outside your control when you chose to site your plant so close to the river? (OK the land was cheap – you got a special deal with the local authority – but was it wise?) You may have chosen to outsource manufacture to a supplier in a poorer country and now depend on them for your supplies. They may ship the product but because it is seized by pirates it doesn't reach its destination – you may therefore need an alternative source of supply! A few years ago we would have thought this highly unlikely, but after 300 years and equipped with modern technology pirates have returned to the seas once again.

Maintenance of Equipment

In a manufacturing environment the process plant, machinery and any other equipment on which process capability depends need to be maintained and for this you will need:

- A list of the equipment on which process capability depends.
- Defined maintenance requirements specifying maintenance tasks and their frequency.
- A maintenance programme which schedules each of the maintenance tasks on a calendar.

- Procedures defining how specific maintenance tasks are to be conducted.
- Procedures governing the decommissioning of plant prior to planned maintenance.
- Procedures governing the commissioning of plant following planned maintenance.
- Procedures dealing with the actions required in the event of equipment malfunction.
- Maintenance logs that record both the preventive and corrective maintenance work carried out.

In a service environment if there is any equipment on which the capability of your service depends, this equipment should be maintained. Maintenance may often be subcontracted to specialists but nevertheless needs to be under your control. If you are able to maintain process capability by bringing in spare equipment or using other available equipment, your maintenance procedures can be simple. You merely need to ensure you have an operational spare at all times. Where this is not possible you can still rely on the Call-out Service if you can be assured that the anticipated down time will not reduce your capability below that which you have been contracted to maintain.

The requirement does not mean that you need to validate all your word-processing software or any other special aids you use. Maintenance means retaining in an operational condition and you can do this by following some simple rules.

There are several types of maintenance; planned maintenance,[①] preventive maintenance,[①] corrective maintenance[①] and predictive maintenance[①].

An effective maintenance system should be one that achieves its objectives in minimizing down time, i.e., the period of time in which the equipment is not in a condition to perform its function. In order to determine the frequency of checks you need to predict when failure may occur. Will failure occur at some future time, after a certain number of operating hours, when being operated under certain conditions or some other time? An example of predictive maintenance is vibration analysis. Sensors can be installed to monitor vibration and thus give a signal when normal vibration levels have been exceeded. This can signal tool wear and wear in other parts of the machine in advance of the stage where nonconforming product will be generated.

The manuals provided by the equipment manufacturer's should indicate the recommended preventive maintenance tasks and the frequency with which they should be performed covering aspects such as cleaning, adjustments, lubrication, replacement of filters and seals, inspections for wear, corrosion, leakage, damage etc.

Another source of data is from your own operations. Monitoring tool wear, corrective maintenance, analysing cutting fluids and incident reports from operators you can obtain a better picture of a machines performance and predict more accurately the frequency of checks, adjustments and replacements. For this to be effective you need a reporting mechanism that causes operators to alert maintenance staff to situations where suspect malfunctions is observed. In performing such monitoring you cannot wait until the end of the production run to verify whether the tools are still producing conforming product. If you do you will have no data to show when the tool started producing nonconforming product and will need to inspect the whole batch.

An effective maintenance system depends on it being adequately resourced. Maintenance resources include people with appropriate skills, replacement parts and materials with the funds to purchase these material and access to support from Original Equipment Manufacturers (OEMs) when needed. If the OEM no longer supports the equipment, you may need to cannibalize old machines or manufacture the parts yourself.

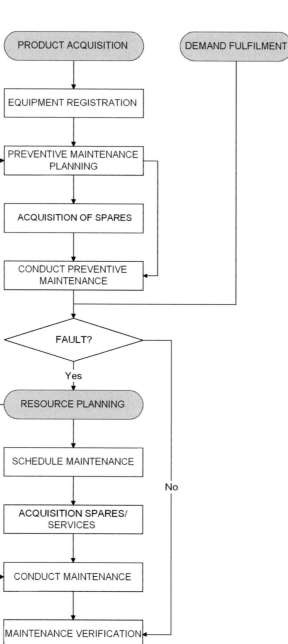

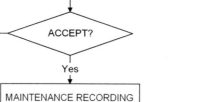

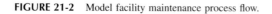

FIGURE 21-2 Model facility maintenance process flow.

This can be a problem because you may not have a new part from which to take measurements. At some point you need to decide whether it is more economical to maintain the old equipment than to buy new. Your inventory control system needs to account for equipment spares and to adjust spares holding based on usage.

For the system to be effective there also has to be control of documentation, maintenance operations, equipment and spare parts. Manuals for the equipment should be brought under document control. Tools and equipment used to maintain the operational equipment should be brought under calibration and verification control. Spare parts should be brought under identity control and the locations for the items brought under storage control. The maintenance operations should be controlled to the extent that maintenance staff should know what to do, know what they are doing and be able to change their performance if the objectives and requirements are not being met. While the focus should be on preventive maintenance, one must not forget corrective maintenance. The maintenance crew should be able to respond to equipment failures promptly and restore equipment to full operational condition in minimum time. The function needs resourcing to meet both the preventive and corrective demands because it is down time that will have most impact on production schedules.

A model facility maintenance process flow is illustrated in Fig. 21-2.

Infrastructure Records

There is no requirement in Clause 6.3 for records of the maintenance of infrastructure (confirmed by published interpretation RFI 003) but there is a requirement in Clause 8.4 for the organization to determine, collect and analyse appropriate data to demonstrate the suitability and effectiveness of the quality management system (which includes the infrastructure) and to evaluate where continual improvement of the effectiveness of the system can be made. It would be difficult to do this without records. There is also a requirement in Clause 4.2.1d for *the quality management system documentation to include records, determined by the organization to be necessary to ensure the effective planning, operation and control of its processes.*

Any records you maintain need to be useful and it would be wise to maintain records particularly of plant, facility and equipment maintenance so that in the event of a problem, a pending prosecution or simply a customer complaint you can carry out a proper investigation. You would want to either demonstrate you acted in a reasonable and responsible manner or find the root cause of a problem – you can't demonstrate you are compliant with Clause 8.5.2 on corrective action unless you have records to analyse but don't go overboard; think what you might need; conduct a process FMEA and this might reveal exactly what records you need to keep.

Chapter 22

Work Environment (6.4)

CHAPTER PREVIEW

This chapter is aimed at all levels of management for it is they who create the work environment that will determine whether the right products and services are developed and produced on time and within budget to the customer's satisfaction.

In the preview to Chapter 20 and in Chapter 9 we pointed out that people are affected by the environment and even when equipped with the right competences may not function as well as expected if the conditions are not appropriate. This means that the physical, social and psychological factors of the work environment need to be managed effectively so that they provide the optimum conditions for worker productivity. In the 2000 version of ISO 9001, the term work environment meant those conditions under which work is performed and the definition in ISO 9000:2005 referred to the physical, social and psychological factors. However, for some unknown reason a note has been added to Clause 6.4 in ISO 9001:2008 which appears to limit the environment affecting conformity of product to the physical factors and typical examples are given as noise, temperature, humidity, lighting or weather. There is no mention of the social and psychological factors that we identified in Chapter 9 (see also *ISO 9001:2008 Commentary* in Chapter 4). As ISO 9000:2005 is invoked in ISO 9001, the former definition still applies therefore rather than creating a conflict ISO 9001:2008 simply creates confusion and the note in Clause 6.4 should not be taken as limiting the work environment. However, there is reluctance in some quarters to introduce the behavioural aspects of management into the management systems assessment process because of the difficulties in obtaining objective evidence in short periods of time allowed for third party audits. Longer periods of observation are often necessary to come to any conclusions about the culture and are therefore omitted from the assessment. Despite the limitations of third party assessment, this should not be cause for management to ignore the social and psychological factors, which have been known to influence productivity since the behavioural school of management emerged in the 1920s with Elton Mayo's experiments in industrial research.

In this chapter we examine the requirements in Clause 6.4 of ISO 9001:2008 and in particular:

- Explore the work environment and the factors affecting worker output;
- Why it is necessary to manage these factors; and
- How to demonstrate you are managing work environment.

The position where the requirements for managing the work environment feature in the managed process is shown in Fig. 22-1. In this case there are two processes. One is

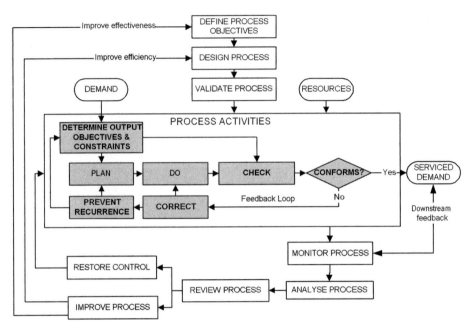

FIGURE 22-1   Where the requirements of Clause 6.4 apply in a managed process.

a work process within the resource management process that has the creation of a physical work environment as its objective and the other is a work process within the mission management process that has the creation of a psychological and social environment as its objective (see Fig. 10-6). The measures in both cases would be an environment acting positively towards the achievement of the planned results. The PDCA in the diagram is planning, creating, checking and improving this environment. The demand comes from the mission management process and the output goes through every process back to the mission management process where it is assessed.

IDENTIFYING AND MANAGING THE WORK ENVIRONMENT (6.4)

The standard requires the organization *to identify and manage the work environment needed to achieve conformity to product requirements.*

What Does this Mean?

ISO 9000:2005 defines *the work environment* as a set of conditions under which people operate and include physical, social and psychological environmental factors. The environment has to be appropriate for the product and for the people producing the product because it affects both.

The social and psychological factors can be considered to be human factors as both are related to human behaviour. However, there are three dimensions to the work environment namely product design, organizational design and job design. The inherent product characteristics will determine its sensitivity to factors in the work environment. A job exists within an organization and there are factors that arise from performing the job such as physical movement, the man–machine interface, the physical environment and the psychological factors of the job. These form the basis of ergonomics (see BS 3138:1992). Even when the ergonomics of the job have been optimized, performance can still be adversely affected by wider influence of factors inherent in the organization caused by the way it functions or the way things get done – its culture and climate (see Chapter 9).

Physical factors of the work environment include space, temperature, noise, light, humidity, hazards, cleanliness, vibration, pollution, accessibility, physical stress and airflow. In addition to visible light, other types of radiation across the whole spectrum impact the physical environment.

Social factors of the work environment are those that arise from interactions between people and include the impact of an individual's family, education, religion and peer pressure as well as the impact of the organization's ethics, culture and climate.

Psychological factors of the work environment are those that arise from an individual's inner needs and external influences and include recognition, responsibility, achievement, advancement, reward, job security, interpersonal relations, leadership, affiliation, self-esteem, and occupational stress. They tend to affect or shape the emotions, feelings, personality, loyalty and attitudes of people and therefore the motivation of people towards the job to which they have been assigned.

While this grouping serves to identify related factors, it is by no means comprehensive or exclusive. Each has an influence on the other to some extent. Relating the identification of such factors to the achievement of conformity of product tends to imply that there are factors of the work environment that do not affect conformity of product. Whether people produce products directly or indirectly, their behaviour affects their actions and decisions and consequently the results of what they do. It is therefore difficult to exclude any factor on the basis that it is not needed to achieve conformity of product in some way or other.

Managing such factors means creating, monitoring and adjusting conditions in which the physical and human factors of the work environment act positively towards achievement of the planned results. Some of the factors affecting the work environment are constraints rather than objectives, i.e., they exist only because we have an objective to achieve. Noise in the workplace occurs because we need to run machines to produce product that satisfies the customer. If we didn't need to run the machines, there would be no noise. However, some of the constraints are of our own making. If the style of management created an environment that was more conducive to good industrial relations, the work force would be more productive.

Why is this Necessary?

This requirement responds to the Leadership and Involvement of People Principles.

The work environment is critical to the product and to worker performance and extends beyond the visible and audible factors commonly observed in a workplace. The physical factors affect both product and producer and all the above factors influence individual behaviour and this has a direct impact on organizational performance and consequently product quality. It is the duty of management to control the physical factors firstly within the levels required by law, secondly within the levels necessary to prevent deterioration of the product and thirdly as necessary for people to perform their jobs as efficiently and effectively as possible. It is also the task of management to create conditions in which personnel are motivated to achieve the results for which they are responsible and therefore remove or contain any de-motivating elements such as friction and conflict in the workplace.

The physical factors of the work environment influence individual behaviour by causing fatigue, distraction, accidents and a series of health problems. There are laws governing many of the physical factors such as noise, air pollution, space and safety. There are also laws related to the employment of disabled people that impact the physical environment in terms of access and ergonomics. The social interactions in the workplace influence interpersonal relationships such as the worker–boss, worker–subordinate, worker–colleague and worker–peer relationship (see Chapter 9).

How is this Demonstrated?

For a solution we can use a similar approach to that taken towards the natural environment. Environmental management is the control over activities, products and processes that cause or could cause environmental impacts. The approach taken is based on the management of cause and effect where the activities, products and processes are the causes or 'aspects', and the resulting effects or potential effects on the environment are the impacts. All effort is focused on minimizing or eliminating the impacts. In the context of the work environment, the causes or aspects would be the physical and human factors and the impacts would be the changes in working conditions. However, unlike management of the natural environment, the effort would not all be focused on reducing or eliminating impacts where a state of zero impact is ideal. In the working environment, the effort should be focused on eliminating negative impact and creating positive or beneficial impacts that also lead to an improvement in performance.

Dealing with the Physical Factors Affecting Product

The physical factors of the work environment that affect product conformity should be identified through design FMEA[?]. This should reveal whether the product is sensitive to temperature, humidity, vibration, light and contamination.

Where material degrades on exposure to light or air, the production processes should be designed to provide the protection required when the material exists the process.

To achieve high performance from electronic components particle and chemical contamination has to be minimized during fabrication and assembly. Cleanrooms are often built in which product is manufactured, assembled and tested. To produce food and drugs to the regulatory standards, high levels of cleanliness and hygiene need to be maintained during production and food preparation. For these and many other reasons, the work environment needs to be controlled. If these conditions apply you should:

- Document the standards that are to be maintained.
- Prohibit unauthorized personnel from entering the areas.
- Provide training for staff who are to work in such areas.
- Provide alarm system to warn of malfunctions in the environment.
- Provide procedures for maintaining the equipment to these standards.
- Maintain records of the conditions as a means of demonstrating that the standards are being achieved.

Dealing with the Physical Factors Affecting People

The physical factors affecting people are more easily dealt with than the human factors primarily because they are more tangible, measurable and controllable. To manage the physical factors they firstly need to be identified and this requires a study of the work environment to be made relative to its influence on the worker. We are not necessarily dealing only with safety issues although these are very important. The noise levels do not need to cause harm for them to be a factor that adversely affects worker performance. Libraries are places of silence simply to provide the best environment in which people can concentrate on reading. No harm arises if the silence is broken!

In dealing with physical factors there is a series of steps you can take to identify and manage these factors:

- Use an intuitive method such as brainstorming to discover the safety related and none safety related factors of the environment such as noise, pollution, humidity, temperature, light, space, accessibility etc.
- Research legislation and associated guidance literature to identify those factors that could exist in the work environment due to the operation of certain processes, use of certain products or equipment. We do X therefore from historical and scientific evidence there will be Y impact. (VDUs, RSI, airborne particles, machinery etc.).
- Determine the standard for each factor that needs to be maintained to provide the appropriate environment.
- Establish whether the standard can be achieved by workspace design, by worker control or by management control or whether protection from the environmental impact is needed (protection of ears, eyes, lungs, limbs, torso or skin).
- Establish what could fail that would breach the agreed standard using FMEA[①] or Hazards Analysis and identify the cause and the effect on worker performance.
- Determine the provisions necessary to eliminate, reduce or control the impact.
- Put in place the measures that have been determined.
- Measure and monitor the working environment for compliance with the standards and implementation of the provisions defined.
- Periodically repeat the previous steps to identify any changes that would affect the standards or the provisions currently in place. Ask – Is the standard still relevant? Are there better methods now available for dealing with this environmental impact?

Dealing with the Human Factors

Managers are often accused of ignoring the human factors but such factors are not easily identified or managed. With physical factors you can measure the light level and adjust it if it's too bright or too dim. You can't measure ethics, culture, climate, occupational stress – all you see are its effects and the primary effect is impacts employee motivation.

Managers need to understand and analyse human behaviour and provide conditions in which employees are motivated to achieve the organization's objectives (see Chapter 9). The standard requires top management to demonstrate commitment to quality and therefore demonstration that it has created an environment conducive to producing conforming product would be addressed by Clause 5.1.

Ergonomics

An employee's body movement in performing a job has important implications on the work environment.

The study of the relationship between a person and his or her job is referred as Ergonomics (see BS 3138) and it deals principally with the relationship between a person and his or her job, equipment and environment and particularly the anatomical, physiological and psychological aspects arising thereon. The layout of the workplace, the distances involved the areas of reach, seating, frequency and type of movement all impact the performance of the worker. These factors require study to establish the optimum conditions that minimize fatigue, meet the safety standards while increasing productivity.

Where people are an integral part of a mechanized process the man–machine interface is of vital importance and has to be carefully considered in process design. The information on display panels should be clear and relevant to the task. The positioning of instruments, and input, output and monitoring devices should allow the operator to easily access information without abnormal movement. The emergency controls should be within easy reach and the operating instructions accessible at the workstation. Legislation and national standards cover many of these aspects.

Key Messages from Part 5

1. No plans will be achieved unless the resources to make them happen are provided. It is therefore incumbent on top management to not only require a management system to be designed but also provide the means for it to function effectively.
2. In the long term, the total resources required to maintain the organization will be less with an effective management system than without but to start with, additional time and skills are required and need to be made available.
3. Individual competence is concerned with the ability of a person to achieve a result whereas training is concerned with the acquisition of skills to perform a task and education is concerned with the acquisition of knowledge. It is therefore not a question of whether a person has the skills and knowledge to do a job but on whether a person is able to achieve the desired outcome.
4. The starting point with competences is to define the outcomes required from a job and then define what makes those performing it successful and agree these with the role or jobholder. These outcomes are the hard results such as products and decisions and the soft results such as behaviours, influence and stamina.
5. When you document your management system you are probably documenting only parts of your processes. The interactions that create problems are probably hidden therefore to make the processes behave like you portray in your documentation, you will need to work on people's competence so that they all reach the stage of unconscious competence.
6. Training activities may degrade the performance, reliability and safety of training equipment and so it should be subject to inspection before and after training exercises.
7. A course that enables the participants to learn by doing, to learn by self-discovery and insight is a *training* course. The participants come away having had a skill changing experience.
8. Maintaining the infrastructure means maintaining output when there is a power cut, a fire, a computer virus, a flood or a gas explosion. Maintaining the infrastructure means making provision for disaster recovery and therefore maintaining business continuity
9. The limitations of third party assessment should not be cause for management to ignore the social and psychological factors which are known to influence productivity.
10. Managing physical, social and psychological environmental factors means creating, monitoring and adjusting conditions in which the physical and human factors of the work environment act positively towards achievement of the planned results

Complying with ISO 9001 Section 7 Requirements on Product Realization

INTRODUCTION TO PART 6

Structure of ISO 9001 Section 7

Product Realization as expressed in Section 7 of ISO 9001 is the Demand Fulfilment Process referred to previously that has interfaces with Resource Management and Demand Creation processes. It is also the Order to Cash process implying that the inputs are orders and the output is cash, therefore it would include the invoicing and banking activities. However the Product Realization requirements include requirements for purchasing, a process that could fit as comfortably under resource management because it is not limited to the acquisition of components but is a process that is used for acquiring all physical resources including services and also human resources such as contract labour. Section 7 also includes requirements for control of measuring devices which would fit more comfortably into Section 8 but it omits the control of nonconforming product which is more to do with handling product than measurement. Product realization does not address demand creation or marketing see text box. Note that Demand Creation is addressed by the standard only through clause 5.2 and that product design is located in Section 7 simply because, it refers to the design of customer specific products. If the products were designed in order to create a demand this work process would be part of Demand Creation.

Limits of Product Realization

By the time you get to product realization the market research has been completed and the demand has been created; a potential customer has been attracted to your organizations and has specified the requirements to be met.

Product realization is driven by customers not markets. The demand creation process precedes product realization and if your organization is market driven, several important processes will be excluded from your quality management system if you simply respond to the product realization requirements. However, you can bring these excluded processes into the QMS under the Customer Focus requirements of Clause 5.2.

Clause 7.6 on the control of measuring equipment is part of section 7 of the ISO 9001 because it is an area of permitted exclusion. As it is primarily concerned with measurement we have addressed the requirements in Part 7 of this book along with the requirements of section 8 of the standard.

Linking Requirements

If we link the requirements together in a cycle (indicating the headings from ISO 9001 in bold type), having marketed the organization's capability and attracted a customer, the cycle commences by the need to **communicate with customers** and **determine the requirements** of customers, of regulators and of the organization relative to the product or service to be supplied. This will undoubtedly involve more **customer communication** and once requirements have been determined we need to **review the requirements** to ensure they are understood and confirm we have the capability to achieve them. If we have identified a need for new products and services, we would then need to **plan product realization** and in doing so use **preventive action** methods to ensure the success of the project and take care of any **customer property** on loan to us. We would undertake product **design and development** and in doing so we would probably need to **identify product, purchase** materials, components and services, build prototypes using the process of **production provision** and **validate** new **processes.** After **design validation** we would release product information into the market to attract customers and undertake more **customer-communication.** As customers enquire about our offerings we would once more **determine the requirements** in order to match customer needs with product offerings and our ability to supply.

Now faced with real customers demanding our products, we would **review the requirements** and confirm we had the capability to supply the product in the quantities and to the delivery schedule required before entering into a commitment to supply. We would then proceed to **plan product realization** once again and undertake **production or service provision.** During production or service delivery we would maintain **traceability** of the product if applicable, perform **measurement and monitoring** and **control the measuring and monitoring devices**. We would **monitor and measure processes** and **monitor and measure products** at each stage of the process. If we found unacceptable variations in the product we would undertake the **control of nonconforming product** and **analyse data** to facilitate **corrective action**. Throughout production or service delivery we would seek the **preservation of product** and take care of **customer property**. Once we had undertaken all the **product verification** and **preserved** the product for delivery, we would ship the product to the customer or complete the service transaction. To complete the cycle **customer communication** would be initiated once more to obtain feedback on our performance.

Here we have linked together all the clauses in Section 7 and many in Section 8 of the standard because the two cannot be separated.

Chapter 23

Planning Product Realization Processes

CHAPTER PREVIEW

This chapter is aimed at those managers operating at the customer interface with responsibility for planning the execution of an order or contract and in particular those involved with project planning, new product planning, production planning, installation planning and other related planning functions.

In Part 3 we addressed the requirements for establishing a quality management system and meeting these requirements should result in a system of processes that will enable the organization to provide products that satisfy customers and other requirements. In many cases, the work processes will have been designed and will form part of the management system. However, the nature and complexity of specific projects, contracts or orders may require these work processes to be tailored or enhanced to suit particular needs. The Demand Fulfilment process is likely to enable the organization to deliver conforming product regardless of the specific features and characteristics of the product. It will generate product specific documents therefore the business process won't change. These product or contract specific documents are what are addressed by these requirements.

In this chapter we examine the requirements in Clause 7.1 of ISO 9001:2008 and in particular:

- Planning and developing product realization processes;
- Creating consistency in process planning;
- Quality objectives and requirement for product;
- Determining the need for documentation;
- Determining the need for resources;
- Determining verification, validation and monitoring activities;
- Determining the criteria for product acceptance;
- Determining the need for records; and
- Documenting product realization planning.

The position where the requirements for product realization planning feature in the managed process is shown in Fig. 23-1. Note 2 in ISO 9001 Clause 7.1 suggesting that the design and development requirements of Clause 7.3 can be applied to processes results in the inclusion of Process Objective, Design and Validation elements from the generic model.

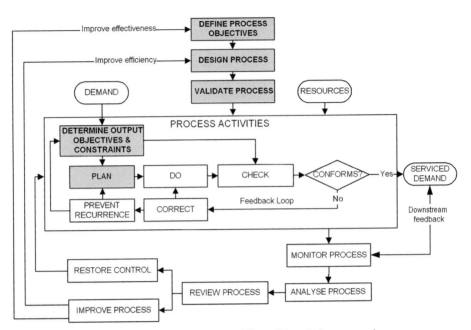

FIGURE 23-1   Where the requirements of Clause 7.1 apply in a managed process.

PLANNING AND DEVELOPING PRODUCT REALIZATION PROCESSES (7.1)

The standard requires the organization *to plan and develop the processes required for product realization.*

What Does this Mean?

The product realization processes are the processes needed to specify, develop, produce and supply the product or service required and would include where applicable, those needed to:

- specify the products and services required by the organization's customers (the sales process);
- plan the provision of the identified products and services (the project, contract or order planning process);
- design the identified products and services so as to meet customer requirements (the design process);
- procure the materials, components, services needed to accomplish the design and/or generate or deliver the product or service (the purchasing process);
- generate the product (the production process);
- supply the product or service (the distribution or service delivery processes);
- install the product on customer premises (the installation process);
- maintain and support the product in service (the product and service support process); and

- provide support to customers (the after sales, technical support or customer support process).

The planning for product realization should define the processes for producing the product features required to meet customer needs.

These processes are all product or service specific and take the input from the customer through a chain of related processes that deliver acceptable products or services to customers. Planning these processes means determining the processes required for a specific project, contract or order and determining their sequence and interrelation. As the nature of planning will vary significantly from organization to organization, the generic term for this type of planning is *product realization planning* thus distinguishing it from specific planning activities such as design planning, production planning, installation planning etc. product realization planning is therefore the overall planning activities needed to meet all requirements for a project, product, contact or order.

Why is this Necessary?

This requirement responds to the Process Approach Principle.

In designing the management system the work processes needed to produce the organization's products and services should have been developed so that planning to meet specific orders does not commence with a blank sheet of paper. These work processes provide a framework that aids the planners in deciding on the specific processes, actions and resources required for specific projects, contracts or orders. The process descriptions may not contain details of specific products, dates, equipment, personnel or product characteristics. These may need to be determined individually for each product – hence the need to plan and develop processes for product realization.

How is this Demonstrated?

Planning Product Realization Processes

There are too many variations in the product realization process to provide much more than an overview.

Where the organization designs products and services in response to customer requirements, design would be part of the Demand Fulfilment process the relationship between the processes might be as shown in Fig. 23-2.

For customer specific products, product planning is driven by customer enquires, orders or contracts. This may require a project planning process in order to provide the customer with a viable proposal. On receipt of a contract or order an order processing process is then needed to confirm and agree customer requirements and the terms and conditions for the supply of product or service. Once the contract or order has been agreed, a project or order-planning process is needed to establish the provisions needed to meet the contract or order requirements.

For proprietary products, product planning is driven by the demand creation process which searches for new opportunities that will result in the development of new products and services and thus its output will lead into project planning which

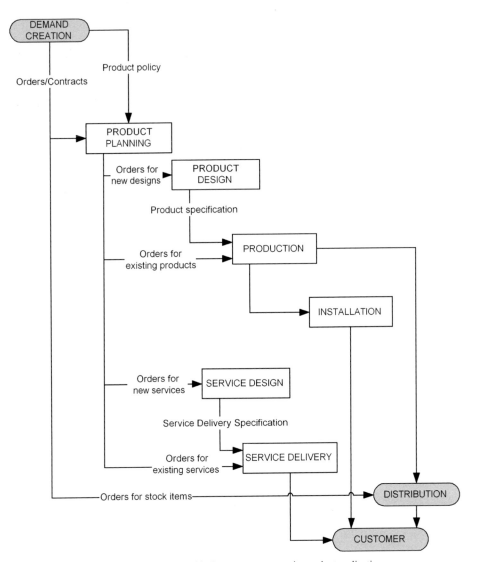

FIGURE 23-2   Relationship between processes in product realization.

subsequently supplies the demand fulfilment process with proven products and service to sell.

Product planning can therefore be driven by customer specific requirements as well as requirements determined by the organization that will reflect or create customer needs and expectations. The product realization planning process therefore includes the sales process and the product planning process. The sales process may not require tailoring for specific enquiries but with major projects for the large procurement agencies, this process often varies depending on the nature of the project and may require careful planning for the organization to be successful. The product planning process will often require tailoring for specific projects or contracts as the nature of the work involved will vary. Where the organization takes orders for existing products or services which the customer selects from a catalogue, no special planning may be needed other than the creation of work orders.

In planning product realization, there are several factors involved – tasks, timing, responsibilities, resources, constraints, dependencies and sequence. The flowcharts for each process that were developed in establishing the management system identify the tasks. The planners job is to establish whether these tasks, their sequence and the process characteristics in terms of throughput, resources, capacity and capability require any modification to meet the requirements of a particular project, contract or order. A typical product planning process is illustrated in Fig. 23-3.

Tools often used in product realization planning are Gantt Charts and PERT Charts. The Gantt chart depicts the tasks and responsibilities on a timescale showing when the

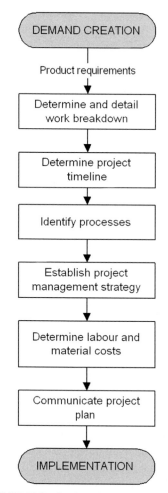

FIGURE 23-3   Product planning process flow.

tasks are to commence and when they are to be completed. PERT charts display the same information but show the relationship between the tasks. These tools are useful in analysing a programme of work, determining resources and determining whether the work can be completed by the required end date using the allocated resources. It is not the purpose here to elaborate on project planning techniques but merely to indicate the scope of the requirement and what is needed to implement it.

A common method of planning projects is to prepare a Project Plan that includes the following depending on whether the project includes new product development or the development of new processes to deliver high volume production:

1. Project objectives;
2. System requirements;
3. Project strategy;
4. Critical success factors and success criteria;
5. Project milestones;
6. Project timeline;
7. Project organization (chart and team responsibilities and authority);
8. Work breakdown structure (major tasks, work packages and deliverables);
9. Resource provision in terms of space, development tools, equipment;
10. Supplier control plan;
11. Information system (strategy, tools and their development);
12. Communication plan (strategy, methods and tasks);
13. Personnel development plan (strategy, education and training for those engaged on the project);
14. Evaluation plan (audits, design reviews and assessments);
15. Project reviews (strategy, project reviews and team reviews); and
16. Contract management.

Developing Product Realization Processes

There is a note that suggests that the design and development requirements may be applied to the development of product realization processes. The development of business processes has been addressed in Chapter 10 but the requirements of Clause 7.3 could certainly be applied to the design and development of the processes for producing the product or delivering the service.

CREATING CONSISTENCY IN PROCESS PLANNING (7.1)

The standard requires planning of the realization processes *to be consistent with the other requirements of the organization's quality management system.*

What Does this Mean?

This means that the processes employed for specific products and services should either be those that form part of the management system, are developments of those that form part of the management system or are new processes that fit into the set of management system processes and meet the same organizational objectives.

Why is this Necessary?

This requirement responds to the Process Approach Principle.

The management system developed to meet the requirements of ISO 9001 is likely to be a generic system, not specific to any particular product, project or contract other than the range of products and services which your organization supplies. By implementing the processes of the management system, product, project or contract specific plans, procedures, specification etc. are generated. However, specific variants, modifications or new activities may be required for particular projects, contracts or orders. It is therefore essential that the provisions made for any particular product, service, project or contract do not conflict with the authorized policies and practices so that the integrity of the system is maintained. Also, if staff are familiar with one way of working, by receiving conflicting instructions staff may apply the incorrect policies and practices to the project.

How is this Demonstrated?

There is often a temptation when planning for specific contracts to change the policies and work processes where they are inflexible, invent new forms, change responsibilities, by-pass known bottlenecks etc. You need to be careful not to develop a mutant management system for specific contracts. If the changes needed are good for the business as a whole, they should be made using the established management system change process.

QUALITY OBJECTIVES AND REQUIREMENT FOR PRODUCT (7.1a)

The standard requires the organization *to determine the quality objectives and requirements for the product.*

What Does this Mean?

This means that for every product or service that is to be supplied there has to be a specification of requirements which if met will deliver a product or service that meets customer requirements.

The quality objectives for the product are those inherent characteristics of the product or service that aims to satisfy customers. For products these would include objectives for functional performance and physical attributes, reliability, maintainability, durability etc. Those for services would include accessibility, responsiveness, promptness, reliability etc. Quality requirements for the product are the inherent characteristics that are required to be met and may equal the quality objectives, but quality objectives may aim higher than the requirements either in an attempt to delight customers or simply to ensure requirements are met. In some cases, the characteristic may be prone to variation due to factors that are not easily controllable. In these cases targets are set higher to be certain of meeting the requirement. In other cases the objective may be specific and quantifiable such as requiring a mean time between failures (MTBF) of 100 000 hours or requiring compliance with EU Directive 90/385/EEC Annex 1. These quality objectives are for specific products whereas the quality objectives referred to in Clause 5.4.1 relate to the whole organization.

Why is this Necessary?

This requirement responds to the Leadership Principle.

The objective of product realization processes is to deliver product or service that meets requirements. It is therefore necessary to establish exactly what requirements the product must satisfy in order to determine whether the processes to be employed are fit for their purpose, i.e., are capable of delivering a product that meets the requirements.

How is this Demonstrated?

A specification should be produced or supplied for each product or service that is designed, produced and delivered. This specification should not only define the characteristics of the product but also should define its purpose or function so that a product possessing the stated characteristics can be verified as being fit for its purpose. It is of little use for a product to meet its specification if the specification does not accurately reflect customer needs.

DETERMINING THE NEED FOR SPECIFIC PROCESSES (7.1b)

The standard requires the organization *to determine the need to establish processes specific to the product.*

What Does this Mean?

As the standard does not define processes other than something that transforms inputs into outputs, the notion of a hierarchy of processes is overlooked. With this definition of a process, the organization itself is a customer-oriented process – it takes inputs from customers and creates products that hopefully satisfy the inputs.

There are two classes of processes – those that are product specific and those that are not. The non-product specific processes are generally the management and support processes such as mission management and resource management. These are used for all the organization's products and services. Design, production and delivery processes are in general product specific because their characteristics may change depending on the nature of the product or service being produced or supplied. However, there are many design, production and delivery processes that do not require customization to deliver the required outputs other than when initially established. An organization that designs electronic circuits can use the same design process regardless of the specific characteristic of particular products. The same is true with respect to production processes where there are many making and moving processes that do not change with the product. Each of the processes has a performance range within which it can process product therefore design processes for electronic circuits would probably be inappropriate for designing ships. Fabrication processes for telephones would probably be inappropriate for washing machines etc. So with the requirement to *determine the need to establish processes specific to the product,* we are not talking about business processes or even work processes but the processes needed to execute a task such as to make a body panel or assemble an engine where the inputs are materials and the output is a finished or semi-finished product.

Why is this Necessary?

This requirement responds to the Process Approach Principle.

When planning for specific products it is necessary to determine whether the intended product characteristics are within the design limits of the existing processes. If the product is similar to existing products no change to the processes may be needed. If the nature of the product is different or if the performance required is beyond the capability of existing processes, new or modified processes will be required. Many problems arise where managers load product into processes without being aware of the process's limitations. Often because people are so flexible, it is assumed that because they were successful at producing the previous product they will be successful with any other products. It is only when the differences are so great as to be glaringly obvious that they stop and think.

How is this Demonstrated?

In planning for a contract or new product or service, the existing processes need to be reviewed against the customer or market requirements. One can then identify whether the system provides an adequate degree of control. Search for unusual requirements and risks to establish whether any adjustment to processes is necessary. This may require you to introduce new activities or provide additional verification stages and feedback loops or prepare contingency plans.

If you have a process hierarchy such as that in the example of Fig. 8-14 new 'processes' might be needed at the task level. In the example, the tasks vary depending on what parts are being made so in effect the new process is reflected in the Manufacturing Plan. In order to make the parts, after machining the materials might need heat treatment, plating or any number of processes.

DETERMINING THE NEED FOR DOCUMENTATION (7.1b)

The standard requires the organization *to determine the need to establish documents specific to the product.*

What Does this Mean?

Documentation specific to the product is any documentation that is used or generated by the product realization processes. Such documents include specifications, drawings, plans, standards, datasheets, manuals, handbooks, procedures, instructions, records, reports etc. that refer to the product or some aspect of the product. Determining the need to establish documentation means that in planning product realization you need to determine the information carriers that will feed each of the processes and be generated by each of the processes.

Why is this Necessary?

This requirement responds to the Process Approach Principle.

The process descriptions will specify the types of information required to operate the process and required to be generated by the process. However, depending on the nature of the product, contract or project, these may need to be customized for the specific product so that they carry the required information to the points of implementation. Information required for one project may not be required for another project. Product configuration, organization structure and locations may all be different and require specific documents that are not used on other projects.

How is this Demonstrated?

A common method for project work is to establish a Work Breakdown Structure (WBS) that identifies all the major packages of work to be carried out. For each major task a work statement is produced that defines the inputs, tasks and outputs required. The outputs are described as a series of deliverables. Some of these will be documents, particularly in the design and planning phases. For less complex projects a list of deliverables may be all that is required, identifying the document by name, the author and delivery date.

DETERMINING THE NEED FOR RESOURCES (7.1b)

The standard requires the organization *to provide resources specific to the product*.

What Does this Mean?

Resources are an available supply of equipment, machines, materials, people etc. that can be drawn on when needed. Therefore this requires detailed planning and logistics management and may require many lists and sub-plans so that the resources are available when required. Inventory management and Information Technology is an element of such planning.

Why is this Necessary?

This requirement responds to the Process Approach Principle.

All businesses are constrained by their resources. No organization has an unlimited capability. It is therefore necessary when planning new or modified products to determine what resources will be required to design, develop, produce and supply the product or service. Even when the requirement is for existing products, the quantity or delivery required might strain existing resources to an extent where failure to deliver becomes inevitable.

How is this Demonstrated?

Successful implementation of this requirement depends on managers having current details of the capability of the process at their disposal. At the higher levels of management, a decision will be made as to whether the organization has the inherent capability to meet the specific requirements. At the lower levels, resource planning focuses on the detail, identifying specific equipment, people, materials, capacity and most important, the time required. A common approach is to use a project-planning tool such as Microsoft Project that facilitates the development of Gantt and PERT Charts and the ability to predict resources levels in terms of manpower and programme time. Other planning tools will be needed to predict process throughput and capability. The type of resources to be determined might include any of the following:

- Special equipment tools, test software and test or measuring equipment.
- Equipment to capture, record and transmit information internally or between the organization and its customers.
- New technologies such as computer aided design and manufacturing (CAD/CAM).
- Fixtures, jigs and other tools.
- New instrumentation either for monitoring processes or for measuring quality characteristics.
- New measurement capabilities.
- New skills required to operate the processes, design new equipment and perform new roles.
- New research and development facilities.
- New handling equipment, plant and facilities.

DETERMINING VERIFICATION AND VALIDATION ACTIVITIES (7.1c)

The standard requires the organization *to determine the required verification, validation, monitoring, measurement, inspection and test activities specific to the product.*

What Does this Mean?

The required verification activities are those activities necessary to establish that a product meets or is meeting the defined and agreed requirements. There are several methods of verification including inspection, test, monitoring, analysis, simulation, observation or demonstration each serving the same purpose but fulfilling it through a different method. All may include measurement but might not be exclusively measurement (*Numbers are the product of counting. Quantities are the product of measurement: Gregory Bateson 1904–1980, U.S. scientist, philosopher*).

Why is this Necessary?

This requirement responds to the Factual Approach Principle.

Provision should be made in the management system for verification of product at various stages through the realization processes. The stage at which verification needs to be performed and the characteristics to be verified at each stage are dependent on the requirements for the particular product. It is therefore necessary to determine the verification required for each product and process.

How is this Demonstrated?

Product Verification

If all the key features and characteristics of your product or service can be verified by a simple examination on final inspection or at the point of delivery, the requirement is easily satisfied. On the other hand if you can't do this, whilst the principle is the same, it becomes more complex.

Generically there are two types of requirements – *Defining requirements* and *Verification requirements*. Defining requirements specify the features and characteristics required of a product, process or service. These may be wholly specified by the customer or by the organization or a mixture of the two. *Verification requirements* specify the requirements for verifying that the defining requirements have been achieved and again may be wholly specified by the customer or by the organization or a mixture of the two. With verification requirements, however, other factors need to be taken into consideration depending on what you are supplying and to whom you are supplying it. In a contractual situation, the customer may specify what is to be verified and how it is to be verified. In a non-contractual situation, there may be statutory legal requirements, compliance with which is essential to avoid prosecution. Many of the national and international standards specify the tests that products must pass rather than specify performance or design requirements, so identifying the verification requirements can be quite a complex issue. It is likely to be a combination of:

- What your customer wants to be verified to meet the need for confidence. (The customer may not demand you demonstrate compliance with all customer requirements, only those which are judged critical.)
- What you need to verify to demonstrate that you are meeting all your customer's defining requirements. (You may have a choice as to how you do this so it is not as onerous as it appears.)
- What you need to verify to demonstrate that you are meeting your own defining requirements. (Where your customer defines the product or service in performance terms, you will need to define in more detail the features and characteristics that will deliver the specified performance and these will need to be verified.)
- What you need to verify to demonstrate that you are complying with the law (product safety, personnel health and safety, conservation, environmental and other legislation).
- What you need to verify to obtain confidence that your suppliers are meeting your requirements.

Verification requirements are not limited to product or service features and characteristics. One may need to consider who carries out the verification, where and when it is carried out and under what conditions and on what quantity (sample or 100%) and standard of product (prototype or production models).

You may find that the only way you can put your product on the market is by having it tested by an independent test authority. You may need a licence to manufacture it, to supply it to certain countries and this may only be granted after independent certification. Some verification requirements only apply to the type of product or service, others to the process or each batch of product and others to each product or service delivery. Some requirements can only be verified under actual conditions of use. Others can be verified by analysis or similarity with other products that have been thoroughly tested.

There are a number of ways of documenting verification requirements:

- By producing defining specifications which prescribe requirements for products or services and also the means by which these requirements are to be verified in-house in terms of the inspections, tests and other means of verification.

- By producing separate verification specifications which define which features and characteristics of the product or service are to be verified and the means by which such verification is to be carried out.
- By producing a quality plan or a verification plan that identifies the verification stages from product conception to delivery and further as appropriate, and refers to other documents that define the specific requirements at each stage.
- By route card referencing drawings and specifications.
- By inspection and test instructions specific to a production line, product or range of products.

In fact you may need to employ one or more of the above techniques to identify all the verification requirements.

Document Verification

It is necessary to verify that all the documentation needed to produce and install the product is compatible; that you haven't a situation where the design documentation requires one thing and the production documents require another or that details in the design specification conflict with the details in the test specification. Incompatibilities can arise in a contract that has been compiled by different groups. For example the contract requires one thing in one clause and the opposite in another. Many of the standards invoked in the contract may not be applicable to the product or service required. Production processes may not be qualified for the material specified in the design – the designer may have specified materials that are unavailable!

In order to ensure compatibility of these procedures, quality-planning reviews need to be planned and performed as the new documentation is produced. Depending on the type of contract, several quality planning reviews may be necessary, each scheduled to occur prior to commencing subsequent stages of development, production, installation or servicing. The quality planning reviews during product development can be held in conjunction with the design stage reviews required in Section 7.3.4 of ISO 9001. At these reviews the technical and programme requirements should be examined to determine whether the existing provisions are adequate, compatible and suitable to achieve the requirements and if necessary additional provisions put in place.

DETERMINING THE CRITERIA FOR PRODUCT ACCEPTANCE (7.1c)

The standard requires the organization *to determine the criteria for product acceptance*.

What Does this Mean?

The criteria for product acceptability are those characteristics that the product or service needs to exhibit for it to be deemed acceptable to the customer or the regulator. These are those standards, references, and other means used for judging compliance with defined requirements. In some cases the requirements can be verified directly such as when a measurable dimension is stated. In other cases the measurements to be made have to be derived, such as in the food industry where there is a requirement for food to be safe for human consumption. Standards are established for levels of contamination, microbes etc. which if exceeded are indicative of food is not safe for human consumption. Another example is in traffic management systems where speed limits are imposed for certain roads because they have been proved to represent a safe driving speed under normal conditions. The requirement is for people to drive safely but this is open to too much interpretation consequently measurable standards are imposed for effective communication and to ensure consistency in the application of the law.

Why is this Necessary?

This requirement responds to the Factual Approach Principle.

In order to verify that the products or services meet the specified requirements there needs to be unambiguous standards for making acceptance decisions. These standards need to be expressed in terms that are not open to interpretation so that any qualified person using them would reach the same decision when verifying the same character-istics in the same environment using the same equipment. In some cases the requirement is expressed definitively and in other cases subjectively. It is therefore necessary to establish how reliable is 'reliable', how safe is 'safe' and how clean is 'clean'. Speci-fications often contain subjective statements such as good commercial quality, smooth finish etc. and require further clarification in order that an acceptable standard can be attained.

How is this Demonstrated?

A common method of determining acceptance criteria is to analysis each requirement and establish measures that will indicate that the requirement has been achieved. In some cases national or international standards exist for use in demonstrating acceptable performance. The secret is to read the statement then ask yourself, *'Can I verify we have achieved this?'* If not, select a standard that is attainable, unambiguous and acceptable to both customer and supplier that if achieved will be deemed as satisfying the intent of the requirement.

The results of some processes cannot be directly measured using gauges, tools, test and measuring equipment[①] and so an alternative means has to be found of determining what is conforming product. The term given to such means is 'Workmanship Criteria', criteria that will enable producers and inspectors to gain a common understanding of what is acceptable and unacceptable. Situations where this may apply in manufacturing are soldering, welding, brazing, riveting, deburring etc. It may also include criteria for finishes, photographs, printing, blemishes and many others. Samples indicating the acceptable range of colour, grain and texture may be needed and if not provided by your customer, those that you provide will need customer approval.

The criteria need to be defined by documented standards or by samples and models that clearly and precisely define the distinguishing features that represent both con-forming and nonconforming product. In order to provide adequate understanding it may be necessary to show various examples of workmanship from acceptable to unaccept-able so that the producer or inspector doesn't strive for perfection or rework product unnecessarily. These standards like any others need to be controlled. Documented standards should be governed by the document control provisions. Samples and models need to be governed by the provision for controlling measuring devices and be subject to periodic examination to detect deterioration and damage. They should be certified as authentic workmanship samples and measures taken to preserve their integrity. Ideally they should be under the control of the inspection authority or someone other than the person responsible for using them so that there is no opportunity for them to be altered without authorization. The samples represent your company's standards, they do not belong to any individual and if used by more than one person you need to ensure consistent interpretation by training the users.

DETERMINING THE NEED FOR RECORDS (7.1d)

The standard requires the organization to determine *the records needed to provide evidence that the realization processes and resulting product meet requirements.*

What Does this Mean?

Countless verification activities will be carried out at various levels of product and service development, production and delivery. These activities will generate data and this data needs to be collected in a form that can be used to demonstrate that processes and products fulfil requirements. This does not mean that every activity needs to be recorded but the manner in which the data is recorded and when it is to be recorded should be determined as part of the planning activity.

Why is this Necessary?

This requirement responds to the Factual Approach Principle.

Without records indicating the results that have been obtained from product and process verification, compliance cannot be demonstrated to those on the scene of the action such as customers, managers and analysts. When investigating failures and plotting performance trends, records are also needed for reference purposes.

While procedures should define the records that are to be produced, these are the records that will be produced if these procedures are used. On particular contracts, only those procedures that are relevant will be applied and therefore the records to be produced will vary from contract to contract. Special conditions in the contract may make it necessary for additional records to be produced.

How is this Demonstrated?

There are two parts to this requirement. One concerning product records and the other concerning process records.

Product Records

By assessing the product requirements and identifying the stage in the process where these requirements will be verified, the type of records needed to capture the results should be determined. In some cases common records used for a variety of products may suffice. In other cases, product specific records may be needed that prescribe the characteristics to be recorded and the corresponding acceptance criteria to be used to indicate pass or fail conditions.

Process Records

The records required for demonstrating process performance should be identified during process development. Continued operation of the process should generate further records that confirm that the process is functioning properly – i.e., meeting the requirements for which it was designed. Periodically, process managers should review their processes and establish that the process continues to function as planned. These reviews should apply to the office processes as well as the shop floor processes so that sales, design and purchasing processes are subject to the same reviews as production, distribution, installation and service delivery.

Process records should indicate the process objectives and exhibit performance data showing the extent to which these objectives are being achieved. These may be in the form of bar charts, graphs, pie charts etc. From the process hierarchy of Fig. 11-3, you will observe that there would be process records for business processes, work processes and for activities but not tasks.

DOCUMENTING PRODUCT REALIZATION PLANNING (7.1)

The standard requires the product realization planning *to be in a form suitable for the organization's method of operations.*

What Does this Mean?

The output of planning can be in a variety of forms depending on the nature of the product, project, contract or service and its complexity. For simple products, the planning output may be a single document. For complex products, the planning output may take the form of a project plan and several supplementary plans each being in the form of a manual with several sections. What it is called is immaterial.

The word 'form' in the requirement is intended to convey manner or type and not a specific document into which data is captured, i.e., a Form (RFI 022).

Why is this Necessary?

This requirement responds to the Process Approach Principle.

The standard does not impose a particular format for the output of the planning activity or insist that such information carriers are given specific labels. Each product is different and therefore the planning outputs need to match the input requirements of the processes they feed.

How is this Demonstrated?

Discrete plans are needed when the work to be carried out requires detailed planning beyond that already planned for by the management system. The system will not specify everything you need to do for every job. It will usually specify only general provisions that apply in the majority of situations. You will need to define the specific activities to be performed, the documentation to be produced and resources to be employed. The contract may specify particular standards or requirements that you must meet and these may require additional provisions to those defined in the documented management system.

The planning outputs are dependent on the work that is required and therefore may include:

- Project plans,
- Product development plans,
- Production plans,
- Procurement plans,
- Reliability and maintainability programme plans,
- Control plans or verification plans,
- Installation plans,
- Commissioning plans,
- Performance evaluation plans.

It is not necessary to produce a separate quality plan if the processes of the management system that are to be utilized are identified in the project plan. Sometimes, the project is so complex that separate quality or quality assurance plans may be needed simply to separate the subject matter into digestible chunks. The disadvantage in giving any document a label with the word *quality* in the title is that it can sometimes be thought of as a document that serves only the Quality Department rather than a document that defines the provisions for managing the various processes that will be utilized on the project. A useful rule to adopt is to avoid giving documents a title that reflect the name of a department wherever possible.

Chapter 24

Customer-Related Processes

CHAPTER PREVIEW

This chapter is aimed at those personnel operating at the customer interface with responsibility for defining product requirements, negotiating contracts and handling customer feedback.

A process for determining product requirements should be designed so that it takes as its input the identified need for a product and passes this through several stages where requirements from various sources are determined, balanced and confirmed as the definitive requirements that form the basis for product realization. The input can either be a customer-specific requirement or the market specification that results from market research (see Chapter 14 on *Customer focus*) or a sales order for an existing product. However, this is not blue sky stuff – remember the product realization process is triggered by a customer placing a demand on the organization either because it has a product to sell or a capability to offer. The output may indeed be presented in several documents – the product requirement specification containing the hardware and software requirements and the service requirement specification containing the service requirements. Alternatively where service is secondary, the requirement may be contained in the contract.

The sub-heading in Clause 7.2.1 of 'Determination of requirements related to the product' within a section on customer-related processes implies that there are other requirements that do not relate to the product that may form part of the customer requirements. However, ISO 9000:2005 defines a product as the result of a process and includes services among these. It is therefore difficult to imagine any aspect of customer requirements that would not relate to the product or service that is being provided.

In this chapter we examine the requirements in Clause 7.2 of ISO 9001:2008 on customer-related processes and in particular:

- Product requirements specified by the customer;
- Requirements necessary for known intended use;
- Statutory and regulatory requirements;
- Organization's product requirement;
- Review of requirements related to the product;
- Providing product information;
- Handling enquiries;
- Handling contracts and orders;
- Handling contract amendments; and
- Customer feedback.

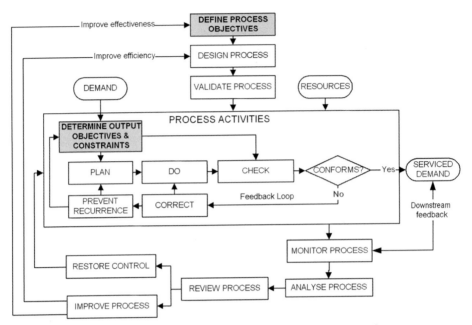

FIGURE 24-1 Where the requirements of Clause 7.2 apply in a managed process.

The position where the requirements on customer-related processes feature in a managed process is shown in Fig. 24-1. The requirement of Clause 7.2.1 equates with establishing the product objectives or product requirements as part of the demand fulfilment process and those of Clauses 7.2.2 and 7.2.3 are also part of this process. One requirement of Clause 7.2.3 is unconnected as it applies to all customer feedback and as such feed into process monitoring but this is monitoring of all processes not only the demand fulfilment process. It would perhaps have been better if this requirement had been placed under customer satisfaction in Section 8.

PRODUCT REQUIREMENTS SPECIFIED BY THE CUSTOMER (7.2.1a)

The standard requires the organization *to determine requirements specified by the customer including requirements for delivery and post-delivery activities.*

What Does this Mean?

Customers are consumers, clients, end users, retailers, purchasers and beneficiaries; therefore requirements specified by the customer need not be limited to the organization that is purchasing the product or the service. ISO 9000:2005 defines requirements as needs or expectations that are stated, generally implied or obligatory and therefore any information that is expressed by the customer as a need or expectation, whether in writing or verbally is a requirement. To determine such requirements means that the needs and expectations that are either stated verbally, or in writing, implied or obligatory have to be resolved, pinned-down and defined so that neither party is in any doubt as to what is required. The requirements for delivery mean requirements pertaining to the delivery dates, methods of shipment, transportation, transmission or other means for conveying the product or service to the customer in a specified condition (confirmed published interpretation RFI 050). Similarly, with post-delivery requirements, these are the requirements pertaining to warranty, the support the customer requires from the organization to maintain, service, assist or otherwise retain the product or service in a serviceable state and to recycle or dispose of the product.

Why is this Necessary?

This requirement responds to the Customer Focus Principle.

The purpose of the requirements is to ensure that you have established the requirements you are obliged to meet before you commence work. This is one of the most important requirements of the standard. The majority of problems down stream can be traced either to a misunderstanding of customer requirements or insufficient attention being paid to the resources required to meet customer requirements. Get these two things right and you are half way towards satisfying your customer needs and expectations.

How is this Demonstrated?

Customers will convey their requirements in various forms. Many organizations do business through purchase orders or simply order over the telephone or by electronic or surface mail. Some customers prefer written contracts others prefer a handshake or a verbal telephone agreement. However, a contract does not need to be written and signed by both parties to be a binding agreement. Any undertaking given by one party to another for the provision of products or services is a contract, whether written or not. The requirement for these provisions to be determined rather than documented, places the onus on the organization to understand customer needs and expectations, not simply react to what the customer has transmitted. It is therefore necessary in all but simple transactions to enter into a dialogue with the customer in order to understand what is required. Through this dialogue, assisted by checklists that cover your product and service offerings, you can tease out of the customer all the requirements that relate to the product. Sometimes, the customer wants one of your products or services but in fact needs another but has failed to realize it. Customer's wants are not needs unless the two coincide. It is not until you establish needs that you can be certain that you can satisfy the customer (see also *Stakeholder requirements* in Chapter 3). There may be situations when you won't be able to satisfy customer's needs because the customer simply does not have sufficient funds to pay you for what is necessary!

Many customer requirements will go beyond end product or service requirements. They will address delivery, quantity, warranty, payment, recycling, disposal and other legal obligations. With every product one provides a service. For instance one may provide delivery to destination, invoices for payment, credit services (if they don't pay on delivery they are using your credit services), enquiry services, warranty services etc. and the principal product may not be the only product either – there may be packaging, brochures, handbooks, specifications etc. With services there may also be products such as brochures, replacement parts and consumables, reports, certificates etc.

In ensuring the contract requirements are adequately defined, you should establish where applicable that:

- there is a clear definition of the purpose of the product or service you are being contracted to supply;
- the conditions of use are clearly specified;
- the requirements are specified in terms of the features and characteristics that will make the product or service fit for its intended purpose;
- the quantity, price and delivery are specified;
- the contractual requirements are specified including warranty, payment conditions, acceptance conditions, customer supplied material, financial liability, legal matters, penalties, subcontracting, licences and design rights, recycling and disposal;
- the management requirements such as points of contact, programme plans, work breakdown structure, progress reporting, meetings, reviews, interfaces are specified; and
- the quality assurance requirements such as quality system standards, quality plans, reports, customer approvals and surveillance, product approval procedures and concessions are specified.

Published interpretation RFI 009 confirms that a specified requirement does not necessarily imply that it has to be documented but it is wise to have the requirement documented in case of a dispute later. The document also acts as a reminder as to what was agreed but it is vital when either of the parties that made agreement move on, leaving their successors to continue the relationship. This becomes very difficult if the agreements were not recorded particularly if your customer representative moves on before you have submitted your first invoice. The document needs to carry an identity and if subject to change, an issue status. In the simple case this is the serial numbered order and in more complicated transactions, it will be a multipage contract with official contract number, date and signatures of both parties.

PRODUCT REQUIREMENTS NECESSARY FOR KNOWN INTENDED USE (7.2.1b)

The standard requires the organization *to determine product requirements not specified by the customer but necessary for known intended use.*

What Does this Mean?

There are two ways of looking at this requirement:

- From the viewpoint of an identified market need.
- From the viewpoint of a specific contract or order.

Market Need

The process of identifying future customer needs and expectations was addressed in Chapter 14 on *Customer focus.* The output of this process will be in the form of a market research report that contains information from which a new product requirement can be developed.

Specific Contract or Order

The customer is not likely to be an expert in your field. The customer may not know much about the inner workings of your product and service offerings and may therefore specify the requirements only in performance terms. In such cases, the onus is on the organization to determine the requirements that are necessary for the product or service to fulfil its intended use. For example, if a customer requires an electronic product to operate close to high voltage equipment, the electronics will need to be screened to prevent harmful radiation from affecting its performance. The customer may not know that this is necessary but during your dialogue, you establish the conditions of use and as a result identify several other requirements that need to be met. These are requirements not specified by the customer but necessary for known intended use.

Why is this Necessary?

This requirement responds to the Customer Focus Principle.

It is necessary to convert the results of market research into definitive product requirements so as to form a basis for new product development.

In the case of specific orders it is important to identify requirements necessary for intended use. For instance, subsequent to delivery of a product, a customer could inform you that your product does not function properly and you establish that it is being used in an environment that is outside its design specification. You would not have a viable case if the customer had informed you that it was going to be placed near high voltage equipment and you took no action.

How is this Demonstrated?

Careful examination of customer needs and expectations is needed in order to identify all the essential product requirements. A useful approach is to maintain a check list or datasheet of the products and services offered which indicates the key characteristics and the limitations, what it can't be used for, what your processes are not capable of. Of course such data needs to be kept within reasonable bounds but it is interesting to note that a manufacturer of refrigerators was successfully sued under product liability legislation for not providing a warning notice that the item was not safe for a person to stand on the top surface. It is therefore important to establish what the customer intends to use the product for, where and how they intend to use it and for how long they expect it to remain serviceable. With proprietary products many of these aspects can be clarified in the product literature supplied with the goods or displayed close to the point of service delivery. With custom designed products and services, a dialogue with the customer is vital to understand exactly what the product will be used for through its design life.

STATUTORY AND REGULATORY REQUIREMENTS (7.2.1c)

The standard requires the organization *to determine statutory and regulatory requirements applicable to the product.*

What Does this Mean?

Almost all products are governed by regulations that constrain or prohibit certain inherent characteristics.

Statutory requirements may apply to the prohibition of items from certain countries, power supply ratings, security provisions, markings and certain notices.

Regulatory requirements may apply to health, safety and environmental emissions, electromagnetic compatibility and these often require accompanying certification of compliance.

If you intend exporting the product or service, it would be prudent to determine the regulations that would apply before you complete the design requirement. Failure to meet some of these requirements can result in no export licence being granted as a minimum and imprisonment in certain cases if found to be subsequently noncompliant.

While there may be no pollution from using the product, there may be pollution from making, moving or disposing of the product and therefore, regulations that apply to production processes are indeed product-related.

Why is this Necessary?

This requirement responds to the Customer Focus Principle.

The customer may not be aware of all regulations that apply but will expect the supplier of the products and services required to be fully aware and have complied with all of them without exception. It is necessary to be fully aware of such statutes and regulations for the following reasons:

- A failure to observe government health and safety regulations could close a factory for a period and suspend your ability to supply customers.
- Health and safety hazards could result in injury or illness and place key personnel out of action for a period.
- Environmental claims made by your customers regarding conservation of natural resources, recycling etc. may be compromised if environmental inspections of your organization show a disregard for such regulations.
- The unregulated discharge of waste gases, effluent and solids may result in public concern in the local community and enforce closure of the plant by the authorities.
- A failure to take adequate personnel safety precautions may put product at risk.
- A failure to dispose of hazardous materials safely and observe fire precautions could put plant at risk.
- A failure to provide safe-working conditions for personnel may result in public concern and local and national inquiries that may harm the reputation of the organization.
- A failure to observe the codes of conduct of a professional body that regulates the professional services provided by an organization may result in disqualification of key staff and clients or their property being harmed or put at risk.

It is therefore necessary to maintain an awareness of all regulations that apply regardless of the extent to which they may or may not relate to the product.

How is this Demonstrated?

In order to determine the applicable statutes and regulations you will need a process for scanning the environment, identifying those that are relevant and capturing them in your management system. The legislators don't know what is relevant to your organization – only you know that so a dialogue with legal experts may be necessary to identify all those regulations that apply (*Scanning the environment* is part of the mission management process see Fig. 10-5).

There are lots of regulations and no guarantees of finding them all. However, you can now search through libraries on the Internet and consult bureaux, trade associations and government departments to discover those that apply to you. Ignorance of the law they say is no excuse.

The requirement also applies to products you purchase that are resold under the original manufacturer's label or re-badged under your label or incorporated into your product. Regulations that would apply to your products also apply to products you have purchased. There may be regulations that only apply to products you have purchased because of their particular form, function or material properties. Such regulations may not apply to your other products. It pays therefore to be vigilant when releasing purchase orders.

It is not simply the product that may have to meet regulations but the materials used in making the product. These materials are a direct consequence of the chosen design solution, therefore in general:

- the product has to be safe during use, storage and disposal;
- the product has to present minimum risk to the environment during production, storage, use and disposal;
- the materials used in the manufacture of the product have to be safe during use, storage and disposal; and
- the materials used in manufacture of the product have to present minimum risk to the environment during use and disposal.

Although you may not have specified a dangerous substance in the product specification, the characteristics you have specified in the product specification may be such that can only be produced by using a dangerous substance.

The conduct of professional staff of organizations providing services, such as architects, lawyers, solicitors, medical practitioners, will be governed by their respective professional body or by law and these regulations are an applicable service requirement. This is confirmed by published interpretation RFI 020.

ORGANIZATION'S PRODUCT REQUIREMENT (7.2.1d)

The standard requires *any additional product requirements considered necessary by the organization to be determined.*

What Does this Mean?

In addition to the requirements specified by the customer and the regulations that apply, there may be requirements imposed by the organization's policies that impinge on the particular products or services that are to be supplied. The product policy may impose certain style, appearance, reliability and maintainability requirements or prohibit use of certain technologies or materials. Other requirements may serve to aid production or distribution that are of no consequence to the customer but necessary for the efficient and effective production, storage and supply of the product.

Why is this Necessary?

This requirement responds to the Leadership Principle.

The requirement is necessary in order that relevant organizational policies and objectives are deployed through the product and service offerings. A failure to identify such constraints at the requirement definition stage could lead to abortive design work or if left undetected, the supply of products or services that harm the organization's reputation. Often, an organization is faced with the task of balancing customer needs with those of other stakeholders. It may therefore be appropriate in some circumstances for the organization to decline to meet certain customer requirements on the grounds that they conflict with the needs of certain stakeholders.

How is this Demonstrated?

The organization's requirements should be defined in technical manuals that are used by designers, production and distribution staff. These will often apply to all the organization's products and services but will, however, need to be reviewed to identify the specific requirements that apply to particular products.

REVIEW OF REQUIREMENTS RELATED TO THE PRODUCT (7.2.2)

Conducting the Review (7.2.2)

The standard requires the organization *to review the requirements related to the product*.

What Does this Mean?

A review of the requirements related to the product means that all the requirements that have been identified through the requirement determination process should be examined together preferably by someone other than those who gathered the information. The review may be quite independent of any order or contract but may need to be repeated should an order or contract for the product be received.

Requirements related to the product or service could include:

- Characteristics that the product is required to exhibit, i.e., the inherent characteristics.
- Price and delivery requirements.
- Procurement requirements that constrain the source of certain components, materials or the conditions under which personnel may work.
- Management requirements related to the manner in which the project will be managed, the product developed, produced and supplied.
- Security requirements relating to the protection of information.
- Financial arrangements for the deposit of bonds, payment conditions, invoicing etc.

- Commercial requirements such as intellectual property, proprietary rights, labelling, warranty, resale, copyright etc.
- Licensing requirements relating to individuals permitted to provide a service such as a pilot's licence, driving licence, professional licence to practice (RFI 020).
- Personnel arrangements such as access to the organization's facilities by customer personnel and vice versa.

Why is this Necessary?

This requirement responds to the Factual Approach Principle.

The process of determining the requirements that relate to the product consists of several stages culminating in a definitive statement of the product requirements. All processes should contain review stages as a means of establishing that the process output is correct and that the process is effective. The review referred to in this requirement is therefore necessary to establish that the output of the requirement determination process is correct (two examples show where this features in the demand creation process see Figs 8-15 and 10-7).

How is this Demonstrated?

The information gathered as a result of determining the various product requirements should be consolidated in the form of a specification, contract or order and then subject to review. The personnel who should review these requirements depend on their complexity and there are three situations that you need to consider.

1. Development of new product to satisfy identified market needs – New product development.
2. Sales against the organization's requirements – Proprietary sales.
3. Sales against specific customer requirements – Custom sales.

New Product Development

In setting out to develop a new product there may not be any customer orders – the need for the product may have been identified as a result of market research and from the data gathered a definitive product requirement is developed. The product requirement review is performed to confirm that the requirements do reflect a product that will satisfy the identified needs and expectations of customers. At the end product level, this review may be the same as the design input review, but there are other outputs from market research such as the predicted quantities, the manner of distribution, packaging and promotion considerations. The review should be carried out by those functions representing the customer, design and development, production, service delivery and in service support so that all views are considered.

Proprietary Sales

In a proprietary sales situation, you may simply have a catalogue of products and services advertising material and a sales office taking orders over the telephone or over the counter. There are two aspects to the review of requirements. The first is the initial review of the requirements and advertising material before they are made available for potential customers to view and the second is where the sales person reviews the customer's request against the catalogue to determine if the particular product is available and can be supplied in the quantity required. We could call these reviews, Requirement Review and Transaction Review. As a customer may query particular features, access to the full product specification or a technical specialist may be necessary to answer such queries.

Custom Sales

In custom sales situation the product or service is being produced or customized for a specific customer and with several departments of the organization having an input to the contract and its acceptability. These activities need coordinating so that you ensure all are working with the same set of information. You will need to collect the contributions of those involved and ensure that they are properly represented at meetings. Those who negotiate contracts on behalf of the company, carry a great responsibility.

One aspect of a contract often overlooked is the shipment of finished goods. You have ascertained the delivery schedule, the place of delivery, but how do you intend to ship it (by road, rail and ship or by air). It makes a lot of difference to the costs. Also delivery dates often mean the date on which the shipment arrives not the date it leaves. You therefore need to build into your schedules an appropriate lead-time for shipping by the means agreed to the required destination. If you are late you may need to employ speedier means but that will incur a premium for which you may not be paid. Your financial staff will therefore need to be involved in the requirement review.

Having agreed the requirements, you need to convey them to their point of implementation in sufficient time for resources to be acquired and put to work. Policy deployment or Quality Function Deployment (QFD)$^{\oslash}$ are tools you can use for this purpose.

Timing of Review (7.2.2)

The standard requires the review *to be conducted prior to the decision or commitment to supply a product to the customer (e.g., submission of a tender, acceptance of a contract or order).*

What Does this Mean?

A tender is an offer made to a potential customer in response to an invitation. The acceptance of a contract is a binding agreement on both sides to honour commitments. Therefore, the period before the submission of a tender or acceptance of a contract or order is a time when neither side is under any commitment and presents an opportunity to take another look at the requirements before legal commitments are made.

Why is this Necessary?

This requirement responds to the Factual Approach Principle.

The purpose of the requirement review is to ensure that the requirements are complete, unambiguous and attainable by the organization. It is therefore necessary to conduct such reviews before a commitment to supply is made so that any errors or omissions can be corrected in time. There may not be opportunities to change the agreement after a contract has been signed without incurring penalties. Customers will not be pleased by organizations that have underestimated the cost, time and work required to meet their requirements and may insist that organizations honour their commitments – after all an agreement is a promise and organizations that break their promises do not survive for long in the market place.

How is this Demonstrated?

The simplest method of implementing this requirement is to make provision in the requirement determination process for a requirement review to take place before tenders are submitted, contracts are signed or orders accepted. In order to ensure this happens staff need to be educated and trained to react in an appropriate manner to situations in which the organization will be committed to subsequently honouring its obligations. At one level this means that staff stop, think and check before accepting an order. At another level, this means that staff seek out someone else to perform the checks so that there is another pair of eyes focused on the requirements. At a high level, this means that a review panel is assembled and the requirements debated and all issues resolved before the authorized signatory signs the contract. One means of helping staff to react in an appropriate manner is to provide forms with provision for a requirement review box that has to be checked or signed and dated before the process may continue. With computer-based systems, provision can also be made to prevent the transaction being completed until the correct data has been entered. This process is needed also for any amendments to the contract or order so that the organization takes the opportunity to review its capability with each change.

Ensuring that Product Requirements are Defined (7.2.2a)

The standard requires that the review of requirements *ensures product requirements are defined*.

What Does this Mean?

This means that the review should verify that all the requirements specified by the customer, the regulators and the organization have been defined, there are no omissions, no errors, no ambiguities and no misunderstandings.

Why is this Necessary?

This requirement responds to the Factual Approach Principle.

During the requirement determination process there are many variables. The time allowed, the competence of the personnel involved, the knowledge of the customer of what is needed and the accessibility of information. A deficiency in any one of these can result in the inadequate determination of requirements. It is therefore necessary to subject these requirements to review to ensure that they are correct before a commitment to supply is made.

How is this Demonstrated?

Some organizations deal with orders that are so predictable that a formal documented review before acceptance adds no value. But however predictable the order, it is prudent to establish that it is what you believe it to be before acceptance. Many have been caught out by the small print in contracts or sales agreements such as the following wording 'This agreement takes precedence over any conditions of sale offered by the supplier' or 'Invoices must refer to the order reference otherwise they will be rejected'.

If the customer is choosing from a catalogue or selecting from a shelf of products, you need to ensure that the products offered for sale are properly described. Such descriptions must not be unrepresentative of the product otherwise you may be in breech of national laws and statutes. In other situations you need some means of establishing that the customer requirements are adequate.

One means of doing this is to use checklists that prompt the reviewers to give proper consideration to important aspects before accepting contracts. Another method is to subject the requirements to an independent review by experts in their field, thus ensuring that a second pair of eyes scans the requirements for omissions, ambiguities and errors.

Resolving Differences (7.2.2b)

The standard requires the review *to ensure that contract or order requirements differing from those previously expressed are resolved.*

What Does this Mean?

Previously expressed requirements are those that may have been included in an invitation to tender issued by the customer. Whether or not you have submitted a formal tender, any offer you make in response to a requirement is a kind of tender. Where a customer's needs are stated and you offer your product, you are implying that it responds to your customer's stated needs. You need to ensure that your 'tender' is compatible with your customer's needs otherwise the customer may claim that you have sold a product that is not 'fit for purpose'.

Why is this Necessary?

This requirement responds to the Customer Focus Principle.

In situations where the organization has responded to an invitation to tender for a contract, it is possible that the contract (when it arrives) may differ from the draft conditions against which the tender was submitted. It is therefore necessary to check whether any changes have been made that will affect the validity of the tender. Customers should indicate the changes that have been made but they often don't – they would if there was a mutually beneficial relationship in place.

How is this Demonstrated?

On receipt of a contract that has been the subject of an invitation to tender, or the subject of an unsolicited offer of product or service, it is prudent to check that what you are now being asked to provide is the same as that which you offered. If the product or service you offer is in any way different than the requirement, you need to point this out to your customer and reach agreement before you accept the order. Try and get the contract changed, but if this is not possible, record the differences in your response to the contract. Don't rely on verbal agreements because they can be conveniently forgotten when it suits one party or the other, or as is more common, the person you conversed with moves on and the new person is unable to act without written agreement – such is the world of contracting!

Ensuring that the Organization has the Ability to Meet Defined Requirements (7.2.2c)

The standard requires that the review *ensures that the organization has the ability to meet defined requirements.*

What Does this Mean?

The organization needs to be able to honour its obligations made to its customers. Checks therefore need to be made to ensure that the necessary resources including plant, equipment, facilities, technology, personnel, competency and time are available or will be available to discharge these obligations when required.

Often the contract for design and development and the contract for production are two separate contracts. They may be placed on the same organization but it is not unusual for the production contract to be awarded to an organization that did not design the product for cost reasons. This requirement is concerned with business capability rather than process capability and addresses the question, do we as a business have the capability to make this product in the quantity required and deliver it in the condition required to the destination required over the time period required and within the price required?

Why is this Necessary?

This requirement responds to the Leadership Principle.

You must surely determine that you have the necessary capability before accepting the contract as to find out afterwards that you haven't the capability to honour your obligations could land you in deep trouble. There may be penalty clauses in the contract or the nature of the work may be such that the organization's reputation could be irrevocably damaged as a result. This requirement demands an analytical approach and objective evidence to show that it's feasible to meet the requirements. When the customer awards such a contract, there is a commitment on both sides and if the supplier is later found to be unable to deliver, the customer suffers, it puts the programme in jeopardy.

How is this Demonstrated?

Prior to accepting a contact to manufacture a product of proven design, a business risk assessment is necessary as a safeguard against programme failure. The risk assessment should answer the following questions:

- What new technical capabilities will be needed for us to make this product?
- Will we be able to develop the additional capabilities within the timescales permitted?
- Can we make this product in the quantities required and in the timescales required?

- In consideration of the timescales and quantities required, can we make this product at a price that will provide an acceptable profit?
- Do we have the slack in our capacity to accommodate this programme at this time?
- Do we have the human resources required?
- If we require additional human resources, can we acquire them and train them to the level of competence required in the timescales?
- Is there a sufficient supply of materials and components available in order to resource the production processes?
- Do we (or our partners) have the capability to transport this product to the required destination and protect it throughout the journey?

If you don't have any of the above, you will need to determine the feasibility of acquiring the relevant licence, the skills, the technology etc. within the timescale. Many organizations do not have staff waiting for the next contract so it is a common practice for companies to bid for work for which they do not have the necessary numbers of staff. However, what they need to ascertain is from where and how quickly they can obtain the appropriate staff. If a contract requires specialist skills or technologies that you don't already possess, you need to consider the probability that you will not be able to acquire them in the timescale. It is also likely that your customer will want an assurance that you have the necessary skills and technologies before the contract is placed. No organization can expect to hire extraordinary people at short notice; in fact all you can expect to be available are average people and you may well have no choice than to accept less than average people. With good management skills and a good working environment you may be able to get these average people to do extraordinary things but it is not guaranteed!

A sales person who promises a short delivery to win an order invariably places an impossible burden on the company. A company's capability is not increased by accepting contracts beyond its current level of capability. You need to ensure that your sales personnel are provided with reliable data on the capability of the organization do not exceed their authority and always obtain the agreement of those who will execute the contractual conditions before their acceptance.

In telephone sales transactions or transactions made by sales personnel alone, the sales personnel need to be provided with current details of the products and services available, the delivery times, prices and procedures for varying the conditions.

Maintaining Records of Product Requirement Reviews (7.2.2)

The standard requires *the results of the review and actions as a consequence of the review to be recorded (see clause 4.2.4).*

What Does this Mean?

The result of the review may be a decision in which case a record of the decision is all that is necessary. However, the result could be a list of actions to be executed to correct the definition of requirement, or a list of concerns that need to be addressed. If the review is conducted with customer's representatives present, records of the review could include modifications, interpretations and correction of errors that may be held back until the first contract amendment. In such cases the review records act as an extension to any contract.

Why is this Necessary?

This requirement responds to the Factual Approach Principle.

For both new product development and order processing, records of the review are necessary as a means of recalling accurately what took place or as a means of reference in the event of a dispute or to distribute to those having responsibility for any actions that have been agreed. During the processing of orders and contracts, records of the requirement review indicate the stage in the process that has been reached and are useful if the process is interrupted for any reason.

How is this Demonstrated?

There should be some evidence that a person with the authority to do so has accepted each product requirement, order or contract. This may be by signature or by exchange of letters or e-mails. You should also maintain a register of all contracts or orders and in the register indicate which were accepted and which were declined. This is useful when assessing the effectiveness of the demand creation process. If you prescribe the criteria for accepting a contract, the signature of the contract or order together with this register can be adequate evidence of requirement review. If requirement reviews require the participation of several departments in the organization, their comments on the contract, minutes of meetings and any records of contract negotiations with the customer will represent the records of product requirement review. It is important, however, to be able to demonstrate that the requirement being executed was reviewed for adequacy, differences in the tender and for supplier capability, before work commenced. The minimum you can have is a signature accepting an assignment to do work or supply goods but you must ensure that those signing the document know what they are signing for. Criteria for accepting orders or contracts can be included in the appropriate procedures. It cannot be stressed too strongly the importance of these actions. Most problems are caused by the poor understanding or poor definition of requirements.

Handling Undocumented Statements of Requirements (7.2.2)

The standard requires that where the customer provides no documented statement of requirement, *the customer requirements are to be confirmed by the organization before acceptance*.

What Does this Mean?

Customers often place orders by telephone or in face-to-face transactions where no paperwork passes from the customer to the organization. Confirmation of customer requirements is an expression of the organization's understanding of the obligations it has committed to honour.

Why is this Necessary?

This requirement responds to the Customer Focus Principle.

Confirmation is necessary because when two people talk, it is not uncommon to find that although they use the same words, they each interpret the words differently. Confirming an understanding will avoid problems later. Either party to the agreement could move jobs leaving the successors to interpret the agreement in a different way.

How is this Demonstrated?

The only way to implement this requirement is for the organization to send a written acknowledgement to the customer confirming the requirements that form the basis of the agreement. In this way there should be no ambiguity, but if later the customer appears to be requiring something different, you can point to the letter of confirmation. If you normally use e-mail for correspondence, obtain an e-mail receipt that it has been read (not merely *received* as it could be overlooked) otherwise always send confirmation by post as e-mails can easily be inadvertently lost or deleted. Keep a copy of the e-mail and the letter and bring them under records control. Saving specific e-mails as text files in an appropriate directory on the server is better than simply keeping them on your local hard drive as a message in Outlook Express or similar program.

Changes to Product Requirements (7.2.2)

The standard requires that where product requirements are changed, *the organization ensure that relevant documents are amended and that relevant personnel are made aware of the changed requirements.*

What Does this Mean?

Product requirements may be changed by the customer, by the regulators or by the organization itself and this may be made verbally or by changing the affected product requirement documents. This requirement means that all documents affected by the change are amended and that the changes are transmitted to those who need to know.

Why is this Necessary?

This requirement responds to the Factual Approach Principle.

When changes are made to product requirements, the documents defining these requirements need to be changed otherwise those using them will not be aware of the changes. Also, changes to one document may have an impact on other related documents and unless these too are changed, the users will be working with obsolete information. It is therefore necessary to promulgate changes in a way that users are able to achieve the desired results.

How is this Demonstrated?

In some organizations *product requirement change control* is referred to as *configuration management*.[1] Once a baseline set of requirements has been agreed, any changes to the baseline need to be controlled such that accepted changes are promptly made, and rejected changes are prevented from being implemented. If there is only one product specification and no related information, configuration management is similar to document control (see Chapter 12). When there are many specifications and related information, configuration management introduces a further dimension of having to control the compatibility between all the pieces of information. Document control is concerned with controlling the information carriers, whereas configuration management

---

[1] Lyon, D.D. (2000). *Practical CM – Best Configuration Management Practices*. Butterworth Heinemann.

is concerned with controlling the information itself. If a system parameter changes there may be a knock-on effect through the sub-systems, equipment and components. The task is to identify all the items affected and as each item will have a product specification, this task will result in a list of affected specifications, drawings etc. While the list looks like a list of documents, it is really a list of items that are affected by the change. The requirement of ISO 9001 makes it appear a simple process but before documents are amended, the impact on each item may vary and have to be costed before being implemented. Major redesign may be necessary, tooling, handling equipment, distribution methods etc. may be affected and, therefore, the process of not only communicating the change, but also communicating the effects and the decisions relating to the change is an essential part of configuration management. It is not so much as who should be informed as what is affected. Identify what is affected and you should be able to identify who should be informed.

CUSTOMER COMMUNICATION (7.2.3)

Providing Product Information (7.2.3a)

The standard requires the organization *to determine and implement effective arrangements for communicating with customers relating to product information.*

What Does this Mean?

Product information could be in the form of advertising material, catalogues, a web site, specifications or any medium for promoting the organization's products and services. Effective arrangements would be the processes that identified, planned, produced and distributed information that accurately describes the product.

Why is this Necessary?

This requirement responds to the Customer Focus Principle.

Customers are only aware of product information that is accessible but whether they receive it or retrieve it (as is the case if the information is posted on the Internet), it must accurately represent the products and services offered otherwise, it is open to misrepresentation and liable to prosecution in certain countries. It is therefore necessary to employ an effective process for communicating production information.

How is this Demonstrated?

This requirement is concerned with the quality of information available to customers and has two dimensions. There is the misleading of customers into believing a product or service provides benefits that it cannot deliver (accuracy) and there is the relationship between information available to customers and information as would need to be to properly represent the product (compatibility).

Accuracy depends on getting the balance right between imaginative marketing and reality. Organizations naturally desire to present their products and services in the best light – emphasizing the strong points and playing down or omitting the weak points. Providing the omissions are not misleading to the customer this is legitimate. What is needed is a product advertising process that ensures product information accurately represents the product and does not infringe advertising regulations and sale of goods laws.

Compatibility depends on maintaining the product information once it has been released. Product information takes many forms and keeping it all compatible is not an easy task. An information control process is therefore needed to ensure that information compatibility is maintained when changes are made.

Handling Enquiries (7.2.3a)

The standard requires the organization *to determine and implement effective arrangements for communicating with customers relating to enquiries.*

What Does this Mean?

Customer enquiries are the result of the effectiveness of the marketing process. If this has been successful, customers will be making contact with the organization to seek more information, clarify price, specification or delivery or request tenders, proposals or quotations.

Why is this Necessary?

This requirement responds to the Customer Focus Principle.

If the personnel receiving a customer enquiry are uninformed or not competent to deal effectively with enquiries, a customer may not receive the treatment intended by the organization and either go elsewhere or be misled. Both situations may result in lost business and dissatisfied customers.

How is this Demonstrated?

An enquiry-handling process is needed as part of the sales process that ensures customers are fed correct information and treated in a manner that maximizes the opportunity of a sale. Enquiries should be passed through a process that will convert the enquiry into a sale. As the person dealing with the enquiry could be the first contact the customer has with the organization, it is vital that they are competent to do the job. Frequent training and monitoring are therefore necessary to prevent customer dissatisfaction. This has become more apparent with telephone sales where a recorded message informs the customer that the conversation may be monitored for quality assurance purposes. As with all processes, you need to establish what you are trying to achieve, what affects your ability to get it right and how you will measure your success. The potential for error is great, whether customers are dealing with someone in person or cycling through a menu during a telephone transaction or on the Internet. Both human and electronic enquiry-handling processes need to be validated regularly to ensure their continued effectiveness. A typical enquiry conversion process flow is illustrated in Fig. 24-2.

Handling Contracts and Orders (7.2.3b)

The standard requires the organization *to determine and implement effective arrangements for communicating with customers relating to contracts or order handling.*

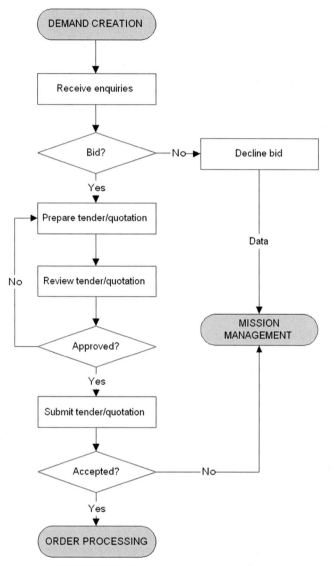

FIGURE 24-2 Enquiry conversion process flow.

What Does this Mean?

When an order or contract is received, several activities need to be performed in addition to the determination and review of product requirements and in each of these activities, communication with the customer may be necessary to develop an understanding that will secure an effective relationship.

Why is this Necessary?

This requirement responds to the Customer Focus Principle.

Customer enquiries may or may not result in orders. However, when an order or a contract is received, it is necessary to pass it through an effective process that will ensure both parties are in no doubt as to the expectations under the contract.

How is this Demonstrated?

A process should be established for handling orders and contracts with the objective of ensuring both parties are in no doubt as to the expectations under the contract before work commences. A typical order processing process is illustrated in Fig. 24-3.

Apart from the requirement determination and reviews stages, there will be:

- order or contract registration – recording its receipt;
- order or contract acknowledgement – informing the customer that the order has been received and that the organization intends or does not intend to offer a bid or supply a product or service;
- requirement determination (as addressed previously);
- requirement review (as addressed previously);
- order or contract negotiation;
- order or contract acceptance; and
- order or contract communication.

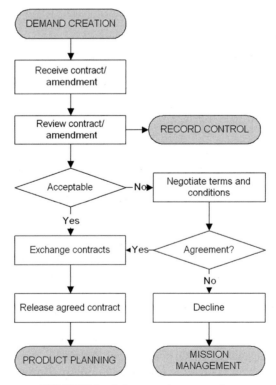

FIGURE 24-3   Order processing process flow.

Handling Contract Amendments (7.2.3b)

The standard requires the organization *to determine and implement effective arrangements for communicating with customers relating to amendments.*

What Does this Mean?

An order or contract amendment is a change that corrects errors, rectifies ambiguities or otherwise makes improvements.

Why is this Necessary?

This requirement responds to the Customer Focus Principle.

The need for an amendment can arise at any time and be initiated by either party to an agreement.

As orders and contracts are primarily a source of reference, it is necessary to ensure that only agreed amendments are made and that any provisional amendments or disagreed amendments are not acted on or communicated as though they were approved. Otherwise, the basis of the agreement becomes invalid and may result in a dissatisfied customer or an organization that cannot recover its costs.

How is this Demonstrated?

There may be several reasons why a customer needs to amend the original contract – customer needs may change, your customer's customer may change the requirement or details unknown at the time of contract may be brought to light. Whatever the reasons you need to provide a process for amending existing contracts under controlled conditions and conveying the amendments to those who need to act on the information.

On contracts where direct liaison with the customer is permitted between several individuals, e.g., a project manager, contract manager, design manager, procurement manager, manufacturing manager, quality assurance manager, it is essential to establish ground rules for amending contracts, otherwise your company may unwittingly be held liable for meeting requirements beyond the funding that was originally predicted. It is often necessary to stipulate that only those changes to contract that are received in writing from the contract authority of either party will be legally binding. Any other changes proposed, suggested or otherwise communicated should be regarded as being invalid. Agreement between members of either project team should be followed by an official communication from the contract authority before binding either side to the agreement.

Having officially made the change to the contract, a means has to be devised to communicate the change to those who will be affected by it. You will need to establish a distribution list for each contract and ensure that any amendments are issued on the same distribution list. Clearly, this list needs to be under control to ensure that no one who would act on the changes is omitted and as a consequence jeopardize delivery of conforming product.

Customer Feedback (7.2.2c)

The standard requires the organization *to determine and implement effective arrangements for communicating with customers relating to customer feedback including customer complaints.*

What Does this Mean?

Customer feedback is any information conveyed by the customer in relation to the quality of the products or services provided. Sometimes this may be positive in the form of compliments, praise or suggestions and other times the feedback could be negative in the form of a complaint or an expression of disapproval.

Why is this Necessary?

This requirement responds to the Customer Focus Principle.

Without an effective process for capturing customer feedback the organization would be missing opportunities for improving its performance, as it is perceived to be by its customers. While not an accurate measure of customer satisfaction, customer feedback provides objective evidence that can be used in such an assessment.

How is this Demonstrated?

It will be easy for you to say 'We don't have any complaints' if your processes are not designed to alert you to customer dissatisfaction. Therefore the absence of complaints should not be indicative that your customers are satisfied (see Chapter 29 on *Customer satisfaction*). You can only handle the customer feedback that you receive and record therefore the first thing to do is to ensure that you have a mechanism for capturing it effectively. Probably the best place is at the point of delivery or first use and a method often used is to include a return card with the product so that the customer can register their view.

Customers may complain about your products and services but not go to the extent of writing a formal complaint. They may also compliment you on your products and services but again not bother to put it in writing. Compliments and complaints may arise in conversation between the customer and your sales and service staff and this is where you need to instil discipline and ensure that they are captured. The primary difference between compliments and complaints is that compliments deserve a thank you and complaints deserve action, therefore the processes for dealing with compliments and complaints will differ.

The complaint handling process should cover the following aspects to be effective:

- A definition of when a message from a customer can be classified as a complaint;
- The method of capturing the customer complaints from all interface channels with the customer;
- The behaviour expected from those on the receiving end of the complaint;
- The registration of complaints in order that you can account for them and monitor progress;
- A form on which to record details of the complaint, the date, customer name etc.;
- A method for acknowledging the complaint in order that the customer knows you care;

- A method for investigating the nature and cause of the complaint;
- A method for replacing product, repeating the service or for compensating the customer;
- A link with other processes to trigger improvements that will prevent a recurrence of the complaint;
- A means of ensuring that the complaint is not closed until the specific issue has been resolved to the customer's satisfaction and plans for preventing recurrence of the complaint have been agreed and their implementation is being tracked.

The compliment handling process should cover the following:

- A definition of when a message from a customer can be classified as a compliment;
- The method of capturing the compliments from all interface channels with the customer;
- The behaviour expected from those on the receiving end of the compliment;
- The registration of compliments in order that you can account for those you can use in your promotional literature;
- A method for keeping staff informed of the compliments made by customers; and
- A method of rewarding staff when compliments result in further business.

Chapter 25

Design and Development

CHAPTER PREVIEW

This chapter is aimed at those personnel responsible for and involved with the design and development of products and services intended for customers. This may involve departmental managers, project managers, development engineers, product and service designers, specialists and personnel involved with design assurance, tests and trials.

Design and development is a process and as such needs to be managed effectively in order to deliver products with features that reflect customer needs and expectations. It should be a process that encourages creativity and innovation rather than inhibit it but with checks and balances to regulate overspend and delay. Any project that is triggered by a customer requirement will have constraints and would only include those journeys into the unknown where the customer is prepared to fund the necessary research.

In this chapter we examine the requirements in Clause 7.3 of ISO 9001:2008 and in particular:

- Control of design and development;
- Planning design and development;
- Inputs to the design and development process;
- Outputs from the design and development process;
- Design reviews;
- Design verification;
- Design validation; and
- Control of design changes.

The headings in this section of the standard are all prefixed by the words "design and development" because in many cases the new product is a development of a previous product. However, if we take this to the extremes, we would have stopped designing automobiles once the first one emerged on to the market in the nineteenth century as all subsequent models from any manufacturer would be developments from the original idea. Another view is that when a design (whether original or not) reaches a prototype stage, all subsequent iterations are developments. In the list above you will notice that the word 'development' has been omitted in the last four items. This is because whatever the stage of design or development of that design, the reviews, the verification, validation and the changes are of the design. The output from the process is a design; the fact that the output might be a development of a previous design does not alter this.

The position where the requirements on design and development feature in the managed process is shown in Fig. 25-1. Due to the structure of the standard, activities normally part

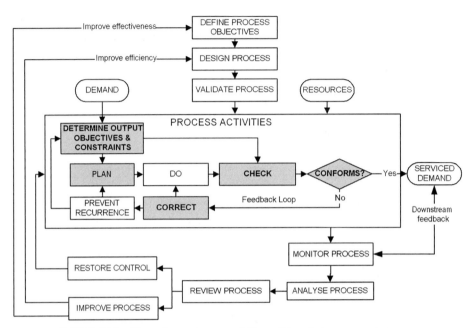

FIGURE 25-1   Where the requirements of Clause 7.3 apply in a managed process.

of design and development such as the activity of design and development itself (the standard does not actually require anything to be designed, only that it be controlled) and corrective action etc. are excluded. Correction is equivalent to design changes.

DESIGN AND DEVELOPMENT CONTROL (7.3.1)

The standard requires the organization *to control design and development of the product.*

What Does this Mean?

Design can be as simple as replacing a part in an existing product with one of a different specification, or as complex as the design of a new product or any of its subsystems. Design can be of hardware, software (or a mixture of both) and can be of new services or modified services.

These requirements apply to the product that will be supplied to the organizations customers including any packaging (see RFI 043). They are not intended to apply to tooling or anything that might come into contact with the product but is not shipped with the product. However, imposing similar controls over the design of such items may be desirable. They also apply in cases where a design has been purchased and the related product manufactured or service delivered (confirmed in published interpretation RFI 049).

Before design commences there is either a requirement or simply an idea. Design is a creative process that creates something tangible out of an idea or a requirement. The controls specified in the standard apply to the product and service design process, but can be applied also to the design of processes that will produce the product. In order to succeed, the process of converting an idea into a design that can be put into production or service has to be controlled. Design is often a process which strives to set new levels of performance, new standards or creates new wants and as such can be a journey into the unknown. On such a journey we can encounter obstacles we haven't predicted which may cause us to change our course but our objective remains constant. Design control is a method of keeping the design on course towards its objectives and as such will comprise all the factors that may prevent the design from achieving its objectives.

To control any design activity be it either product or service design, there are 10 primary steps you need to take in the design process:

1. Establish the customer needs.
2. Convert the customer needs into a definitive specification of the requirements.
3. Plan for meeting the requirements.
4. Organize resources and materials for meeting the requirements.
5. Conduct a feasibility study to discover whether accomplishment of the requirements is feasible.
6. Conduct a project definition study to discover which of the many possible solutions will be the most suitable.
7. Develop a specification that details all the features and characteristics of the product or service.
8. Produce a prototype or model of the proposed design.
9. Conduct extensive trials to discover whether the product, service or process that has been developed meets the design requirements and customer needs.
10. Feed data back into the design and repeat the process until the product or service is proven to be fit for the task.

Control of design and development does not mean controlling the creativity of the designers – it means controlling the process through which new or modified designs are produced so that the resultant design is one that truly reflects customer needs. It therefore controls the inputs, the selection of components, standards, materials, processes, techniques and technologies and the outputs.

Why is this Necessary?

This requirement responds to the Process Approach Principle.

Without control over the design and development process several possibilities may occur:

- Design will commence without an agreed requirement.
- Costs will escalate as designers pursue solutions that go beyond what the customer really needs.
- Costs will escalate as suggestions get incorporated into the design without due consideration of the impact on development time and cost.
- Designs will be released without adequate verification and validation.
- Designs will be expressed in terms that cannot be implemented economically in production or use.

The bigger the project, the greater the risk that the design will overrun budget and timescale. Design control aims to keep the design effort on course so that the right design is released on time and within budget.

How is this Demonstrated?

Control of the design and development requires the application of the same principles as any other process. The standard actually identifies the controls that need to be applied to each design but there are other controls needed in order to apply the requirements of ISO 9001 Clause 4.1. Typical product design and service design process flow charts are illustrated in Figs 25-2 and 25-3, respectively. (Note that the process flow reflects traditional sequential design and not the more modern agile software development)

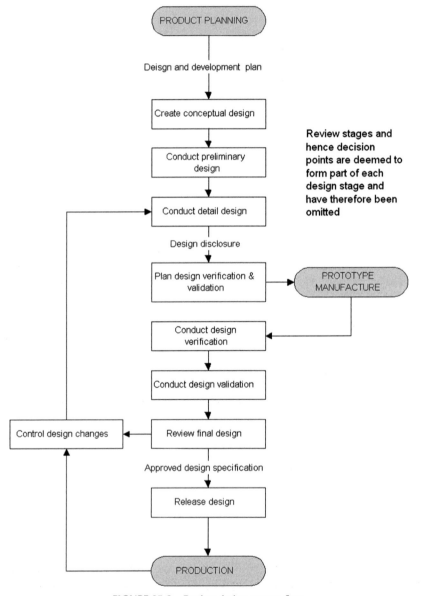

FIGURE 25-2   Product design process flow.

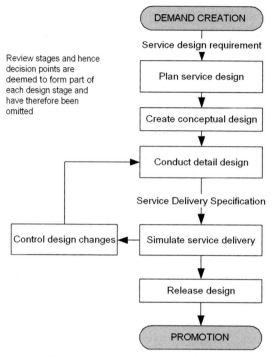

FIGURE 25-3 Service design process flow.

The design process is a key process in enabling the organization to achieve its objectives. These objectives should include goals that apply to the design process (see Chapter 16 on *Establishing quality objectives*). Consequently there need to be:

- Objectives for the design process;
- Measures for indicating achievement of these objectives;
- A defined sequence of sub-processes or tasks that transform the design inputs into design outputs;
- Links with the resource management process so that human and physical resources are made available to the design process when required;
- Review stages for establishing that the process is achieving its objectives; and
- Processes for improving the effectiveness of the design process.

DESIGN AND DEVELOPMENT PLANNING (7.3.1)

Preparing the Plans (7.3.1)

The standard requires the organization *to plan design and development of the product.*

What Does this Mean?

Planning the design and development of a product means determining the design objectives and the design strategy, the design stages, timescales, costs, resources and responsibilities needed to accomplish them. Sometimes, the activity of design itself is considered to be a planning activity but what is being planned in such cases is the product and not the design.

Why is this Necessary?

This requirement responds to the Leadership Principle.

The purpose of planning is to determine the provisions needed to achieve an objective. In most cases, these objectives include not only a requirement for a new or modified product but also requirements governing the costs and product introduction timescales (Quality, Cost and Delivery or QCD). Remove these constraints and planning becomes less important but there are few situations when cost and time are not the constraints. It is therefore necessary to work out in advance whether the objective can be achieved within the budget and timescale. One problem with design is that it is often a journey into the unknown and the cost and time it will take cannot always be predicted. It may in fact result in disaster and either a complete reassessment of the design objective or the technology of the design solution. This has been proven time and again with major international projects such as Concorde, the Channel Tunnel and the International Space Station. Without a best guess some projects would not get underway so planning is a vital first step to get the funding and secondly to define the knowns and unknowns so that risks can be assessed and quantified.

How is this Demonstrated?

You should prepare a design and development plan for each new design and also for any modification of an existing design that radically changes the performance of the product or service. For modifications that marginally change performance, control of the changes required may be accomplished through the design change process.

Design and development plans need to identify the activities to be performed, by whom they will be performed and when they should commence and should be completed. One good technique is to use a network chart (often called a PERT chart), which links all the activities together. Alternatively a bar chart may be adequate. There does need to be some narrative in addition as charts in isolation rarely convey everything required.

Design and development is not complete until the design has been proven as meeting the design requirements, so in drawing up a design and development plan you will need to cover the planning of design verification and validation activities. The plans should identify as a minimum:

- The design requirements;
- The design and development programme showing activities against time;
- The work packages and names of those who will execute them. (Work packages are the parcels of work that are to be handed out either internally or to suppliers);
- The work breakdown structure showing the relationship between all the parcels of work;
- The reviews to be held for authorizing work to proceed from stage to stage;
- The resources in terms of finance, people and facilities;
- The risks to success and the plans to minimize them; and
- The controls that will be exercised to keep the design on course.

Planning for all phases at once can be difficult as information for subsequent phases will not be available until earlier phases have been completed. So, your design and development plans may consist of separate documents, one for each phase and each containing some detail of the plans you have made for subsequent phases.

Your design and development plans may also need to be subdivided into plans for special aspects of the design such as reliability plans, safety plans, electromagnetic compatibility plans and configuration management plans. With simple designs there may be only one person carrying out the design activities. As the design and development plan needs to identify all design and development activities, even in this situation you will need to identify who carries out the design, who will review the design and who will verify the design. The same person may perform both the design and the design verification activities. However, it is good practice to allocate design verification to another person or organization because it will reveal problems overlooked by the designer. On larger design projects you may need to employ staff from various disciplines. The responsibilities of all these people or groups need to be identified and a useful way of parcelling up the work is to use work packages that list all the activities to be performed by a particular group. If you subcontract any of the design activities, the supplier's plans need to be integrated with your plans and your plan should identify which activities are the supplier's responsibility. While purchasing is dealt with in Clause 7.4 of the standard, the requirements also apply to design activities.

Stages of Design and Development Process (7.3.1a)

The standard requires *the stages of design and development to be determined.*

What Does this Mean?

A stage in design and development is a point at which the design reaches a phase of maturity. There are several common stages in a design process. The names may vary but the intent remains the same.

- Feasibility stage: The stage during which studies are made of a proposed objective to determine whether practical solutions can be developed within time and cost constraints. This stage usually terminates with a design brief.
- Conceptual design stage: The stage during which ideas are conceived and theories tested. This stage usually terminates with a preferred concept in the form of a design requirement.
- Design definition stage: The stage during which the architecture or layout takes form and the risks assessed and any uncertainty resolved. This stage usually terminates with design requirement specifications for the components comprising the product, service or process.
- Detail design stage: The stage during which final detail characteristics are determined and methods of production or delivery established. This stage usually terminates with a set of specifications for the construction of prototypes. Process design and development will also commence as soon as engineering drawings and tooling requirements are released.

- Development stage: The stage during which the prototype is proven using models or simulations and refined. This stage usually terminates with a set of approved specifications for the construction, installation and operation of the product. Process validation also occurs during this phase even though it is referred to under the heading *Control of production* in ISO 9001.

Why is this Necessary?

This requirement responds to the Process Approach Principle.

Any endeavour is more easily accomplished when undertaken in small stages. By processing a design through several iterative stages, a more robust solution will emerge than if the design is attempted in one cycle.

How is this Demonstrated?

In drawing up your design and development plans you need to identify the principal activities and a good place to start is with the list of 10 steps detailed in Chapter 23 of which the last five are explained further above. Any more detail will in all probability be a breakdown of each of these stages initially for the complete design and subsequently for each element of it. If dealing with a system you should break it down into subsystems, equipment, assemblies and so on. It is most important that you agree the system hierarchy and associated terminology early on in the development programme otherwise you may well cause both technical and organizational problems at the interfaces.

Planning Review Verification and Validation Activities (7.3.1b)

The standard requires *the review, verification and validation activities appropriate to each design and development stage to be determined.*

What Does this Mean?

Each design stage is a process that takes inputs from the previous process and delivers outputs to the next stage. Within each process are verification, validation and review points that feedback information into the process to produce a further iteration of the design. At the end of each stage the output needs to be verified, validated and reviewed before being passed on to the next stage. The further along the design cycle, the more rigorous and complex the verification, validation and review stages will need to be. The verification stages are those stages where design output of a stage is checked against the design input for that stage to ensure that the output is correct. The validation stages occur sequentially or in parallel to confirm that the output is the right output by comparing it with the design brief or requirement. The review stages are points at which the results of verification and validation are reviewed to confirm the design solution, recommend change and authorize or halt further development. A note added to the 2008 edition of ISO 9001 reminds us that review verification and validation have distinct purposes and can be conducted and recorded separately or in any combination.

Why is this Necessary?

This requirement responds to the Factual Approach Principle.

The checks necessary to select and confirm the design solution need to be built into the design process so that they take place when they will have the most beneficial effect on the design. Waiting until the design is complete before verification, validation or review will in all probability result in extensive rework and abortive effort.

How is this Demonstrated?

The stages of verification, validation and review should be identified in the design and development plan, but at each stage there may need to be supplementary plans to contain more detail of the specific activities to be performed. This may result in a need for a separate design verification plan.

The design verification plan should be constructed so that every design requirement is verified and the simplest way of confirming this is to produce a verification matrix of requirement against verification methods. Another matrix in a similar form is a Quality Function Deployment Chart (QFD)[1]. You need to cover all the requirements, those that can be verified by test, by inspection, by analysis, by simulation or demonstration or simply by validation of product records. For those requirements to be verified by test, a test specification will need to be produced. The test specification should specify which characteristics are to be measured in terms of parameters, limits and the conditions under which they are to be measured.

The verification plan needs to cover some or all of the following details as appropriate:

- A definition of the product design standard that is being verified.
- The objectives of the plan (you may need several plans covering different aspects of the requirement).
- Definition of the specifications and procedures to be employed for determining that each requirement has been achieved.
- Definition of the stages in the development phase at which verification can most economically be carried out.
- The identity of the various models that will be used to demonstrate achievement of design requirements. (Some models may be simple space models, others laboratory standard or production standard depending on the need.)
- Definition of the verification activities that are to be performed to qualify or validate the design and those which need to be performed on every product in production as a means of ensuring that the qualified design standard has been maintained.
- Definition of the test equipment, support equipment and facilities needed to carry out the verification activities.
- Definition of the timescales for the verification activities in the sequence in which the activities are to be carried out.
- Identification of the location of the verification activities.
- Identification of the organization responsible for conducting each of the verification activities.
- Reference to the controls to be exercised over the verification activities in terms of the procedures, specifications and records to be produced, the reviews to be

conducted during the programme and the criteria for commencing, suspending and completing the verification operations. (Provision should also be included for dealing with failures, their remedy, investigation and action on design modifications.)

As part of the verification plan, you should include an activity plan that lists all the planned activities in the sequence they are to be conducted and use this plan to progressively record completion and conformance. The activity plan should make provision for planned and actual dates for each activity and for recording comments such as recovery plans when the programme does not proceed exactly as planned. It is also good practice to conduct test reviews before and after each series of tests so that corrective measures can be taken before continuing with abortive tests (see also under the heading *Design validation*).

The designers and those performing the verification activities should approve the verification plan. Following approval the document should be brought under document control. Design verification is often a very costly activity and so any changes in the plan should be examined for their effect on cost and timescale. Changes in the specification can put back the programme by months whilst new facilities are acquired, new jigs, cables, etc. procured. However, small your design, the planning of its verification is vital to the future of the product. Lack of attention to detail can rebound months (or even years) later during production.

Determining Responsibilities and Authority for Design and Development Activities (7.3.1c)

The standard requires *the responsibilities and authorities for design and development activities to be determined.*

What Does this Mean?

To cause the activities in the design and development plan to happen, they have to be assigned to either a person or an organization. Once assigned and agreed by both parties, the assignee becomes responsible for delivering the required result. The authority delegated in each assignment conveys a right to the assignee to make decisions affecting the output. The assignee becomes the design authority for the items designed but this authority does not extend to changing the design requirement – this authority is vested in the organization that delegated or sponsored the design for the item.

Why is this Necessary?

This requirement responds to the Leadership Principle.

Responsibility for design activities needs to be defined so that there is no doubt as to who has the right to take which actions and decisions. Authority for design activities needs to be delegated so that those who are responsible have the right to control their own output. Also the authority responsible for the requirements at each level of the design needs defining so that there is a body to which requests for change can be routed. Without such a hierarchy, there would be anarchy resulting in a design that failed to fulfil its requirements.

How is this Demonstrated?

Within the design and development plan the activities need to be assigned to a person, group or organization equipped with the resources to execute them. Initially the feasibility study may be performed by one person or one group but as the design takes shape, other personnel are required and possibly other external organizations may be required to undertake particular tasks or products.

One way of assigning responsibilities is to use the work package technique. With this approach you can specify not only what is to be done but estimate the required hours, days months or years to do it and then obtain the group's acceptance and consequently commitment to the task.

One of the difficulties with assigning design work is ensuring that those to whom the work is assigned understand the boundary conditions, i.e., what is included and what is excluded (see below under the heading *Organizational interfaces*).

You also need to be careful that work is not delegated or subcontracted to parties about whom you have little knowledge. In subcontracts, clauses that prohibit subcontracting without your approval need to be inserted, thereby enabling you to retain control.

Managing Organizational Interfaces (7.3.1)

The standard requires *the interfaces between different groups involved in design and development process to be managed to ensure effective communication and clarity of responsibilities.*

What Does this Mean?

Where there are many different groups of people working on a design they need to work together to produce an output that meets the overall requirement when all outputs are brought together. To achieve this each party needs to know how the design work has been allocated and to which requirements each party is working so that if there are problems, the right people can be brought together.

Why is this Necessary?

This requirement responds to the Leadership Principle.

If the interfaces between design groups are not properly managed, there are likely to be technical problems arising from groups changing interface requirement without communicating the changes to those affected. Political problems might arise from groups assuming the right to do work or make decisions that have been allocated to other groups. Cost overruns might arise from groups not communicating their difficulties when they are encountered.

How is this Demonstrated?

You should identify where work passes from one organization to another and the means used to convey the requirements such as work instructions, work package descriptions or contracts. Often in design work, the product requirements are analysed to identify further requirements for constituent parts. These may be passed on to other groups as input requirements for them to produce a design solution.

In doing so these groups may in fact generate further requirements in the form of development specifications to be passed to other groups and so on. Some of these transactions may be in-house but some might be subcontracted. Some organizations only possess certain design capabilities and subcontract most of the hardware, software or specialist service components to specialists such as the IT architecture and network design. In this way they concentrate on the business they are good at and get the best specialist support through competitive tenders. These situations create organizational interfaces that require effective information control processes.

In managing the organizational interfaces you will need to:

- define the customer and the supplier in the relationship;
- define the product requirements that the supplier is to meet (the objectives and outputs);
- define the work that the supplier is to carry out with the budget and time constraints;
- define the responsibility and authority of this work (who does what, who approved what);
- define the process used for conveying information and receiving feedback;
- define the reporting and review requirements for monitoring the work;
- conduct regular interface review meetings to check progress and resolve concerns; and
- periodically review the interface control process for its effectiveness.

One mechanism of transmitting technical interface information is to establish and promulgate a set of baseline requirements that are to be used at commencement of design for a particular phase. Any change to these requirements should be processed by a *change control board or* similar body and following approval a change to the baseline is made. This baseline listing becomes a source of reference and if managed properly ensures that no designer is without the current design and interface information.

Interfaces should be reviewed along with other aspects of the design at regular design reviews scheduled prior to the completion of each phase or more often if warranted. Where several large organizations are working together to produce a design, an *interface control board* or similar body may need to be created to review and approve changes to technical interfaces. Interface control is especially difficult with complex projects. Once underway an organization like a large ship gains momentum and takes some time to stop. The project manager may not know of everything that is happening. Control is largely by information and it can often have a tendency to be historical information by the time it reaches its destination. So it is important to control changes to the interfaces. If one small change goes unreported, it may cause months of delay correcting the error – such as two tracks of a railway or two ends of a tunnel being misaligned.

Ensuring that Plans are Updated as the Design Progresses (7.3.1)

The standard requires *planning output to be updated, as appropriate, as the design and development progresses.*

What Does this Mean?

Planning commences before work is performed but as work progresses and the unknown becomes the known, its direction may need to change and therefore the plans need to change.

Why is this Necessary?

This requirement responds to the Factual Approach Principle.

The design and development plan is a source of reference for those on the project – it communicates the work to be performed, who is to perform it and when. It should form the basis on which the design and development costs have been estimated and the work is proceeding. Therefore, if the basis for the plan changes, the plan needs to change so that it reflects the design that is being produced and provides legitimacy for the actions and decisions being carried out. If the plan is not updated, those in possession of it may waste valuable effort in performing work that is no longer required or may not be able to provide the resources when needed and therefore additional costs may be incurred.

How is this Demonstrated?

Some design planning needs to be carried out before any design commences, but it is an iterative process and therefore the design plans may be completed progressively as more design detail emerges. It is not unusual as design gets underway that problems are encountered which require a change in direction. When this occurs the original plans should be changed. The design and development plan should be placed under control after it has been approved. When a change in the plan is necessary you should use the document change request mechanism to change the design and development plan and not implement the change until the request has been approved. In this way you remain in control.

DESIGN AND DEVELOPMENT INPUTS (7.3.2)

Determining and Recording Design Inputs (7.3.2)

The standard requires *inputs relating to product requirements to be, determined and records maintained (see Clause 4.2.4).*

What Does this Mean?

The design inputs are the requirements governing the design of the intended product. They include all the requirements determined from an analysis of customer and regulatory requirements and the organization's requirements. It may appear that this requirement duplicates those addressed in Clause 7.2 of the standard, but as the design is decomposed into subsystems, equipment, components, materials and processes, design inputs are the inputs into the design of each of these levels and will therefore become more specific through the hierarchy. The records to be maintained are the resultant specifications that describe these requirements.

Why is this Necessary?

This requirement responds to the Factual Approach Principle.

The design input requirements constitute the basis for the design without which there is no criteria to judge the acceptability of the design output.

How is this Demonstrated?

Design inputs should reflect the customer, regulator and organization's needs and be produced or available before any design commences.

To identify design input requirements you need to identify the following factors:

- The purpose of the product or service.
- The conditions under which it will be used, stored and transported.
- The skills and category of those who will use and maintain the product or service.
- The countries to which it will be sold and the related regulations governing sale and use of products.
- The special features and performance characteristics which the customer requires the product or service to exhibit (including life, reliability, durability and maintainability).
- The constraints in terms of timescale, operating environment, cost, size, weight or other factors.
- The standards with which the product or service needs to comply.
- The products or services with which it will directly and indirectly interface and their features and characteristics.
- The documentation required of the design output necessary to manufacture, procure, inspect, test, install, operate and maintain a product or service.

Organizations have a responsibility to establish their customer requirements and expectations. If you do not determine conditions that may be detrimental to the product and you supply the product as meeting the customer needs and it subsequently fails, the failure is your liability. If the customer did not provide reasonable opportunity for you to establish the requirements, the failure may be the customer's liability. If you think you may need some extra information in order to design a product that meets the customer needs, you must obtain it or declare your assumptions. A nil response is often taken as acceptance in full.

In addition to customer requirements there may be industry practices, national standards, company standards, the experience gained from previous designs and other sources of input to the design input requirements to be taken into account. The result of competitive analysis should be used to ensure that product design requirements are not putting the product at a distinct disadvantage even before design commences. You should provide design guides or codes of practice that will assist designers identify the design input requirements that are typical of your business.

The design output has to reflect a product that is producible or a service that is deliverable. The design input requirements may have been specified by the customer and consequently not have taken into account your production capability. The product of the design may therefore need to be producible within your current production capability using your existing technologies, tooling, production processes, material handling equipment etc.

Having identified the design input requirements, you need to document them in a specification that when approved is brought under document control. The requirements should not contain any solutions at this stage so as to provide freedom and flexibility to the designers. If the design is to be subcontracted, this makes for fair competition and removes from you the responsibility for the solution. Where specifications contain solutions, the supplier is being given no choice and if there are delays and problems the supplier may have a legitimate claim against you.

Defining Functional and Performance Requirements (7.3.2a)

The standard requires design inputs *to include functional and performance requirements.*

What Does this Mean?

Functional requirements are those related to actions that the product is required to perform with or without external stimulus. Performance requirements relate to the results or behaviours required by such actions under stated conditions. Normally a product's characteristics are stated in physical, functional and performance terms rather than functional and performance but no matter. The intent of the requirement is that all characteristics that the product is required to exhibit should be included in the design input requirements and expressed in terms that are measurable.

Why is this Necessary?

This requirement responds to the Process Approach Principle.

All the characteristics need to be stated otherwise the resultant design may not reflect a product that fulfils the conditions for intended use. Two products may possess the same physical and functional characteristics but perform differently due to the individual arrangement of their component parts and the materials and processes used in their construction.

How is this Demonstrated?

From the statement of product purpose, the conditions of use and the skills of those who will use it, the most obvious characteristics can be derived and divided into physical, functional and performance requirements. The physical characteristics might include size, mass, appearance and material properties. Functional characteristics might include speed, power, capacity and a wide range of characteristics that give the product distinctive features. Performance characteristics might include reliability, maintainability, durability, flammability, portability, safety etc. Sustainability may be a customer requirement, therefore material and component selection criteria may need to be specified.

Defining Statutory and Regulatory Requirements (7.3.2b)

The standard requires design inputs *to include applicable statutory and regulatory requirements.*

What Does this Mean?

At the end product level, the applicable statutory and regulatory requirements are those addressed by Clause 7.2.1c. However, as the design unfolds additional statutory and regulatory requirements may become applicable as specific subsystems, equipment, components, materials and processes are identified.

Why is this Necessary?

This requirement responds to the Process Approach Principle.

The statutory and regulatory requirements that apply are dependent on a range of factors that emerge during the design process once it is known what type of device is required. These regulations need to be identified in the design input so that the resultant design is proven to meet them before commitment to production is granted.

How is this Demonstrated?

Statutory and regulatory requirements are those that apply in the country to which the product or service is to be supplied. Whilst some customers have the foresight to specify these, others often don't. Just because such requirements are not specified in the contract doesn't mean you don't need to meet them. Further detail on *statutory and regulatory requirements* is provided in Chapter 24.

Deriving Information from Previous Designs (7.3.2c)

The standard requires the design inputs *to include applicable information derived from previous similar designs.*

What Does this Mean?

Most designs are a development of that which was designed previously. It is rare for a design to be completely new. Even if the product concept is new, it may contain design solutions used previously. The history of these previous designs contains a wealth of information that may be applicable to the application that is currently being considered.

Why is this Necessary?

This requirement responds to the Continual Improvement Principle.

Using the lessons learnt from previous designs is corrective action – preventing the recurrence of problems that have occurred in the past. If previous design history is not utilized, the problems may recur.

How is this Demonstrated?

In principle the design history of a product should be archived and made available to future designers. Design history can be placed in a database or library that is accessible to future designers. A rather old way of doing this was for companies to create design manuals containing data sheets, fact sheets and general information sheets on design topics, which is a sort of design guide that captured experience. Companies should still be doing this but many will by now have converted to electronic storage media with the added advantage of a search engine. Information will also be available from trade associations, libraries and learned societies. Often professional journals, published literature and even newspapers can contain useful information for designers. In your model of the design process you need to install a research process that is initiated at some stage in the design of a system, subsystem, equipment or component. The database or libraries need to structure the information in a way that it will return relevant data on previous designs. One advantage of submitting the design to a review by those not involved in the design is that they bring their experience to the review and identify approaches that did not work in the past, or put forward more effective ways of doing such things in the future.

Within the design input requirements, such information would appear either as preferred solutions or non-preferred solutions, either directly or by reference to learned papers, standards, guide etc.

Identifying Other Essential Requirements (7.3.2d)

The standard requires *design inputs to include any other requirements essential for design and development.*

What Does this Mean?

In addition to the requirements identified there may be requirements that are dictated because of the organizational policies, national and international politics as was addressed under Clause 7.2.1d.

Why is this Necessary?

This requirement responds to the Process Approach Principle.

The organization may wish to maintain a certain profile or reputation through its designs and therefore may impose requirements that may impact the design input requirements.

How is this Demonstrated?

One specific series of requirement that may not emerge from the forgoing are technical interface requirements. Some of these may need to be written around a particular supplier. However, within each development specification the technical interfaces between systems, subsystems, equipment etc. should be specified so that when all these components are integrated they function properly. In some situations it may be necessary to generate separate interface specifications defining requirements that are common to all components of the system. In a large complex design, minor details of a component may be extremely important in the design of another component. Instead of providing designers with specifications of all the components, it may be more economical (as well as more controllable) if the features and characteristics at the interface between components are detailed in separate interface specifications.

Reviewing Design Input Requirements (7.3.2)

The standard requires design inputs *to be reviewed for adequacy and for the requirements to be complete, unambiguous and not in conflict with other requirements.*

What Does this Mean?

Adequacy in this context means that the design input requirements are a true reflection of the customer needs.

Why is this Necessary?

This requirement responds to the Factual Approach Principle.

The determination of design inputs results in information that needs to be reviewed prior to its release otherwise incorrect information may enter the design process. It is prudent to obtain customer agreement to the design requirements before commencing the design. In this way you will establish whether you have correctly understood and translated customer needs. It is advisable also to hold an internal design review at this stage so that you may benefit from the experience of other staff in the organization.

How is this Demonstrated?

The review of the design input requirements needs to be a systematic review, not a superficial glance. Design work will commence on the basis of what is conveyed in the requirements or the brief, although you should ensure that there is a mechanism in place to change the information should it become necessary later. In fact such a mechanism should be agreed at the same time as agreement to the requirement is reached.

In order to detect incomplete requirements you either need experts on tap or checklists to refer to. It is often easy to comment on what has been included but difficult to imagine what has been excluded. It is also important to remove subjective statements.

Ambiguities arise where statements imply one thing but the context implies another. You may also find cross-references to be ambiguous or in conflict. To detect the ambiguities and conflicts you need to read statements and examine diagrams very carefully. The same items shown on one diagram may be shown differently in another. There are many other aspects you need to check before being satisfied they are fit for use. Any inconsistencies you find should be conveyed to the appropriate person with a request for action. Any changes to correct the errors should be self-evident so that you do not need to review all the information again.

DESIGN AND DEVELOPMENT OUTPUTS (7.3.3)

Documenting the Design and Development Output (7.3.3)

The standard requires that *the outputs of design and development be in a form suitable for verification against the design and development inputs.*

What Does this Mean?

Design output is the product of the design process and will therefore comprise information and/or models and specimens that describe the design in all its detail, the calculations, assumptions and the rationale for the chosen solution. It is not simply the specifications or drawings because should the design need to be changed, the designer may need to revisit the design data to modify parameters and assumptions. By requiring the design output to be in a form suitable for verification, the characteristics of the

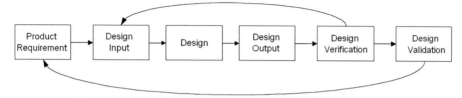

FIGURE 25-4 Relationship between design verification and design validation.

product need to be expressed in measurable terms. One would therefore expect form, fit and function to be specified in units of measure with allowable tolerances or models and specimens to be capable of use as comparative references.

It is interesting to note that the requirement omits validation. This is because design outputs are verified against design inputs whereas, the design is validated against the original product requirement using a product or simulation that accurately reflects the design, thereby by-passing the design input and output as illustrated in Fig. 25-4.

Why is this Necessary?

This requirement responds to the Process Approach Principle.

Unless the design output is expressed in a form that enables verification, it will not be possible to verify the design with any certainty.

How is this Demonstrated?

The design input requirements should have been expressed in a way that would allow a number of possible solutions. The design output requirements should therefore be expressed as *all* the inherent features and characteristics of the design that reflect a product that will satisfy these requirements. It should therefore fulfil the stated or implied needs, i.e., be fit for purpose.

Product Specifications

Product specifications should specify requirements for the manufacture, assembly and installation of the product in a manner that provides acceptance criteria[①] for inspection and test. They may be written or CAD generated specifications, engineering drawings, diagrams, inspection and test specifications and schematics. With complex products you may need a hierarchy of documents from system drawings showing the system instal-lation to component drawings for piece-part manufacture. Where there are several documents that make up the product specification there should be an overall listing that relates documents to one another.

Service specifications should provide a clear description of the manner in which the service is to be delivered, the criteria for its acceptability, the resources required including the numbers and skills of the personnel required, the numbers and types of facilities and equipment necessary and the interfaces with other services and suppliers.

In addition to the documents that serve product manufacture and installation or service delivery, documents may also be required for maintenance and operation. The product descriptions, handbooks, operating manuals, user guides and other documents which support the product or service in use are as much a part of the design as the other product requirements. Unlike the manufacturing data, the support documents may be published either generally or supplied with the product to the customer. The design of such documentation is critical to the success of the product as poorly constructed handbooks can be detrimental to sales.

The requirements within the product specification need to be expressed in terms that can be verified. You should therefore avoid subjective terms such as 'good quality components', 'high reliability', and 'commercial standard parts' etc. as these requirements are not sufficiently definitive to be verified in a consistent manner.

Design Calculations

Throughout the design process, calculations will need to be made to size components and determine characteristics and tolerances. These calculations should be recorded and retained together with the other design documentation but may not be issued. In performing design calculations it is important that the status of the design on which the calculations are based is recorded. When there are changes in the design these calculations may need to be repeated. The validity of the calculations should also be examined as part of the design verification activity. One method of recording calculations is in a designer's logbook that may contain all manner of things and so the calculations may not be readily retrievable when needed. Recording the calculations in separate reports or in separate files along with the computer data will improve retrieval.

Design Analyses

Analyses are types of calculations but may be in the form of comparative studies, predictions and estimations. Examples are stress analysis, reliability analysis, failure modes analysis, and hazard analysis. Analyses are often performed to detect whether the design has any inherent modes of failure and to predict the probability of occurrence. FMEA is addressed in Chapter 37 as it is a preventive measure. The analyses assist in design improvement and the prevention of failure, hazard, deterioration and other adverse conditions. Analyses may need to be conducted because the end-use conditions may not be reproducible in the factory. Assumptions may need to be made about the interfaces, the environment, the actions of users etc. and analysis of such conditions assists in determining characteristics as well as verifying the inherent characteristics.

Design is an iterative process therefore the above analyses are not performed once but after each iteration of the design so that by the end of the design process any inherent weaknesses will have either been eliminated, reduced or contained by component redundancy, derating, error-proofing or warning notices on the product or in the user documentation.

Ensuring that Design Output Meets Design Input Requirements (7.3.3a)

The standard requires that design and development output *meets the design and development input requirements.*

What Does this Mean?

The characteristics of the resultant design should be directly or indirectly traceable to the design input requirements. In some cases a dimension may be stated in the design input which is easily verified when examining the design specifications, drawings etc.

In other cases the input requirement may be stated in performance terms that are translated into a number of functions which when energized provide the required result. In other cases a parameter may be specified above or below the design input requirement so as to allow for variation in production.

Why is this Necessary?

This requirement responds to the Factual Approach Principle and needs no explanation.

How is this Demonstrated?

The techniques of design verification can be used to verify that the design output meets the design input requirements. However, design verification is often an iterative process. As features are determined, their compliance with the requirements should be checked by calculation, analysis or test on development models. Your development plan should identify the stages at which each requirement will be verified so as to give warning of non-compliance as early as possible.

Providing Information for Purchasing, Production and Service Provision (7.3.3b)

The standard requires that design and development output *provide appropriate information for purchasing, production and service provision.*

What Does this Mean?

British Standard BS 7000 defines a design as *a set of instructions (specifications, drawings, schedules etc.) necessary to construct a product.* Therefore, the inherent characteristics of the product that facilitates procurement, production and servicing need to be defined. Tooling for production is considered to be part of the production process but information within the design output is needed to enable tooling to be designed.

The instructions needed to produce, inspect, test, protect, preserve, install and maintain the product may be produced by the designers but are strictly outputs of the production, installation and servicing processes that are derived from inputs that comprise the design description.

Why is this Necessary?

This requirement responds to the Systems Approach Principle.

A design description alone will not result in its realization unless information is provided for procuring the materials and components, preparing the product for production and maintaining the product in service.

How is this Demonstrated?

Products should be designed to facilitate procurement, manufacture, storage, transportation, installation and servicing and therefore additional characteristics to those required for end use may be necessary. Examples include, geometric tolerances, specific part numbers, part marking, assembly aids, error-proofing, lifting points, transportation and storage protection. Techniques used to identify such design provisions are as follows:

- FMEA (see Chapter 37);
- Producibility Analysis;
- Testability Analysis; and
- Maintainability Analysis.

Defining Acceptance Criteria (7.3.3c)

The standard requires design and development output to *contain or reference product acceptance criteria.*

What Does this Mean?

Acceptance criteria are the requirements that, if met, will deem the product acceptable. It means that characteristics should be specified in measurable terms with tolerances or limits. These limits should enable all production versions to perform to the product specification, providing such limits are well within the limits to which the design has been tested. It means that every requirement should be stated in such a way that it can be verified – that there is no doubt as to what will be acceptable and what will be unacceptable.

Why is this Necessary?

This requirement responds to the Factual Approach Principle.

Where product characteristics are specified in terms that are not measurable or are subjective, they lend themselves to misinterpretation and variation such that no two products produced from the same design will be the same and will exhibit consistent performance.

How is this Demonstrated?

A common method used to ensure characteristics are stated in terms of acceptance criteria[i] is to define them by reference to product standards. These standards may be developed by the organization or may be of national or international status. Standards are employed to enable interchangeability, repeatability and to reduce variety.

Where there are common standards for certain features, these may be contained in a standards manual. Where this method is used it is still necessary to refer the standards in the particular specifications to ensure that the producers are always given full criteria. Some organizations omit common standards from their specifications. This makes it difficult to specify different standards or to subcontract the manufacture of the product or operation of a service without handing over proprietary information.

Specifying Essential Characteristic (7.3.3d)

The standard requires design and development output *to specify the characteristics of the product that are essential to its safe and proper use.*

What Does this Mean?

Certain characteristics will be critical to the installation, operation or maintenance of the product. These are sometimes called Critical to Quality characteristics (CTQ). These can be divided into two types. Those characteristics that the product needs to exhibit in order to function correctly and those characteristics that are exhibited when the product is put together, used or maintained incorrectly.

Why is this Necessary?

This requirement responds to the Customer Focus Principle.

Alerting assemblers, users and maintainers to CTQ characteristics increases their sensitivity, provides the awareness to plan preventive measures and thus reduces the probability of an incident or accident.

How is this Demonstrated?

The design output data should identify by use of symbols or codes the CTQ characteristics. This will enable the manufacturers to determine the measures needed to ensure no variation from specification when the characteristics are initially produced and ensure no alteration of these characteristics during subsequent processing.

Drawings should also indicate the warning notices required, where such notices should be placed and how they should be affixed. Examples that indicate improper function or potential danger are red lines on tachometers to indicate safe limits for engines, audible warnings that signal unsafe loading e.g., a stall warning onboard an airliner or 'No Step' notices on the flying surfaces of an aircaft wing or warnings on computers to indicate an incorrect command etc. In some cases it may be necessary to mark dimensions or other characteristics on drawings to indicate that they are critical and employ special procedures for dealing with any variations. In passenger vehicle component design, certain parts are regarded as safety-critical because they carry load or need to behave in a certain manner under stress. Others are not critical because they carry virtually no load so there can be a greater tolerance on deviations from specification.

Failure Modes and Effects Analysis and Hazard Analysis are techniques that aid the identification of characteristics crucial to the safe and proper functioning of the product.

Approval of Design Outputs (7.3.3)

The standard requires design and development outputs to *be approved prior to release.*

What Does this Mean?

Although the requirement for design outputs to be approved prior to release may appear to duplicate the requirement of Clause 4.2.3 on document control, there is a subtle difference. Document approval is not the same as design approval and design release is not the same as document issue. When a design is approved it is the description of that design in whatever form that is approved. Design approval therefore applies to all the documents, models, and specimens etc. that constitute the design description, not as separate entities but as a whole.

Why is this Necessary?

This requirement responds to the Factual Approach Principle.

While design documents should pass into the document control process for issue, several iterations may be needed before the design output is complete and ready for release. It is important that these iterations are under control otherwise the full impact of changes in one component may not be reflected in other components. When the design is ready for verification it is released.

How is this Demonstrated?

The requirements in both previous and current versions of ISO 9001 do not recognize that design reviews are not only concerned with reviewing a design but are also concerned with granting design approval. This oversight resulted in design approval being planted under design outputs and the inevitable consequence that it would be interpreted as document approval.

There are sound reasons for separating design approval from document approval. Design approval should proceed through three stages:

- Design information should be approved before being presented to a design review – in this way the reviewers only work with information that has been checked and found acceptable.
- Designs should be approved before the design is subject to verification – in this way prototypes are produced, or simulated using a complete set of design information that has been found acceptable.
- Designs should be verified before being subject to validation – in this way trials are only conducted on models representative of those that will enter production.

The design output may consist of many documents each of which fulfils a certain purpose. It is important that these documents are reviewed and verified as being fit for their purpose before release using the documentation controls developed for meeting Section 4.5 of ISO 9001. In the software industry, where documentation provides the only way of inspecting the product prior to installation, document inspections called Fagan Inspections, (after Michael Fagan of IBM) are carried out not only to identify the errors, but to collect data on the type of error and the frequency of occurrence. By analysing this data using statistical techniques the results assist in error removal and prevention.

Design documentation reviews can be made effective by providing data requirements for each type of document as part of the design and development planning process. The data requirement can be used both as an input to the design process and as acceptance criteria[1] for the design output documentation review. The data requirements would specify the input documents and the scope, content and format required for the output document. Contracts with procurement agencies often specify deliverable documents and by invoking formal data requirements in the contract, the customer is then assured of the outputs.

As design documents are often produced at various stages in the design process, they should be reviewed against the input requirements to verify that no requirements have been overlooked and that the requirements have been satisfied.

DESIGN AND DEVELOPMENT REVIEW (7.3.4)

Planning Design Reviews (7.3.4)

The standard requires that *at suitable stages, systematic reviews of design and development be performed in accordance with planned arrangements.*

What Does this Mean?

ISO 9000:2005 defines a review as an activity undertaken to determine the suitability, adequacy and effectiveness of the subject matter to achieve established objectives. This is the way the term should be interpreted in the context of ISO 9001. When the term review is used in another context, it simply means to have another look at something. The ISO 9000:2005 definition implies that a design review is an activity undertaken to determine the suitability, adequacy and effectiveness of a design to meet the design requirement. Suitability means 'Has an appropriate design solution been developed?' Adequacy means 'Does the design solution meet all the design requirements?' and effectiveness means 'Have we got the right design objective?' Design reviews are therefore *not* document reviews.

Systematic reviews are those that cover the complete design from the high level down to the smallest component and all the associated requirements in a logical manner. The review has to be stage-by-stage, methodical with purpose. Systematic reviews probe the design solution and the interfaces between all components for design weaknesses and delves into the detail to explore how requirements are fulfilled.

Suitable stages are at the transition between the various phases of design maturity in the design process. In simple terms designs begin with a conceptual phase, proceed through a definition phase and end with a detail design phase. Development commences with a detail design and proceeds through several iterations involving verification and validation and may continue through several enhancements before the design becomes obsolete and a new design idea is conceived.

The planned arrangements are the stages in the design and development cycle that reviews have been planned and the expected inputs and outputs and the acceptance criteria[①] for proceeding to the next stage.

Why is this Necessary?

This requirement responds to the Leadership Principle.

A design represents a considerable investment by the organization. There is therefore a need for a formal mechanism for management and the customer (if the customer is sponsoring the design) to evaluate designs at major milestones. The purpose of the review is to determine whether the proposed design solution is compliant with the design requirement and should continue or should be changed before proceeding to the next phase. It should also determine whether the documentation for the next phase is adequate before further resources are committed. Design review is that part of the design control process which measures design performance, compares it with pre-defined requirements and provides feedback so that deficiencies may be corrected before the design is released to the next phase.

How is this Demonstrated?

Review Schedules

A schedule of design reviews should be established for each product, process or service being developed. In some cases there will need to be only one design review. After completion of all design verification activities but depending on the complexity of the design and the risks, you may need to review the design at some or all of the following intervals:

- Design requirement review: To establish that the design requirements can be met and reflect the needs of the customer before commencement of design.
- Conceptual design review: To establish that the design concept fulfils the requirements before project definition commences.
- Preliminary design review: To establish that all risks have been resolved and development specifications have been produced for each sub-element of the product or service before detail design commences.
- Critical design review: To establish that the detail design for each sub-element of the product or service complies with its development specification and that product specifications have been produced before manufacture of the prototypes.
- Design validation readiness review: To establish the configuration of the baseline design and readiness before commencement of design validation.
- Final design review: To establish that the design fulfils the requirements of its development specification before preparation for its production.

Design Review Input Data

The input data for the review should be distributed and examined by the review team well in advance of the time when a decision on the design has to be made. A design review is not a meeting. However, a meeting will often be necessary to reach a conclusion and to answer questions of the participants. Often analysis may need to be performed on the input data by the participants in order for them to determine whether the design solution is the most practical and cost-effective way of meeting the requirements.

Conducting Design Reviews (7.3.4a and b)

The standard requires design reviews *to be conducted to evaluate the ability of the results of the design and development to fulfil requirements, identify problems and propose required actions.*

What Does this Mean?

Design reviews occur at the end of a design phase when there are results to review. This means that every phase needs an objective, the achievement of which is evaluated at the review. The results of the design may be concepts, models, calculations, drawings, specifications or any output which describes the maturity of the design at a particular stage. During the initial phases, the key performance characteristics will be evaluated and at subsequent design reviews further definition enables the design to be evaluated against more definitive requirements until all requirements are fulfilled. Each review may reveal design weaknesses that need to be resolved before proceeding to the next phase.

Why is this Necessary?

This requirement responds to the Factual Approach Principle.

It would be folly to proceed with a design that possesses significant weaknesses and therefore the design review provides an opportunity to identify these weaknesses early on and take action to eliminate them before compounding the errors.

How is this Demonstrated?

Although design documents may have been through a vetting process, the purpose of the design review is not to review documents but to subject the design to an independent board of experts for its judgement as to whether the most satisfactory design solution has been chosen. By using a design review methodology, flaws in the design may be revealed before it becomes too costly to correct them. Design reviews also serve to discipline designers by requiring them to document the design logic and the process by which they reached their conclusions, particularly the options chosen and the reasons for rejecting others.

The experiences of previous designs provide a wealth of information of use to designers that can alert them to potential problems. In compiling this information designers can feed off the experience of others not only in the same organization but also in different organizations and industries. By using technical data available from professional institutions, associations, research papers etc. checklists can be complied that aid the evaluation of designs.

Participants at Design Reviews (7.3.4)

The standard requires participants in design reviews *to include representatives of functions concerned with the design and development stage(s) being reviewed.*

What Does this Mean?

Representatives of functions concerned with a design include not only the designers but also those sponsoring the design such as the customers, marketing personnel or upper management, those that will be responsible for transforming the design into a product or service, those responsible for maintaining the product, using the product or disposing of the product in fact any party that has an interest in the quality of the design solution.

Why is this Necessary?

This requirement responds to the Involvement of People Principle.

Design reviews should be performed by the management or the sponsor rather than the designers, in order to release a design to the next phase of development. The designer has had one look at the design and when satisfied presents the design to an impartial body of experts so as to seek approval and permission to go-ahead with the next phase. Designers are often not the budget holders, or the sponsors. They often work for others. Even in situations where there is no specific customer or sponsor or third party, it is good practice to have someone else look at the design. A designer may become too close to the design to spot errors or omissions and so will be biased towards the standard of his or her own performance. The designer may welcome the opinion of someone else because it may confirm that the right solution has been found or that the requirements can't be achieved with the present state of the art. If a design is inadequate and the inadequacies are not detected before production commences the consequences may well be disastrous. A poor design can lose a customer, a market or even a business so the advice of independent experts should be valued.

How is this Demonstrated?

The review team should have a collective competency greater than that of the designer of the design being reviewed. For a design review to be effective it has to be conducted by someone other than the designer. The requirement for participants to include representatives of all functions concerned with the design stage means those who have an interest in the results.

The review team should comprise as appropriate, representatives of the purchasing, manufacturing, servicing, marketing, inspection, test, reliability, QA authorities etc. as a means of gathering sufficient practical experience to provide advance warning of potential problems with implementing the design. The number of people attending the design review is unimportant and could be as few as the designer and his or her supervisor provided that the supervisor is able to impart sufficient practical experience and there are no other personnel involved at that particular design stage. There is no advantage gained in staff attending design reviews that add no value in terms of their relevant experience, regardless of what positions they hold in the company. The representation at each review stage may well be different.

The chairman of the review team should be the authority responsible for placing the development requirement and should make the decision as to whether design should proceed to the next phase based on the evidence substantiated by the review team.

Design Review Records (7.3.4)

The standard requires *the results of the reviews and actions arising from the review to be recorded.*

What Does this Mean?

The results of the design review are not simply minutes of a meeting but all the evidence that has been accumulated in evaluating a particular design, identifying problems and determining actions required to resolve them.

Why is this Necessary?

This requirement responds to the Factual Approach Principle.

The results of the design review should be documented in a report rather than minutes of a meeting because they represent objective evidence that may be required later to determine product compliance with requirements, investigate design problems and compare similar designs. Even when no problems are found, the records of the review provide a baseline that can be referred to when making subsequent changes.

How is this Demonstrated?

The report should have the agreement of the full review team and should include:

- The criteria against which the design has been reviewed.
- A list of the documentation that describes the design being reviewed and any evidence presented which purports to demonstrate that the design meets the requirements.
- The decision on whether the design is to proceed to the next stage.
- The basis on which confidence has been placed in the design.
- A record of any uncompleted corrective actions from previous reviews.
- The recommendations and reasons for corrective action – if any.
- The members of the review team and their roles.

DESIGN AND DEVELOPMENT VERIFICATION (7.3.5)

Performing Design Verification (7.3.5)

The standard requires *design and development verification to be performed in accordance with planned arrangements to ensure the output meets the design and development inputs.*

What Does this Mean?

ISO 9000:2005 defines verification as confirmation, through the provision of objective evidence that specified requirements have been fulfilled. There are two types of verification, those verification activities performed during design and on the component parts to verify conformance to specification and those verification activities performed on the completed design to verify performance against the design input. There should be design requirements or published standards for each product in the hierarchy down to component and raw material level. Each of these design requirements represents acceptance criteria[©] for verifying the design output of each stage. Verification may take the form of a document review, laboratory tests, alternative calculations, similarity analyses or tests and demonstrations on representative samples, prototypes etc. In all these cases the purpose is to prove that the design is right, i.e., it meets the requirements.

The reference to planned arrangements again means that verification plans should be adhered to.

Why is this Necessary?

This requirement responds to the Factual Approach Principle.

Verification is fundamental to any process and unless the design is verified, there will be no assurance that the resultant design meets the requirements.

How is this Demonstrated?

Timing

The standard does not state when design verification is to be performed although verification of the design after launch of product into production would not be a wise thing to do. The stages of verification will therefore mirror the design review schedule but may include additional stages. Design verification needs to be performed when there is a verifiable output.

Verification Process

During the design process many assumptions may have been made and will require proving before commitment of resources to the replication of the design. Some of the requirements such as reliability and maintainability will be time-dependent. Others may not be verifiable without stressing the product beyond its design limits. With computer systems, the wide range of possible variables is so great that proving total compliance would take years. It is, however, necessary to subject a design to a series of tests and examinations in order to verify that all the requirements have been achieved and that features and characteristics will remain stable under actual operating conditions. The tests that confirm features and characteristics under nominal operating conditions are called *verification tests* whereas, those test that confirm features and characteristics at the operating extremes are called *qualification tests*. These differ from other tests because they are designed to establish the design margins and prove the capability of the design (see Design validation).

The design verification process should provide for the following:

- Test specifications to be produced that define the features and characteristics that are to be verified for design verification and acceptance.
- Test plans to be produced that define the sequence of tests, the responsibilities for verification their conduct, the location of the tests and test procedures to be used.
- Test procedures to be produced that describe how the tests specified in the test specification are to be conducted together with the tools and test equipment to be used and the data to be recorded.
- All measuring equipment to be within calibration during the tests.
- The test sample to have successfully passed all planned in-process and assembly inspections and tests prior to commencing verification tests.
- The configuration of the product in terms of its design standard, deviations, nonconformities and design changes to be recorded prior to and subsequent to the tests.
- Test reviews to be held before tests commence to ensure that the product, the facilities, tools, documentation and personnel are in a state of operational readiness for verification.
- Test activities to be conducted in accordance with the prescribed specifications, plans and procedures.
- The results of all tests and the conditions under which they were obtained to be recorded.
- Deviations to be recorded, action taken and the product subject to re-verification prior to continuing with the tests.
- Test reviews to be performed following verification tests to confirm that sufficient objective evidence has been obtained to demonstrate that the product fulfils the requirements of the test specification.

Development Models

Many different types of models may be needed to aid product development, test theories, experiment with solutions etc. However, when the design is complete, prototype models representative in all their physical and functional characteristics to the production models may need to be produced.

If design is proven on uncontrolled models then it is likely that there will be little traceability to the production models. Production models may therefore contain features and characteristics that have not been proven. The only verification that needs to be performed on production models is for those features and characteristics that are subject to change due to the variability in manufacturing, either of raw materials or of assembly processes.

When building prototypes, the same materials, locations, suppliers, tooling and processes should be used as will be used in actual production so as to minimize the variation.

Development tests will not yield valid results if obtained using uncontrolled measuring equipment[T], therefore the requirements of Clause 7.6 on measuring devices apply to the design process. Design is not complete until the criteria for accepting production versions have been established. Products need to be designed so as to be testable during production using the available production facilities. The proving of production acceptance criteria[T] is therefore very much part of design verification.

Development Tests

Where tests are needed to verify conformance with the design specification, development test specifications will be needed to specify the test parameters, limits and operating conditions. For each development test specification there should be a corresponding development test procedure that defines how the parameters will be measured using particular test equipment and taking into account any uncertainty of measurement[T]. Test specifications should be prepared for each testable item. Whilst it may be possible to test whole units, equipment or subsystems you need to consider the procurement and maintenance strategies for the product when deciding which items should be governed by a test specification. Two principal factors to consider are:

- Testable items sold as spare parts.
- Testable items the design and/or manufacture of which are subcontracted.

If you conduct trials on parts and materials to prove reliability or durability, these can be considered to be verification tests. For example, you may conduct tests in the laboratory on metals for corrosion resistance or on hinges for reliability and then conduct validation tests under actual operating conditions when these items are installed in the final product.

Verifying Compliance with Regulations

Having designed the product to meet the applicable statutes and regulations you need to plan for verifying that they have been met. Verification of compliance can be accomplished through discrete checks combined with other tests, inspections and analyses, however, it may be more difficult to demonstrate compliance through the records alone. In some cases tests such as pollution tests, safety tests, proof loading tests, electromagnetic compatibility tests, pressure vessel tests etc. are so significant that separate tests and test specifications are the most effective method of verifying compliance with regulations.

Alternative Design Calculations

Verification of some characteristics may only be possible by calculation rather than by test, inspection or demonstration. In such cases the design calculations should be checked either by being repeated by someone else or by performing the calculations by an alternative method. When this form of verification is used the margins of error permitted should be specified in the verification plan.

Comparing Similar Designs

Design verification can be a costly exercise. One way of avoiding unnecessary costs is to compare the design with a similar one that has been proven to meet the same requirements. This approach is often used with designs that use a modular construction. Modules used in previous designs need not be subject to the range of tests and examinations necessary if their performance has been verified either as part of a proven design or has been subject to such in-service use that will demonstrate achievement of the requirements. Care has to be taken when using this verification method that the requirements are the same and that evidence of compliance is available to demonstrate compliance with the requirements. Marginal differences in the environmental conditions and operating loads can cause the design to fail if it was operating at its design limit when used in the previous design.

Recording Design Verification Results (7.3.5)

The standard requires *the results of the verification and any required actions to be recorded.*

What Does this Mean?

The results of design verification comprise:

- the criteria used to determine acceptability;
- data testifying the standard of the design being subject to verification;
- the verification methods;
- data testifying the conditions, facilities and equipment used to conduct the verification;
- the measurements;
- analysis of the differences between planned and achieved results; and
- actions to be taken on the differences.

Why is this Necessary?

This requirement responds to the Factual Approach Principle.

Any decision to proceed either to the next stage of development or into production or operations needs to be based on fact and the records of design verification provide the facts. As the definition of verification explains, it is the provision of objective evidence that requirements have been fulfilled and the records of verification constitute this objective evidence.

How is this Demonstrated?

In planning design verification, consideration needs to be given to the output, its format and content. The basic content is governed by the design specification but the data to be recorded both before, during and after verification need to be prescribed (see *Determining the need for records* in Chapter 23). Some data may be generated electronically and other data may be collected from observation. Often there are lots of different pieces of evidence that need to be collected, collated and assembled into a dossier in a secure format. These factors need to be sorted out before commencing verification so that all the necessary information is gathered at the time. After verification a report of the activities may also be necessary to explain the results, possible causes of any variation and recommendations for action for presentation at a design review.

DESIGN AND DEVELOPMENT VALIDATION (7.3.6)

Performing Validation (7.3.6)

The standard requires design and development validation *to be performed in accordance with planned arrangements to confirm that resulting product is capable of fulfilling the requirements for the specified application or intended use where known.*

What Does this Mean?

ISO 9000:2005 defines validation as confirmation through the provision of objective evidence that requirements for a specific intended use or application have been fulfilled. Specified requirements are often an imperfect definition of needs and expectations and therefore to overcome inadequacies in the manner in which requirements can be specified, the resultant design needs to be validated against intended use or application.

Design validation (also known as design qualification) is a process of evaluating a design to establish that it fulfils the intended use requirements. It goes further than design verification in that validation tests and trials may stress the product of such a design beyond operating conditions in order to establish design margins of safety and performance. Design validation can also be performed on mature designs in order to establish whether they will fulfil different user requirements to the original design input requirements. An example is where software designed for one application can be proven fit for use in a different application or where a component designed for one environment can be shown to possess a capability that would enable it to be used in a different environment. Multiple validations may therefore be performed to qualify a design for different applications.

Why is this Necessary?

This requirement responds to the Factual Approach Principle.

Merely requiring that the design output meets the design input would not produce a quality product or service unless the input requirements were a true reflection of the customer needs and this is not always possible. If the input is inadequate the output will be inadequate: 'garbage-in-garbage-out' to use a common software expression. Let's face it; if design validation were performed effectively, all the product recalls would be down to errors in manufacturing.

How is this Demonstrated?

Design validation may take the form of qualification tests which stress the product up to and beyond design limits - beta tests where products are supplied to several typical users on trial in order to gather operational performance data, performance trials and reliability and maintainability trials where products are put on test for prolonged periods to simulate usage conditions.

Other examples are beta tests or public testing conducted on software products where tens or hundreds of products are distributed to designated customer sites for trials under actual operating conditions before product launch. Sometimes, commercial pressures force termination of these trials and premature launch of products in order to beat the competition.

As the cost of testing vast quantities of equipment would be too great and take too long, qualification tests particularly on hardware are usually performed on a small sample. The test levels are varied to take account of design assumptions, variations in production processes and the operating environment.

Products may not be put to their design limits for some time after their launch into service, probably far beyond the warranty period. Customer complaints may appear years after product launch. When investigated this may be traced back to a design fault which was not tested for during the verification programme. Such things as corrosion, insulation, resistance to wear, chemicals, climatic conditions etc. need to be verified as being within the design limits. The design validation process should follow the same pattern as addressed previously under design verification.

Following qualification tests, your customer may require a demonstration of performance in order to accept the design. These tests are called design acceptance tests. They usually consist of a series of functional and environmental tests taken from the qualification test specification supported by the results of the qualification tests. When it has been demonstrated that the design meets all the specified requirements, a Design Certificate can be issued. The design standard that is declared on this certificate is the standard against which all subsequent changes should be controlled and from which production versions should be produced.

Demonstrations

Tests exercise the functional properties of the product. Demonstrations on the other hand, serve to exhibit usage characteristics such as access, maintainability including interchangeability, repairability and serviceability. Demonstrations can be used to prove safety features such as the crash tests filmed at high speed. When the film is played at normal speed, the crumpling of the steel and movement of the dummy against the air bag show up characteristics that prove whether the safety features behave as intended. However, one of the most important characteristics that need to be demonstrated is producibility. Can you actually make the product economically in the quantities required? Does production yield a profit or do you need to produce 50 to yield 10 good ones? The demonstrations should establish whether the design is robust. Designers may be selecting components at the outer limits of their capability. A worst-case analysis should have been performed to verify that under worst-case conditions, i.e., when all the components fitted are at the extreme of their tolerance range, the product will perform to specification. Analysis may be more costly to carry out than a test and by assembling the product with components at their tolerance limits you may be able to demonstrate economically the robustness of the design.

Product Approval

A product approval process is often required in large-scale production situations such as the automotive and domestic appliances sectors. When one considers the potential risk involved in assembling unapproved products into production models, it is hardly surprising that the customers impose such stringent requirements. The process provides assurance that the product meets all design criteria and is capable of production in the qualities required without unacceptable variation. It is intended to validate that products made from production materials, tools and processes meet the customer's engineering requirements and that the production process has the potential to produce product meeting these requirements during an actual production run at the quoted production rate.

The process commences following design and process verification during which a production trial run using production standard tooling, suppliers, materials etc. produces the information needed to make a submission for product approval. Until approval is granted, shipment of production product may not be authorized. If any of the processes change then a new submission is required. Shipment of parts produced to the modified specifications or from modified processes would not be authorized until customer approval is granted.

Timing of Validation (7.3.6)

The standard requires validation *to be completed wherever applicable prior to the delivery or implementation of the product.*

What Does this Mean?

As indicated previously, validation trials may take some considerable time but until confidence in the capability of a design to fulfil intended use requirements is known, any decision to launch into production or into operation involves risk. The requirement, however, recognizes that it may not be possible or practical to hold production until all the results of validation have been obtained and assessed.

Why is this Necessary?

This requirement responds to the Systems Approach Principle.

There are some characteristics such as safety and reliability that need to be demonstrated before launching into production otherwise unsafe or unreliable products might be put onto the market. One has only to scan the recall programmes accessible on the Internet to notice that many products are indeed launched into production with major faults (see Appendix B). Some failures may be due to the quality of conformity but there are also some that are due to design weaknesses that should have been detected in the verification and validation programmes.

It is thought that in the USA alone, there are 30 million product recalls every year. Probably the biggest recall of all time occurred in April 1996 when Ford USA recalled up to 9 million vehicles that may have been equipped with a faulty ignition switch. In July 1999, General Motors USA recalled 1.1 million vehicles that may have had anti-lock brake problems. Launching into production without sufficient evidence that the decision satisfies all stakeholders can therefore be very costly.

How is this Demonstrated?

The simplest approach is to wait until all the evidence from verification and validation trials have been assessed before launching into production or going operational. In practice it depends on knowing what the risks are and therefore is a balance between risk and the impact any delay in production launch or going operational may have. It would therefore be prudent to conduct a risk assessment in such circumstances. However, it should be noted that there is no mean time between failure (MTBF) until you actually have a failure, so you need to keep on testing until you know anything meaningful about the product's reliability.

Recording Results. (7.3.6)

The standard requires *the results of the validation and subsequent follow-up actions to be recorded.*

What Does this Mean?

The results of validation are similar to those required for verification except that duration of testing and trials is important in quantifying the evidence. The results should not only indicate that the product meets intended use requirements but also satisfies market need. If by the time you have the validation results, the anticipated demand for the product has declined, it might not be prudent to launch into production.

Why is this Necessary?

This requirement responds to the Factual Approach Principle.

The reasons for recording validation results are the same as for verification results but the decision based on the results is far more significant. Going into production or going operational with the wrong product will be disastrous. There have been several examples of this over the years. The Ford Edsel was a classic example as was the Sinclair C5 motorized tricycle. There was no market for such products and sales did not materialize in the quantities anticipated.

How is this Demonstrated?

As with design verification, consideration of the output, its format and content needs to be given early in the design phase so that the correct data is captured during validation trials.

CONTROL OF DESIGN AND DEVELOPMENT CHANGES (7.3.7)

Identification and Recording of Design Changes (7.3.7)

The standard requires design and development changes to *be identified and records maintained.*

What Does this Mean?

This clause covers two different requirements involving two quite different control processes. Design changes are simply changes to the design and can occur at any stage in the design process from the stage at which the requirement is agreed to the final certification that the product or process design is proven. Development changes can occur at any time in the life cycle of the design that extends until the product or process is obsolete. Following design certification, changes to the product or process to incorporate design changes are generally classed as 'modifications'.

Changes to design documents are not design changes unless the characteristics of the product are altered. Changes in the presentation of design information or to the system of measurement (Imperial units to metric units) are not design or development changes.

Why is this Necessary?

This requirement responds to the Process Approach Principle.

You need to control design changes to permit desirable changes to be made and to prohibit undesirable changes from being made. Change control during the design process is a good method of controlling costs and timescales because once the design process has commenced every change will cost time and effort to address. This will cause delays whilst the necessary changes are implemented and provide an opportunity for additional errors to creep into the design. 'If it's not broke don't fix it!' is a good maxim to adopt during design. In other words don't change the design unless it already fails to meet the requirements or you have discovered the requirements to be wrong. Designers are creative people who love to add the latest devices, the latest technologies, to stretch performance and to go on enhancing the design regardless of the timescales or costs. One reason for controlling design changes is to restrain the otherwise limitless creativity of designers in order to keep the design within the budget and timescale.

How is this Demonstrated?

Product Design Changes

The imposition of change control is often a difficult concept for designers to accept. They would prefer change control to commence after they have completed their design rather than before they have started. They may argue that until they have finished there is no design to control. They would be mistaken. Designs proceed through a number of stages (as described previously under the heading *Design reviews*). Once the design requirements have been agreed, any changes in the requirements should be subject to formal control. When a particular design solution is complete and has been found to meet the requirements at a design review, it should be brought under *change control*. Between design reviews the designers should be given complete freedom to derive solutions to the requirements. Between the design reviews there should be no change control on incomplete solutions.

Design changes will result in changes to documentation but not all design documentation changes are design changes. This is why design change control should be treated separately from document control. You may need to correct errors in the design documentation and none of these may materially affect the product. The mechanisms you employ for such changes should be different from those you employ to make changes that do affect the design. By keeping the two types of change separate you avoid bottlenecks in the design change loop and only present the design authorities with changes that require *their* expert judgement.

Identifying and Recording Product Design Changes

The documentation for design changes should comprise the change proposal, the results of the evaluation, the instructions for change and traceability in the changed documents to the source and nature of the change. You will therefore need:

- A Change Request form which contains the reason for change and the results of the evaluation – this is used to initiate the change and obtain approval before being implemented. An example of a design change proposal is shown in Fig. 25-5.
- A Change Notice that provides instructions defining what has to be changed – this is issued following approval of the change as instructions to the owners of the various documents that are affected by the change. A change Notice is probably unnecessary for process changes.
- A Change Record that describes what has been changed – this usually forms part of the document that has been changed and can be either in the form of a box at the side of the sheet (as with drawings) or in the form of a table on a separate sheet (as with specifications). For processes, the change record could be incorporated into the document that describes the process be it a specification, flow chart, or control plan.

Where the evaluation of the change requires further design work and possibly experimentation and testing, the results for such activities should be documented to form part of the change documentation.

At each design review a design baseline should be established which identifies the design documentation that has been approved. The baseline should be recorded and change control procedures employed to deal with any changes. These change procedures should provide a means for formally requesting or proposing changes to the design. For complex designs you may prefer to separate proposals from instructions and have one form for proposing design changes and another form for promulgating design changes after approval. You will need a central registry to collect all proposed changes and provide a means for screening those that are not suitable to go before the review board (either because they duplicate proposals already made or because they may not satisfy certain acceptance criteria[①] which you have prescribed). On receipt, the *change proposals* should be identified with a unique number that can be used on all related documentation that is subsequently produced. The change proposal needs to:

- identify the product of which the design is to be changed;
- state the nature of the proposed change;

DESIGN CHANGE PROPOSAL				**DCP/**	

ORIGIN					Class
Name	Company	Dept	Tel:	Date	Priority
System		Subsystem		Item	

SPECIFICATIONS CONCERNED					
Specification	Title			Date	Revision

REASON FOR APPLICATION

[ ] Hazards	[ ] Production costs	[ ] Noncompliance
[ ] Incompatibility	[ ] Maintenance costs	[ ] Other
[ ] Potential failure	[ ] Disposal costs	[ ] Other

CHANGE DESCRIPTION

COST ESTIMATE (+Additions or -Savings)

Phase	Hardware	Software	Tools/test Equipment	Publications	Planning
Development					
Production					
Operation					
Maintenance					

EFFECT OF CHANGE

On customers and suppliers

On other systems/equipments/parts

On programmes

On transportation and distribution networks

DECISION of ENGINEERING CHANGE BOARD

Accepted [ ]	Rejected [ ]	Hold [ ]
Reasons		
Authorized by		Date

FIGURE 25-5 Sample design change proposal.

- identify the principal requirements, specifications, drawings or other design documents which are affected by the change;
- state the reasons for the change either directly or by reference to failure reports, nonconformity reports, customer requests or other sources; and
- provide for the results of the evaluation, review and decision to be recorded.

Identifying and Recording Product Modifications

As *modifications* are changes to products resulting from design changes, the identity of modifications needs to be visible on the product that has been modified. If the issue status of the product specification changes, you will need a means of determining whether the product should also be changed. Not all changes to design documentation are design changes that result in product changes and not all product changes are modifications. (Nonconformities may be accepted which change the product but not the design.) Changes to the drawings or specifications that do not affect the form, fit or function of the product are usually called 'alterations' and those which affect form, fit or function are 'modifications'. Alterations should come under 'document control' whereas design changes should come under 'configuration control'[①]. You will therefore need a mechanism for relating the modification status of products to the corresponding drawings and specifications. Following commencement of production, the first design change to be incorporated into the product will usually be denoted by a number such as Mod 1 for hardware and by Version or Release number for software. The practices for software differ in that versions can be incremented by points such as 1.1, 1.2 etc. where the second digit denotes a minor change and the first digit a major change. This modification notation relates to the product whereas, issue notation relates to the documentation that describes the product. You will need a modification procedure that describes the notation to be used for hardware and software.

Within the design documentation you will need to provide for the attachment of modification plates on which to denote the modification status of the product.

Prior to commencement of production, design changes do not require any modification documentation, the design changes being incorporated in prototypes by rework or rebuild. However, when product is in production, instructions will need to be provided so that the modification can be embodied in the product. These modification instructions should detail:

- The products that are affected by part number and serial number.
- The new parts that are required.
- The work to be carried out to remove obsolete items and fit new items or the work to be carried out to salvage existing items and render them suitable for modification.
- The markings to be applied to the product and its modification label.
- The tests and inspections to be performed to verify that the product is serviceable.
- The records to be produced as evidence that the modification has been embodied.

Modification instructions should be produced after approval for the change that has been granted and should be submitted to the change control board or design authority for approval before release.

Review and Evaluation of Changes (7.3.7)

The standard requires *the changes to be reviewed and approved before implementation, including the evaluation of the effect of changes on constituent parts and delivered product.*

What Does this Mean?

A change to a design that has not proceeded beyond a design review or verification stage is still in progress and therefore requires no approval. When a design is reviewed or verified it means that any change to the information on which that decision was taken needs to be evaluated for its effect on the design and any product produced from that design which may be in production or in service.

Why is this Necessary?

This requirement responds to the Factual Approach Principle.

By controlling change you control cost so it is a vital organ of the business and should be run efficiently. The requirement for changes to be approved before their implementation emphasizes the importance of this control mechanism. The requirement for evaluation of changes for impact on product in service is necessary to transfer any benefits from the change to customers. This is especially important if the reason for change was to improve safety, reliability or regulatory compliance.

How is this Demonstrated?

Following the commencement of design you will need to set up a change control board or panel comprising those personnel responsible for funding the design, administering the contract and for accepting the product. All change proposals should be submitted to such a body for evaluation and subsequent approval or disapproval before the changes are implemented. Such a mechanism will give you control of all design changes. By providing a two-tier system you can also submit all design documentation changes through such a body. They can filter the alterations from the modifications and the minor changes from the major changes.

The change proposals need to be evaluated:

- to validate the reason for change;
- to determine whether the proposed change is feasible;
- to judge whether the change is desirable;
- to determine the effects on performance, costs and timescales;
- to determine the impact of the change on other designs with which it interfaces and in which it is used;
- to examine the documentation affected by the change and consequently programme their revision; and
- to determine the stage at which the change should be embodied.

The evaluation may need to be carried out by a review team, by suppliers or by the original proposer, however, regardless of who carries out the evaluation, the results should be presented to the change control board for a decision.

During development there are two decisions the board will need to make:

- whether to accept or reject the change; and
- when to implement the change in the design documentation.

If the board accepts the change, the changes to the design documentation can either be submitted to the change control board or processed through your document control procedures. With CAD systems there is no reason why changes cannot be incorporated immediately following their approval. One does not need to accumulate design changes for incorporation into the design when design validation has been completed.

During production the change control board will need to make four decisions:

- Whether to accept or reject the change;
- When to implement the change in the design documentation;
- When to implement the modification in new product; and
- What to do with existing product in production, in store and in service.

The decision to implement the modification will depend on when the design documentation will be changed and when new parts and modification instructions are available. The modification instructions can either be submitted to the change control board or through your document control procedures. The primary concern of the change control board is not so much the detail of the change but its effects, its costs and the logistics in its embodiment. If the design change has been made for safety or environmental reasons you may need to recall product in order to embody the modification. Your modification procedures need to provide for all such cases. With safety issues, there may be regulatory procedures that need to be implemented to notify customers, recall product, implement modifications and to release modified product back into service.

Verification and Validation of Design Changes (7.3.7)

The standard requires design changes *to be verified and validated as appropriate before implementation*.

What Does this Mean?

Design changes are no different than original designs in that they need to be verified and validated.

Why is this Important?

This requirement responds to the Factual Approach Principle.

When a change is made to an approved design, any previous review, verification and validation may be invalidated by the change. Re-verification and validation may therefore be necessary.

How is this Demonstrated?

Depending on the nature of the change, the verification may range from a review of calculations to a repeat of the full design verification programme. The changes may occur before the design has reached the validation phase and therefore not warrant any change to the validation programme. It is therefore necessary when evaluating a design change to determine the extent of any verification and validation that may need to be repeated. Some design changes warrant being treated as projects in their own right, recycling the full design process. Other changes may warrant verification on samples only or verification may be possible by an analysis of the differences with a proven design.

In some cases the need for a design change may be recognized during production tests and in order to define the changes required you might wish to carry out trial modifications or experiments. Any changes to the product during production should be carried out under controlled conditions, hence the requirement that approval of design changes be given before their implementation. To allow such activities as trial modifications and experiments to proceed you will need a means of controlling these events. If the modification can be removed in a way that will render the production item in no way degraded, you can impose simple controls for the removal of the modification. If the item will be rendered unserviceable by removing the modification, alternative means may need to be determined otherwise you will sacrifice the product. It is for this reason that organizations provide development models on which to try out modifications.

Chapter 26

Purchasing

CHAPTER PREVIEW

This chapter is aimed at those personnel at the supplier interface who are responsible for or involved with the purchasing of products and services that directly or indirectly impact the end product supplied to customers. This may include purchasing managers, designers, project managers, supplier quality assurance and those involved with supplier development and goods inwards.

Most organizations depend on the support of other organizations to produce product that satisfies their customers. This support takes many forms but typically includes the supply of materials, components, semi-finished products, equipment, labour, professional services and laboratory services. Whether embodied in the end product or simply used in its design, development, production or delivery, the quality of these supplies will impact the quality of the product provided to customers.

In this chapter we examine the requirements in Clause 7.4 of ISO 9001:2008 and in particular:

- The provisions needed to ensure purchased product conforms to specified requirements;
- Control of suppliers;
- Evaluation and selection of suppliers;
- Results of supplier evaluation;
- Documents describing products to be purchased;
- Adequacy of purchasing requirements;
- The provisions needed to verify purchased product meets requirements; and
- Legitimizing verification on supplier's premises.

The position where the requirements on purchasing feature in the managed process is shown in Fig. 26-1. Due to the structure of the standard, activities normally part of purchasing such as the activity of purchasing itself (the standard does not actually require anything to be purchased) is excluded. The demand on the process is from interfacing processes requiring resources that need to be purchased and the serviced demand is purchased resource that satisfies that demand.

PURCHASING PROCESS (7.4.1)

Ensuring Purchased Product Conforms to Specified Requirements (7.4.1)

The standard requires *the organization to ensure that purchased product conforms to specified purchase requirements.*

ISO 9000 Quality Systems Handbook
Copyright © 2009, David Hoyle. Published by Elsevier Ltd. All rights reserved.

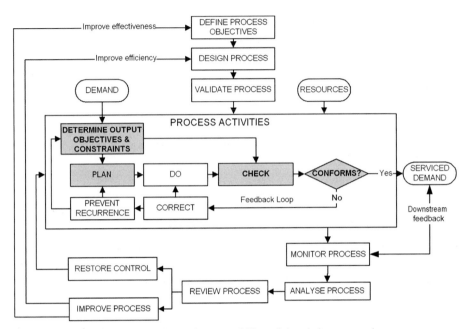

FIGURE 26-1   Where the requirements of Clause 7.4 apply in a managed process.

What Does this Mean?

ISO 9000:2005 defines a supplier as an organization or person that provides a product and in ISO 9001 a product can be services, hardware, software or processed materials. A supplier may therefore be a producer, distributor, retailer, vendor, contractor, subcontractor or service provider. Purchased product includes any product or service that is purchased, freely given or otherwise acquired that will in any way have an effect on conformity of the deliverable product to customers (confirmed by published interpretation RFI 042 & 047). Specified purchase requirements are those requirements that are specified by the customer, the organization or by statutes and regulations that apply to purchased product. This would include any requirements limiting the conditions or the source of supply.

Why is this Necessary?

This requirement responds to the Factual Approach Principle.

   All organizations have suppliers of one form or another in order to provide products and service to their customers. Some of them directly or indirectly impact the product being supplied to the organization's customers and others may have no impact at all such as office supplies. From the scope of the standard we draw the conclusion that the requirement is not intended to apply to products and service that have no impact on the customer but why would you not want to manage such purchasing activities as effectively as other purchasing activities? This is not to say that one should apply the same controls and constraints to all purchases but one needs to apply those controls and constraints that are appropriate to the use of purchased product.

How is this Demonstrated?

Once the make or buy decision has been made, control of any purchasing activity follows a common series of activities.

There are four key processes in the procurement cycle which should be managed effectively.

- The specification process which starts once the need has been identified and ends with a request to purchase.
- The evaluation process which starts following the request to purchase and ends with the placement of the order or contract.
- The surveillance process which starts following placement of order or contract and ends on delivery of supplies.
- The acceptance process which starts following delivery of supplies and ends with entry of supplies onto the inventory and/or payment of invoice.

Whatever you purchase the processes will be very similar although there will be variations for purchased services such as subcontract labour, computer maintenance, consultancy services etc. Where the purchasing process is relatively simple, one route may suffice but where the process varies you may need separate routes so as to avoid *all* purchases, regardless of value and risk, going through the same process and incurring unnecessary costs and delay. A typical procurement process flow is illustrated in Fig. 26-2.

Control of Suppliers (7.4.1)

The standard requires *the type and extent of control applied to the supplier and the purchased product to be dependent on the effect of the purchased product on subsequent product realization or the final product.*

What Does this Mean?

The requirement contains two quite separate requirements – one applying to the product and the other applying to the supplier.

Regarding the product, the type of control refers to whether the controls should act before, during or after receipt of product. The extent of control refers to whether it is remote or on the suppliers' premises and whether product is accepted on the basis of supplier data or is to be evaluated before authorizing delivery.

Regarding the supplier the type of control refers to whether or not to qualify the supplier and the extent of control refers to the degree to which the organization is involved with the supplier in managing the purchase.

Purchased product can have varying degrees of impact on the processes and products of the organization ranging from no impact to critical impact. A product with a critical impact would warrant stringent control over its purchase, whereas a product with negligible or no impact may warrant no more than a simple visual check on delivery to verify receipt of the right product.

Why is this Necessary?

This requirement responds to the Process Approach Principle.

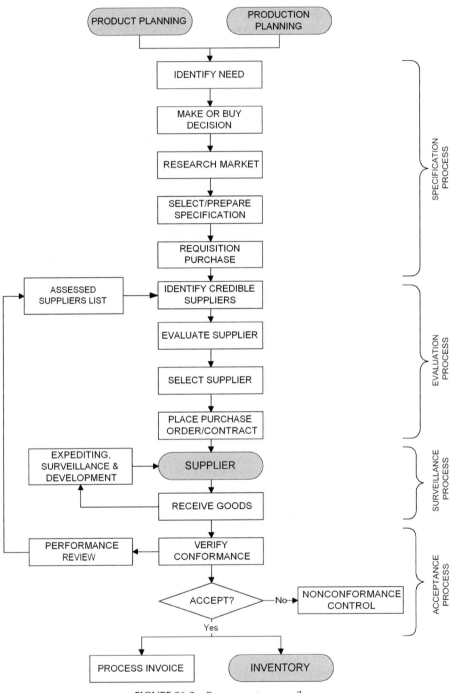

FIGURE 26-2 Procurement process flow.

As it would not be prudent to exercise no control over suppliers, it would also be counterproductive to impose rigorous controls over every supplier and purchased product. A balance has to be made on the basis of risk to the processes in which the purchased product is to be used and the final product into which the purchased product may be installed.

How is this Demonstrated?

Selecting the Degree of Control

You need some means of verifying that the supplier has met the requirements of your order and the more unusual and complex the requirements, the more control will be required. The degree of control you need to exercise over your suppliers depends on the confidence you have in their ability to meet your requirements. If you have high confidence in a particular supplier you can concentrate on the areas where failure is more likely. If you have no confidence you will need to exercise rigorous control until you gain sufficient confidence to relax the controls. The fact that a supplier has gained ISO 9001 registration for the products and services you require should increase your confidence, but if you have no previous history of their performance it does not mean they will be any better than the supplier you have used for years which is not registered to ISO 9001. Your purchasing process needs to provide the criteria for selecting the appropriate degree of control and for selecting the activities you need to perform.

With suppliers of proprietary products, your choices are often limited because you have no privileges. Control over your suppliers is therefore exercised by the results of receipt inspection or subsequent verification. If your confidence in a supplier is low you can increase the level of verification and if high you can dispense with receipt verification and rely on in-process controls to alert you to any deterioration in supplier performance.

In determining the degree of control to be exercised you need to establish the nature and degree of risks and avoid imposing requirements which would be too burdensome in relation to the risks concerned. There are several options available from low to high risk. Imposing ISO 9001 on your suppliers is not necessarily the best solution in all situations.

Quality Assurance by Product Verification on Receipt Using Standard Resources The quality of the product or service can be verified on receipt using your normal inspection and test techniques. Sampling inspection on receipt should be used when statistical data is unavailable to you or you don't have the confidence for permitting ship-to-line. This is the least costly of methods and usually applies where achievement of the requirements is measurable by examination of the end product in your facilities. You would not normally impose ISO 9001 on such suppliers until you had confidence to remove verification on receipt.

Quality Assurance by Product Verification on Receipt Using Additional Resources The quality of the product can be verified on receipt providing you acquire additional equipment or facilities. This is more costly than the previous method but may be economic if there is high utilization of the equipment. You would not normally impose ISO 9001 on such suppliers until you had confidence to remove verification on receipt or if the burden of receipt verification because too onerous and the next level was a not practical for some reason. This is equivalent to Modules E in the EU Conformity Assessment legislation.

Quality Assurance by Product Verification on Supplier's Premises   The quality of the product can be verified by witnessing the final acceptance on the suppliers' premises. If you don't possess the necessary equipment or skill to carry out product verification, this method may be an economic compromise and should yield as much confidence in the product as the previous methods. You do, however, need to recognize that your presence on the supplier's premises may affect the results. They may omit tests that are problematical or your presence may cause them to be particularly diligent, a stance that may not be maintained when you are not present. This is a level where ISO 9001 applies but may not be necessary if full confidence can be gained by being on site.

Quality Assurance by Product Verification by Third Party   The verification of the product could be contracted to a third party such as a part evaluation laboratory. This can be very costly and is usually only applied with highly complex products and where safety is of paramount importance. The supplier should be provided the option of seeking ISO 9001 certification or paying for the independent product verification. This is equivalent to Modules B and C in the EU Conformity Assessment legislation.

Quality Assurance by Control of Production, Installation and Servicing   The product characteristics are such that they cannot be verified by examination of the end product alone and can only be verified by having complete confidence in the suppliers manufacturing, installation and services processes. Such is the case if nonconformity of critical characteristics is unlikely to be detected before use. In such cases you need to rely on what the contractor tells you. To gain sufficient confidence you can impose quality management system requirements, require certain manufacturing and verification documents to be submitted to you for approval and carry out periodic audit and surveillance activities either directly of using third parties. This is a level where ISO 9001 can be invoked in contracts supplemented by customer specific requirements as necessary. This is equivalent to Module D in the EU Conformity Assessment legislation.

Quality Assurance by Control of Design   The product characteristics are such that they cannot be verified by control of production alone and can only be verified by having complete confidence in the suppliers design processes. In such cases you need to rely on what the contractor tells you and to gain sufficient confidence you can impose quality management system requirements, require certain design and verification documents to be submitted to you for approval and carry out periodic audit and surveillance activities either directly or using third parties. This is a level where ISO 9001 can be invoked in contracts supplemented by customer specific requirements as necessary. This is equivalent to Modules H in the EU Conformity Assessment legislation.

In order to relate the degree of inspection to the importance of the item, you should categorize purchases as follows:

- If the subsequent discovery of nonconformity before use will *not* cause design, production, installation or operational problems of any nature, a simple identity, carton quantity and damage check may suffice. An example of this would be mechanical fasteners.

- If the subsequent discovery of nonconformity before use will cause *minor* design, production, installation or operational problems, you should examine the features and characteristics of the item on a sampling basis. An example of this would be electrical, electronic or mechanical components.
- If the subsequent discovery of nonconformity before use will cause *major* design, production, installation or operational problems then you should subject the item to a complete test to verify compliance with all prescribed requirements. An example of this would be an electronic unit.

These criteria would need to be varied depending on whether the items being supplied were in batches or separate. However, these are the kinds of decisions you need to take in order to apply practical receipt verification procedures.

Defining Supplier Controls

When carrying out supplier surveillance you will need a plan which indicates what you intend to do and when you intend to do it. You will also need to agree the plan with your supplier. If you intend witnessing certain tests the supplier will need to give you advanced warning of commencing such tests so that you may attend.

The quality plan would be a logical place for such controls to be defined. Some companies produce a Quality Assurance Requirement Specification to supplement ISO 9001 and also produce a Supplier Surveillance Plan. In most other cases the controls may be defined on the reverse side of the purchase order as standard conditions coded and selected for individual purchases. However, don't impose onerous requirements on simple purchases. Requiring test samples for literature you purchase or ISO 9001 certification from a bookseller is rather ludicrous. Make provision for the relevant conditions to be selected by the buyer otherwise you run the risk of suppliers ignoring requirements that might be relevant.

Evaluation and Selection of Suppliers (7.4.1)

The standard requires the organization *to evaluate and select suppliers based on their ability to supply product in accordance with the organization's requirements and to establish criteria for selection, evaluation and re-evaluation.*

What Does this Mean?

In searching for a supplier you need to be confident that the supplier can provide the product or service you require. This means that the decision to select a supplier should be based on knowledge about that supplier's capability to meet your requirements. The decision should be based on facts gathered as a result of an evaluation against criteria that you have established.

Why is this Necessary?

This requirement responds to the Process Approach Principle.

It would be foolish to select a supplier without first verifying that it was able to meet your requirements in some way or other. Failure to check out the supplier and its products may result in late delivery of the wrong product. It may also mean that you might not know immediately that the product does not meet your requirement and discover much later that it seriously impacts your commitments to your customer.

How is this Demonstrated?

Selection Process

The process for selection of suppliers varies depending on the nature of the products and services to be procured. The more complex the product or service, the more complex the process. You either purchase products and services to your specification (custom) or to the supplier's (proprietary). For example you would normally procure stationery, fasteners or materials to the *supplier's specification* but procure an oil platform, radar system or road bridge to *your specification*. There are grey areas where proprietary products can be tailored to suit your needs and custom made products or services that primarily consist of proprietary products configured to suit your needs. There is no generic model, each industry seems to have developed a process to match its own needs. However, we can treat the process as a number of stages some of which do not apply to simple purchases as shown in Table 26-1. At each stage the number of potential suppliers is whittled down to end with the selection of what is hoped to be the most suitable that meets the requirements. With 'custom' procurement this procurement cycle may be exercised several times. For instance there may be a competition for each phase of project feasibility, project definition, development and production. Each phase may be funded by the customer. On the other hand, a supplier may be invited to tender on the basis of previously demonstrated capability but has to execute project feasibility, project definition and development of a new version of a product at its own cost. Supplier capability will differ in each phase. Some suppliers have good design capability but lack the capacity for quantity production, others have good research capability but lack development capability.

TABLE 26-1 Supplier Evaluation and Selection Stages

Stage	Purpose	Proprietary	Tailored	Custom
Preliminary supplier assessment	To select credible suppliers	✓	✓	✓
Pre-qualification of suppliers	To select capable bidders		✓	✓
Qualification of suppliers	To qualify capable bidders			✓
Request for Quotation (RFQ)	To obtain prices for products or services	✓	✓	
Invitation to Tender (ITT)	To establish what bidders can offer			✓
Tender or Quote Evaluation	To select a supplier	✓	✓	✓
Contract Negotiation	To agree terms and conditions	✓	✓	✓

You need to develop a supplier evaluation and selection process and in certain cases this may result in several closely related procedures for use when certain conditions apply. Do not try to force every purchase through the same selection process. Having purchasing policies that require three quotations for every purchase regardless of past performance of the current supplier is placing price before quality. Provide flexibility so that process complexity matches the risks anticipated. Going out to tender for a few standard nuts and bolts would seem uneconomical. Likewise, placing an order for £1 million of equipment based solely on the results of a third party ISO 9001 certification would seem reckless.

Preliminary Supplier Assessment

The purpose of the preliminary supplier assessment is to select a *credible* supplier and not necessarily to select a supplier for a specific purchase. There are millions of suppliers in the world, some of which would be happy to relieve you of your wealth given half a chance, and others that take pride in their service to customers and are a pleasure to have as partners. You need a process for gathering intelligence on potential suppliers and for eliminating unsuitable ones so that the buyers do not need to go through the whole process from scratch with each purchase. The first step is to establish the type of products and services you require to support your business, and then search for suppliers that claim to provide such products and services. In making your choice, look at what the supplier says it will do and what it has done in the past. Is it the sort of firm that does what it says it does or is it the sort of firm that says what you want to hear and then conducts its business differently? Some of the checks needed to establish the credibility of suppliers are time consuming and would delay the selection process if undertaken only when you have a specific purchase in mind. You will need to develop your own criteria but typically unsuitable suppliers may be those that:

- are unlikely to deliver what you want in the quantities you may require;
- are unable to meet your potential delivery requirements;
- cannot provide the after-sales support needed;
- are unethical;
- do not comply with the health and safety standards of your industry;
- do not comply with the relevant environmental regulations;
- do not have a system to assure the quality of supplies;
- are not committed to continuous improvement; and
- are financially unstable.

You may also discriminate suppliers on political grounds such as a preference for supplies from certain countries or a requirement to exclude supplies from certain countries.

The supplier assessment will therefore need to be in several parts:

Technical assessment This would check the products, processes or services to establish that they are what the supplier claims them to be. Assessment of design and production capability may be carried out at this stage or be held until the pre-qualification stage when specific contracts are being considered.

Quality system assessment This would check the certification status of the management system, verifying that any certification was properly accredited. For non-ISO 9001 registered suppliers, a quality management system assessment may be carried out at this stage either to ISO 9001 or the customer's standards.

Financial assessment This would check the credit rating, insurance risk, stability etc.

Ethical assessment This would check probity, conformance with professional standards and codes.

These assessments do not need to be carried out on the supplier's premises. Much of the data needed can be accumulated from a supplier questionnaire and searches through directories and registers of companies. You can choose to rely on assessments carried out by accredited third parties to provide the necessary level of confidence. (The Directories of Companies of Assessed Capability that are maintained by the Accreditation Agencies can be a good place to start.) The assessments may yield suppliers over a wide range and you may find it beneficial to classify suppliers as follows:

Class A   ISO 9001 certified and demonstrated capability – This is the class of those certified suppliers with which you have done business for a long time and gathered historical evidence that proves their capability.

Class B   Demonstrated capability – This is the class of those suppliers you have done business with for a long time and warrant continued patronage on the basis that it's better to deal with those suppliers you know than those you don't. They may not even be contemplating ISO 9001 certification, but *you* get a good product, a good service and no hassle.

Class C   ISO 9001 certified and no demonstrated capability – This is the class of those certified suppliers with whom you have done no business. This may appear a contradiction because ISO 9001 certification is obtained on the basis of demonstrated capability, but you have not established their capability to meet *your* requirements.

Class D   Capable with additional assurance – This is the class of first time suppliers with which you have not done sufficient business to put in class B and where you may need to impose ISO 9001 requirements or similar to gain the confidence you need.

Class E   Unacceptable performance that can be neutralized – This class is for those cases where you may be able to compensate for poor performance if they are sole suppliers of the product or service.

Class F   No demonstrated capability – This is the class of those suppliers you have not used before and therefore have no historical data.

Class G   Demonstrated unacceptable performance – This is the class of those suppliers that have clearly demonstrated that their products and services are unacceptable and it is uneconomic to compensate for their deficiencies.

Caution is advised on the name you give to this list of suppliers. All have been assessed (even if it is limited to a literature survey) but all may not have been visited or used. Some organizations refer to it as an Approved Suppliers List (ASL) or Assessed Vendor List (AVL) but if you include unacceptable suppliers you cannot call it an ASL. If the data is stored in a database, the fields can be protected to prevent selection of unacceptable suppliers.

If your requirements vary from product to product, suppliers approved for one product may not be approved for others. If your procurement requirements do not vary from product to product, you may well be able to maintain an AVL. Most will meet your minimum criteria for doing business with your company but may not be capable of meeting specific product or service requirements. Others you will include simply because they do supply the type of product or service you require but their credibility is too low at present to warrant preferred status. In the process you have eliminated the 'cowboys' or 'rogues' – there is no point in adding these to the list because you have established that they won't change in the foreseeable future.

Pre-Qualification of Suppliers

Pre-qualification is a process for selecting suppliers for known future work. The design will have proceeded to a stage where an outline specification of the essential parameters has been developed. You know roughly what you want but not in detail. Pre-qualification is undertaken to select those suppliers that can demonstrate that they have the capability to meet your specific requirements on quality, quantity, price and delivery. A supplier may have the capability to meet quality, quantity and price requirements but not have the capacity available when you need the product or service. One that has the capacity may not offer the best price and one that meets the other criteria may not be able to supply product in the quantity you require.

A list of potential bidders can be generated from the ASL by searching for suppliers that match given input criteria specific to the particular procurement. However, the evidence you gathered to place suppliers on your ASL may now be obsolete. Their capability may have changed and therefore you need a sorting process for specific purchases. If candidates are selected that have not been assessed, an assessment should be carried out before proceeding any further.

Once the list is generated a Request for Quotation (RFQ) or Invitation to Tender (ITT) can be issued depending on what is required. RFQs are normally used where price only is required. This enables you to disqualify bidders offering a price well outside your budget. ITTs are normally used to seek a line-by-line response to technical, commercial and managerial requirements. At this stage you may select a number of potential suppliers and require each to demonstrate its capability. You know what they do but you need to know if they have the capability of producing a product with specific characteristics and can control its quality.

When choosing a bidder you also need to be confident that continuity of supply can be assured. One of the benefits of ISO 9001 certification is that it should demonstrate that the supplier has the capability to supply certain types of products and services. However, it is not a guarantee that the supplier has the capability to meet *your* specific requirements. Suppliers that have not gained ISO 9001 certification may be just as good. If the product or service you require can only be obtained from a non-registered contractor. Using an ISO 9001 registered supplier should enable you to reduce your supplier controls, so by using a non-ISO 9001 registered supplier you will need to compensate by performing more quality assurance activities yourself, or employ a third party to do so.

Depending on the nature of the work you may require space models, prototypes, process capability studies, samples of work as evidence of capability. You may also make a preliminary visit to each potential bidder but would not send out an evaluation team until the qualification stage.

Qualification of Suppliers

Of those potential bidders that are capable, some may be more capable than others. Qualification is a stage executed to compile a short list of bidders following pre-qualification. A detail specification is available at this stage and production standard models may be required to qualify the design. Some customers may require a demonstration of process capability to grant production part approval.

During this stage of procurement a series of meetings may be held depending on the nature of the purchase. A pre-bid meeting may be held on the customer's premises to enable the customer to clarify the requirements with the bidders. A mid-bid meeting or pre-award assessment may be held on the supplier's premises at which the customer's Supplier Evaluation Team carries out a capability assessment on site. This assessment may cover:

- an evaluation of the product;
- an audit of design and production plans to establish that, if followed, they will result in compliant product;
- an audit of operations to verify that the approved plans are being followed;
- an audit of processes to verify their capability; and
- an inspection and test of product (on-site or off-site) to verify that it meets the specification.

The result of supplier qualification is a list of capable suppliers that will be invited to bid for specific work.

ISO 9001 certification was supposed to reduce the amount of supplier assessments by customers and it has in certain sectors. However, the ISO 9001 certification whilst focused on a specific scope of registration is often not precise enough to give confidence to customers for specific purchases.

The evaluation may qualify two or three suppliers for a specific purchase. The tendering process will yield only one winner but the other suppliers are equally suitable and should not be disqualified because they may be needed if the chosen supplier fails to deliver.

Invitation to Tender

Once the bidders have been selected, an Invitation to Tender (ITT) needs to be prepared to provide a fixed baseline against which unbiased competitive bids may be made. The technical, commercial and managerial requirements should be finalized and subject to review and approval prior to release. It is important that all functions with responsibilities in the procurement process review the tender documentation. The ITT will form the basis of any subsequent contract.

The requirements you pass to your bidders need to include as appropriate:

- The tender conditions, date, format, content etc.
- The terms and conditions of the subsequent contract.
- A specification of the product or service that you require that transmits all of the relevant requirements of the main contract (see *Purchasing specifications*).
- A specification of the means by which the requirements are to be demonstrated (see *Purchasing specifications*).
- A statement of work which you require the supplier to perform – It might be design, development, management or verification work and will include a list of required

deliverables such as project plans, quality plans, production plans, drawings, test data etc. (You need to be clear as to the interfaces both organizationally and technically.)

- A specification of the requirements which will give you an assurance of quality – This might be a simple reference to ISO 9001, but as this standard does not give you any rights or flow down your customer's requirements you will probably need to amplify the requirements (see *Quality management system requirements*).

In the tendering phase each of the potential suppliers is in competition, so observe the basic rule that what you give *one* must be given to *all*! It is at this stage that your supplier conducts the tender review defined in Clause 7.2.2 of ISO 9001.

Tender or Quote Evaluation

On the due date when the tenders should have been received, record those that have been submitted and discard any submitted after the deadline. Conduct an evaluation to determine the winner – the supplier that can meet all your requirements (including confidence) for the lowest price. The evaluation phase should involve all your staff that were involved with the specification of requirements. You need to develop scoring criteria so that the result is based on objective evidence of compliance.

The standard does not require that you only purchase from 'approved suppliers'. It does require that you maintain records of the results of supplier evaluations but does not prohibit you from selecting suppliers that do not fully meet your purchasing require-ments. There will be some suppliers that fully meet your requirements and others that provide a product with the right functions but quality, price and delivery may be less than you require. If the demonstrated capability is lacking in some respects you can adjust your controls to compensate for the deficiencies. Obviously not a preferred option but you may have no choice. Deming advocated in the 4th of his 14 points the orga-nizations should '*Move toward a single supplier for any one item, on a long-term relationship of loyalty and trust*'.

In some cases your choice may be limited to a single source because no other supplier may market what you need. On other occasions you may be spoilt for choice. With some proprietary products you are able to select particular options so as to tailor the product or service to your requirements. It remains a proprietary product because the supplier has not changed anything just for you. The majority of products and services you will purchase from suppliers, however, is likely to be from catalogues. The designer may have already selected the item and quoted the part number in the specification. Quite often you are buying from a distributor rather than the manufac-turer and so need to ensure that *both* the manufacturer and the distributor will meet your requirements.

Contract Negotiation

After selecting a 'winner' you may need to enter contract negotiations in order to draw up a formal subcontract. It is most important that none of the requirements are changed without the supplier being informed and given the opportunity to adjust the quotation. It is at this stage that your supplier conducts the requirement review defined in Clause 7.2.2 of ISO 9001. It is pointless negotiating the price of products and services that do not meet your needs. You will just be buying a heap of trouble! Driving down the price may also result in the supplier selling their services to the highest bidder later and leaving you high and dry!

Satisfying Regulatory Requirements

The first step in meeting this requirement is to establish a process that will identify all current regulatory requirements pertaining to the part or material. You need to identify the regulations that apply in the country of manufacture and the country of sale. This may result in two different sets of requirements. For example a part may be manufactured in Mexico and sold in California or made in UK and sold in India. In one case the regulations on recycling materials may be tougher in the country of sale and in the other case, there may be restrictions prohibiting sale of vehicles containing materials from a particular country. It is difficult to keep track of changes in import and export regulations but using the services of a legal department or agency will ease the burden.

In order to ensure compliance with this requirement you need to impose on your suppliers through the purchase order, the relevant regulations and through examination of specifications, products and by on-site assessment, verify that these regulations are being met. It is not sufficient to merely impose the requirement on your supplier through the purchase order. You can use the certified statements of authorized independent inspectors as proof of compliance instead of conducting the assessment yourself. However, such inspections may not extend to the product being supplied and therefore a thorough examination by your technical staff will be needed. Once deemed compliant, you need to impose change controls in the contract that prohibit the supplier changing the process or the product without your approval. This may not be possible when dealing with suppliers supplying product to their specification or when using offshore suppliers where the system of law enforcement cannot be relied on. In such cases you will need to accurately define the product required and carry out periodic verification for continued compliance.

Criteria for Periodic Evaluation

For one-off purchases periodic re-evaluation would not be necessary. Where a commitment from both parties is made to supply products and services continually until terminated, some means of re-evaluation is necessary as a safeguard against deteriorating standards.

The re-evaluation may be based on supplier performance, duration of supply, quantity, risk or changes in requirements and conducted in addition to any product verification that may be carried out. Suppliers are no different than customers in that their performance varies over time. People, organizations and technologies change and may impact the quality of the service obtained from suppliers. The increasing trend for customers to develop partnerships with suppliers has led to supplier development pro-grammes where customers work with suppliers to develop their capability to improve process capability, delivery schedules or reduce avoidable costs. These programmes replace re-evaluations because they are ongoing and any deterioration in standards is quickly detected. In addition the effect of recent mergers, acquisitions and affiliations on the effectiveness of the quality system should be verified.

Results of Supplier Evaluation (7.4.1)

The standard requires *the records of evaluations and any necessary follow-up actions to be maintained.*

What Does this Mean?

Records of evaluations are documents containing the results of the evaluation. This is not the Approved Supplier List or AVL used to select suppliers but the objective evidence that was used to make the decision as to whether a supplier should be listed in such a document.

Why is this Necessary?

This requirement responds to the Factual Approach Principle.

Records of supplier evaluation are necessary in order to select suppliers on the basis of facts rather than opinion. They are also necessary for comparisons between competing suppliers as a mere listing provides little information on which to judge acceptability. Even in small organizations these records are required according to published interpretation RFI 046. However, this appears to conflict with the requirement in Clause 4.2.1d where it requires the quality management system to include records, determined by the organization to be necessary to ensure the effective planning, operation and control of its processes. Therefore if you can demonstrate you can control your suppliers without records of evaluations and any necessary follow-up actions there appears to be a good case for not maintaining any records in this situation.

How is this Demonstrated?

Although records of evaluations are not the same as a list of approved suppliers, there is a need for both. The list identifies which suppliers have been or not been evaluated and the records support the decision to include suppliers in the list or qualify suppliers in different categories.

Evaluation Records

Evaluation records can be classified in three groups:

- Initial evaluation for supplier selection;
- Supplier performance monitoring; and
- Re-evaluation to confirm approval status.

The initial evaluation records would include the evaluation criteria, the method used, the results obtained and the conclusions. They may also include information relevant to the supplier such as supplier history, advertising literature, catalogues and approvals. These records may not contain actions and recommendations because the evaluation may have been carried out under a competitive tender. The actions come later, when re-evaluations are performed and continued supply is decided.

You should monitor the performance of all your suppliers and classify each according to prescribed guidelines. Supplier performance will be evident from audit reports, surveillance visit reports and receipt inspections carried out by you or the third party if one has been employed. You need to examine these documents for evidence that the supplier's quality management system is controlling the quality of the products and services supplied. You can determine the effectiveness of these controls by periodic review of the supplier's performance. What some firms call 'vendor rating'. By collecting data on the performance of suppliers over a long period you can measure their effectiveness and rate them on a suitable scale. In such cases you should measure at least three characteristics, quality, delivery and service. Quality would be measured by the ratio of defective products to conforming products. Delivery would be measured by the number of days early or late and service would be measured by the responsiveness to actions requested by you. The output of these reviews should be in the form of updates to the list of assessed suppliers.

Re-evaluation records would include all the same information as the initial evaluation but in addition contain follow-up actions and recommendations, the supplier's response and evidence that any problems have been resolved.

Listing Suppliers

It is important that you record those suppliers that should not be used due to previously demonstrated poor performance so that you don't repeat the mistakes of the past. Assessing suppliers is a costly operation. Having established that a supplier has or hasn't the capability of meeting your requirements you should enter their details in a database. The database should be made available to the purchasing authority thereby avoiding the necessity of re-assessments each time you wish to place an order. The database of assessed suppliers should not only identify the name and address of the company but also provide details of the products and service that have been evaluated. This is important because the evaluation performed to place suppliers on the list will have only covered particular products and services. Other products and services offered by the supplier may not have been acceptable. Some firms operate several production lines each to different standards. A split between military products and civil products is most common. Just because the military line met your requirement doesn't mean that the other production line will also meet your requirements. Calling it a List of Assessed Suppliers does not imply that it only lists approved suppliers – it allows you to include records of all suppliers with which you have done business and classify them accordingly. By linking purchases with the List of Assessed Suppliers you can indicate usage status, e.g., current, dormant or unused.

You will need a procedure for generating and managing the database of Assessed Suppliers adding new suppliers, changing data and reclassifying suppliers that no longer meet your criteria.

PURCHASING INFORMATION (7.4.2)

Describing Products to be Purchased (7.4.2)

The standard requires purchasing information to *describe the product to be purchased*.

What Does this Mean?

Purchasing information is the information that identifies the product or service to be purchased and which is used to make purchasing decisions. Not all of this information may be conveyed to the supplier. Some information may be needed by buyers to select the correct product or service required.

Why is this Necessary?

This requirement responds to the Process Approach Principle.

The supplier needs to know what the organization requires before it can satisfy the need and although the standard does not specifically require the information to be recorded, you need to document purchasing requirements so that you have a record of what you ordered. This can then be used when the goods and the invoice arrive to confirm that you have received what you ordered. The absence of such a record may prevent you from returning unwanted or unsatisfactory goods.

How is this Demonstrated?

The essential purchasing information must be communicated to suppliers so that they know what you require, but it is not essential to submit your purchasing documents to your suppliers. In fact many purchases will be made from catalogues by telephone, quoting reference number and quantity required. Providing you have a record and can compare this with the goods received and the invoice, you are protected against paying for goods you didn't order.

Product Identification

The product or service identification should be sufficiently precise as to avoid confusion with other similar products or services. The supplier may produce several versions of the same product and denote the difference by suffixes to the main part number. To ensure you receive the product you require you need to carefully consult the literature provided and specify the product in the same manner as specified in the literature or as otherwise advised by the supplier.

Purchasing Specifications

If you are procuring the services of a supplier to design and/or manufacture a product or design and/or deliver a service, you will need specifications which detail all the features and characteristics which the product or service is to exhibit. The reference number and issue status of the specifications need to be specified in the event that they change after placement of the purchase order. This is also a safeguard against the repetition of problems with previous supplies. These specifications should also specify the means by which the requirements are to be verified so that you have confidence in any certificates of conformity that are supplied. For characteristics that are achieved using special processes you need to ensure that the supplier employs qualified personnel and equipment. Products required for particular applications need to be qualified for such applications and so your purchasing documents will need to specify what qualification tests[①] are required.

Quality Management System Requirements

Management system requirements are only necessary inclusions in purchasing information when the quality of the product cannot be verified on receipt or when confidence in the product and the supplier is needed to permit the supplier to ship direct into stock or onto the production line.

Management system requirements can be invoked in your purchasing documents whether or not your supplier is registered to a management system standard, but doing so may cause difficulties. If the firm is not registered they may not accept the requirement or may well ignore it, in which case you will need to compensate by invoking surveillance and audit requirements in the order. If your purchasing documents do not reference ISO 9001 or its equivalent and you have taken alternative measures to assure the quality of the supplies, you need not invoke quality management requirements in the order. There is little point in imposing ISO 9001 on non-registered suppliers when ordering from a catalogue. It only makes sense when the supplier is prepared to make special arrangements for your particular order – arrangements which may well cost you more for no added value. Remember that the requirements in Clause 7.4.2 apply only where appropriate.

Adequacy of Purchasing Requirements (7.4.2)

The standard requires the organization *to ensure the adequacy of specified purchase requirements prior to their communication to the supplier.*

What Does this Mean?

The adequacy of purchasing information is judged by the extent to which it accurately reflects the requirements of the organization for the products concerned. Communication of such requirements to the supplier can be verbal or through documentation and processed by post or electronically.

Why is this Necessary?

This requirement responds to the Factual Approach Principle.

The acceptance of an order by a supplier places it under an obligation to accept product or service that meets the stated requirements. It is therefore important that such information is deemed adequate before being released to the supplier.

How is this Demonstrated?

Prior to orders being placed the purchasing information should be checked to verify that it is fit for its purpose. This is a requirement of Clause 4.2.3a (Published interpretation RFI 001). The extent to which you carry out this activity should be on the basis of risk and if you choose not to review and approve all purchasing information, your procedures should provide the rationale for your decision. In some cases orders are produced using a computer and transmitted to the supplier directly without any evidence that the order has been reviewed or approved. The purchase order does not have to be the only purchasing document. If you enter purchasing data onto a database, a simple code used on a purchase order can provide traceability to the approved purchasing documents.

You can control the adequacy of the purchasing data in at least four ways:

- Provide the criteria for staff to operate under self-control.
- Classify orders depending on risk and only review and approve those that present a certain risk.

- Select those orders that need to be checked on a sample basis.
- Check everything they do (this is Theory X[©] management and not recommended).

A situation where staff operate under self-control would be in the case of telephone orders where there is little documentary evidence that a transaction has taken place. There may be an entry on a computer database showing that an order has been placed with a particular supplier. So how would you ensure the adequacy of purchasing requirements in such circumstances? There follows a number of steps you can take:

- Provide buyers with read only access to approved purchasing data in the database.
- Provide buyers with read only access to a list of approved suppliers in the database.
- Provide a computer file containing details of purchasing transactions with read and write access.
- Provide a procedure that defines the activities, responsibilities and authority of all staff involved in the process.
- Train the buyers in the use of the database.
- Route purchase requisitions only to trained buyers for processing.

The above approach is suitable for processing routine orders, however, where there are non-standard conditions a more variable process needs to be developed.

VERIFICATION OF PURCHASED PRODUCT (7.4.3)

Ensuring Purchased Product Meets Requirements (7.4.3)

The standard requires the organization *to establish and implement the inspection or other activities necessary for ensuring that purchased product meets specified purchase requirements.*

What Does this Mean?

Verification is one of the fundamental elements of the control loop and in this case the verification serves to ensure that the output from the purchasing process meets the purchasing requirement. Verification may be achieved by several means before or after product is delivered or by building confidence in the source of supply so that product may enter the organization without any physical inspection. The requirement does not state when such verification should be performed and clearly it can be before, during and after receipt of the product. The standard leaves it to the organization's discretion to choose the timing that is appropriate to its operations.

Why is this Necessary?

This requirement responds to the Factual Approach Principle.

When you purchase items as individuals it is a natural act to inspect what you have purchased before you use it. To neglect to do this may result in you forfeiting your rights to return it later if found defective or nonconforming. When we purchase items on behalf of our employers we may not be as tenacious, so the company has to enforce its own verification policy as a way of protecting itself from the mistakes of its suppliers. Another reason for product verification is that it is often the case that characteristics are not accessible for inspection or test after subsequent processing. Characteristics that have not been verified prior to or on receipt may never be verified. If the end product passes the prescribed tests we tend to assume that the quality of individual components must be acceptable but there is a risk that it is not and inherent nonconformities have yet to be revealed.

How is this Demonstrated?

There are several ways of verifying that purchased product meets requirements and these were outlined previously under the heading *Selecting the degree of control*. Assessments by third parties alone would not give sufficient confidence to remove all receiving inspection for deliveries from a particular supplier. You need to examine product as well as the system until you have gained the confidence to reduce inspection and eventually remove it.

Timing of Verification Activities

If you have verified that product conforms to the specified requirements before it arrives you can receive product into your company and straight onto the production line. An example of this is where you have performed acceptance tests or witnessed tests on the supplier's premises. You may also have obtained sufficient confidence in your supplier that you can operate a 'just-in-time' arrangement but you must be able to show that you have a continuous monitoring programme that informs you of the supplier's performance.

If you have not verified that product meets requirements before it arrives you need to install a 'gate' through which only conforming items may pass. You need to register the receipt of items and then pass them to an inspection station equipped to determine conformance with your purchasing requirements. If items would normally pass into stores following inspection, as a safeguard you should also make provision for the store-person to check that all items received have been through inspection, rejecting any that have not. By use of labels attached to items you can make this a painless routine. If some items are routed directly to the user, you need a means of obtaining written confirmation that the items conform to the prescribed requirements so that at receipt inspection you can provide evidence that:

- nothing comes into the company without being passed through inspection;
- nothing can come out of inspection into stores without it being verified as conforming.

If the user is unable to verify that requirements have been met, you will need to provide either evidence that it has passed your receipt inspection or has been certified by the vendor.

Receiving Inspection

The verification plans should prescribe the acceptance criteria[i] for carrying out receipt inspection. The main aspects to cover are as follows:

- Define how the receipt inspection personnel obtain current purchasing requirements.
- Categorize all items that you purchase so that you can assign levels of receipt inspection based on given criteria.

- For each level of inspection, define the checks that are to be carried out and the acceptance criteria to be applied.
- Where dimensional and functional checks are necessary, define how the receipt inspection personnel obtain the acceptance criteria and how they are to conduct the inspections and tests.
- Define the action to be taken when the product, the packaging or the documentation is found to be acceptable.
- Define the action to be taken when the product, the packaging or the documentation is found to be unacceptable.
- Define the records to be maintained.

Evaluation of Supplier's Statistical Data

If the supplier supplies statistical data from the manufacturing process that indicates that quality is being controlled, then an analysis of this data based on assurances you have obtained through site evaluation can provide sufficient confidence in part quality to permit release into the organization. Where you have required your suppliers to send a certificate of conformity (C of C) testifying the consignment's conformity with the order, you cannot omit all receiving checks. Once supplier capability has been verified, the C of C allows you to reduce the frequency of incoming checks but not to eliminate them. The C of C may need to be supported with test results therefore you would need to impose this requirement in your purchasing documents. However, take care to specify exactly what test results you require and in what format you require them presented because you could be provided with attribute data[①] when you really want variables data[①].

Purchased Labour

This requirement poses something of a dilemma when purchasing subcontract labour because clearly it cannot be treated the same as product. You still need to ensure, however, that the labour conforms with your requirements before deployment to the job. Such checks will include verification that the personnel provided could demonstrate competence and they are who they say they are and the legitimate work permits where applicable. These checks can be made on the documentary evidence provided such as certificates, but you will probably wish to monitor their performance because it is the effort you have purchased not the people. You will not be able to verify whether they are entirely suitable until you have evaluated their performance so you need to keep records of the personnel and their performance during the tenure of the contract.

Dealing with Product Audits on Supplier's Premises

Within your procedures you need to provide a means of identifying which items have been subject to inspection at the supplier's premises and the receipt inspection action to be taken depending on the level of that inspection. In one case, your representative on the supplier's premises may have accepted the product. In another case, your representative may have accepted a product from the same batch but not the batch that has been delivered. Alternatively your representative may have only performed a quality audit to gain a level of confidence. You need to specify the inspection to be carried out in all such cases. Even when someone has performed inspection at the supplier's premises, if there is no evidence of conformity the inspections are of little value. The fact that an inspection was carried out is insufficient. There has to be a statement of what was checked, what results were obtained and a decision as to whether conformity had been achieved. Without such evidence you may need to repeat some of the inspections carried out on the supplier's premises.

Third Party Assessments or Part Evaluation

In cases where you don't have the skills or the resources to verify products on supplier sites it can be outsourced to a competent organization such as a part evaluation laboratory. This is what is known as Third Party Assessment. You could use the third party to undertake specialist assessments that support your own incoming inspection. This would give confidence that the components you receive are built under adequately controlled conditions.

Purchased Product Verification Records

Clause 7.4.3 of ISO 9001 does not require purchased product verification records (confirmed by published interpretation RFI 002). However, evidence of conformity with the acceptance criteria[①] is required to be maintained by Clause 8.2.4 and you need to think how you will demonstrate you are monitoring purchasing processes (Clause 8.2.3) and providing information relating to suppliers (Clause 8.4) without purchased product verification records.

Legitimizing Verification on Supplier's Premises (7.4.3)

The standard requires *that where the organization or its customer intends to perform verification activities at the supplier's premises, the organization is to state the intended verification arrangements and method of product release in the purchasing information.*

What Does this Mean?

If you choose a verification method other than receipt inspection that involves a visit to the supplier's premises, the supplier has a right to know and the proper vehicle for doing this is through the purchasing information such as a contract or order.

Why is this Necessary?

This requirement responds to the Factual Approach Principle.

The supplier needs to know if you or your customers intend to enter its premises to verify product before shipment so that they may make the necessary arrangements and establish that the proposed methods are acceptable to them.

How is this Demonstrated?

Verification by the Organization

The acceptance methods need to be specified at the tendering stage so that the supplier can make provision in the quotation to support any of your activities on site. When you visit a supplier you enter its premises only with their permission. The product remains their property until you have paid for it and therefore you need to be very careful how you behave. The contract or order is likely to only give access rights to products and areas related to your order and not to other products or areas. You cannot dictate the methods the supplier should use unless they are specified in the contract. It is the results in which you should be interested not the particular practices unless you have evidence to demonstrate that the steps they are taking will affect the results.

Verification by Your Customer

In cases where your customer requires access to your suppliers to verify the quality of supplies, you will need to transmit this requirement to your supplier in the purchasing information and obtain agreement. Where a firm's business is wholly that of contracting to customer requirements, a clause giving their customers certain rights will be written into their standard purchasing conditions. If this is an unusual occurrence, you need to identify the need early in the contract and ensure it is passed on to those responsible for preparing subcontracts. You may also wish to impose on your customer a requirement that you are given advanced notice of any such visits so that you may arrange an escort. Unless you know your customer's representative very well it is unwise to allow unaccompanied visits to your suppliers. You may for instance have changed, for good reasons, the requirements that were imposed on you as the main contractor when you prepared the subcontract and in ignorance your customer could inadvertently state that these altered requirements are unnecessary.

When customers visit your suppliers or inspect product on receipt, they have the right to reserve judgement on the final acceptance of the product because it is not under their direct control and they may not be able to carry out all the test and inspections that are required to gain sufficient confidence. Customer visits are to gain confidence and not to accept product. The same rules apply to you when you visit your suppliers. The final decision is the one made on receipt or sometime later when the product is integrated with your equipment and you can test it thoroughly in its operating environment or equivalent.

Chapter 27

Production and Service Provision

CHAPTER PREVIEW

This chapter is aimed at those personnel responsible for or involved with producing product or providing services to customers. This may include production managers, manufacturing managers, service managers, call centre managers and personnel engaged in planning, producing, inspecting, packing and shipping product or planning, delivering, checking services and handling post-delivery issues such as installation, warranty and maintenance.

In this chapter we examine the requirements in Clause 7.5 of ISO 9001:2008 and in particular:

- Planning production and service provision,
- Availability of information that describes the product,
- Availability of work instructions,
- Use of suitable equipment,
- Use and of monitoring and measuring devices,
- Implementation of monitoring and measurement activities,
- Release processes,
- Delivery processes,
- Post-delivery processes,
- Validation of processes,
- Identification and traceability,
- Customer property,
- Preservation of product.

The headings in this section of ISO 9001 are a little confusing. The title of the first section 'Control of production and service provision' implies that the subsequent sections serve some other purpose to the control of production and service provision when in fact they are wholly part of it. You would not be controlling product if you were using unvalidated processes, unidentified product and uncontrolled customer property. Therefore, in this chapter the heading 'Control of production and service provision' has been omitted in order to place all the topics at the same level in the structure of this chapter. The order remains the same as in the standard which means that it is not in a PDCA sequence.

The position where the requirements on production and service provision feature in the managed process is shown in Fig. 27-1. Due to the structure of the standard, activities normally part of production and service provision such as correcting

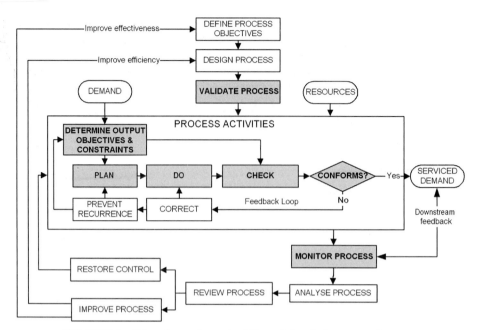

FIGURE 27-1 Where the requirements of Clause 7.5 apply in a managed process.

nonconformities etc. are excluded. Note that unlike design and purchasing requirements, Clause 7.5 does actually require the organization to carry out the activity to which it relates, i.e., production, hence the 'DO' box is shaded.

PLANNING PRODUCTION AND SERVICE PROVISION (7.5.1)

The standard requires the organization *to plan and carry out production and service provision under controlled conditions.*

What Does this Mean?

The process referred to in this section of the standard is the result producing process, the process of implementing or replicating the design. It is the process that is cycled repeatedly to generate product or to deliver service. It differs from the design process in that it is arranged to reproduce product or service to the same standard each and every time. The design process is a journey into the unknown whereas the production process is a journey along a proven path with a predictable outcome. The design process requires control to keep it on course towards an objective, the production process requires control to maintain output of a prescribed standard.

There are two ways in which product quality can be controlled: by controlling the product that emerges from the producing processes or by controlling the processes through which the product passes. Process control relies on control of the elements that drive the process, whereas product control relies on verification of the product as it emerges from the process. In practice it is a combination of these that yields products of consistent quality. If you concentrate on the process output to the exclusion of all else, you might find there is a high level of rework of the end product. If you concentrate on the process using the results of the product verification, you will gradually reduce rework until all output products are of consistent quality. It will therefore be possible to reduce dependence on output verification (Deming's 3rd of his 14 points).

Controlled conditions are conditions under which the outputs are predictable and are capable of being changed by a measurable degree. If the factors that affect process outputs could not be identified and changed, the process is not under control. There are usually eight factors that affect the control of any process.

- The quality of the people – competence to do the job with the required proficiency when required (Clause 6.2.1).
- The quality of the physical resources – capability of plant, machinery, equipment and tools (Clause 7.5.1c).
- The quality of the physical environment – level of temperature, cleanliness and vibration (Clause 6.4).
- The quality of the human environment – degree of physical stress, physiological stress and motivation (Clause 6.4).
- The quality of the information – degree of accuracy, currency, completeness, usability and validity (Clauses 7.5.1a and b).
- The quality of materials – adequacy of physical properties, their consistency and purity (Clause 7.4).
- The quantity of resources – time, money, information, people, materials, components, equipment etc. (Clause 6.1).
- The quality of measurement – units, values, timing and integrity (Clauses 7.5.1d and e and 7.6).

If you can't identify and control these factors you are not in control of the process.

Only three of the eight factors are addressed by Clause 7.5.2 of the standard. The others are addressed by other clauses of the standard meaning that the list of items in Clause 7.5.1 is not all that you would need to do to provide controlled conditions. Although the process you are controlling may not necessarily provide the resources needed, to remain in control you need the ability to verify you are getting what you need to produce the requirement process outputs.

The use of the term *provision* in the requirement is not significant to its meaning. The term *process* would have been more consistent with the principles on which the standard is based.

Why is this Necessary?

This requirement responds to the Leadership Principle.

Controlled conditions enable the organization to achieve its objectives. If operations were carried out in conditions in which there were no controls, the outputs would be the result of chance and totally unpredictable.

How is this Demonstrated?

The planning of production and servicing processes requires three levels:

- identifying which processes are required to produce products and deliver services;
- designing, commissioning and qualifying these processes for operational use; and

- routing the product through the appropriate qualified processes or running the service delivery process.

Production Planning

In order to identify the production processes required to produce a particular product you need production requirements in the form of product specifications which define the features and characteristics of the product which are to be achieved. By studying these specifications you will be able to determine the processes required to turn raw materials and bought out components into a finished product.

The next stage is to design the processes that have been identified. In many cases, existing processes may well satisfy the need but process approval may be required if the tolerance on product characteristics is much less than the currently demonstrated process capability.

The plans that route an item through the various processes from raw material to finished product are often called route cards or shop travellers. You may need separate plans for each process and each part with an overall plan which ensures that the product goes through the right processes in the right sequence. The number of plans is usually determined by the manner in which the specifications are drawn up. You may have drawings for each part to be made or one drawing covering several parts.

Unless products, processes and facilities are developed in parallel, the product will be unlikely to reach the market when required. This requires product and process development to proceed simultaneously with facility development and as a consequence the term 'simultaneous engineering'[①] or 'concurrent engineering' has emerged to optimize the relationship between design and manufacturing functions. It is not a case of designing only those products for which facilities exist, but designing those products that will give you a competitive edge and laying down facilities that will enable you to fulfil that promise.

To ensure that the processes are carried out under controlled conditions the production plans need to:

- identify the product in terms of the specification reference and its issue status;
- define the quantity required;
- define which section is to perform the work;
- define each stage of manufacture and assembly;
- provide for progress through the various processes to be recorded so that you know what stage the product has reached at any one time;
- define the special tools, processing equipment, jigs, fixtures and other equipment required to produce the product (general purpose tools and equipment need not be specified because your staff should be trained to select the right tool for the job);
- define the methods to be used to produce the product either directly or by reference to separate instructions;
- define the environment to be maintained during production of the product if anything other than ambient conditions;
- define the process specifications and workmanship standards to be achieved;
- define the stages at which inspections and test are to be performed and the methods to be used;
- define any special handling, packaging, marking requirements to be met; and
- define any precautions to be observed to protect health, safety and environment.

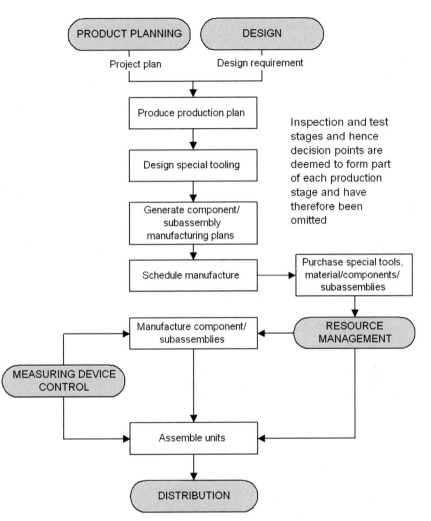

FIGURE 27-2 Typical production process flow.

These plans create a basis for ensuring that work is carried out under controlled conditions, but the staff, equipment, materials, processes and documentation must be up to the task before work commences. A simple production process is illustrated in Fig. 27-2. The shaded boxes indicate interfaces external to the production process. The variables are too numerous to illustrate the intermediate steps.

Service Planning

The variety of different types of services makes it impractical to lay down any prescription for how they should be planned apart from some fundamental provisions. A typical service delivery process flow is illustrated in Fig. 27-3.

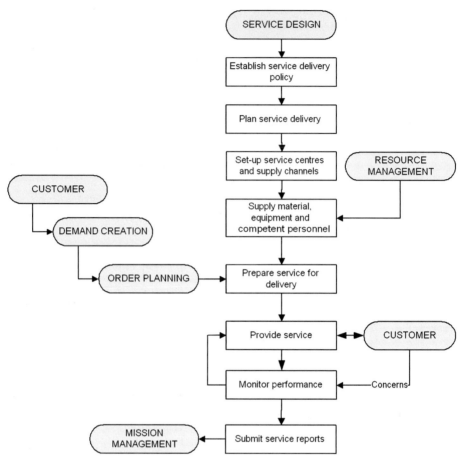

FIGURE 27-3   Typical service delivery process.

To ensure that the processes are carried out under controlled conditions the service delivery plans need to:

- define the service objectives and targets;
- define the stages in the process to achieve these objectives;
- define the inputs and outputs for each stage in the process;
- define the resources required to generate these inputs and outputs;
- define the methods to be used to operate any equipment and generate the inputs and outputs;
- define the methods of measuring stage outputs;
- define the methods of ensuring the integrity of these measurements;
- define the information and resources required for each stage to be performed as planned;
- define the actions to be taken when problems are encountered;
- define the information provided to service users that indicates the operation and availability of the service; and
- define the precautions to be observed to protect health, safety and environment.

AVAILABILITY OF INFORMATION THAT DESCRIBES THE PRODUCT (7.5.1a)

The standard requires the organization *to control production and service provision through the availability of information that describes the characteristics of the product.*

What Does this Mean?

This information tells you what to make or provide and the criteria the output must meet for it to be fit for its purpose. The information is the input to the production or service delivery process usually coming out of the design process but may be direct from customers. It may take the form of definitive specifications, drawings, layouts or any information that specifies the physical and functional characteristics that the product or service is required to meet.

Why is this Necessary?

This requirement responds to the Process Approach Principle.

Information is one of the key factors needed to control processes. Without information specifying the product to be produced or the service to be delivered, there is no basis on which to commence production.

How is this Demonstrated?

In order to ensure the right information is available, there needs to be a communication channel opened between the product design and production process or the service design and the service delivery process as appropriate. Along this channel needs to pass all the information required to produce and accept the product or deliver the service. Provision also needs to be made for transmitting changes to this information in such a manner that the recipients can readily determine what the changes are and why they have been made. Often the design information is reissued, identified with a revision code leaving the recipient to work out what has changed and whether it affects what has gone before. The change management process therefore needs to take into account the factors that affect the effectiveness of the interfacing processes.

AVAILABILITY OF WORK INSTRUCTIONS (7.5.1b)

The standard requires the organization *to control production and service provision through the availability of work instructions where necessary.*

What Does this Mean?

Information is one of the key factors needed to control processes. In addition to information specifying the product characteristics to be achieved, information relating to how, when and where the activities required to convert the inputs into useable outputs are to be performed will be needed. This type of information is often termed as work instructions. There are two forms of work instructions – instructions that inform people what work to do and when to do it and instructions that inform people how to do work – the latter are often called procedures. This topic is addressed in more detail in Chapter 11 under *Documents that ensure effective operation and control of processes.*

The requirement implies that work instructions may not always be required. In routine operations such as serving food in a fast food outlet, the workers do not need instructions telling them what work to do and when to do it because that has been conveyed during their training. After being deemed competent they will not need to refer to work instructions again except for reassurance but such instructions may well exist in the training manuals. Some people take instructions from their managers, others take instruction from the situation, the equipment or other signals they receive.

Why is this Necessary?

This requirement responds to the Process Approach Principle.

As people are not normally mind readers, they are unlikely to know what is required unless the work instructions come direct from the customer or the supplier. Once they know what is required they may need further instructions on how to carry out the work because it is not intuitive or learnt through education and training. Work instructions may also be necessary to ensure consistent results. In fact any operation that requires tasks to be carried out in a certain sequence to obtain consistent results should either be specified through work instructions or procedures or developed into a habit.

How is this Demonstrated?

Work instructions can take many forms:

- Schedules indicating when work should be complete.
- Plans indicating what work is to be performed and what resources are available.
- Specifications indicating the results the work should achieve.
- Process descriptions indicating the stages through which work must pass.
- Procedures indicating how work should be performed and verified.

There are instructions for specific activities and instructions for specific individuals – whether they are contractors or employees is not important – the same requirements apply. As each employee may perform different jobs, they may each have a different set of instructions that direct them to specific sources of information. Therefore, it is unnecessary to combine all instructions into one document although they could all be placed in the same binder for easy access.

By imposing formal controls you safeguard against informality that may prevent you from operating consistent, reliable and predictable processes. The operators and their supervisors may know the tricks and tips for getting the equipment or the process to operate smoothly. You should discourage informal instructions because you cannot rely on them being used by others when those who know them are absent. If the tips or tricks are important, encourage those who know them to bring them to the process owner's attention so that changes can be made to make the process run smoothly all the time.

If you have a manufacturing process that relies on skill and training then instructions at the workstation are unnecessary. For example, if fixing a tool in a tool holder on a lathe is a skill, learnt during basic training, you don't need to provide instructions at each workstation where normal tool changes take place. However, if the alignment of the tool is critical and requires knowledge of a setting up procedure, then either documented instruction or training is necessary. Even for basic skills you can still provide standard machinery data books that are accessible near the workstation. There is merit in not providing basic textbooks to operators because the information is soon outdated and operators relying on such data instead of consulting the authorized data may inadvertently induce variation into the process.

USE OF SUITABLE EQUIPMENT (7.5.1c)

The standard requires the organization *to control production and service provision through the use of suitable equipment.*

What Does this Mean?

Equipment is one of the key factors needed to control processes. Suitable equipment is equipment of the right type and capability to fulfil the requirements for which it is needed. Although the term equipment is used, the intent is to imply any physical resource that is needed to achieve the process objectives. It means that the resources have to be serviceable and capable of the performance, accuracy and precision required, i.e. maintained, calibrated, qualified, verified or otherwise approved as appropriate.

Why is this Necessary?

This requirement responds to the Process Approach Principle.

Process outputs cannot be achieved unless the physical resources that are essential to perform the work are fit for their purpose. In any other state, the human resources would be used to compensate for the inadequacies of the equipment – a state that can be sustained in some circumstances but not for long without degrading the quality of the work.

How is this Demonstrated?

The equipment should be selected or designed during the planning process. In selecting such equipment you should determine whether it is capable of producing, maintaining or handling conforming product in a consistent manner. You also need to ensure that the equipment is capable of achieving the specified dimensions within the stated tolerances. Process capability studies can reveal deficiencies with equipment that are not immediately apparent from inspection of the first off.

There may be documentation available from the supplier of the equipment that adequately demonstrates its capability; otherwise you may need to carry out qualification[①] and capability tests to your own satisfaction. In the process industries the plant is specially designed and so needs to be commissioned and qualified by the user. Your procedures need to provide for such activities and for records of the tests to be maintained.

When equipment or plant is taken out of service either for maintenance or for repair, it should not be re-introduced into service without being subject to formal acceptance tests which are designed to verify that it meets your declared standard operating conditions. Your procedures need to provide for such activities and for records of the tests to be maintained.

Drawings should be provided for jigs, fixtures, templates and other hardware devices and they should be verified as conforming with these drawings prior to use. They should also be proven to control the dimensions required by checking the first-off to be produced from such devices. Once these devices have been proven they need checking periodically to detect signs of wear or deterioration. The frequency of such checks should be dependent on usage and the environment in which they are used.

Tools which form characteristics such as crimping tools, punches, press tools etc. should be checked prior to first use to confirm they produce the correct characteristics and then periodically to detect wear and deterioration. Tools that need to maintain certain temperatures, pressures, loads etc. in order to produce the correct characteristics in materials should be checked to verify that they will operate within the required limits.

Steel rules, tapes and other indicators of length should be checked periodically for wear and damage and although accuracy of greater than 1 mm is not normally expected, the loss of material from the end of a rule may result in inaccuracies that affect product quality.

While you may not rely entirely on these tools to accept product, the periodic calibration or verification of these tools may help prevent unnecessary costs and production delays. While usage and environment may assist in determining the frequency of verification hardware checks, these factors do not affect software. Any bugs in software have always been there or were introduced when it was last modified. Software therefore needs to be checked prior to use and after any modifications have been carried out, so you cannot pre-determine the interval of such checks.

USE OF MONITORING AND MEASURING EQUIPMENT (7.5.1d)

The standard requires the organization *to control production and service provision through the availability and use of measuring and monitoring equipment.*

What Does this Mean?

Measurement is one of the key factors needed to control processes. This means providing the equipment needed to measure product features and monitor process performance and also providing adequate training and instruction for this equipment to be used as intended.

Why is this Necessary?

This requirement responds to the Process Approach Principle.

Product quality can only be determined if the equipment needed to measure and monitor product and process characteristics are available and used.

How is this Demonstrated?

When designing the process for producing product or delivering service you should have provided stages at which product or service features are verified. You may also need to install monitoring devices that indicate when the standard operating conditions have been achieved and whether they are being maintained. The equipment used to perform measurements needs to be available where the measurements are to be performed. The monitoring devices need to be accessible to process operators for information on the performance of the process to be obtained. The monitoring equipment may be located in inaccessible places providing the signals are transmitted to the operators controlling the process.

IMPLEMENTATION OF MONITORING AND MEASUREMENT ACTIVITIES (7.5.1e)

The standard requires the organization *to control production and service provision through the implementation of monitoring and measurement.*

What Does this Mean?

Measurement is one of the key factors needed to control processes. Monitoring and measurement is the means by which product and process characteristics are determined. The specifications define the target values and the process description or plan defines when measurements should be taken to ascertain whether the targets have been met.

Why is this Necessary?

This requirement responds to the Factual Approach Principle.

In order to control product quality the achieved characteristics need to be measured and the process operating conditions need to be monitored. All controls need a verification stage and a feedback loop. You cannot control production processes without performing some kind of verification.

How is this Demonstrated?

Controlled conditions in production include in-process monitoring and in-process inspection and test. In a service delivery process they may include inspections of information, personnel and facilities as well as a review of process outputs.

The production of some products can be controlled simply by inspection after the product has been produced. In other cases, such as with the continuous production of food and drugs, you may need to monitor certain process parameters to be sure of producing conforming product. By observing the variability of certain parameters using control charts, you can determine whether the process is under control within the specified limits.

The purpose of monitoring the process is firstly to establish its capability of producing product correctly and consistently. Secondly it is to alert the process operators to conditions that indicate that the process is becoming incapable of producing the product correctly and consistently.

Process monitoring can be achieved by observing sensors installed in the production process that measure key process parameters or, samples can be taken at discrete intervals and prescribed measurements taken. In both cases the measurements should be recorded for subsequent analysis and any decision made to allow the process to continue or to stop should also be recorded together with the reasons for the decision. The data to be recorded should be specified in advance on the forms or computer screens provided at the workstation. This will give personnel a clear indication of what to record, when and where to record it. It also simplifies auditing if data is required in all boxes on a form or computer screen. A blank box would then indicate an unusual occurrence that should be checked. The forms should also indicate the accept or reject limits so that the operator can easily judge when the process is out of control.

Operators should be trained to both operate the plant and control the process. As added assurance you should take samples periodically and subject them to a thorough examination. The sampling plan should be defined and documented and operators trained to determine what causes the results they observe. *Operators should therefore understand what causes the dots on the chart to vary.* Process control comes about by operators knowing what results to achieve, by knowing what results are being achieved and by being able to correct performance should the results not be as required. They need to understand what is happening during processing to cause any change in the results as they are being monitored. In the process specification you will need to define the parameters to be observed and recorded and the limits within which the process is to be controlled.

RELEASE PROCESSES (7.5.1f)

The standard requires *the implementation of product release activities*.

What Does this Mean?

Release activities are decision points where process output is confirmed as complete and moved onto the next stage in a process or to another process. These are sometimes called 'gates' through which product has to pass before being deemed acceptable for further processing or delivery.

Why is this Necessary?

This requirement responds to the Process Approach Principle.

Within the chain of processes from customer requirement to delivery of product there are many interfaces between processes. Wherever decision points between processes are absent, errors may pass from one process to another resulting in customer dissatisfaction.

How is this Demonstrated?

In the production process there are several release points:

- Release from component storage, where components and materials are held pending completion of the input requirement.
- Release from job set-up, where work is held pending verification of job set-up.
- Release from in-process inspection, where product is held pending completion or rework, sample tests, repair etc.
- Release from final inspection, where product is held pending completion of all inspection stages, sampling etc.
- Release from finished product storage, where product is held pending receipt of customer order or customer instruction.

Release from Storage Areas

The content of storage areas should be known at all times in order that you can be confident that only that which is in storage areas is of a known condition. Storage areas containing conforming items should be separate from those containing nonconforming items. It follows therefore that when an item is taken from a storage area the person taking it should be confident it is conforming. If free access is given to add and remove items in such areas, this confidence is lost. If at any time the controls are relaxed, the whole stock becomes suspect.

There is often a need to supply items as free issue because the inadvertent loss of small value items is less than the cost of the controls to prevent such loss. This practice can be adopted only if the identity of the items can be determined wholly by visual inspection by the person using them.

There are, however, issues other than quality that will govern the control of items in stock. Inventory control is a vital part of any business. Stock ties up capital, so the less stock that is held the more capital the firm has available to apply to producing output.

A common solution that satisfies both inventory control and quality control is to institute a stock requisition system. Authorization of requisitions may be given by a person's supervisor or can be provided via a work order. If someone has been authorized to carry out a particular job, this should authorize the person to requisition the items needed. Again for inventory control reasons you may wish to impose a limit on such authority, requiring the person to seek higher authority for additional items above a certain value. If an operator is requesting additional items you need to know why as the process was intended to deliver certain output from the inputs supplied.

Held Product

In continuous production, product is inspected by taking samples from the line that are then examined whilst the line continues producing product. In such cases you will need a means of holding product produced between sampling points until the results of the tests and inspections are available. You will also need a means of releasing product when the results indicate that the product is acceptable. So a Product Release Procedure or Held Product Procedure may be necessary.

Release from Job Set-Ups

In setting up a job prior to commencing a production run, you need to verify that all the requirements for the part are being met. You will therefore need job set-up instructions so as to ensure that each time the production of a particular part commences, the process is set up against the same criteria. In addition, process parameters may change whenever there is material changeover, a job change or if significant time periods lapse between production runs.

Documentation verifying job set-ups should include instructions to perform the set-up and records that demonstrate that the set-up has been performed as required. This requires that you record the parameters set, the sample size and retain the control charts used that indicate performance.

Release from Inspection

Every verification is a stage where product is verified as either conforming or non-conforming. Provisions are needed for signifying when product is ready for release either to the next stage in the process, into quarantine store awaiting decision or back into the process for rework or completion. Often this takes the form of an inspection label appropriately annotated. With services, release is often signified by an 'In-service' notice a 'Ready' indicator, illuminated sign or other indicator. In many cases, however, it is only 'Out of Service' notices that are posted.

DELIVERY PROCESSES (7.5.1f)

The standard requires *the implementation of delivery activities.*

What Does this Mean?

Delivery is an activity that serves the shipment or transmission of product to the customer and is one part of the distribution process. Delivery may include preparation for delivery such as packing, notification, transportation, customs, arrival at destination and unpacking on customer premises. In the consumer sector, this may involve agents, wholesalers, and retailers before the end user received the goods. In the service delivery process this means the fulfilment of the service and may include transmission of information and payment mechanisms. In the consumer goods market there may be intermediaries such as agents, wholesalers, retailers, resellers etc. which exist to distribute product to the customer.

Why is this Necessary?

This requirement responds to the Process Approach Principle.

The process of moving goods from producer to customer is an important process in the management system. Although good product design, economic production and effective promotion are vital for success, these are useless if the customer cannot access the product and take ownership. It is necessary to control delivery activities because conforming product may be degraded by the manner in which it is protected during transit. It may also be delayed by the manner in which it is transported. You may be under an obligation to supply product by certain dates or within so many days of order and as a consequence control of the delivery process is vital to honour these obligations.

How is this Demonstrated?

A typical distribution process is illustrated in Fig. 27-4 indicating that like any process it needs to be designed and that the origin of the process inputs is the demand creation process.

The distribution or marketing channel promotes the physical flow of goods and services along with ownership title, from producer to consumer or business user.[1] Often

---

[1] Boone Louis, E., and Kurtz David, L. (2001). *Contemporary marketing.* 10th Edition, Harcourt College Publishers.

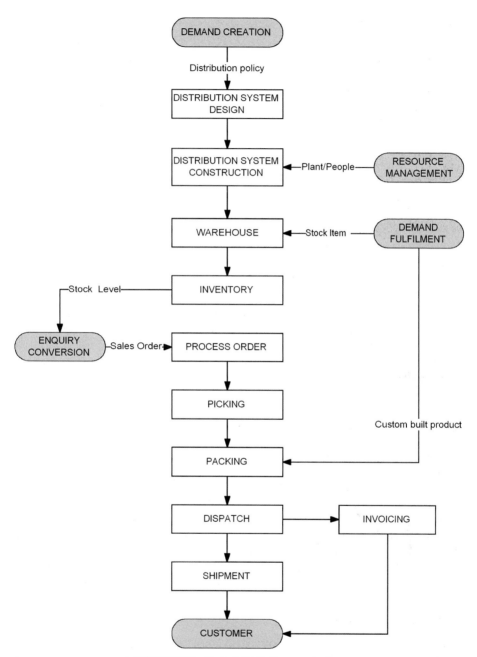

FIGURE 27-4 Typical distribution process flow.

the logistics for moving goods to outlets where consumers are able to purchase them is a business in its own right but nevertheless starts out in the demand creation process when determining the distribution strategy (see Fig. 10-7). There are several different distribution channels depending on the type of goods and the market into which they are to be sold as illustrated in Fig. 27-5.

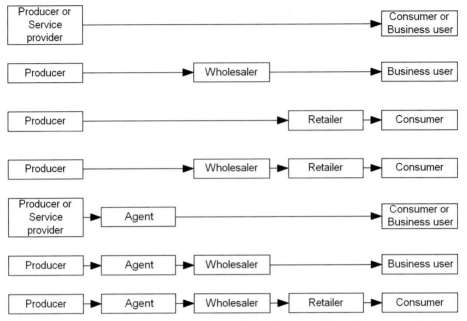

FIGURE 27-5 Distribution channels (Source: Adapted from Boonze and Kurtz *Contemporary Marketing* Fig. 13-2).

Delivery takes place between each of the parties in the distribution chain and for each party there are several aspects to the delivery process:

- Preparation of product such as cleaning and preservation,
- Packing of product,
- User information,
- Product certification,
- Labelling and transit information,
- Handling,
- Customer notification,
- Transportation,
- Tracking.

Preparation and packing of product is addressed under preservation later in this chapter as the methods also apply to internal processing. However, within the delivery process there will be specific packing stages that are different in nature to internal packing stages.

Sometimes, delivery is made electronically using a modem and telephone line. The product may be a software package or a document stored in electronic form. Protection of the product is still required but takes a different form. You need to protect the product against loss and corruption during transmission or downloading.

When shipping consumer goods it is necessary to include user information such as operating instructions, handbooks, warranty and return instructions.

Customers may require product certificates testifying the fulfilment of contracts or order requirement. Customs may require certain legal information on the outside of the package otherwise the consignment will be held at the port of entry and customers will be none too pleased.

The type of transport employed is a key factor in getting shipments to customers on time.

On-Time Delivery

To guarantee shipment on time, you either need to maintain an adequate inventory of finished goods, for shipment on demand or utilize only predictable processes and obtain sufficient advanced order information from your customer. Without sufficient lead-time on orders you will be unlikely to meet the target. There will be matters outside your control and matters over which you need complete control. It is the latter that you can do something about and take corrective action should the target not be achieved.

Firstly you need to estimate the production cycle time during the production trial runs in the product and process validation phase, assess risk areas and build in appropriate contingencies. An assessment of your supplier's previous delivery performance will also enable you to predict their future performance. When new processes become stabilized over long periods and the frequency of improvement reduces as more and more problems are resolved, you will be able to reduce lead-time.

Your planning and delivery procedures need to record estimated and actual delivery dates and require the data to be collected and analysed through delivery performance monitoring. When targets are not met you should investigate the cause under the corrective action procedures and formulate corrective action plans. Where the cause is found to be a failure of the customer to supply some vital information or equipment, it would be prudent not to wait for the periodic analysis but react promptly.

Customer Notification

A means for notifying the customer of pending delivery is often necessary. Your organization might be linked with the customer electronically so that demands are transmitted from the customer to trigger the delivery process. However, the customer may need to change quantities and delivery dates due to variations in production. This does not mean that the changes will always be to shorten delivery times but on occasions the delivery times may need to be extended owing to problems on the assembly line or as a result of problems with other suppliers. The customer may not have made provision to store your product so needs to be able to urgently inform you to hold or advance deliveries. If the customer reduced the quantity required from that previously demanded, you could be left with surplus product and consequently need protection through the contract for such eventualities.

POST-DELIVERY PROCESSES (7.5.1f)

The standard requires *the implementation of post-delivery activities*.

What Does this Mean?

Post-delivery activities are those performed after delivery of the consignment to the customer and may include:

- Servicing,
- Warranty claims,

- Technical support,
- Maintenance,
- Logistics,
- Installation,
- Disposal or recycling.

Why is this Necessary?

This requirement responds to the Customer Focus Principle.

Control of post-delivery activities is just as important as pre-delivery if not more so as the customer may be losing use of the product and want prompt resolution to the problems encountered. Post-delivery performance is often the principal reason why customers remain loyal or choose a competitor. Even if a product does give trouble, a sympathetic, prompt and courteous post-delivery service can restore confidence.

How is this Demonstrated?

The wide range of post-delivery services makes a detailed analysis impractical in this book. However, there are some simple measures that can be taken that would apply to all types of post-delivery activities.

- Define the nature and purpose of the post-delivery service.
- Define post-delivery policies that cover such matters as handling complaints, offering replacement product, service or installation staff conduct.
- Establish conditions of post-delivery contracts with customers.
- Specify objectives and measures for each feature of the service such as response time, resolution time.
- Communicate the policies and objectives and ensure their understanding by those involved.
- Define the stages in the process needed to achieve these objectives.
- Identify the information needs and ensure control of this information.
- Identify and provide the resources to deliver the service.
- Install verification stages to verify achievement of stage outputs.
- Provide communication channels for feeding intelligence into production and service design processes.
- Determine methods for measuring process performance.
- Measure process performance against objectives.
- Determine the capability of the process and make changes to improve performance.
- Determine process effectiveness and pursue continual improvement.

VALIDATION OF PROCESSES (7.5.2)

The standard requires the organization *to validate any production and service provision where the resulting output cannot be verified by subsequent measurement or monitoring.*

What Does this Mean?

Many processes do not present any difficulty in the verification of the output against the input requirements regardless of the tools, personnel, facilities or other means used to carry out the process. The resultant features and characteristics are relatively easily determined. However, there are some processes where the output is totally dependent on the personnel, the equipment and the facilities and what is more, cannot be fully verified by examination of the output at any stage of assembly. Among such processes are welding, soldering, adhesive bonding, casting, forging, forming, heat treatment, protective treatments and inspection and test techniques such as X-ray examination, ultrasonics, environmental tests, mechanical stress tests. The standard only requires process validation where, as a consequence of not being able to verify the output, deficiencies become apparent only after the product is in use.

In service industries, special processes include correctness of financial or legal documents, software, professional advice etc. In such cases, these processes are not separated for special treatment because all processes in the business may fall into this category.

Why is this Necessary?

This requirement responds to the Factual Approach Principle.

If any of these factors on which the performance of a process depends is less than adequate, deficiencies may not become apparent until long after the product is installed, used or enters service. Normally, product characteristics are verified before release but when this is not possible without destroying the product, the process needs to be qualified as capable of only producing conforming product.

How is this Demonstrated?

You will only have to demonstrate conformity with this requirement for processes that satisfy the criteria in Clause 7.5.2 of the standard. This is reinforced by published interpretation RFI 023. However, quite why this is the case is puzzling as according to ISO 9000:2005, validation means *'confirmation, through the provision of objective evidence that the requirements for a specific intended use or application have been fulfilled'*. Therefore confirming that a process fulfils its purpose would appear to be validation and apply to all processes.

To limit the potential for deficiencies to escape detection before the product is released, measures should be taken that ensure the suitability of all equipment, personnel, facilities and prevent varying conditions, activities or operations. A thorough assessment of the processes should be conducted to determine their capability to maintain or detect the conditions needed to consistently produce conforming product. The limits of capability need to be determined and the processes applied only within these limits.

You should produce and maintain a list of manufacturing processes that have been validated as well as a list of the personnel who are qualified to operate them. In this way you can easily identify an unqualified process, an unauthorized person or an obsolete list if you have neglected to maintain it.

Where process capability relies on the competence of personnel, personnel operating such processes need to be appropriately educated and trained and undergo examination of their competency. If subcontracting manufacturing processes you need to ensure that the supplier only employs qualified personnel and has qualified process equipment and facilities.

Where there is less reliance on personnel but more on the consistency of materials, environment and processing equipment, the particular conditions need to be specified. Where necessary restrictions should be placed on the use of alternative materials, equipment and variations in the environment. Operating instructions should be used that define the set-up, operation and shutdown conditions and the sequence of activities required to produce consistent results. The resultant product needs to be thoroughly tested using such techniques that will enable the performance characteristics to be measured. This may involve destructive tests to measure tensile and compressive strength, purity, porosity, adhesion, electrical properties etc. In production, samples should be taken at set frequencies and the tests repeated.

In production you need to ensure that only those personnel, equipment, materials and facilities that were qualified are employed in the process otherwise you will invalidate the qualification and inject uncertainty into the results.

The records of qualified personnel using special processes should be governed by the training requirements. Regarding the equipment, you will need to identify the equipment and facilities required within the process specifications and maintain records of the equipment. This data may be needed to trace the source of any problems with product that was produced using this equipment. To take corrective action you will also need to know the configuration of the process plant at the time of processing the product. If only one piece of equipment is involved, the above records will give you this information but if the process plant consists of many items of equipment which are periodically changed during maintenance, you will need to know which equipment was in use when the fault was likely to have been generated.

IDENTIFICATION AND TRACEABILITY (7.5.3)

Identifying Product (7.5.3.1)

The standard requires the organization *to identify the product by suitable means throughout product realization where appropriate.*

What Does this Mean?

The requirements for product identification are intended to enable products and services with one set of characteristics to be distinguishable from products or services with another set of characteristics.

The option of applying this requirement 'where appropriate' implies that there are situations where product identity is unnecessary. There are of course situations where attaching an identity to a product would be impractical such as for liquids or items too small but the product nevertheless has an identity that is conveyed through the packaging and associated information. In the food industry, the biscuits on the conveyor might not carry an identity but the box into which they are packed does as does the instruction that ordered the biscuits to be produced. Thus identifying a product by suitable means might require the product to be labelled, or might require the container to be labelled. Services are somewhat different. Many are not identified other than by the nature of what the organization does by generic categories such as investment, mortgage, financial planning services of banks. Where there are differences for instance in interest rates, the 'products' are given different names such as Instant Access Account, 90 Day account and so on.

Why is this Necessary?

This requirement responds to the Process Approach Principle.

Product identity is vital in many situations to prevent inadvertent mixing, to enable re-ordering, to match products with documents that describe them and to do that basic of all human activities – to communicate. Without codes, numbers, labels, names and other forms of identification we cannot adequately describe the product or service to anyone else or be certain we are looking at the right product. The product must be identified in one way or another otherwise it cannot be matched to its specification.

How is this Demonstrated?

Separate product identity is necessary where it is not inherently obvious. If products are so dissimilar that inadvertent mixing would be unlikely to occur, a means of physically identifying the products is probably unnecessary. 'Inherently obvious' in this context means that the physical differences are large enough to be visible to the untrained eye. Functional differences, therefore, no matter how significant as well as slight differences in physical characteristics such as colour, size, weight and appearance would constitute an appropriate situation for documented identification procedures.

Identifying product should start at the design stage when the product is conceived. The design should be given a unique identity, a name or a number and that should be used on all related information. When the product emerges into production, the product should carry the same number or name but in addition it should carry a serial number or other identification to enable product features to be recorded against specific products. If verification is on a go or no go basis, product does not need to be serialized. If measurements are recorded some means has to be found of identifying the measurements with the product measured. Serial numbers, batch numbers and date codes are suitable means for achieving this. This identity should be carried on all records related to the product.

Apart from the name or number given to a product you need to identify the version and the modification state so that you can relate the issues of the drawing and specifications to the product they represent. Products should either carry a label or markings with this type of information in an accessible position or bear a unique code number that is traceable to such information.

You may not possess any documents that describe purchased product. The only identity may be marked on the product itself or its container. Where there are no markings, information from the supplier's invoice or other such documents should be transferred to a label and attached to the product or the container. Information needs to be traceable to the products it represents.

The method of identification depends on the type, size, quantity or fragility of the product. You can mark the product directly (provided the surface is not visible to the end user unless of course identity is part of the brand name) tie a label to it or the container in which it is placed. You can also use records remote from the product providing they bear a unique identity that is traceable to the product.

Marking products has its limitations because it may damage the product, be removed or deteriorate during subsequent processing. If applied directly to the product, the location and nature of identification should be specified in the product drawings or referenced process specifications. If applied to labels which are permanently secured to the product, the identification needs to be visible when the product is installed so as to facilitate checks without its removal. If the identity is built into the forging or casting, it is important that it is legible after machining operations. One situation which can be particularly irritating to customers is placing identification data on the back of equipment and then expecting the customer to state this identity when dealing with a service call thus causing delay while the customer dives under the desk to locate the serial number and drops the telephone in the panic!

Verification Status (7.5.3)

The standard requires the organization *to identify the status of the product with respect to measurement and monitoring requirements throughout product realization.*

What Does this Mean?

Product status with respect to monitoring and measurement means an indication as to whether the product conforms or does not conform to specified requirements. Thus identifying product status enables conforming product to be distinguishable from nonconforming product.

Why is this Necessary?

This requirement responds to the Factual Approach Principle.

Measurement does not change a product but does change our knowledge of it. Therefore it is necessary to identify which products conform and which do not so that inadvertent mixing, processing or delivery is prevented.

How is this Demonstrated?

The most common method of denoting product status is to attach labels either to the product or to containers holding the product. *Green labels* for acceptable good and *Red labels* for reject goods. Labels should remain affixed until the product is either packed or installed. Labels should be attached in a way that prevents their detachment during handling. If labels need to be removed during further processing, the details should be transferred to inspection records so that at a later date the status of the components in an assembly can be checked through the records. At dispatch, product status should be visible. Any product without status identification should be quarantined until re-verified and found conforming. Once a product has passed through the product realization process and is in use, it requires no product status identity unless it is returned to the production process for repair or other action.

It should be possible when walking through a machine shop for example, to identify which products are awaiting verification, which have been verified and found conforming and which have been rejected. If by chance, some product was to become separated from its parent batch, it should still be possible to return the product to the location from whence it came. A machine shop is where this type of identification is essential – it is where mix-ups can occur. In other places, where mix-ups are unlikely, verification status identification does not need to be so explicit.

Identifying product status is not just a matter of tying a label on a product. The status should be denoted by an authorized signature, stamp, mark or other identity which is applied by the person making the accept or reject decision and which is secure from misuse. Signatures are acceptable as a means of denoting verification status on paper records but are not suitable for computerized records. Secure passwords and 'write only' protection has to be provided to specific individuals. Signatures in a workshop environment are susceptible to deterioration and illegibility that is why numbered inspection stamps with unique markings evolved. The ink used has to survive the environment and if the labels are to be attached to the product for life, it is more usual to apply an imprint stamp on soft metal or bar code.

Small and fragile products should be held in containers and the container sealed and marked with the product status. Large products should either carry a label or have a related inspection record.

In some situations the location of a product can constitute adequate identification of product status. However, these locations need to be designated as Awaiting Inspection, Accepted Product or Reject Product or other such labels as appropriate to avoid the inadvertent placement of items in the wrong location. The location of product in the normal production flow is not a suitable designation unless an automated transfer route is provided.

When a service is out of service, tell your customers. Services that rely on products should carry a label or a notice when accessed. A bank cash machine is one example where a notice is displayed when the machine is out of service. In some cases customers may need to be informed by letter or telephone.

With software the verification status can be denoted in the software as a comment or on records testifying its conformance with requirements.

With documentation you can either denote verification status by an approval signature on the document or by a reference number, date and issue status that is traceable to records containing the approval signatures.

If you use stamps, you will need a register to allocate stamps to particular individuals and to indicate which stamps have been withdrawn. When a person hands in his stamp it is good practice to avoid using the same number for 12 months or so to prevent mistaken identity in any subsequent investigations.

Traceability

Where traceability is a requirement the standard requires *the organization to control the unique identification of the product and maintain records.*

What Does this Mean?

Traceability is a process characteristic. It provides the ability to trace something through a process to a point along its course either forwards or backwards through the process and determine as necessary, its origin, its history and the conditions to which it was subjected. Traceability may be a requirement of the customer, legislation or statutes or simply a requirement of the organization in order to conduct investigations when events do not proceed as planned.

Why is this Necessary?

This requirement responds to the Factual Approach Principle.

One needs traceability to find the root cause of problems. If records cannot be found which detail what happened to a product then nothing can be done to prevent its recurrence. Although the standard only requires traceability when required by contract or law, it is key to enabling corrective action.

In situations of safety or national security it is necessary to be able to locate all products of a batch in which a defective product has been found so as to eliminate them before there is a disaster. It is also very important in the aerospace, automobile, medical devices and food and drugs industries – in fact, any industry where human life may be at risk due to a defective product being in circulation.

Traceability is also important to control processes. You may need to know which products have been through which processes and on what date if a problem is found sometime later. The same is true of test and measuring equipment. If on being calibrated a piece of test equipment is found to be out of calibration then it is important to track down all the equipment which has been validated using that piece of measuring equipment. This in fact is a requirement of ISO 9001 Clause 7.6 but no requirement for traceability is specified.

How is this Demonstrated?

Providing traceability can be an onerous task. Some applications require products to be traced back to the original ingot from which they were produced. Traceability is achieved by coding items and their records such that you can trace an item back to the records at any time in its life. The chain can be easily lost if an item goes outside your control. For example, if you provide an item on loan to a development organization for investigation and it is returned sometime later, without a certified record of what was done to it, you have no confidence that the item is in fact the same one, unless it has some distinguishing features. The inspection history may also be invalidated because the operations conducted on the item were not certified. Traceability is only helpful when the chain remains unbroken. It can also be costly to maintain. The system of traceability that you maintain should be carefully thought out so that it is economic. There is little point in maintaining an elaborate traceability system for the once in a lifetime event when you need it, unless your very survival or society's survival depends on it. However, if there is a field failure, in order to prevent recurrence you will need to trace the component back through the supply chain to establish which operation on which component was not per-formed correctly simply to rule out any suggestion that other products might be affected.

The conventions you use to identify product and batches need to be specified in the product specifications and the stage at which product is marked specified in the relevant procedures or plans. Often such markings are automatically applied during processing, as is the case with printed circuits, mouldings, ceramics, castings, products etc. Process setting up procedures should specify how the marking equipment or tools are to be set up.

If you do release a batch of product prior to verification being performed and one out of the batch is subsequently found to be nonconforming, you will need to retrieve all others from the same batch. This may not be as simple as it seems. In order to retrieve a component which has subsequently been assembled into a printed circuit board, which has itself been fitted into a unit along with several other assemblies, not only would you need a good traceability system but also one that is constantly in operation.

It would be considered prudent to prohibit the premature release of product if you did not have an adequate traceability system in place. If nonconformity will be detected by the end product tests, allowing production to commence without the receipt tests being available may be a risk worth taking. However, if you lose the means of determining conformity by premature release, don't release the product until you have verified it as acceptable.

CUSTOMER PROPERTY (7.5.4)

Care of Customer Property (7.5.4)

The standard requires the organization *to exercise care with customer property while it is under the organization's control or being used by the organization.*

What Does this Mean?

Customer property is any property owned or provided by the customer and can include intellectual property and personal data. The product being supplied may have been produced by a competitor, by the customer or even by your own firm under a different contract. Customer property is any property supplied to you by your customer and not only what is to be incorporated into product to be supplied to customers. Customer owned tooling and returnable packaging also constitutes customer supplied product. The property being used may be supplied by the customer such as tools, software, and equipment or made available for the organization's use such as test and development facilities on customer premises.

> **Data Loss**
>
> The spate of data loss since the widespread use of laptop computers and portable storage media by public bodies shows how vulnerable our systems are in the hands of those carrying responsibility for it without the added risk of an external threat.

Documentation is not considered customer property because it is normally freely issued and ownership passes from customer to supplier on receipt. However, if the customer requires the documentation to be returned at the end of the contract, it should be treated as customer property.

Why is this Necessary?

This requirement responds to the Customer Focus Principle.

Anything you use that does not belong to you should be treated with due care particularly if it has been supplied for your use and is expected to be returned in good condition.

How is this Demonstrated?

For customer property that is used on your own premises you should maintain a register containing the following details:

- Name of product, part numbers, serial numbers and other identifying features.
- Name of customer and source of product if different.
- Delivery note reference, date of delivery.
- Receipt inspection requirements.
- Condition on receipt including reference to any rejection note.
- Storage conditions and place of storage.
- Maintenance specification if maintenance is required.
- Current location and name of custodian.
- Date of return to customer or embodiment into supplies.
- Part number and serial number of product embodying the customer supplied product.
- Dispatch note reference of assembly containing the product.

These details will help you keep track of the customer supplied product whether on embodiment loan^{①} or contract loan^{①} and will be useful during customer audits or in the event of a problem with the item either before or after dispatch of the associated assembly.

There might also be a need for a definition of responsibilities – a table showing which of the three parties (customer, supplier and your organization) is responsible for product acquisition, verification, repair, return to supplier, defect investigation etc. and what the associated financial liabilities are.

Identification of Customer Supplied Property (7.5.4)

The standard requires the organization *to identify customer property provided for use or incorporation into the product.*

What Does this Mean?

Identifying customer property means attaching labels or other means of identification that denote its owner.

Why is this Necessary?

This requirement responds to the Process Approach Principle.

If customer property carries an identity that distinguishes it from other product, it will prevent inadvertent disposal or unauthorized use.

How is this Demonstrated?

Customer property may carry suitable identification but if not, labels, containers or other markings may be necessary to distinguish it from organization owned property. As customer property may have been supplied by the organization originally as in the case of a repair service, labels indicating the owner should suffice. In a vehicle service area for instance, a label is attached to the car keys rather than labelling the car itself.

When deciding the type of marking, consideration needs to be given to the conditions of use. Markings may need to be permanent in order to be durable under the anticipated conditions of use. It would be wise to seek guidance from the customer if you are in any doubt as to where to place the marking or how to apply it. Metal identification plates stamped with the customer's identity, date of supply, contract and limitations of use are durable and permanent.

Verification of Customer Supplied Property (7.5.4)

The standard requires the organization *to verify customer property provided for use or incorporation into the product.*

What Does this Mean?

Customers may supply product purchased from other suppliers for installation in an assembly purchased from your organization.

Why is this Necessary?

This requirement responds to the Factual Approach Principle.

Product needs to be verified before incorporation into the organization's product regardless of its source firstly, to establish the condition of the item on receipt in the event that it is damaged, defective or is incomplete and secondly, to verify that it is fit for the intended purpose before use. If you fail to inspect the product on receipt you may find difficulty in convincing your customer later that the damage was not your fault.

How is this Demonstrated?

When property is received from a customer it should be processed in the same way as purchased product so that it is registered and subject to receipt inspection. The inspection you carry out may be limited if you do not possess the necessary equipment or specification, but you should reach an agreement with the customer as to the extent of any receipt inspection before the product arrives. You also need to match any delivery note with the product because the customer may have inadvertently sent you the wrong product. Unless you know what you are doing it is unwise to energize the product without proper instructions from the customer.

Protection of Customer Supplied Property (7.5.4)

The standard requires the organization *to protect customer property provided for use or incorporation into the product.*

What Does this Mean?

Protection means safeguarding against loss, damage, deterioration and misuse.

Why is this Necessary?

This requirement responds to the Customer Focus Principle.

As the property will either be returned to the customer on completion of contract or will be incorporated into your products, it is necessary to protect the product from conditions that may adversely affect its quality.

How is this Demonstrated?

Where the customer supplied property is in the form of products that could be inadvertently degraded, they should be segregated from other products to avoid mixing, inadvertent use, damage or loss. Depending on the size and quantity of the items and the frequency with which your customer supplies such products you may require a special storage areas. Wherever the items are stored you should maintain a register of such items, preferably separate from the store e.g. in inventory control or the project office. The authorization for releasing customer supplied property from stores may need to be different for inventory control reasons. You also need to ensure that such products are insured. You will not need a corresponding purchase order and they may not therefore be registered as stock or capital assets. If you receive customer supplied property very infrequently, you will need a simple process that is only activated when necessary rather than being built into the inventory control process. Under such circumstances it is easy to lose these products and forget they are someone else's property. You need to alert staff to take extra care especially if they are high value items which cannot readily be replaced.

Maintenance of Customer Supplied Property (7.5.4)

The standard requires the organization *to maintain customer property provided for use or incorporation into the product.*

What Does this Mean?

Maintenance of customer supplied property means retaining the property in the condition in which it was originally provided.

Why is this Necessary?

This requirement responds to the Process Approach Principle.

Customer supplied property that is issued for incorporation into supplies does not often require maintenance – however, items for use in conjunction with the contract may be retained for such duration that maintenance is necessary.

How is this Demonstrated?

If the property requires any maintenance you should be provided with a maintenance specification and the appropriate equipment to do the job. Maintenance may include both preventive and corrective maintenance but you should clarify with your customer which it is. You may have the means for preventive maintenance such as lubrication and calibration but not for repairs. Always establish your obligations in the contract regarding customer supplied property, because you could take on commitments for which you are not contractually covered if something should go wrong. You need to establish who will supply the spares and re-certify the equipment following repair.

Reporting Problems to the Customer (7.5.4)

The standard requires *occurrence of any customer property that is lost, damaged or otherwise found to be unsuitable for use to be reported to the customer and records maintained.*

What Does this Mean?

While customer property is on your premises, it may be damaged, develop a fault or become lost. Also when using customer property on customer premises, events may occur that result in damage or failure to the property.

Why is this Necessary?

This requirement responds to the Customer Focus Principle.

It is necessary to record and report any damage, loss or failure to the customer so that as owners they may decide the action that is required. Normally, the organization does not have responsibility to alter, replace or repair customer property unless authorized to do so under the terms of the contract.

How is this Demonstrated?

The customer is responsible for the product they supply wherever it came from in the first place. It is therefore very important that you establish the condition of the product before you store it or use it. In the event that you detect that the product is damaged, defective or is incomplete, you should place it in a quarantine area and report the condition to the customer. Even if the product is needed urgently and can still be used, you should obtain the agreement of your customer before using inferior product, otherwise you may be held liable for the consequences.

You could use your own reject note or nonconformity report format to notify the customer of a defective product but these are not appropriate if the product is lost. You also need a customer response to the problem and so a form that combined both a statement of the problem and of the solution would be more appropriate.

PRESERVATION OF PRODUCT (7.5.5)

The standard requires the organization *to preserve the product during internal processing and delivery to the intended destination in order to maintain conformity to requirements* and goes on to require these measures *to include identification, handling, packaging, storage and protection a*s applicable.

What Does this Mean?

These requirements are concerned with conformity control.[①] They apply to service operations that involve the supply of product such as maintenance.

Identification in this context means identifying the product through the packaging and can apply at any stage in the product realization process. Handling refers to the manner by which product is moved by hand or by machine. Packaging refers to the materials employed to protect the product during movement and storage refers to the place where product is held pending use, shipment or further processing. Protection applies at all stages in product realization.

Why is this Necessary?

This requirement responds to the Process Approach Principle.

As considerable effort will have gone into producing a conforming product, it is necessary to protect it from adverse conditions that could change the physical and functional characteristics. In some cases preservation is needed immediately the characteristics have been generated (e.g., surface finish). In other cases, preservation is only needed when the product leaves the controlled environment (e.g., food, chemicals and electronic goods). Preservation processes need to be controlled in order that product remains in its original condition until required for use.

How is this Demonstrated?

Determination of preservation requirements commences during the design phase or the manufacturing or service planning phase by assessing the risks to product quality during its manufacture, storage, movement, transportation and installation. Packaging design should be governed by the requirements of Clause 7.3 although if you only select existing designs of packaging these requirements would not be applicable.

The preservation processes should be designed to prolong the life of the product by inhibiting the effect of natural elements. While the conditions in the factory can be measured, those outside the factory can only be predicted.

Having identified there is a risk to product quality you may need to prepare instructions for the handling, storage, packing, preservation and delivery of particular items. In addition to issuing the instructions you will need to reference them in the appropriate work instructions in order that they are implemented when necessary. Whatever the method, you will need traceability from the identification of need to implementation of the provisions and from there to the records of achievement.

Identification

Packages for export may require different markings than those for the home market. Those for certain countries may need to comply with particular laws. Unless your customer has specified labelling requirements, markings should be applied both to primary and secondary packaging as well as to the product itself. Markings should also be made with materials that will survive the conditions of storage and transportation. Protection can be given to the markings while in storage and in transit but this cannot be guaranteed while products are in use. Markings applied to the product therefore need to be resistant to cleaning processes both in the factory and in use. Markings on packaging are therefore essential to warn handlers of any dangers or precautions they must observe. Limited Life Items should be identified so as to indicate their shelf life. The expiry date should be visible on the container and provisions should be made for such items to be removed from stock when their indicated life has expired.

While a well-equipped laboratory can determine the difference between different products and materials the consumer needs a simple practical method of identification and labelled packets is often a reliable and economic alternative. For products that start to deteriorate when the packaging seal is broken, the supplier's responsibility extends beyond delivery to the point of use. In such cases markings need to be applied to the containers to warn the consumers of the risks.

Handling

Handling provisions serve two purposes both related to safety. Protection of the product from the individual and protection of the individual handling the product. This latter condition is concerned with safety and addressed through other provisions, however, the two cannot and should not be separated and handling procedures should address both aspects.

Handling product can take various forms depending on the hazard you are trying to prevent from happening. In some cases notices on the product will suffice, such as 'LIFT HERE' or 'THIS WAY UP' or the notices on batteries warning of acid. In other cases you will need to provide special containers, or equipment. There follows a short list of handling provisions that your procedures may need to address:

- Lifting equipment,
- Pallets and containers,
- Conveyors and stackers,
- Design features for enabling handling of product,
- Handling of electrostatic sensitive devices,
- Handling hazardous materials,
- Handling fragile materials.

Packaging

Packing processes should be designed to protect the product from damage and deterioration under the conditions that can be expected during storage and transportation. You will need a means of identifying the packaging and marking requirements for particular products and of determining processes for the design of suitable packaging including the preservation and marking requirements. Depending on your business you may need to devise packages for various storage and transportation conditions, preservation methods for various types of product and marking requirements for types of product associated with their destination.

Unless your customer has specified packaging requirements, there are several national standards that can be used to select the appropriate packaging, marking and preservation requirements for your products. Your procedures should make provision for the selection to be made by competent personnel at the planning stage and for the requirements thus selected to be specified in the packing instructions to ensure their implementation.

Packing instructions should not only provide for protecting the product but also for including any accompanying documentation such as:

- assembly and installation instructions,
- licence and copyright notices,
- certificates of conformity,

- packing list identifying the contents of the container,
- export documents,
- warranty cards.

The packing instructions are likely to be one of the last instructions you provide and probably the last operation you will perform for a particular consignment. This also presents the last opportunity for you to make mistakes! They may be your *last* mistakes but they will be the *first* your customer sees. The error you made on component assembly probably won't be found, but the slightest error in the packaging, the marking, the enclosures will almost certainly be found therefore this process needs careful control. It may not be considered so skilled a process but all the same it is vital to your image.

Storage

In order to preserve the quality of items that have passed receipt verification they should be transferred to stockrooms in which they are secure from damage and deterioration. You need secure storage areas for several reasons:

- For preventing personnel from entering the stockrooms and removing items without authorization.
- For preventing items from losing their identity. (Once the identity is lost it is often difficult, if not impossible to restore complete identification without testing material or other properties.)
- For preventing vermin damaging the stock.
- For preventing climatic elements causing stock to deteriorate.

While loss of product may not be considered a quality matter, it is if the product is customer property or if it prevents you from meeting your customer requirements. Delivery on time is a quality characteristic of the service you provide to your customer and therefore secure storage is essential.

To address these requirements you will need to identify and specify the storage areas that have been established to protect product pending use or delivery. Although it need be only a brief specification, the requirements to be maintained by each storage area should be specified based on the type of product, the conditions required preserving its quality, its location and environment. Products that require storage at certain temperatures should be stored in areas that maintain such temperatures. If the environment in the area in which the room is located is either uncontrolled or at a significantly higher or lower temperature, an environmentally controlled storage area will be required.

All items have a limit beyond which deterioration may occur and therefore temperature, humidity, pressure, air quality, radiation, vibration etc. may need to be controlled. At some stage, usually during design or manufacturing or service planning, the storage conditions need to be defined and displayed. In many cases dry conditions at room temperature are all that is necessary but problems may occur when items requiring non-standard conditions are acquired. You will need a means of ensuring that such items are afforded the necessary protection and your storage procedures need to address this aspect. It is for this reason that it is wiser to store items in their original packaging until required for use. If packets need to be opened to verify product identity etc. the packaging design is already noncompliant.

Any area where product is stored should have been designated for that purpose in order that the necessary controls can be employed. If you store product in undesignated areas then there is a chance that the necessary controls will not be applied. Designation can be accomplished by placing notices and markers around the area to indicate the boundaries where the controls apply.

Each time the storage controller retrieves an item for issue, there is an opportunity to check the condition of stock. However, some items may have a slow turnover in certain storage areas, e.g., where spares are held pending use. It is also necessary to plan and carry out regular checks of the overall condition of the stockroom for damage to the fabric of the building. Rainwater may be leaking onto packaging and go undetected until that item is removed for use.

Some items such as electrolytic capacitors and two part adhesives may deteriorate when dormant. Others such as rubber materials, adhesive tape and chemicals deteriorate with the passage of time regardless of use. These are often referred to as 'Shelf Life Items' or 'Limited Life Items'. Dormant electronic assemblies can deteriorate in storage and in the unlikely event that product would remain in storage for more than one year, provision should be made to retest equipment periodically or prior to release.

The assessment interval should depend on the type of building, the stock turnover, the environment in which the stock is located and the number of people allowed access. The interval may vary from storage area to storage area and should be reviewed and adjusted as appropriate following the results of the assessment.

Segregation

Segregation is vital in many industries where products can only be positively identified by their containers. It is also important to prevent possible mixing or exposure to adverse conditions or cross contamination. Examples where segregation makes sense are:

- Toxic materials,
- Flammable materials,
- Limited life items,
- Explosives.

Segregation is not only limited to the product but also to the containers and tools used with the product. Particles left in containers and on tools – no matter how small, can cause blemishes in paint and other finishes as well as violate health and safety regulations. If there are such risks in your manufacturing process then procedures need to be put in place that will prevent product mixing.

Segregation may also be necessary in the packaging of products not only to prevent visible damage but also to prevent electrical damage as with electrostatic sensitive devices. Segregation may be the only way of providing adequate product identity as is the case with fasteners.

Cleaning

Where applicable, preservation processes should require that the product be cleaned before being packed and preservative applied. In other cases the product may need to be stored in sealed containers in order to retard decay, corrosion and/or contamination.

Key Messages from Part 6

1. Product realization is driven by customers not markets. The demand creation process precedes product realization. If your organization is market driven, several important processes will be excluded from your quality management system if you simply respond to the product realization requirements.

2. All organizations are constrained by their resources; none has an unlimited capability but success depends on managers having knowledge of current capability of the process in order for their plans to be viable.

3. A sales person who promises a short delivery to win an order invariably places an impossible burden on the company. A company's capability is not increased by accepting contracts beyond its current level of capability.

4. Design is often a process which strives to set new levels of performance and as such can be a journey into the unknown. On such a journey we can encounter obstacles we haven't predicted which may cause us to change our course but our objective remains unchanged.

5. Design control is a method of keeping the design on course towards its objectives; it does not mean controlling the creativity of the designers; it means controlling the process through which new or modified designs are produced so that the resultant design is one that truly reflects customer needs.

6. Change control during the design process is a good method of controlling costs and timescales because once the design process has commenced every change will cost time and effort to address.

7. One reason for controlling design changes is to restrain the otherwise limitless creativity of designers in order to keep the design within the budget and timescale.

8. Design changes will result in changes to documentation but not all design documentation changes are design changes. By keeping the two types of change separate you avoid bottlenecks in the design change loop and only present the design authorities with changes that require *their* expert judgment.

9. A change to a design that has not proceeded beyond a design review or verification stage is still in progress and therefore requires no approval. When a design is reviewed or verified it means that any change to the information on which that decision was taken needs to be evaluated for its effect on the design and any product produced from that design.

10. It would be counterproductive to impose rigorous controls over every supplier and purchased product. A balance has to be made on the basis of risk to the processes and products in which the purchased product is to be used.

11. You should discourage informal instructions because you cannot rely on them being used by others when those who know them are absent.

12. There will be matters outside your control and matters over which you need complete control. It is the latter that you can do something about and take corrective action should the target not be achieved.

13. Operators should understand what causes the dots on the chart to vary. Process control comes about by operators knowing what results to achieve, by knowing what results are being achieved and by being able to correct performance should the results not be as required.

14. According to ISO 9000:2005, validation means "*confirmation, through the provision of objective evidence that the requirements for a specific intended use or application have been fulfilled*" Therefore confirming that a process fulfils its purpose would appear to be validation and apply to all processes not simply production processes.

15. Measurement does not change a product but does change our knowledge of it.

16. It would be considered prudent to prohibit the premature release of product if you did not have an adequate traceability system in place.

17. Although the standard only requires traceability when required by contract or law, it is key to finding the root cause of problems and undertaking corrective action.

18. Determination of preservation requirements commences during the design phase or the manufacturing or service planning phase by assessing the risks to product quality during its manufacture, storage, movement, transportation and installation

19. Mistakes you make at the packing stage may be your *last* mistakes but they will be the *first* your customer sees.

20. While loss of product may not be considered a quality matter delivery on time is a quality characteristic of the service you provide to your customer and therefore secure storage is essential.

Complying with ISO 9001 Section 8 Requirements on Measurement, Analysis and Improvement

INTRODUCTION TO PART 7

Structure of ISO 9001 Section 8

There is a sort of logic in the structure of the requirements in this section but there are some gaps. It would have assisted understanding if the same terms as used in Clause 8.1 had been used in the headings of Clauses 8.2–8.5. In that way the relationships would have been more obvious. The general requirements of Clause 8.1 are amplified by Clauses 8.2–8.5 so the requirements in Clause 8.1 are not separate to those in Clauses 8.2–8.5 with the exception of Clause 8.3 on the control of nonconforming product and those on statistical techniques. This latter requirement is stated once because it applies to all monitoring, measurement and analysis processes. Clause 8.3 on the other hand appears in Section 8 not because it has anything to do with monitoring, measurement, analysis and improvement but because its inclusion in Section 7 would imply that it could be excluded from the management system – quite illogical! (see ISO 9001 Clause 1.2).

It should also not to be assumed that Section 8 includes all requirements on measurement, analysis and improvement as such requirements are also addressed in Clauses 5.5.2, 5.5.3, 5.6, 6.1, 7.3.5, 7.3.6, 7.3.7, 7.4.3 and 7.6. Rather than address the requirements of Clause 7.6 in Part 6 of this book we have included them in Part 8 among the other requirements on measurement.

Linking the Requirements

We can link the requirements of Section 8 together in a number of separate cycles (indicating the headings from ISO 9001 in bold italic type).

During the design and development of the business and work processes we would undertake a review of the established practices to identify potential problems and undertake **preventive action** to prevent occurrence of such problems. Before implementing the management system processes or any changes thereto, we would perform **internal audits** (or undertake **process validation**) to determine whether these processes met the relevant requirements of the standard, enabled the organization to fulfil its **policies** and **objectives** and produce the required products and services. Any potential problems discovered would be subject to **preventive action** to prevent occurrence of such problems. At appropriate stages in the production process we would **monitor and measure product** for compliance with specified requirements, periodically undertake **product audits** to establish the effectiveness of the process controls, initiate **control of nonconforming product** on detecting nonconformity and undertake **corrective action** to eliminate the cause of nonconformity when process targets had not been met. After introducing new or changed practices and periodically thereafter, we would perform **internal audits** to determine whether the planned arrangements were being implemented as intended and undertake **corrective action** to bring about **improvement** by better control. Periodically we would **monitor and measure processes** for their ability to achieve planned results and undertake **process audits** to establish whether the achieved results arose from implementing the planned arrangements and if necessary, undertake corrective action to reduce variation and bring about **improvement** by better control. We would also **analyse data** resulting from these reviews and bring about **improvements** by better utilization of resources. Sometime after establishing the **organization's purpose,** setting **policies** and **objectives** that were **customer focused** and installed the enabling **processes** we would collect and **analyse data** in order to monitor **customer satisfaction** and undertake **corrective action** to bring about **improvement** by better control.

One observes from this consolidation that the order in which the clauses are mentioned is not remotely the same as the order they are addressed in the standard, that some clauses in the standard appear several times and others are drawn from other sections thus demonstrating that you cannot treat the clauses in isolation or in any particular sequence.

Quality Assurance

Both customers and managers have a need for an assurance of quality because they are not in a position to oversee operations for themselves. They need to place trust in the producing operations, thus avoiding constant intervention.

ISO 9000:2005 defines *quality assurance* as part of quality management focused on providing confidence that quality requirements will be fulfilled.

This concept has some similarity with the concept of the financial audit which provides assurance of financial probity by establishing that:

1. The plan for controlling the organization's finances is sound and will, if followed, ensure all financial transactions can be accounted for
2. The plan is being followed as prescribed

Quality assurance activities do not control quality, they establish the extent to which quality will be, is being or has been controlled. All quality assurance activities are post-event activities (i.e., after the event being assessed) and off-line and serve to build confidence in results, in claims, in predictions, etc. If a person tells you they will do a certain job for a certain price in a certain time, can you trust them or will they be late, overspent and over limits? The only way to find out is to gain confidence in their capability and that is what quality assurance activities are designed to do. Quite often, the means to provide the assurance need to be built into the process, such as creating records, documenting plans, documenting specifications, reporting reviews etc. Such documents and activities also serve to control quality as well as assure it. ISO 9001 provides a basis for obtaining an assurance of quality, if you are the customer, and a basis for controlling quality, if you are the supplier.

The following steps can obtain an assurance of quality

- Acquire the documents that declare the organization's plans for achieving quality.
- Produce a plan that defines how an assurance of quality will be obtained, i.e. a quality assurance plan.
- Organize the resources to implement the plans for quality assurance.
- Establish whether the organization's proposed product or service possesses characteristics that will satisfy customer needs.
- Assess operations, products and services of the organization and determine where and what the quality risks are.
- Establish whether the organization's plans make adequate provision for the control, elimination or reduction of the identified risks.
- Determine the extent to which the organization's plans are being implemented and risks contained.
- Establish whether the product or service being supplied has the prescribed characteristics.

There are internal and external audits. External audits may be performed by customers or regulators or by independent third parties. Internal audits may be performed by dedicated auditors from finance or quality departments or by personnel engaged by the manager responsible for the results of the process. Internal audit goes beyond checking that procedures have been followed, into determining whether what was planned to be done has been done and whether the right things were planned to be done.

Measurement

Measurement, analysis and improvement processes are vital to the achievement of quality. Until we measure using devices of known integrity, we know little about a process or its outcomes. Before we measure we need standards, targets, requirements etc. we can use to judge the results of measurement, in fact there is little point measuring anything unless we have a clear idea of what we are looking for. The target value is therefore vital but arbitrary values demotivate personnel. Targets should always be focused on purpose so that through the chain of measures from corporate objectives to component dimensions there is a soundly based relationship between targets, measures, objectives and the purpose of the organization, process or product. But if we measure using instruments that are unfit for purpose, we will be misled by the results. With the results of valid measurement we can make a judgement on the basis of facts. The facts will tell us whether we have met the target. Analysis of the facts will tell us whether the target can be met using the same methods or better methods or whether the target is the right target to aim for.

Measurement tells us whether there has been a change in performance. Change is a constant. It exists in everything and is caused by physical, social or economic forces. When we measure the same parameter on different entities we expect slight variation. However, if we measure the same parameter using the same device we might not expect there to be a change, but the inaccuracies inherent in the measuring system will lead to a variation in readings. To understand change we need to understand its cause. Some change is represented by variation about a norm and is predictable – it is a natural phenomenon of a process and when it is within acceptable limits it is tolerable. Other change is represented by erratic behaviour and is not predictable but its cause can be determined and eliminated through measurement, analysis and improvement.

Invariably, managers do monitor and measure outputs but often react to the results and start tinkering with the process when what they observed was not random but the effect of a problem with an assignable cause. What is not often measured is the capability of the process to deliver the required outputs. This is now commonplace in manufacturing especially among the mass producers such as white goods, automotive and, aerospace etc. but even in these industries it is only common in the making processes. There is little evidence that process capability is even considered in the management and innovative processes but the requirements of this section are meant to apply to all processes.

Coming to Terms with Correction, Corrective Action and Preventive Action

Within Section 8 of ISO 9001 there are requirements for control of nonconformity, corrective action and preventive action and are all related to product nonconformity. There is considerable confusion about these terms because (a) both definitions of corrective and preventive action include the word *prevent* and (b) we tend to think of correction as fixing a problem or correcting an error so when the word corrective and action are put together we naturally think it is referring to the actions we take when correcting errors but this is not how the terms are used in the ISO 9000 family of standards. There are also three states of nonconformity (see text box) that add to the confusion and in the following explanations we specify which type of nonconformity applies.

Types of Nonconformity

Potential nonconformity is when conditions are present that may cause a nonconformity if no action is taken. For example (1) a process that will produce nonconforming product if not arrested or (2) a design that produces a product that will become nonconforming under certain conditions of use.

Actual nonconformity is a verified non-fulfilment of a requirement. For example (1) when measuring a product characteristic the observed value is different to the required value or (2) when subjecting a product to the specified operating conditions it fails to function as required.

Suspect nonconformity is when there is the possibility that conditions could have caused nonconformity. For example (1) a product from a batch in which some have been found nonconforming; (2) a product that possesses some of the same characteristics as the nonconforming product or (3) a product that has been accepted using equipment subsequently found out of calibration or (4) product is mishandled but shows no obvious signs of damage or (5) a product with no indication of verification status.

Correction

Correction is the term used to describe the action of removing an actual or suspect nonconformity in a product before its acceptance. Correction does not stop the nonconformity recurring. As correction is applied before a product is completed, actions intended to restore, recover or remedy the situation are inappropriate as a conforming condition has not been reached. Examples are:

- While observing the performance of a process you notice that the values are drifting towards the upper limits. You adjust the process and bring it back under control. You have corrected the process and avoided an occurrence of product nonconformity but not a future occurrence; you have merely delayed its occurrence.

- You are delivering a service and the customer points out an error which you correct immediately.
- Sometimes a product may be inadvertently submitted for verification before all operations have been completed and in such cases correction involves normal operations to complete the item
- A product can be made to conform by continuation of processing; this type of correction is called rework.
- An unknown state can be corrected by carrying out verification and declaring the product conforming or nonconforming.

Remedial Action

Remedial action is the term used to describe the action of removing an actual nonconformity in a product that was previously deemed conforming. Examples are

- If a conforming product has been damaged or in some other way becomes nonconforming, action to remove the detected nonconformity is a *remedial action.*
- An automated cash machine in a bank has malfunctioned and the supervisor investigates, locates the problem and resets the operating conditions.
- Repair is a remedial action that restores an item to an acceptable condition but unlike rework, it may involve changing the product so that it differs from the specification but fulfils the intended use

Recovery Action

Recovery action is the action of seeking out products with the same characteristics as those found nonconforming. It is not a corrective action because it removes a nonconformity not the cause of the nonconformity. Recall action is a recovery action even though the nonconformity might not have been exhibited.

Corrective Action

Corrective action is the term used to describe the pattern of activities that traces the symptoms of a actual or suspect nonconformity to its cause, produces solutions for preventing its recurrence, implements the change and monitors that the change has been successful. Such an action prevents the recurrence of the nonconformity. A problem has to exist for you to take corrective action. Corrective action uses Root Cause Analysis to discover and eliminate the actual causes. **Containment action** removes an immediate cause thus allowing production to continue until product or process design change removes the root cause. Examples are:

- The root cause of nonconformity is found to be a deficiency in design practices. The action to introduce new design practices is a corrective action.
- If the conditions for nonconformity are present because they have been detected in other similar products but have not yet resulted in failure, the action of preventing failure is a corrective action. This argument is based on the premise that a nonconformity is a non-fulfilment of a requirement, therefore a product that has not exhibited failure but possess the potential for failure is a nonconforming product not a potentially nonconforming product i.e. we know the nonconformity exists, its just a matter of time before it causes a failure.
- If a process is already running and will produce nonconforming output either now or in the future it is not capable. Any action whether undertaken now or before a nonconforming product is produced is a corrective action because there is an existing flaw in the process that has not been eliminated by design. The action taken simply puts the process back to where it should have been (see also preventive action below). However, by taking corrective action on the process your action becomes a preventive action on the product.

Preventive Action

Preventive action is the term used to describe the action of eliminating a cause of a potential nonconformity. Preventive action is triggered by a discovery (what if this trend continues or what if these conditions exist?) and serves to prevent a problem from occurring for the first time in a product or process. It uses search and remove techniques such as FMEA, HACCP etc. to discover and eliminate the potential causes. Examples are:

- Monitoring performance and observing an undesirable trend, then taking action to remove the root cause before a failure occurs is taking *preventive action* even though there is no evidence that the occurrence could be imminent. However, it is correcting a process nonconformity and preventing a product nonconformity.

- Risk analysis, failure modes analysis, hazards analysis, stress analysis, reliability predictions or any similar type of analysis performed to identify design weaknesses is a preventive action *if* action is taken as a result of the analysis.
- Any action taken to ensure success. Research, planning, training, preparation, organizing and resourcing are all activities which if done well will prevent nonconformities arising.

Chapter 28

Monitoring, Measurement, Analysis and Improvement Processes

CHAPTER PREVIEW

This chapter is aimed at all those managers responsible for the outputs of processes. This may include process owners, departmental managers and those personnel designing and operating business and work processes.

Measurement, analysis and improvement are strictly sub-processes within each business process. In many cases these processes do not exist as separate work processes let alone business processes for every process should include monitoring, measurement, analysis and improvement activities. However, parent processes will often capture data from monitoring and measurements within sub-processes. This may happen when assessing a variety of data from individual processes to determine customer satisfaction or for discovering common cause problems and subsequently devising company wide improvement programmes.

The danger in treating these activities as separate processes is that they may be made to serve objectives that have not been derived from the parent process objective.

Beware of Reductionism

If the manager of an internal department starts to think of its analysis operations as a separate business unit, it might begin to promote its analysis capability beyond its original remit resulting in the resource allocation being depleted.

A Managing Director decides to outsource the internal quality audit function and the new supplier becomes so focused on undercutting the competition that the information yielded by the audits is of little value other than to keep the ISO 9001 certification body content.

The Managing Director being unaware of the part played by internal quality audit tasks the Financial Director with a new job of determining whether the organization's policies and strategies are being implemented as planned thus re-establishing the internal quality audit function under a new manager.

In this chapter we examine the requirements in Clause 8.1 of ISO 9001:2008 and in particular the processes:

- to demonstrate conformity such as design verification and validation;
- to ensure conformity of management system such as internal audit and validation of processes; and
- to continually improve the effectiveness of the management system such as management review.

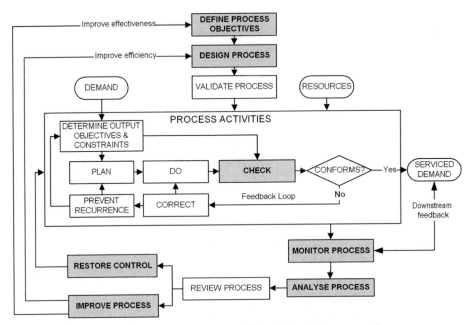

FIGURE 28-1   Where the requirements of Clause 8.1 apply in a managed process.

The position where requirements on monitoring, measurement, analysis and improvement processes feature in a managed process is shown in Fig. 28-1.

PROCESSES TO DEMONSTRATE CONFORMITY (8.1)

The standard requires the organization *to plan and implement the monitoring, measurement, analysis and improvement processes needed to demonstrate conformity to product requirements.*

What Does this Mean?

If a standard, target or performance indicator defines what we should be doing, measurement tells us what we are doing. Therefore without measurement we really know very little about our performance or the quantity or quality of an output. The control of quality depends on an ability to measure quality be it the quality of products, services, processes, systems, organizations or simply the quality of actions and decisions. Without measurement we won't even know whether we are getting better, getting worse or staying the same. The size of the performance gap shown in Fig. 2-1 will not be known.

The requirement for monitoring, measurement, analysis and improvement processes to be planned means that these activities should not be left to chance.

The monitoring processes that are needed to demonstrate conformity of product are addressed in the standard by:

- The monitoring of customer satisfaction (8.2.1).
- Monitoring of product (8.2.4) including design and development review (7.3.4).

These are the processes that keep operations and operating conditions under periodic or continual observation in order to be alerted to events before they occur. This is so that action can be taken to prevent nonconformity. Typical monitoring processes are those employed to check that machines are functioning correctly, that processes are under statistical control, that there are no serious bottlenecks or other conditions that may cause abnormal performance.

The measurement processes that are needed to demonstrate conformity of product are addressed in the standard by:

Measurement of product (8.2.4) including:

- Design and development verification (7.3.5);
- Design and development validation (7.3.6);
- Verification of purchased product (7.4.3); and
- Control of monitoring and measuring devices (7.6)

These are the processes that determine the characteristics of product in order to establish whether the product meets defined requirements. Typical measurement processes involve physical measurement of a product using inspection, test or demonstration techniques, or non-physical measurement of services using observation or examination techniques. These processes are found in design, production and service delivery operations.

The analysis processes to demonstrate conformity of product are addressed in the standard by:

- Design and development review (7.3.4) (This is both a monitoring and an analysis process.)
- Corrective action (8.5.2).
- Preventive action (8.5.3).

These are those processes that convert product data into knowledge from which decisions of conformity against prescribed standards can be made. Typical analysis processes are chemical and microbiological analyses to ascertain the properties or composition of a substance, stress analysis to ascertain the load bearing capacity of a structure. Other types of analysis may be used to compare current designs with previously validated designs, determine time-dependent characteristics, assess vulnerability, susceptibility and durability.

The improvement processes needed to demonstrate conformity of the product are those processes that eliminate the cause of nonconformities and prevent their occurrence. These include the corrective action processes and preventive action processes. (These are dealt with under the relevant headings.)

Why is this Necessary?

This requirement responds to the Factual Approach Principle.

Monitoring, measurement, analysis and improvement processes are necessary to control the quality, cost and delivery of output. Where processes needed to demonstrate conformity of the product are performed, they should form part of a plan, the intention of which is to discover whether the product conforms to requirements. Having discovered that information, it should also be planned that the information is passed through analysis processes on to the decision-makers so that decisions on product acceptability are made on the basis of fact and not opinion. Also as part of the plan, it is necessary to make provision for dealing with unacceptable product, either making it conform or preventing its unintended use or further processing.

How is this Demonstrated?

The Measurement Process[i]

Measurement begins with a definition of the measure, the quantity that is to be measured, and it always involves a comparison of the measure with some known quantity of the same kind. If the measure is not accessible for direct comparison, it is converted or 'transduced' into an analogous measurement signal. As measurement always involves some interaction between the measure and the observer or observing instrument, there is always an exchange of energy, which, although in everyday applications is negligible, can become considerable in some types of measurement and thereby limit accuracy.[1] So for instance if we want to know whether food is safe to eat, rather than having people sample it, we might count the bugs in it and if the bug count is below a certain level (the standard) we deem it to be safe for human consumption. Therefore, providing the standards are expressed in measurable terms we can measure conformity.

Any measuring requirement for a quantity requires the measurement process to be capable of accurately measuring the quantity with consistency. For this to happen, the factors that affect the result need to be identified and a process designed that takes into account the variations in these factors and delivers a result that can be relied on as being accurate within defined limits.

To measure something we need the following:

- Sensor (a detecting device that can be human with or without measuring instruments).
- Converter if the sensor is not human (a device for converting the signal from the sensor into a form that the human senses can detect).
- Transmitter where the measurement is done remotely (a device for transmitting the signal to a receiver for analysis).

Sensors need to be accurate, precise, reliable and economic. Sensors that tell lies are of use only to those who wish to deceive. It is too easy to look at a clock, a speedometer, a thermometer or any other instrument and take it for granted that it is telling you the truth. We often put more credence into the readings we get from instruments than we do from our own sensors but both can be equally inaccurate.

Measurement Uncertainty[i]

There is uncertainty in all measurement processes. There are uncertainties attributable to the measuring equipment[i] being used, the person carrying out the measurements and the environment in which the measurements are carried out. When you make a measurement with a calibrated instrument you need to know the specified limits of permissible error (how close to the true value the measurement is). If you are operating

---

[1] Measurement (1994–1999) *Britannica® CD 99 Multimedia Edition*©. Encylopædia Britannica, Inc.

under stable environmental conditions, you can assume that any calibrated equipment will not exceed the limit of permissible error. Stable conditions exist when all variation is under statistical control. This means that all variation is due to common causes only and none due to special causes. In other cases you will need to estimate the amount of error and take this into account when making your measurements. Test specifications and drawings etc. should specify characteristics in true values, i.e., values that do not take into account any inherent errors. Your test and inspection procedures, however, should specify the characteristics to be measured taking into account all the errors and uncertainties that are attributable to the equipment, the personnel and the environment when the measurement system is in statistical control. This can be achieved by tightening the tolerances in order to be confident that the actual dimensions are within the specified limits.

Controlling Variation in Measurement Processes

Measurement processes must be in statistical control so that all variation is due to common cause and not special cause variation[①]. It is often assumed that the measurements taken with a calibrated equipment are accurate and indeed they are if we take account of the variation that is present in every measuring system and bring the system under statistical control. Variation in measurement processes arises due to bias, repeatability, reproducibility, stability and linearity.

Bias is the difference between the observed average of the measurements and the reference value.

Repeatability is the variation in measurements obtained by one appraiser using one measuring equipment to measure an identical characteristic on the same part.

Reproducibility is the variation in the average of the measurements made by different appraisers using the same measuring instrument when measuring an identical characteristic on the same part.

Stability is the total variation in the measurements obtained with a measurement system on the same part when measuring a single characteristic over a period of time.

Linearity is the difference in the bias values through the expected operating range of the measuring equipment[①].

It is only possible to supply parts with identical characteristics if the measurement processes as well as the production processes are under statistical control. In an environment in which daily production quantities are in the range of 1000–10 000 units, inaccuracies in the measurement processes that go undetected can have a disastrous impact on customer satisfaction and consequently profits.

Calibration is a process that has evolved to enable us to establish the accuracy and precision of our measuring devices. It ensures the integrity of measurement but is not limited to physical devices although much of what is written about calibration is concerned with such things. The characteristics of bias, repeatability, reproducibility, stability and linearity in a measuring system are just as relevant when the measuring device is a school examination, a training assessment, a driving test or any other device for assessing the knowledge, skill or competence of personnel to achieve prescribed standards. What value are the results of an examination if a written answer to a question receives widely different marks from different examiners? What value is a college degree if not all recipients are graded using the same measuring process of known integrity?

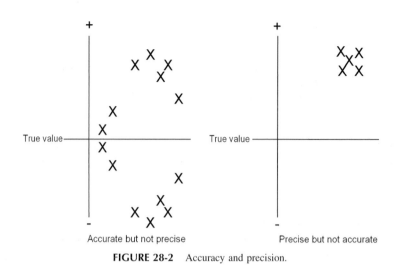

FIGURE 28-2   Accuracy and precision.

Accuracy and Precision in Measurement Processes

Accuracy and precision are often perceived as synonyms but they are quite different concepts. Accuracy is the difference between the average of a series of measurements and the true value. Precision is the amount of variation around the average. So you can have a measuring device that gives a large variation around the true value with repeated measurements but whose average is the true value (see Fig. 28-2).

Alternatively you could have a device which gives small variation with repeated measurements around a value which is wide of the true value. The aim is to obtain both accuracy and precision. The difference in accuracy and precision can cause expensive errors. You should not assume that the result you have obtained is both accurate and precise unless the device has been calibrated immediately prior to use and the results of its accuracy and precision provided.

The concept of accuracy and precision is also used in operating processes when determining process capability (see Chapter 31).

Planning the Monitoring, Measurement and Analysis Processes

Whether monitoring or measuring, the processes are very similar and consist of a uniform series of elements.[2]

- The characteristic to be measured or monitored.
- The units of measure.
- The standard or specification to be achieved.
- The sensor for detecting variance.
- The human and physical resources for collecting, analysing, transmitting and presenting the data.
- Interpretation and verification of results.

---

[2] Juran, J.M. (1995) *Managerial Breakthrough*. Second Edition, McGraw-Hill.

- Decision on the action needed.
- Taking action.

Some of these elements were addressed under *Control of measuring and monitoring equipment* in Chapter 32.

In planning the monitoring and measurement processes needed to demonstrate conformity of the product the first thing to do is to identify the characteristics that need to be achieved and then determine how and where they are going to be verified. It is also necessary to determine the conditions that affect the achievement of these characteristics such as key process parameters and establish how these will be monitored.

A useful approach is to develop for each product a Verification Matrix that identifies the requirement to be achieved, the level at which the requirement is achieved and the method to be employed. Some requirements may be verified during design verification and not require confirmation during production because they are inherent features of the design. Other requirements may need to be reconfirmed on each product due to variations in materials or processes. Some characteristics may be only accessible for verification at the component or sub-assembly levels whereas others can be verified at the end product level.

Another useful approach that is used in the automotive industry is a control plan. The aim of the control plan is to ensure that all process outputs will be in a state of control by providing process monitoring and control methods that control product and process characteristics. Apart from product identification data, the type of information contained in a control plan for each manufacturing operation is as follows:

- Operation description,
- Characteristic,
- Specification,
- Measurement technique,
- Sample size and frequency,
- Control method (control chart, 100% inspection, functional test, check sheet etc.),
- Reaction plan (what to do in the event of a nonconformity).

For each production process, there should be a process specification that defines the standard operating conditions that need to be maintained and the means by which variation in these conditions is to be detected. Various instruments may be needed to provide a visual indication of operating conditions and the specification should also indicate the frequency of checks if they are located remote from the operator.

Determining Measurement Methods

Techniques for establishing and controlling process capability are essentially the same – the difference lies in what you do with the results. Firstly you need to know if you can make the product or deliver the service in compliance with the agreed specification. For this you need to know if the process is capable of yielding conforming product. Statistical Process Control (SPC) techniques will give you this information. Secondly you need to know if the product or service produced by the process actually meets the requirements. SPC techniques will also provide this information. However, having obtained the results you need the ability to change the process in order that all product or service remains within specified limits and this requires either real-time or off-line process monitoring to detect and correct variance. To verify process capability you periodically rerun the analysis by measuring output product characteristics and establishing that the results demonstrate that the process remains capable.

When carrying out quality planning you will be examining intended product characteristics and it is at this stage that you will need to consider how achievement is to be measured and what tool or technique is to be used to perform the measurement. When you have chosen the tool you need to describe its use in the control plan.

PROCESSES TO ENSURE CONFORMITY OF MANAGEMENT SYSTEM (8.1)

The standard requires the organization *to plan and implement the monitoring, measurement, analysis and improvement processes needed to ensure conformity of the quality management system.*

What Does this Mean?

Conformity of the management system means that the system has been designed with the capability of implementing the defined policies and fulfilling the established objectives and targets and is being operated in a manner consistent with these policies, objectives and targets.

The monitoring processes to ensure conformity of the management system are addressed in the standard by:

- Reporting on management system performance (5.5.2).
- Review of system adequacy (5.6).
- Internal audit (8.2.2).

(Note: System adequacy is interpreted as a state where the system delivers the required outputs. They may not be produced efficiently or be the right outputs but they are the ones required.)

More generally in this context the monitoring processes are those that keep operations and operating conditions under periodic or continual observation. The reason for doing this is to be alerted to events before, during or immediately after they occur so that action can be taken to prevent or minimize their effects. Typical monitoring processes are those employed in the utility industries where advanced warning of potential or actual problems is needed to maintain the supply of electricity, water and gas to consumers. Management monitor expenditure or rate of expenditure in order to forecast whether their current and future commitments can be met. Operators monitor machines to check that they are functioning correctly, that processes are under statistical control. Distribution staff may monitor on-time delivery and ring the alarm bells when this drops below the norm.

The measurement processes needed to ensure conformity of the management system are those processes that measure the performance of the business processes. Every process should not only contain provision for measuring output but for measuring whether the process objectives are being achieved. Audits are one means of verifying that the management system conforms in both its design and implementation. Such audits would not only determine if procedures were being followed but whether the desired results were being achieved and any regulations complied with. (Audits are dealt with separately under the relevant heading.)

The analysis processes to ensure conformity of the management system are addressed in the standard by:

- Internal audit (8.2.2). This is in the context of audits of implementation.
- Analysis of data (8.4).

More generally the analysis processes needed to ensure conformity of the management system are those processes that convert process data into knowledge from which decisions of conformity against prescribed standards can be made. Examples include the analysis of variation in processes, process capability studies, behavioural analysis, organizational analysis, workflow analysis, analysis of constraints, risk analysis, hazard analysis and defect analysis.

The improvement processes needed to ensure conformity of the management system are the corrective action processes that reduce variation in process performance and restore the status quo. (Corrective action is dealt with separately under the relevant heading.)

Why is this Necessary?

This requirement responds to the Factual Approach Principle.

As the management system is the means by which the organization achieves its objectives, it is necessary to ensure that the processes are well designed and are operating properly. Much of the focus in previous versions of the standard was on measuring product and auditing procedures. It has been widely accepted that both these techniques are not only ineffective in ensuring customer satisfaction but are also uneconomic. A more effective technique is to monitor and measure processes and strive to reduce variation in these processes. By setting up processes that are capable of delivering conforming output and then monitoring, measuring and improving these processes so that process capability is assured, less dependence needs to be put on measuring product and auditing procedures and as a result less resources are utilized.

How is this Demonstrated?

These requirements are expanded in the other sections of Clause 8 of the standard and therefore the methods for implementing them are addressed under the appropriate heading.

PROCESSES TO CONTINUALLY IMPROVE THE EFFECTIVENESS OF THE MANAGEMENT SYSTEM (8.1)

The standard requires the organization *to plan and implement the monitoring, measurement, analysis and improvement processes needed to continually improve the effectiveness of the quality management system.*

Case Study – Measuring System Effectiveness

We measure the effectiveness of our QMS by the level of customer complaints and nonconformities arising from inspections and audits and in the past the external auditors have been satisfied with our approach. ISO 9001 appears to imply that we need to determine how well the QMS is enabling us to achieve our objectives. Therefore, if we are failing to achieve our objectives will the auditors judge that the QMS is ineffective and thus identify a major nonconformity?

ISO 17021 now requires third party auditors to determine whether the QMS is effective in enabling the organization to achieve its objectives, key performance targets and compliance with legislation therefore auditors will want to see evidence of a link between the system and the results.

If you are only looking at the level of customer complaints and nonconformities arising from inspections and audits, the way you measure the effectiveness of your QMS needs to change.

Firstly, the objectives the QMS is designed to achieve as a minimum should be derived from the needs and expectations of customers. Unless you can demonstrate the link, the external auditors should issue a nonconformity against Clause 5.4.1.

Secondly, you should be measuring your performance against these objectives. Unless you can demonstrate the link, between what you are measuring and the objectives the QMS is designed to achieve the external auditors should issue a nonconformity against either Clause 8.2.1, 8.2.3 or 8.2.4. Also if you can't demonstrate the integrity of the measurement methods you are using, the external auditors should issue a nonconformity against Clause 7.6.

Thirdly, you should be taking action to improve performance and this should be undertaken in line with the established processes. Unless you can demonstrate the link between the actions you are taking or plan to take and the performance as measured the external auditors should issue a nonconformity against clause 8.5.2.

Fourthly, unless you can demonstrate the link between the preventive actions taken when designing the QMS, the objectives and what you deduce was the cause of the failure to achieve your objectives, the external auditors should issue a nonconformity against Clause 8.5.3.

The external auditors will therefore not issue a major nonconformity if evidence is found that you are failing to achieve your objectives, but should do so, if you can't satisfy the four conditions above.

What Does this Mean?

The monitoring and measurement processes to continually improve the effectiveness of the management system are addressed in the standard by:

- Management review (5.6).
- Review of requirements related to product (7.2.2).
- Monitoring and measurement of processes (8.2.3).

Continual improvement of the effectiveness of the management system is addressed in more detail against Clause 8.5.1 of the standard. However, it is important to distinguish between improving the management system and improving the effectiveness of the management system. The management system can be improved by correcting deficiencies in process design and implementation but this is improving conformity – making the process perform as it should do. Effectiveness is about doing the right things therefore improving the effectiveness of the management system means making the objectives, standards or targets established for activities, tasks or processes meet the needs of the organization to accomplish its purpose or mission. The objectives, targets and standards focus action therefore if they are not the right objectives, targets and standards the action, no matter how well it is performed, will be the wrong action for the organization. It may be the right action in another organization or in different circumstances. This is why Clause 7.2.2 appears in the list above.

The only analysis process that serves continual improvement in system effectiveness is the process for scanning the environment for changes that impact the objectives and targets that have been established. This is addressed by Clause 5.6.2f.

The improvement process for improving the effectiveness of the management system is the *change management process*. This process would be triggered by the results of scanning the environment for changes as stated above and would manage the change through the organization.

When planning production the characteristics susceptible to variation will be identified. The techniques need to detect and control this variation also need to be identified. The techniques chosen will depend on what is to be measured, the quantities, the accuracy and precision of measurement and probability of error.

Why is this Necessary?

This requirement responds to the Continual Improvement Principle.

Many organizations are managed as a series of functions. However, the structure tends to dictate the way the objectives are set and therefore each function sets its objectives based on what it believes is necessary to meet the organization's purpose and mission. When functional objectives are based on quotas such as numbers of orders processed, number of designs produced, number of products produced, the challenge is to do more rather than do better. Effectiveness is not a measure of meeting quotas but fulfilling purpose. It is therefore necessary to monitor and measure the effectiveness of the system as a whole – not the functions. This is why it is necessary to manage the system as a series of processes not as a series of functions.

There is a right time and a wrong time to determine the methods of measurement. Doing it when you have a thousand units to measure is definitely the wrong time. Therefore, when planning production, it is prudent to establish measurement methods and specifically the methods by which variation in product characteristics and process capability is to be detected.

How is this Demonstrated?

Improvement Processes

A process for continual improvement would be a series of interrelated activities, resources and behaviours that bring about an improvement in performance. Such a process would operate at several different levels in the organization rather than at one level. It would be impractical to have a single continual improvement process through which all improvements are channelled. It is more likely that continual improvement processes form part of other processes but the principles on which each is based would be the same.

Each managed process has the following three improvement mechanisms.

- One that brings about improvement by better control – reducing variation about a mean.
- One that brings about improvement by greater efficiency – reducing resources, doing more for less.
- One that brings about improvement by greater effectiveness – raising standards, doing the right things, being ahead of stakeholder needs and expectations.

Quality improvement is addressed in more detail in Chapter 35. The secret is to make continual improvement endemic in the organization – don't go in for razzmatazz, treat every improvement as routine – an opportunity for improving control, finding better ways of doing things or finding different things to do that will delight customers. The way to do this is to install review and improvement cycles in each process, in each stage of each process and within each tasks of each stage of each process. In Fig 35-6 there are three improvement cycles, each dealing with one type of improvement. Build these into your processes and you will bring about continual improvement.

Determining Statistical Methods

The standard requires the monitoring, measuring and improvement processes to include the determination of applicable methods such as statistical techniques and the extent of their use.

What Does it Mean?

There are various ways of monitoring, measuring and analysing things and some methods are better than others. One field of data collection and analysis that can be used to demonstrate conformity of the product is to use statistical techniques. Any technique that uses statistical theory to reveal information is a statistical technique. However, there is a difference between statistical theory and statistics. Any set of figures that are intended to describe an entity or phenomena can be regarded as statistics. When one makes a prediction from these figures that attributes a particular quality to an entity or phenomena one needs to apply statistical theory for the prediction to be valid. Techniques such as Pareto analysis, histograms, correlation diagrams and matrix analysis are regarded as statistical techniques but although numerical data is used, there is no probability theory involved. These techniques are used for problem solving not for making product acceptance decisions. Other techniques such as SPC, reliability prediction and maintainability prediction use probability theory to provide a result which may not be absolute fact but which is the most probable result that can be deduced from the facts about a product or a number of products.

Why is it Necessary

This requirement responds to the Factual Approach Principle.

Statistical techniques are ways of collecting, analysing, presenting, and interpreting facts and figures. These might include probabilities, distributions and relationships etc. that are used to draw conclusions, make comparisons or make decisions. It is important that people have sufficient understanding so as not to feel helpless or frightened when exposed to statistical information.

How is this Demonstrated?

Identifying the Need for Statistical Techniques

The standard does not require you to use statistical techniques but identify the need for them. Within your procedures you will therefore need a means of determining when statistical techniques will be needed to determine product characteristics and process capability. One way of doing this is to use checklists when preparing customer specifications, design specifications and verification specifications and procedures. These checklists need to prompt the user to state whether the product characteristics or process capability will be determined using statistical techniques and if so which techniques are to be used.

Control Plans or similar documents will identify the process stages and from each stage there will be a process output – something that is measurable or countable. The tools used to measure these characteristics with the accuracy and precision required should be identified together with the methods used to capture and assess the data. These might be statistical tools such as check sheets, x bar-R charts, x-MR charts etc.

Defining Statistical Techniques

Where statistical techniques are used for establishing, controlling and verifying process capability and product characteristics, procedures need to be produced for each application. You might for instance need a Process Control Procedure, Process Capability Analysis Procedure, Receipt Inspection Procedure, and Reliability Prediction Procedure etc. The procedures need to specify when and under what circumstances the techniques should be used and provide where applicable, detail instruction on the sample size, collection, sorting and validation of input data, the plotting of results and application of limits. Guidance will also need to be provided to enable staff to analyse and interpret data, convert data and plot the relevant charts as well as make the correct decisions from the evidence they have acquired. Where computer programs are employed, they will need to be validated to demonstrate that the results being plotted are accurate. You may be relying on what the computer tells you rather than on any direct measurement of the product.

Awareness of Statistical Techniques

A basic understanding of statistical concepts that is well founded and frequently refreshed will pay dividends because it will heighten the sensitivity of managers and staff to variation. Statistically aware personnel are more likely to question results, not only when presented with a graph or chart but when observing operations, listening to conversations or reading reports.

Equipping personnel with the awareness necessary can be accomplished by establishing one or more teams of those who need knowledge of statistical concepts to perform their jobs as well as possible. Tutors, mentors, or coaches are appointed to train and lead the teams. By exposure to case studies and the techniques in action followed by classroom instruction and then mentoring coupled with recommended reading, the teams gradually become aware of statistical concepts and their fear or helplessness subsides so they emerge feeling empowered. It is essential that the mentor observes behaviour and has the respect of those they observe. No one is too high in an organization to escape this awareness. Perhaps at the upper levels, the examples and case studies need to be tailored to situations in which the executives might find themselves. But it is also important for them to realize the consequences of failing to observe variation and taking action even with the most mundane of jobs.

The staff assigned to quality planning need an even wider appreciation of statistical concepts and it is probably useful to have an expert in your company on whom staff can call from time to time. If the primary technique is SPC then you should appoint an SPC Coordinator who can act as mentor and coach to the other operators of SPC techniques.

Use of Statistical Techniques

With modern tools and techniques, the technical expertise has been taken out of statistics – staff simply put in the figures and out come the charts often leading to complacency. A wider appreciation of the concepts and heightened awareness should enable observers to question why the dots on the chart vary or remain the same and to reveal significant process problems. This is what we mean by empowerment (see Chapter 9). All managers need a basic appreciation but those in production ought to be able to apply the techniques their staff use so that they can detect when they are not being applied correctly. A manager who can look at a set of figures or perform a simple calculation and immediately detect that there is something amiss is a valuable asset in any organization. Simply plotting measurements will reveal whether flinching[①] has been used.

Auditing Use of Statistical Techniques

Auditors need to be able to determine whether the right techniques are being applied and whether the techniques are being applied as directed. Remember that the auditor's task is to determine whether the system is effective, so the ability to differentiate between the uses of inappropriate techniques is essential. Putting a tick in a box opposite a requirement for SPC charts should imply that the charts have been produced correctly not simply that the charts exist. When looking at an SPC chart the auditor should be able to determine whether the pattern of dots on the chart is indicative of the state of the process as reported by the operator or manager and that they know what they are doing.

Chapter 29

Customer Satisfaction

CHAPTER PREVIEW

This chapter is aimed at all those managers and other personnel at the customer interface responsible for gathering and analysing data on customer perceptions.

Customer satisfaction is the motive behind ISO 9001, therefore it becomes of paramount importance that the organization has up-to-date intelligence on the perceptions of current customers about its products and services. Any organization that is not monitoring customer perceptions or is ignoring them will ultimately fail for customers are the life blood of every organization. It is important also to use a factual approach to collecting and analysing such data. This will influence tactical and perhaps strategic decisions about product features (quality of design) and production/delivery processes (quality of conformance).

Customer satisfaction should be one of the key performance indicators[①] and data on customer perceptions will serve to validate not only the business outputs but also the assumptions made about customer requirements that were inputs to product realization.

In this chapter we examine the requirements in Clauses 8.2.1 and 8.4 of ISO 9001:2008 in terms on monitoring and analysis of data to determine customer satisfaction.

The position where the requirements on customer satisfaction feature in the managed process is shown in Fig. 29-1. In this case the downstream feedback is data from customers. The data gathered will be as a result of outputs primarily from the demand fulfilment process but there may be some feedback on the customer interface with the demand creation process.

MONITORING CUSTOMER SATISFACTION (8.2.1)

The standard requires the organization *to monitor information relating to customer perception as to whether the organization has met customer requirements and requires the methods of obtaining and using this information to be determined.*

What Does this Mean?

By combining definitions of the terms 'customer satisfaction' and 'requirement' ISO 9000:2005 defines customer satisfaction as *the customer's perception of the degree to which the customer's stated or implied needs or expectations have been fulfilled.*

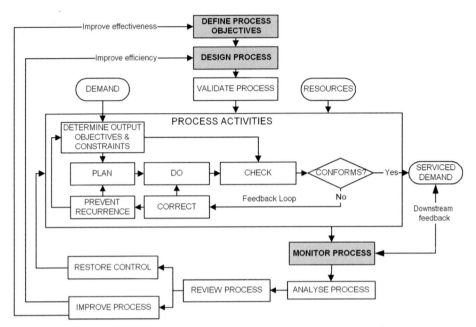

FIGURE 29-1   Where the requirements of Clause 8.2.1 apply in a managed process.

In order to satisfy customers you therefore have to go beyond the stated requirements. Customers are people who differ in their perceptions as to whether the transaction has been satisfactory. The term perception is used because satisfaction is a subjective and human condition unlike acceptance that is based on objective evidence. Customers may accept a product but not be wholly satisfied with it or the service they have received. Whether or not you have done your utmost to please the customer, if the customer's perception is that you have not met their expectations, they will not be satisfied. You could do exactly the same for two customers and find that one is ecstatic about the products and the services you provide and the other is dissatisfied.

As the ISO 9000:2005 definition of a customer is an organization or person that receives a product and includes consumer, client, end user, retailer, beneficiary and purchaser, the search for perceptions needs to go beyond the immediate customer to which your products are supplied (as confirmed by published interpretation RFI 034).

Information relating to customer perception is any meaningful data from which a judgement can be made about customer satisfaction and would include compliments, complaints, sales statistics, survey results etc. The requirement refers to the monitoring of customer perception rather than the measurement of customer satisfaction. One difference is that monitoring involves systematic checks on a periodic or continuous basis, whereas measurement may be a one-off event.

Why is this Necessary?

This requirement responds to the Factual Approach Principle.

The primary purpose of the management system is to enable the organization to achieve its objectives one of which will be the creation and retention of satisfied customers. It therefore becomes axiomatic that customer satisfaction needs to be monitored.

How is this Demonstrated?

There are several ways of monitoring information relating to customer perceptions ranging from unsolicited information to customer focus meetings.

Repeat Orders

The number of repeat orders (e.g., 75% of orders are from existing customers) is one measure of whether customers are loyal but this is not possible for all organizations particular those that deal with consumers and do not capture their names. Another measure is the period over which customers remain loyal (e.g., 20% of our customers have been with us for more than 10 years). A marked change in this ratio could indicate increasing success or impending disaster.

Competition

Monitoring what the competition is up to is an indicator of your success or failure. Do they follow your lead or are you always trying to catch up? Monitoring the movement of customers to and from your competitors is an indicator of whether your customers are being satisfied.

Referrals

When you win new customers find out why they chose your organization in preference to others. Find out how they discovered your products and services. It may be from advertising or may be your existing customers referred them to you.

Demand

Monitoring the demand for your products and services relative to the predicted demand is also an indicator of success or failure to satisfy customers. It could also be an indicator of the effectiveness of your sales promotion programme, therefore analysis is needed to establish which it is.

Effects of Product Transition

When you launch a new product or service, do you retain your existing customers or do they take the opportunity to go elsewhere?

Customer Surveys

The most important part of a customer survey is to ask the right questions and Hill, Self and Ross suggest that the survey will provide a measure of satisfaction only if the questionnaire covers those things the customer was looking for in the first place.[1] There is an eight-step process:

---

[1] Hill Nigel, Self Bill, and Ross Greg (2002). *Customer Satisfaction measurement for ISO 9000:2000*. Butterworth Heinemann.

1. Identify the customer's requirements (needs and expectations);
2. Determine a representative sample of customers;
3. Determine which type of survey is appropriate;
4. Design the questionnaire;
5. Conduct the survey;
6. Analyse the data;
7. Presenting the results; and
8. Taking action on the results.

There are several types of survey that can be used.

- Personal interview;
- Telephone interview; and
- Self-completion questionnaires.

The self-completion form relies on responses to questionnaires and seeks to establish customer opinion on a number of topics ranging from specific products and services to general perceptions about the organization. The questionnaires can be sent to customers in a mail shot, included with a shipment or filled in before a customer departs (as with hotels and training courses). These questionnaires are somewhat biased because they only gather information on the topics perceived as important to the organization.

It should be noted that self-completion questionnaires by themselves are not an effective means of gathering customer opinion. Customers don't like them and are not likely to take them seriously unless they have a particular issue they want to bring to your attention. It is much better to talk face to face with your customer using an interview checklist. Think for a moment how a big customer like Ford and General Motors would react to thousands of questionnaires from their suppliers. They would either set up a special department just to deal with the questionnaires or set a policy that directs staff not to respond to supplier questionnaires. Economics alone will dictate the course of action that customers will take.

The personal form of survey is conducted through interview such as a customer service person approaching a customer with a questionnaire while the customer is on the organization's premises. This may apply to hotels, airports, entertainment venues and large restaurants. With this method there is the opportunity for dialogue and capturing impromptu remarks that hide deep-rooted feelings about the organization.

User Surveys

If your immediate customer is not the end user you could seek the opinion of users to learn of their experiences with your products and get a better understanding of their needs and expectations. These surveys might yield opportunities for future uses, modifications or applications of your products. Depending on the type of product you supply you might need to do a pilot survey to test its validity before running it on the user population.

Lost Business Analysis

An analysis of lost business may provide an insight into your tendering practices although it is often difficult to get reliable data. A potential customer to which you have submitted your tender or quotation may cooperate but they are often not interested as they don't perceive it of being any benefit to them; in other words you are seen as wasting their time. However, it depends on how you approach it and what your relationship has been with that particular customer. Certainly a questionnaire through the post or by e-mail is unlikely to be answered. Lost business analysis needs a personal touch.

Delivery Feedback

If the arrangement with your customer is for scheduled deliveries over a long period of time, you might receive immediate feedback without requesting it. Otherwise you set up a delivery receipt mechanism that provides the recipient to comment on their experience with the delivery. Some organizations include a Customer feedback card with the delivery, on-line sales often have a feedback link that customers can follow to records their experience.

The problem with delivery feedback is that many satisfied customers won't bother to register feedback so you can't measure customer satisfaction by how many feedback returns you receive. You can only assess those you do receive and ignore the proportion of the potential number of returns. As with all customer surveys, a 3% response should be regarded as pretty good.

Dealer Reports

If you ship product to a dealer, you will have a different relationship than you have with the dealer's customer and may get useful feedback.

Warranty Claims

Warranty claims are tangible therefore they can be assessed objectively. They provide an immediate indicator of customer satisfaction particularly if the product represents a substantial investment for the customer. On the other hand, if the product is a throw away item, you may only receive warranty claims from disgruntles customers; others might not consider the effort worthwhile.

Focus Meetings

A personal form of obtaining information on customer satisfaction is to arrange to meet with your customer. Seek opinions from the people within the customer's organization such as from Marketing, Design, Purchasing, Quality Assurance and Manufacturing departments etc. Target key product features as well as delivery or availability, price and relationships. This form is probably only suitable where you deal with other organizations.

Complaints

As stated previously, if your processes have not been designed to alert you to customer dissatisfaction, you may be under the illusion that as you have not received any complaints, your customer must be satisfied. The process for handling customer complaints is addressed under *Customer feedback* in Chapter 24. Here the topic is *monitoring* and therefore you should be looking at the overall number of complaints, the upward or downward trend and the distribution of complaints by type of customer, location and nature of complaint. Coding conventions could be used to assign complaints to various categories covering the product (or parts thereof) packaging, labelling, advertising, warranty, support etc. Any complaint, no matter how trivial, is indicative of a dissatisfied customer. The monitoring methods need to take account of formal complaints submitted in writing by the customer and verbal complaints given in conversation via telephone or meeting. Everyone who comes into contact with customers should have a method of capturing customer feedback and communicating it reliably to a place for analysis.

Compliments

Compliments are harder to monitor because they can vary from a passing remark during a sales transaction to a formal letter. Again, all personnel who come into contact with customers should have a non-intrusive method for conveying to the customer that the compliment is appreciated and will be passed on to the staff involved.

Performance Indicators

In general, the previous indicators provide data about the organization as a whole rather than specific product. In addition it is necessary and probably much easier to gather data direct from customers about specific product such as:

- Performance of delivered parts;
- Customer disruptions including the end user;
- Delivery performance against schedule; and
- Customer notifications related to quality or delivery issues.

ANALYSIS OF CUSTOMER SATISFACTION DATA (8.4a)

The standard requires analysis of data *to provide information relating to customer satisfaction*.

What Does this Mean?

This requirement seeks to take the data generated by the monitoring process addressed in Clause 8.2.1 and through analysis produce meaningful information on whether customers are in fact satisfied with the products and services offered by the organization.

Why is this Necessary?

This requirement responds to the Factual Approach Principle.

Customer satisfaction is not something one can monitor directly by installing a sensor. One has to collect and analyse data to draw conclusions.

How is this Demonstrated?

We can now look at those ways by which data can be collected relative to the different techniques of monitoring customer satisfaction:

- Repeat orders: This data can be collected from the order processing process.
- Competition: This data is more subjective and results from market research.

- Referrals: This data can be captured from sales personnel during the transaction or later on follow-up calls.
- Demand: This data can be collected from sales trends.
- Effects of product transition: This data can be collected from sales trends following new product launch.
- Surveys: This data can be collected from survey reports.
- Focus meetings: This data can be collected from the meeting reports.
- Complaints: This data can be collected from complaints recorded by customers or by staff on speaking with customers.
- Compliments: This data can be collected from written compliments sent in by customers or by staff on speaking with customers.

As indicated above there are several sources of data, several ways in which it can be collected and several functions involved. Provisions need to be made for transmitting the data from the processes where it can be captured to the place where it is to be analysed. It is evident that sales and marketing personnel are involved and as information on customer perceptions is vital for these functions to manage their own operations effectively, it may be appropriate to locate the analysis process within one of these departments. In some organizations, customer support groups are formed to provide the post-sales interface with customer and in such cases they would probably perform the analysis.

The Process

The integrity of your process for determining customer satisfaction is paramount otherwise you could be fooling yourselves into believing all is well when it is far from reality. The process therefore needs to be free from bias, prejudice and political influence.

In defining the process you will need to:

- Determine the sources from which information is to be gathered;
- Determine the method of data collection – the forms, questionnaires and interview checklists to be used;
- Determine the frequency of data collection;
- Devise a method for synthesizing the data for analysis;
- Analyse trends;
- Determine the methods to be used for computing the customer satisfaction index;
- Establish the records to be created and maintained;
- Identify the reports to be issued and to whom they should be issued; and
- Determine the actions and decisions to be taken and those responsible for the actions and decisions.

Pareto analysis can be used to identify the key areas on which action is necessary. For example it may turn out that 80% of the sales come from repeat orders indicating a slow down in the number of new customers. Also 80% of the complaints may be from one market sector that generates only 20% of the sales –which is an indication that 80% of customers may be satisfied. Alternatively, 80% of the compliments may come from 20% of the customers but as they represent 80% of the sales it may prove very significant. The important factor is to look for relationships that indicate major opportunities and not insignificant opportunities for improvement. Use the results to derive the business plans, product development and process development plans for current and future products and services.

Frequency of Measurement

Frequency needs to be adjusted following changes in products and services and major changes in organization structure such as mergers, downsizing, plant closures etc. Changes in fashion and public opinion should also not be discounted. Repeating the survey after the launch of new technology, new legislation or changes in world economics affecting the industry may also affect customer perception and consequently satisfaction.

Trends

To determine trends in customer perception you will need to make regular measurements and plot the results preferably by particular attributes or variables. The factors will need to include quality characteristics of the product or service as well as delivery performance and price. The surveys could be linked to your improvement programmes so that following a change, and allowing sufficient time for the effect to be observed by the customer, customer feedback data could be secured to indicate the effect of the improvement.

Customer dissatisfaction will be noticeable from the number and nature of customer complaints collected and analysed as part of your corrective action procedures. This data provides objective documentation or evidence and again can be reduced to indices to indicate trends.

By targeting the final customer using data provided by intermediate customers, you will be able to secure data from the users but it may not be very reliable. A nil return will not indicate complete satisfaction so you will need to decide whether the feedback is significant enough to warrant attention. Using statistics to make decisions in this case may not be a viable approach because you will not possess all the facts!

Customer Satisfaction Index

A customer satisfaction index that is derived from data from an independent source would indeed be more objective. Such schemes are in use in North America, Sweden and Germany. A method developed by a Professor Claes Fornell has been in operation for 15 years in Sweden and is now being used at the National Quality Research Center of the University of Michigan Business School. The American Customer Satisfaction Index (ACSI) covers seven sectors, 40 industries and some 189 companies and government agencies. It is sponsored by the ASQ and the University of Michigan Business School with corporate sponsorship from Federal Express, Sears Roebuck, Florida Power and Light and others. The index was started in 1994[2] and using data obtained from customer interviews, sector reports are published indicating a CSI for each listed organization thereby providing a quantitative and independent measure of performance useful to economists, investors and potential customers.

---

[2] American Society for Quality (2001). http://acsi.asq.org/.

Chapter 30

Internal Audit

CHAPTER PREVIEW

This chapter is aimed at all those managers responsible for the outputs of processes for it is they who should be verifying that the processes are efficient and effective. However, it is often the case that auditors from quality and/or finance departments undertake these activities on behalf of the process owners or of top management.

In this chapter we examine the requirements in Clause 8.2.2 of ISO 9001:2008 on internal audit and in particular:

- Auditing for conformance with planned arrangements;
- Auditing for compliance;
- Auditing for effective implementation and maintenance;
- Planning audits;
- Defining audit criteria, scope, frequency and methods;
- Selection of auditors;
- Audit procedures;
- Taking action following the audit; and
- Follow-up audits.

The position where the requirements on internal audit feature in the managed process is shown in Fig. 30-1. It is referred to as process monitoring and analysis in the diagram because that aligns with the intent of the audits. There are different types of audits such as system audits, process audits, product audits and conformity audits etc. but in general they are all outside the process operation flow and carried out on processes at different levels. As the internal audit requirements also include remedial and corrective action requirements, they also tick the review and improvement box.

AUDITING FOR CONFORMITY WITH PLANNED ARRANGEMENTS (8.2.2a)

The standard requires the organization *to conduct internal audits at planned intervals to determine whether the quality management system conforms to the planned arrangements (see Clause 7.1).*

What Does this Mean?

ISO 9000:2005 defines an audit as *a systematic, independent and documented process for obtaining audit evidence and evaluating it objectively to determine the extent to*

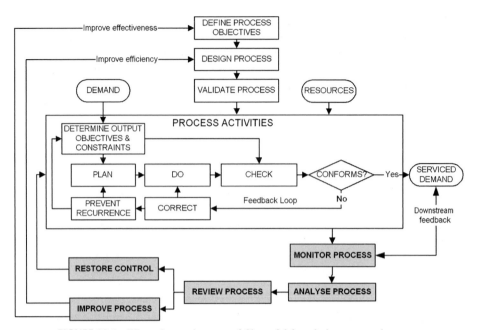

FIGURE 30-1 Where the requirements of Clause 8.2.2 apply in a managed process.

which agreed criteria are fulfilled. Audits are a means of verification and as such involve monitoring and measurement. Both these words are used in the standard and it seems as though they are intended to convey different concepts but are actually part of the same concept.

Planned intervals can be any time after completion of product realization planning and at any time these plans change.

As the requirement refers to Clause 7.1, the planned arrangements referred to are the plans for product realization and these will include the plans for determining the product features that will meet customer needs and the plans for determining processes for producing those product features.

> **Rationale for Audits**
>
> Changing personnel, priorities, operating environment and markets all put pressure on the system as formally established and if left unattended over time, adherence to policies and practices, which the organization has deemed necessary to accomplish its mission, begins to decline.
>
> Internal audits are carried out to determine the continued effectiveness of the management system and can act as a safeguard against deterioration in standards.

Conformity of the quality management system to the planned arrangements means that audits should verify whether the quality management system is capable of enabling the organization to implement the plans for product realization.

Why is this Necessary?

This requirement responds to the Factual Approach Principle.

The purpose of quality audits is to establish, by an unbiased means, factual information on quality performance. Quality audits are the measurement component of the quality system. Having established a quality system it is necessary to install measures that will inform management whether the system is capable of enabling the organization to design and produce the products it is committed to supply. One of the most common deficiencies as Deming points out[1] is that "a goal without a method for reaching it is useless", but it is a common practice for management to set goals like undertaking a commitment to meet a customer requirement without knowing whether the processes in place are capable of fulfilling those goals. A project audit is needed to reveal such deficiencies.

How is this Demonstrated?

An audit may be triggered by an Invitation to Tender, the award of a contract, or an undertaking that requires the development of new or modified products, processes and services. We can call this type of audit a project audit.

The project audit should establish in stages whether:

- The existing product realization processes are capable of designing and producing product with the features required to meet customer requirements;
- The plans for redesigning existing products will be effective, if implemented:
- The processes for producing the product features have been designed;
- Priorities for action have been set; and
- The risks to success have been quantified;
- The information, resources, criteria and methods for effective operation and control of these processes have been identified, developed and provided.
- The necessary monitoring, measurement, analysis and improvement processes have been designed and installed.

By judicious implementation of these project audits, the integrity of the management system will be confirmed.

AUDITING FOR CONFORMITY TO ISO 9001 (8.2.2a)

The standard requires the organization *to conduct internal audits at planned intervals to determine whether the quality management system conforms to the requirements of this International Standard.*

What Does this Mean?

ISO 9001 contains a series of requirements for which there are numerous solutions depending on the nature of the organization and its markets, products and services. As the organization, its processes and products change, periodic confirmation that the system conforms to the standard is necessary to verify it remains in conformity.

Why is this Necessary?

This requirement responds to the Factual Approach Principle.

---

[1] Deming, W. Edwards (1982) *Out of the crisis.,* MITC.

The argument for this requirement is that all organizations should have control over their own operations and not rely on external Certification Bodies to detect system nonconformities. It would therefore be logical for organizations to perform internal audits against the same standards as the external bodies and only use the external bodies as confirmation that they meet the requirements of the standard. Having performed an initial assessment against the requirements of ISO 9001, one should only need to repeat the assessment when the system changes.

How is this Demonstrated?

As the requirement is placed under the heading *Internal audit*, the Certification Body audit cannot be a substitute even though it is performed with the same purpose in mind. There is a school of thought that believes a Certification Body audit will ultimately reduce to a confirmation that the internal conformity audit has been carried out by competent personnel and has demonstrated that the system is effective. This is not beyond the bounds of possibility. All it needs is for the Certification Body to recognize the competence and impartiality of the internal auditors, reduce their certification costs and a new auditing regime will emerge.

ISO 19011 provides guidance on planning and conducting internal audits but it does not describe how to do the audit so you have a free choice as to the methodologies you adopt. If you have taken the approach to ISO 9001 as described in *A practical guide to using the standards* in Chapter 1 and undertaken system development as described in Chapters 10 and 11, the system will have been designed so that it will have all necessary features to conform to the requirements of ISO 9001 but it needs to be verified. One such methodology is a requirement driven conformity audit and another is a results driven conformity audit.

Requirements Driven Conformity Audit

The requirements driven approach starts with a table of requirements and a description of the evidence that should exist which demonstrated conformity with these requirements. There are three steps to this audit:

1. Start by preparing an ISO 9001:2008 Exposition or Compliance Table an extract of which is shown in Table 30-1.
2. Appoint someone else to verify that the Exposition is complete and the evidence indicated in the table is valid as a response to the requirement. This is a desk audit, checking that documentation describes arrangements that satisfy the requirement of ISO 9001.
3. Verify that the evidence referred to in the Exposition actually exists. This is an implementation audit, checking that the system is actually functioning as described, that people are taking the prescribed actions and producing the required evidence.

To build this table, examine each requirement of the standard and translate it into a subject statement, then explain where in the system description evidence of compliance can be found. Through hyperlinks you can link to the location in the system description or other documents that demonstrates compliance (i.e., documents that express intensions and documents that record activities carried out or results achieved). As there are over 260 requirements, this will be a large and complicated matrix so one

TABLE 30-1 Extract of an ISO 9001 Exposition or Compliance

Req. No.	Clause	Subject	Evidence of compliance
	4	**Quality management system**	
	4.1	**General requirements**	
1.	4.1	Establishing a documented quality management system	The Business Management System Description represents the documented management system
2.	4.1	Continually improving the effectiveness of the QMS	Continually improving the effectiveness of the BMS is accomplished through *process reviews*
3.	4.1a	Identifying processes	Process are identified as a result of analysing the stakeholder needs
4.	4.1b	Determining process sequence and interaction	Process sequence and interaction is depicted by the *System Model*
5.	4.1c	Determining criteria and methods for operation and control of processes	The criteria for operation and control of processes are defined by the performance measures related to the process objectives
6.	4.1d	Availability of resources necessary to support the operation and control of processes	The resources necessary to support the operation and control of processes are made available through *Resource Management* processes
7.	4.1.d	Availability of information necessary to support the operation and control of processes	The information necessary to support the operation and control of processes is made available through the Business Management System Description
8.	4.1e	Monitoring, measurement and analysis of processes	All processes are monitored by periodic reviews. Process measurements are undertaken, the data analysed and process performance and efficiency recorded
9.	4.1f	Restoring the status quo	Any action necessary to achieve planned results is recorded in *Performance Log*
10.	4.1f	Continual process improvement	Any action necessary to improve the manner by which the planned results is achieved is recorded in the *Improvement Plan*
11.	4.1	Process management	Processes are managed by the MD
12.	4.1	Outsourcing processes	No processes are outsourced

solution is to compile the matrix as a series of layers. The first layer would list the ISO 9001 requirements as a series of clauses against the core business processes. For each process a list of applicable requirements can then be identified and cross-referenced to the process stage where it is implemented.

A report can be printed and the assumptions, conclusions and recommendations added. It won't have escaped your notice that the requirements are expressed in such a way that a single response is almost impossible and also the only way of collecting evidence of conformity with some requirements is to interview people. For example with the requirement for a quality policy there are 10 separate requirements that could elicit a different response if you asked each of the questions in Table 30-2.

Once a requirements driven conformity audit has been completed and its has been established that the system has been designed to meet all requirements of ISO 9001 it only needs to be checked again if the requirements of the standard change or the system design changes.

Results Driven Conformity Audit

If we begin with the end in mind, we would start an audit by wanting to know if the system was effective, then establish that it conformed to ISO 9001. One of the problems with the requirement driven approach is that the end is conformity and not effectiveness. A results driven conformity audit starts with what the organization is achieving and finished with checking for conformity with ISO 9001.

TABLE 30-2 Sample Conformity Audit

Requirement	Response
Is there a quality policy	Yes
It is appropriate to the purpose of the organization	No evidence found as there is no published statement of the organization's purpose or mission
Does it include a commitment to comply with requirements	Yes
Does it include a commitment to improve the effectiveness of the quality management system	Yes
Does it provide a framework for establishing and reviewing quality objectives	No because the statements are motherhood statements not definable values or principles
Is it communicated within the organization	Yes
Is it understood within the organization	Only 5 out of 20 understood how it applied to what they do
Is it reviewed	Yes
Is it reviewed for continuing suitability	No because there are no criteria for determining its suitability
Is all this directed by top management	No because the Quality Manager produced the statement and top management were simply presented with it to approve

1. Firstly look at what results are being achieved relative to the organization's declared objectives.
2. Establish whether these results are consistent with the intent of ISO 9001, i.e., consistently conforming product, satisfied customers, compliance with relevant statutes and regulations and continual improvement.
3. Discover what processes are delivering these results.
4. Determine whether these processes are being managed effectively (see process audit below).
5. Lastly check that what is being done demonstrated conformity with the requirements of ISO 9001.

In many cases, it may not be possible to get beyond Step 1 simply because the organization cannot demonstrate what it is achieving as there is no simple way of showing performance. Secondly, it might not be possible to separate the objectives of ISO 9001 from the masses of data the organization presents and therefore not get beyond Step 2.

However, once a results-driven audit has been completed and all deficiencies resolved (which might take a considerable time relative to other audits) subsequent audits will be a fraction of the time. If there has been no change in the organization's objectives and no adverse change in its performance when Step 2 is completed, the system remains compliant and therefore there is no justification for checking conformity with any clauses of ISO 9001. If there are changes that will affect the integrity of the system, an analysis needs to be carried out to establish the impact of these changes and repeat Steps 1–5 above on the processes affected by these changes.

AUDITING FOR EFFECTIVE IMPLEMENTATION AND MAINTENANCE (8.2.2b)

The standard requires the organization *to conduct internal audits at planned intervals to determine whether the quality management system is effectively implemented and maintained.*

What Does this Mean?

Effective implementation should not be confused with system effectiveness. The evaluation of system effectiveness serves to explore better ways of doing things, whereas, an evaluation of effective implementation serves to explore whether the processes are being run as intended and delivering the required outputs, i.e., people are doing what they are required to do and the results are having the desired effect. A process may be run as intended but not achieve the desired results indicating a design weakness in the process.

Effectively maintained means that the processes continue to remain capable despite changes in the quantity, condition or nature of the human, physical and financial resources.

In the past it has been assumed that if people were found to be following the procedures as documented, the system was effective. Conversely, if the people were not found to be following the procedures, the system was somehow ineffective. But this was not the reason for the system, i.e., it was not the purpose of the system to force people to follow procedures. The purpose was to ensure results, therefore a system is effective only if it can be demonstrated that the desired results are being achieved.

Why is this Necessary?

This requirement responds to the Factual Approach Principle.

In order to manage the organization effectively, it is necessary to know whether the system for achieving the organization's objectives does the job for which it has been designed. Management also needs a system that does not collapse every time something changes – the system has to be robust. It has to cope with changes in personnel, changes in customer requirements, changes in the environment and in resources. Consequently to remain robust the system has to be effectively maintained.

How is this Demonstrated?

The management system comprises a series of interacting processes and each contributes to its overall effectiveness.

This requirement can be met in one of three ways:

- by system implementation audits;
- by process audits conducted by personnel external to the process; and
- by process audits conducted by personnel operating the process.

System Implementation Audit

There are two ways for conducting the system implementation audit:

- by planning a series of audits that will cover the entire system in one cycle;
- by analysing the results of process audits and determining effectiveness by correlation.

If the system implementation option is chosen, you should not need to build into each process an audit mechanism. The analysis option is only an option when each process has a build-in audit mechanism.

Process Audit

The traditional practice has been for processes to embody provisions for product measurement and process monitoring that is results-oriented. The measures are related to what is being processed – not the extent to which the rules are being followed. The internal audit has taken on this role but it misses the crucial question of establishing the efficiency of the process. Results were not matched with the effort used to produce them. A process is not efficient if it achieves the required results by wasting resources. It is therefore necessary to periodically examine whether:

- the activities are being performed as planned;
- the resources are being effectively utilized.

Such audits would examine activities not only to verify that the prescribed actions and decisions have been taken but also to verify the time and effort taken to perform them. The plans and specifications should define targets for time and effort so that the audit is against agreed targets and does not become a witch-hunt. The audits could be performed by personnel external to the process, such as internal auditors or managers or in fact be performed by the supervisors of the process.

If we assume the system has been developed along the lines expressed in Chapter 10, the business and work processes to be audited will be those the organization manages in order to satisfy its stakeholders. On this premise you would take the following steps in conducting a process audit:

1. Identify the agreed process outputs, measures and targets.
2. Identify the key stages of each process on which delivery of these outputs depend (see Figs 10-5 to 10-8 for examples).
3. Establish for each stage that:
   a. The outputs, measures (or indicators) and targets of each stage have been defined and are consistent with the process outputs (i.e., they are leading and not lagging measures);
   b. The activities being carried out are those necessary to deliver the stage outputs and are consistent with the prescribed policies and procedures;
   c. The materials and equipment being used are those that have been specified and are being used under controlled conditions;
   d. The competences required to produce the stage output have been specified and that the competency of those producing these outputs has been assessed under controlled conditions and found acceptable;
   e. The information being used to produce the stage outputs is that which has been defined and that it is being used under controlled conditions; and
   f. Reviews are carried out to establish whether the stage outputs are being controlled and are meeting the targets.
4. Establish that the provisions made for preventing failure such as error proofing are being effective.
5. Establish that process performance is being measured using the agreed indicators, that results are being recorded and reported and that appropriate action is being taken when targets are not met.
6. Establish that the methods of measurement are soundly based and consistent with the accuracy and precision required for the process outputs.
7. Establish that reviews are carried out to determine whether there are better ways of delivering the process outputs and that recommended improvements have been implemented or planned.
8. Establish that reviews are carried out to verify the relevance of the process outputs, measures and targets to current stakeholder needs and that changes have been implemented or are planned.

Product Audit

As the whole system is required to be audited it would appear logical to perform product audits to establish whether the end product conforms to specification and process audits to establish whether the product resulted from implementing the planned arrangements. The various activities specified by the planned arrangements should result in an output that conforms in full with the specified requirements. However, there is variation in all processes and although the processes may be deemed capable, incidents can occur that escape detection. The product audit is performed to verify that output emerging from the process meets the specified requirements and therefore determines whether the controls in place are effective. Errors that are detected indicate that the controls are not effective as they should have been removed before output was released.

As product audits are performed to verify that the planned controls have been effective it follows that some policy decisions need to be taken regarding:

- When products audits are to be carried out – often this is after product release but before shipment. However, product audits are also appropriate on product returned from customers;
- Who is to perform product audits – often this is a group independent of production so as to be unbiased or having freedom of action and decision;
- What the sampling criteria are to be – often this might be 1 in 10, 1 in 100, 1 in 1000 etc. depending on the production rate and product complexity;
- What happens to the batch from which the sample is taken while the audit takes place – often the batch is held pending the results of the audit;
- What the acceptance criteria[①] are to be – these must address the same characteristics as specified by the customer with the agreed limits. It might be necessary to operate the product under actual or simulated conditions in order to measure some of the characteristics. However, some characteristics might not have been specified for manufacture such as the degree of surface abrasions to be expected from a crankshaft that has run 24 hours under a specific load. Some additional criteria might be therefore needed.

Clearly it might be necessary to develop strip down instructions because it may not be appropriate to simply reverse the assembly instructions. Forms will be needed on which to record the measurements and any observations such as unexpected wear marks, contamination, parts incorrectly oriented, parts missing, low torque etc.

Product auditors need to be extremely vigilant. It's a job for an analytical person – someone who is not daunted by tedium, repetition and attention to detail and it needs to be a person who is cautious, intellectual, works slowly and is good at problem solving. Never put the product auditor under pressure – if you do you will encourage him or her to miss something important. There is no point in employing product auditors and ignoring the results. You will probably want to put some of your best people onto this job.

Documenting the evidence is vital simply because someone or something created it and that someone or something is still around and may be creating more of it as the report is written. In the modern world of digital photography, here is a tool that seems designed for product auditors. It costs virtually nothing to capture exactly what was observed and even date the picture so there is no doubt. But because its easily done, it can be over done.

Traceability is important also. The report should provide sufficient information to identify all products produced in the same batch and likely to possess the same characteristics. If you can't isolate all related products you run the risk of releasing suspect nonconforming product to customers.

PLANNING AUDITS (8.2.2)

The standard requires the audit programme *to be planned taking into consideration the status and importance of the processes and areas to be audited as well as the results of previous audits.*

What Does this Mean?

The audit programme is defined in ISO 9000:2005 as a set of one or more audits planned for a specific time frame and directed towards a specific purpose. The *programme* will therefore have dates on which the audits are to be conducted. As the programme should be directed at the purpose of the audits, there may be a need for different types of audit programmes depending on whether the audits are of the quality system, contracts, projects, processes, products or services. It would therefore be expected that all audits in a particular audit programme would serve the same purpose. The audit programme would also be presented as a calendar chart showing where and when the audits will take place.

Case Study: A Different Approach to Internal Audit

Nippon Koei UK Co. Ltd (NKUK), based in Reading, England, is a firm of consulting engineers providing services in the developing world to clients in the water & wastewater, environment and transportation sectors. It is a wholly owned subsidiary of the long-established Nippon Koei Co. Ltd of Japan.

When it was formed in 2000, NKUK management decided to create a management system that met two basic requirements. The first was to provide a simple and flexible basis, reflecting good management practices, for running the business and keeping control of a wide variety of projects. The second requirement was to obtain ISO 9001:2000 certification.

The directors of the new company had prior knowledge of ISO 9000-based systems. They were particularly concerned about the lack of management value from internal auditing as typically practiced in connection with ISO 9001 and its predecessors. Their perception – from long experience – was one of auditors chasing 'auditees' to catch up with internal audit schedules for the main purpose of retaining certification.

The task, then, was to find a way of meeting the internal audit requirements that not only satisfied the certification body assessor but which also satisfied NKUK's management's requirement for business added-value.

The answer lay in re-examining the Clause 8.2.2 of ISO 9001 and noting that its requirements call for checks to be made of:

1. Planning – specifically of operating processes;
2. Performance against business objectives (for both outcomes and processes); and
3. Conformance to ISO 9001.

We were able to demonstrate to the assessors that the first two areas of requirement are covered by implementing a system of continual management review of operating processes. This approach has the benefit that any anomalies are apparent to the responsible manager in a more timely way than is the case with conventional 'after-the-fact' audits.

And the third requirement of conformance to ISO 9001 is handled separately by means of a simple desk-check against the standard following any change to the system.

This example was provided by Jim Wade of Advanced Training with permission of Nippon Koei UK Co. Ltd and is a practical application of the approach that is advocated in this chapter.

Status has three meanings in this context – the first to do with the relative position of the process or area in the scheme of things; the second to do with the maturity of the process; the third to do with the performance of process.

On the importance of the process, you need to establish to whom is it important – to the customer, the managing director, the public or your immediate superior? You also need to establish the importance of the activity relative to the effect of non-compliance with the planned arrangements[1]. Importance also applies to what may appear minor decisions in the planning or design phase but if the decisions are incorrect it could result in major problems down stream. If not detected, getting the units of measure wrong can have severe consequences particularly if the customer specified dimensions in metric units and the purchase order has them specified in imperial units. Rather than check the figures or the units of measure, audits should verify that the appropriate controls are in place to detect such errors before it is too late.

The status and importance of the activities will determine whether the audit is scheduled once a month, once a year or left for three years. Any longer and the activity might be considered to have no value in the organization.

Why is this Necessary?

This requirement responds to the Process Approach Principle.

There is little point in conducting in-depth audits on processes that add least value. There is also little point auditing processes that have only just commenced operation. You need objective evidence of compliance and that may take some time to be collected. Where the results of previous audits have revealed a higher than average performance in an area (such as zero nonconformities on more than two occasions) the frequency of audits may be reduced. However, where the results indicate a lower than average performance (such as a much higher than average number of nonconformities) the frequency of audits should be increased.

How is this Demonstrated?

An audit programme should be developed for each type of audit as indicated above. Therefore one might prepare audit programmes for the following:

- Project audits (for new developments);
- Product audits (for existing products and services);
- Process audits (for all the organization's processes); and
- Conformity audits (for all regulations and standards that apply).

The conformity audits could focus on *all* regulatory requirements, not simply those of ISO 9001, because they serve the same purpose – to verify conformity with standards.

This requirement focuses on the criteria for choosing the areas, activities, processes etc. to audit and therefore following initial audits to verify that the system is in place and functioning as planned, subsequent audits should be scheduled depending on status and importance.

An audit of one requirement of a policy, standard, process, procedure, contract etc. in one area only will not be conclusive evidence of compliance if the same requirements are also applicable to other areas. Where operations are under different managers but performing similar functions you cannot rely on the evidence from only one area – management style, commitment and priorities will differ. In order to ensure that a particular audit programme is comprehensive you will need to draw up a matrix showing the areas or processes or products etc. to be audited and the dates when the audits are to be carried out. Supporting each audit programme an analysis of the status and importance should be performed and the key aspects to be audited identified. The programme also has to include shift working so that auditors need to be very flexible. One audit per year covering 10% of the quality system in 10% of the organization is hardly comprehensive. However, there are cases where such an approach is valid. If sufficient confidence has been acquired after conducting a comprehensive series of audits over some time, the audit programme can be adjusted so that it targets only those areas where change is most likely, thus auditing more stable areas less frequently.

The management system will contain many provisions, not all of which may be verified on each audit. This may either be due to time constraints or work for which the provisions apply not being scheduled. It is therefore necessary to record those aspects that have or have not been audited and devise the programme so that over a one to three year cycle all provisions are audited in all areas at least once.

As the programme is executed, periodic reviews should be undertaken using data from a variety of sources to confirm that the audit programme is effective. The data should be analysed to determine the status and importance of the areas being audited remains unchanged and if necessary the programme modified to take account of changes in status and importance. This might result in altering the frequency of audits, the depth of the audits and areas covered.

There is no requirement to document the results of any analysis carried out to determine the status and importance of the areas to be audited (Published interpretation RFI 036) but it might useful to do so that the next time you prepare the audit programme you can look back and review the reasons for the decisions that were taken and determine whether they remain valid or should change.

DEFINING AUDIT CRITERIA, SCOPE, FREQUENCY AND METHODS (8.2.2)

The standard requires *the audit criteria, scope, frequency and methods to be defined.*

What Does this Mean?

The audit criteria are the standards for the performance being audited. They may include policies, procedures, regulations or requirements. Examinations without such a standard are surveys but not audits.

The scope of the audit is a definition of what the audit is to cover – the boundary conditions including the areas, locations, shifts, processes, departments etc.

The frequency is the interval over which the audit is to be repeated and can be daily, weekly, monthly, quarterly, annually or longer.

The methods are the manner by which the audit is to be planned, conducted, reported and completed.

In defining these aspects, they may be described through procedures, standards, forms and guides.

Why is this Necessary?

This requirement responds to the Process Approach Principle.

Without defined methods of auditing it is likely that each auditor will choose a different way of performing the audit – some will be good and some not so good. In order to run an effective management system, auditing should aim for best practice and therefore defining auditing methods enables best practice to be defined for the benefit of the organization.

How is this Demonstrated?

For each audit the auditor should as a matter of routine always define the standard against which the audit is to be carried out and the scope of the audit. The frequency of the audit should be defined in the audit programme and the method within the auditing procedures.

SELECTION OF AUDITORS (8.2.2)

The standard requires the selection of auditors and the conduct of audits *to ensure objectivity and impartiality of the audit process and for auditors not to audit their own work.*

What Does this Mean?

This requirement means that auditors should be selected on the basis of their objectivity and impartiality – such that their association with the work being audited should not influence their judgement. The requirement suggests that anyone auditing their own work may be influenced to overlook, hide or ignore facts pertinent to the audit.

Why is this Necessary?

This requirement responds to the Leadership Principle.

If personnel have personally produced a product, they are more likely to be biased and oblivious to any deficiencies than someone totally unconnected with the product. They may be so familiar with the product that they are blind to its full strengths and weaknesses. A second pair of eyes often catches the errors overlooked by the first pair of eyes (see *Rickover on inspection* in Chapter 6). However, auditors are human and if there is a personal relationship between the auditor and the auditee, the judgement of the auditor may be prejudiced. Depending on the nature of any problems found, the auditor being a friend, relation or confidant of the auditee, may be reluctant to or may be persuaded not to disclose the full facts if the findings indicate a serious deficiency. Even a customer may fail to exercise objectivity when it is found that the cause of problems is the inadequacy of the customer requirement!

How is this Demonstrated?

Apart from the requirement for auditors not to audit their own work, any other competent person with experience in the industry could be selected as an auditor.

The requirement for objectivity and impartiality does not mean that one must rule out supervisors, managers, friends, relations or internal customers as auditors. These conditions do not necessarily mean such a person cannot be objective and impartial – there is simply an inherent risk. This risk is overcome by the selection being made on a person's character and track record. It would be foolish to limit the selection of auditors to those who are totally independent because in some small organizations there may be no one who fits this criterion. The difficulty arises in demonstrating subsequent to the audit, that the selected auditor exercised objectivity and impartiality. In organizations that observe a set of shared values, where honesty and trust are prevalent and frequently reinforced, it should not be necessary to demonstrate that the selected auditors meet this criterion. For other organizations a solution is for the auditors to be selected on the basis of having no responsibility for the work audited and no personal relationship with any of the auditees concerned.

By being divorced from the audited activities, the auditor is unaware of the pressures, the excuses, the informal instructions handed down and can examine operations objectively without bias and without fear of reprisals. To ensure their objectivity and impartiality, auditors need not be placed in separate organizations. Although it is quite common for quality auditors to reside in a quality department, it is by no means essential. There are several solutions to retaining impartiality:

- Auditors can be from the same department as the activities being audited, provided they do not perform the activities being audited;
- Separate independent quality audit departments could be set up staffed with trained auditors; and
- Audits could be carried out by competent personnel at any level.

A competent person in this regard is one that can satisfy certain acceptance criteria[i]. A common trap that many fall into is to put people through an ISO 9001 training course and by magic, they become qualified auditors. It is true that such people have been exposed to the requirements of ISO 9001 but they could have had equal exposure simply by reading the document and many of these one to two days courses are no more than tutored reading. An in-depth understanding is what is required and this comes from application followed by examination. Training courses that offer audit simulation or 'live audits' provide a more effective learning environment. Slide shows teach little if anything. Examinations that test memory are not effective. An examination that explains a situation or presents objective evidence and asks the auditor to study the evidence, draw conclusions and explain his/her rationale is more conducive to yielding competent ISO 9001 auditors. However, a more effective method is where the auditor is observed asking questions of people at their place of work, assessing the answers and evidence presented, drawing conclusions and writing a report of their findings.

AUDIT PROCEDURES (8.2.2)

The standard requires *a documented procedure to be established to define the responsibilities and requirements for planning and conducting audits and for establishing records and reporting results.*

What Does this Mean?

This means that the 'who, when, where, what, why and the how' of planning, conducting and reporting audits should be defined and documented. It also means that the methods of maintaining audit records following the audit should also be defined and documented.

Although the requirement calls for *a* documented procedure, it is not essential that your documentation be limited to a single document, because you may need policies, forms, guides and standards to describe the audit process.

Why is this Necessary?

This requirement responds to the Process Approach Principle.

Auditing is a process and as such should be documented in order to achieve consistency, to record best practice and to provide the basis for improvement.

How is this Demonstrated?

The audit procedures should cover the following:

- Preparing the annual audit programme;
- The selection of auditors and team leader if necessary;
- Planning particular type of audits;
- Conducting the audit;
- Recording observations;
- Determining corrective actions;
- Reporting and recording audit findings;
- Implementing corrective actions;
- Confirming the effectiveness of corrective actions;
- The forms on which you plan the audit;
- The forms on which you record the observations and corrective actions;
- Any warning notices you send out of impending audits, overdue corrective actions and escalation actions; and
- The content of the detailed and summary audit report.

Audit Plans

The detail plan for each audit may include dates if it is to cover several days but the main substance of the plan will be what is to be audited, against what requirements and by whom. At the detail level, the specific requirements to be checked should be identified based on risks, past performance and when it was last checked. *Overall plans* are best presented as programme charts and *detail plans* as checklists. Audit planning should not be taken lightly. Audits require effort from auditees as well as the auditor so a well-planned audit designed to quickly discover pertinent facts is far better than a rambling audit that jumps from area to area looking at this or that without any obvious direction.

Audit Checklists

Although checklists may be considered a plan, in the context of an audit they should be considered only as an aid in preparing the auditor to follow trails that may lead to the discovery of pertinent facts. However, there is little point in drawing up a checklist then putting it aside. Its rightful place is after the audit to verify that there is evidence indicating:

- those activities that were compliant;
- those activities that did not comply;
- those activities that were not checked; and
- those activities where there were opportunities for improvement.

Audits of practice against procedure or policy should be recorded as they are observed and you can either do this in note form to be written up later or directly on to observation forms especially designed for the purpose. Some auditors prefer to fill in the forms after the audit and others during the audit. The weakness with the former approach is that there may be some dispute as to the facts if presented sometime later. It is therefore safer to get the auditee's endorsement to the facts at the time they are observed. In other types of audits there may not be an auditee present. Audits of process documentation against policy can be carried out at a desk. One can check whether the documents of the quality system address the relevant clauses of the standard at a desk without walking around the site, but you can't check whether the system is documented unless you examine the operations in practice as there may be activities that make the system work that are not documented. Further guidance is provided in ISO 19011.

Audit Findings Report

An Audit Findings Report is illustrated in Fig. 30-2. You could use a Nonconformity Report but this is usually designed to collect product data so it may not be suitable, apart from the title being a little provocative. The Audit Findings Report simply conveys findings not the conclusions of the audit, i.e., that would be addressed in a summary report. The form illustrated in Fig. 30-2 does not have provision for corrective action unlike so many Audit Reports simply because, there is another form for this (see Fig. 30-2). What is stated on so many Audit Reports as corrective action is action to correct errors because it invariably does not get to the root cause. Many such forms contain actions to correct documentation that should have been correct. The authors of such reports obviously believed that by correcting the documents they were preventing recurrence but all they were doing was making good. Most auditors do not explore the reason why the documents were incorrect. Also the Audit Findings Report in Fig. 30-2 has provision for an impact assessment. Very few Audit Reports do this, but Clause 6.2.2d clearly requires personnel to be aware of the consequences their activities and auditors are no exception.

Audit Summary Report

The Audit Summary Report summarizes the audit in terms of:

- What was audited when;
- Who was involved;
- Why the audit was conducted;
- What was found;
- What was or will be done about it;
- What impact this might have had on the organization if nothing was done;

AUDIT FINDINGS REPORT

Process audited			**SAR/**	
Auditor	Location			Date

AUDIT OBJECTIVES	x	AUDIT CRITERIA
To find evidence of conformity		
To verify the integrity of performance measurement		
To find evidence of applied values and policies		
To find opportunities for improvement		
To verify that the system description is up to date		

AUDIT FINDING

	As relevant define:
	1. What was found
	2. Where was it found
	3 What was wrong with it
	4 Why this needs action

PROVISIONAL IMPACT ASSESSMENT

Impact on Customer and Business

PROPOSED ACTION

Remedial action	As relevant define:
	1. Action to remedy the problem
	2. Action to seek out and correct others
	3. Action to contain the problem

Target date				
Root Cause Analysis and Corrective Action Required	YES	NO	If YES, CAR required by:	
Related NCRs/CARs/SARs				
Findings agreed by:		Actions agreed by:		Date

EFFECTIVENESS OF ACTIONS

All actions taken and effective? ☐Yes ☐No	Reviewed by:	Date closed

FIGURE 30-2 Sample audit findings report.

The summary is intended for top management as they don't need the detail. If this is a separate report to a more detailed report, it might get their attention. One way to deter top management is to place a thick report on their desk.

Audit detail report

Records of an audit simply define what was recorded, whereas the results of an audit go much further. There follows a comprehensive list of topics that should be addressed in the Audit Report.

1. General location of the audit (Site, plant etc.);
2. Audit objectives (what the audit intends to achieve);
3. Audit criteria (the basis for determining success);
4. Audit strategy (basic approach to conducting the audit or summary of the audit process);
5. Audit plan (showing the processes that were audited with dates and times and those not audited (but within scope);
6. Processes or parts thereof not covered but within the audit scope and the reasons for exclusion;
7. Annotated audit checklists (showing what was checked and not checked in each process);
8. Audit team (names and position where appropriate);
9. Personnel interviewed on each process with dates;
10. Audit Findings Reports including the agreed proposals for remedial and corrective action;
11. Obstacles encountered that impact the conclusions;
12. Any unresolved issues between the auditors and the auditees;
13. Audit conclusions (whether audit objectives had been achieved relative to the success criteria);
14. Recommendations for improvement;
15. Follow-up action plan; and
16. Report distribution list.

AUDIT RECORDS (8.2.2)

The standard requires *records of the audits and their results to be maintained (see Clause 4.2.4)*.

What Does this Mean?

The records of the audit would be records of the audit programme for the organization, audit plans for individual audits, the audit reports and records of follow-up audits. Records of the results of the audit should be included in these records. These are explained further below.

Why is this Necessary?

Without adequate records of audits there is no sound basis on which to review performance of the audit process, the audit programme, individual audits and individual auditors, neither will there be objective evidence on which to base decisions for changing any of these.

How is this Demonstrated?

Presenting a pile of audit records is one thing and easily accomplished, demonstrating you have carried out an effective audit and have an effective audit programme is quite another.

The Audit Programme as a Record

The audit programme should be retained from year to year as a record of the processes, functions, departments and locations audited. It might be useful in establishing the last time an area was audited. This means that the programme should be maintained showing planned and executed audits. It is not much use if all it shows is what you planned to audit as plans can be abandoned.

In addition the results of the audit programme reviews should be recorded and these records maintained as evidence that you are managing the audit programme rather than simply doing the same thing over and over regardless of the impact it is having.

Specific Audit Records

For a specific audit the following records should be maintained:

- Audit plans (if not included in the Audit Reports);
- Audit checklists (if not included in the Audit Reports);
- Audit reports (summary and detailed reports);
- Audit findings reports (if not included in the Audit reports); and
- Remedial and corrective action reports (if not contained in the audit findings report).

Auditor Records

Records should also be maintained for individual audits including:

- Audit experience (type and number of audits conducted, position on audit team, duration, dates etc.);
- Auditor competence assessment; and
- Auditor development (training, coaching etc.).

TAKING ACTION (8.2.2)

The standard requires *management responsible for the area audited to ensure that any necessary corrections and corrective actions are taken without undue delay to eliminate the detected nonconformities and their causes.*

What Does this Mean?

Management responsible for the area audited are those who have the authority to cause change. An auditor may have interviewed a supervisor and found opportunities for improvement but the supervisor may not have the authority to agree to any changes – the auditor would therefore need to report the findings to the person who is authorized to take action.

Action without undue delay means that management are expected to act before the detected problem impacts subsequent results.

Eliminating nonconformities and their causes means that managers should:

- Take action to correct the particular nonconformity;
- Search for other examples of nonconformity and establish how widespread the problem is; and
- Establish the route cause of the nonconformity and prevent its recurrence.

Why is this Necessary?

This requirement responds to the Leadership Principle.

There is simply no point in conducting audits and finding problems if management does not intend to take action to prevent such problems impacting results. It often arises that problems detected by internal audits are perceived by management to have no impact on results, so it delays taking any action.

How is this Demonstrated?

To ensure actions are implemented without undue delay, the auditor needs to be sure that a failure to act will in fact impact performance of the process or the system. Management will not implement actions that have no effect on performance even if the action required restores conformity with ISO 9001. It is therefore sensible for the auditor to explain the impact of the detected nonconformity within the audit report – possibly by using a classification convention from critical to minor. But it would be more effective if the potential impact was actually stated and this requires the auditor to have a greater knowledge of the requirements be they in a contract, standard, policy, procedure or work instruction and the consequences of failing to implement them.

Unless the auditee is someone with responsibility for taking the corrective action, the auditee's manager should determine the actions required. If the action required is outside that manager's responsibility, the manager and not the auditor should seek out the appropriate authority and secure a proposal. Your policy manual should stipulate management's responsibility for taking action without undue delay.

A proposed action may not remove the non-compliance; it may be palliative leaving the problem to recur again at some future time. Target dates should be agreed for all actions and the dates should be met as evidence of commitment. Third party auditors will search your records for this evidence so you will need to impress on your managers the importance of honouring their commitments. The target dates also have to match the magnitude of the deficiencies. Small deficiencies which can be corrected in minutes should be dealt with at the time of the audit otherwise they will linger on as sores and show a lack of discipline. Others which may take 10–15 minutes should be dealt with within a day or so. Big problems may need months to resolve and require an orchestrated programme to be implemented. The actions in all cases when implemented should remove the problem, i.e., restore compliance. An action should not be limited to generating another form or procedure because it can be rejected by another manager thereby leaving the deficiency unresolved.

FOLLOW-UP AUDITS (8.2.2)

The standard requires *follow-up activities to include the verification of the actions taken and the reporting of verification results.*

What Does this Mean?

A follow-up audit is an action taken after the audit to verify that agreed actions from the audit have been completed as planned.

Why is this Necessary?

This requirement responds to the Factual Approach Principle.

Verification of actions completed is a normal activity of any managed process. The audit remains open until completion of actions has been confirmed.

How is this Demonstrated?

Follow-up action is necessary to verify that the agreed action has been taken and verify that the original nonconformity has been eliminated. Follow-up audits may be carried out immediately after the planned completion date for the actions or at some other agreed time. However, unless the audit is carried out relatively close to the agreed completion date, it will not be possible to ascertain if the action was carried out without undue delay.

The auditor who carries out the follow-up audit need not be the same that carried out the initial audit. In fact there is some merit in using different auditors in order to calibrate them.

When all the agreed nonconformities have been eliminated the audit report can be closed. The audit remains incomplete until all actions have been verified as being completed. Should any action not be carried out by the agreed date, the auditor needs to make a judgement as to whether it is reasonable to set a new date or to escalate the slippage to higher management. For minor problems, when there are more urgent priorities facing the managers, setting a new date may be prudent. However, you should not do this more than once. Not meeting the agreed completion date is indicative either of a lack of commitment or poor estimation of time and both indicate that there may well be a more deep routed problem to be resolved.

Chapter 31

Measurement and Monitoring of Products and Processes

CHAPTER PREVIEW

This chapter is aimed at all those managers responsible for the outputs of processes and in particular those monitoring and measuring processes and their outputs such as operators, supervisors, inspectors, process engineers and managers depending on the level of the process within the process hierarchy.

The positions where the requirements on measurement and monitoring feature in the managed process are shown in Fig. 31-1. Process validation includes process capability studies.

In this chapter we examine the requirements in Clauses 8.2.3 and 8.2.4 of ISO 9001:2008 and in particular:

- Measuring and monitoring methods (8.2.3),
- Process capability (8.2.3),
- Verifying conformity with product requirements (8.2.4),
- Evidence of conformity (8.2.4),
- Identifying the person(s) authorizing release (8.2.4),
- Product release approval (8.2.4).

MEASURING AND MONITORING METHODS (8.2.3)

The standard requires the organization *to apply suitable methods for monitoring and where applicable, measurement of the quality management system processes.*

What Does this Mean?

In all managed processes there will be stages where outputs are verified against inputs – these are product controls. There also need to be stages where the performance of the process itself is measured – these are process controls.

Monitoring is an on-going activity and, as stated at the beginning of this chapter, is the periodic or continual observation of operations to detect events before they occur so that action can be taken to prevent nonconformity. Measurement on the other hand implies that standards have been set and performance against those standards is being verified. For processes this may involve a range of parameters that are defined as being critical for the process to consistently deliver the correct results.

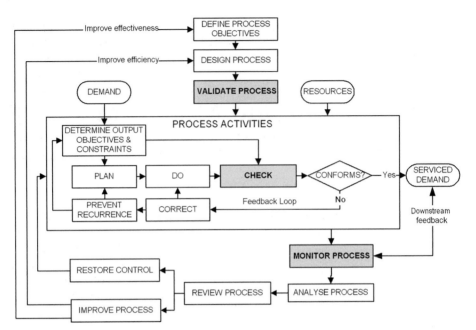

FIGURE 31-1   Where the requirements of Clause 8.2.3 apply in a managed process.

Why is this Necessary?

This requirement responds to the Factual Approach Principle.

It is processes that produce products and therefore measuring products tells us whether the products are correct but does not tell us whether the process is achieving its objectives. Parameters such as the rate at which products are produced, the variability between product characteristics, the resources used by the process and the effect of the process on its environment are parameters that require process measurement.

How is this Demonstrated?

There are several methods available for monitoring and measuring processes. The simplest monitoring method is visual observation by a person trained to detect variations that signal something is not quite right with a process. With industrial processes instruments may be installed to give the observer a visual indicator of performance. Data may be recorded on control charts so that the observer can tell when the performance is deteriorating. With

> **Measurement Maxims**
>
> What gets measured gets managed or
> You can only manage what you can measure.
> You get what you measure.
> Measurement affects that which is being measured.
> Accurate measurement requires accurate data.
> What gets reported gets attention.
> The more layers through which the data passes the less likely its integrity will be preserved.
> Measurement without standards is curiosity.
> Measurement against arbitrary standards creates trouble for someone.

management and administration processes, the observer (often the manager or supervisor) has formal or informal standards against which processes are monitored. Staff sometimes work in an environment in which they can't predict how their manager will react – an environment where the manager has not communicated the standards of performance expected. In other cases the staff work in a predictable environment – where the manager monitors performance against a standard known to the staff. In the former case the processes are not managed – they are reactive. In a managed process, monitoring would only look for the unpredictable where the immediate reaction is 'Why did that happen?' or 'That shouldn't have happened'. Management and administrative processes also lend themselves to management by data rather than observation. Provided that targets have been set, data can be collected, analysed and results produced to show whether performance is on track.

Managers should not only monitor performance against target but also monitor process efficiency. Firstly, they need to know if the processes are running as planned. The people are doing what managers stated they should do – following the policy, following the procedures and instructions as documented. They determine this by audit (see Chapter 8 on *A process approach*). The results of the audit will tell them the extent to which polices and procedures are being followed, not necessarily whether the results are acceptable. There are several outcomes for which data needs to be presented to top management.

- They are doing it OK, and are following the documented procedures. If this is the outcome then can they do it by consuming fewer resources? (Time, money, materials, people etc.).
- They are doing it OK, but are not following the documented procedures. If this is the case the documented procedures are ineffective so need to be re-designed.
- They are not doing it OK, and are following the documented procedures. If this is the outcome then it shows that there are either problems with individual competences or problems with the work environment.
- They not doing it OK and are not following the documented procedures. If this were the case, would it make any difference if they did follow the procedures? – if not, the process needs to be re-designed.

If they are doing it OK, and are following the documented procedures, the question they must ask is, are they using the right success measures, are they aiming at the right targets, do they have the right objectives? The answers to these come out of market research, benchmarking[①] or simply customer focus meetings.

Whatever is being monitored or measured, a soundly based method needs to be used to sense the variance from target, transmit the data, analyse it and compute factual results. The integrity of the measurements need to be sound, i.e., there should be no filtering, screening or reduction of the raw data that is not planned and the personnel performing the measurement should be competent to do so. The methods used also need to be appropriate to the process and in proportion to the impact of process failure on customer satisfaction. For example, you would not use monitoring equipment that gave an accuracy far greater than was needed to control process output, you would not spend weeks finding out whether every action of every procedure had been adhered to before determining whether the process is achieving its objectives.

PROCESS CAPABILITY (8.2.3)

The standard requires the monitoring and measurement methods *to demonstrate the ability of the processes to achieve planned results.*

What Does this Mean?

The planned results are the performance requirements at corporate and system level that relate to a particular process. They are whatever the process has been designed to achieve but should include needs, expectations and obligations from outside the process that impinge on the process.

To demonstrate ability means that either by observation or through validated records, evidence should be available which shows that the process is capable of achieving the planned results. A process is in control when the average spread of variation coincides with the nominal specification for a parameter. The range of variation may extend outside the upper and lower limits but the proportion of output within the limits can be predicted. This situation will remain as long as the process remains in statistical control. A process is in statistical control when the source of inherent variation is from common causes only, i.e., a source of variation that affects all the individual values of the process output and appears random. Common cause of variation results in a stable and repeatable distribution of results over time. When the source of variation causes the location, spread and shape of the distribution to change, the process is not in statistical control. These sources of variation are due to special or assignable causes and must be eliminated before commencing with process capability studies. It is only when the performance of a process is predictable that its capability to meet customer expectations can be assessed. (see also *Taking action of process variation* in this Chapter).

Why is this Necessary?

This requirement responds to the Factual Approach Principle.

Monitoring and measurement should not be performed merely to acquire information; it should either be performed to establish that processes are achieving the results for which they have been designed or for establishing that the results for which the processes have been designed are not adequate to meet the organization's objectives. Without objectives and targets, information from monitoring and measurement is merely interesting but not justification for action. Action should only be taken to bring performance in line with requirements or to change the requirements.

How is this Demonstrated?

In monitoring and measuring processes there are three questions that the process manager needs to be able to answer:

- How do you know the process is performing as planned?
- How do you know the process is achieving the results in the best way?
- How do you know that the results being achieved are those necessary to fulfil organizational goals?

Achieving Objectives

Provisions need to be put in place for each process that enables operations to be monitored and its performance to be measured against process objectives and targets. Objectives and targets need to be established for each process and these should be derived from an analysis of the factors that affect the ability of the process to deliver the desired results.

For example, the purpose of a purchasing process might be to acquire physical resources needed to fulfil the organization's objectives. An analysis of these objectives may reveal a need for a reliable supply of raw materials of consistent quality to be delivered in accordance with schedules that change monthly. A purchasing objective might be to secure a supply of the specific raw materials from a supplier who is capable of meeting the delivery schedules together with raw material quality, service quality and cost targets. These may include targets for response to changes in delivery schedules and raw material specification. Clearly a critical factor is the ability of the purchasing process to respond to change. Not only would the supplier have to respond to change but the internal stages of the purchasing process would also have to respond to change. Monitoring this process would involve observing how changes were received and transmitted through all the stages in the process and alerting staff to bottlenecks or other factors that could jeopardize meeting the targets. Measuring this process would involve checking that deliveries were being made on time and that the variation in cost and quality was also on target. Product measures would look at specific deliveries, whereas process measures would look at all deliveries for evidence that the purchasing objective was being met.

A process may be designed to deliver outputs that meet specification therefore a measure of performance is the ratio of conforming output to total output. If the ratio is less than 1 the process is not capable. Most processes fall into this category because some defective output is often produced, but it is possible to design processes so that they only produce conforming output. This does not mean perfect output but output that is within the limits defined for the process. The target yield for a process may be 97% implying that 100% is not feasible, therefore a yield of 98% is good and a yield of 96% bad – it depends what standards have been set.

This type of measurement requires effective data collection, transmission and analysis points so that information is routed to analysts to determine performance and for results to be routed to decision makers for action. With a process such as order processing, in addition to each order being checked the process should be monitored to establish it is meeting the defined objectives for processing time, customer communication etc. and that there is no situation developing that may jeopardize achievement of the order processing objectives. Therefore, every process will have at least two verification stages – one in-line for verifying output quality and another for verifying process performance against objectives.

The object of process capability studies is to compute the indices and then take action to reduce common cause variation by preventive maintenance, mistake proofing, operator training, revision to procedures and instructions etc. The inherent limitations of attribute data prevent their use for preliminary statistical studies since specification values are not measured. Attribute data have only two values, (conforming or non-conforming, pass or fail, go or no-go, present or absent) but they can be counted, analysed and the results plotted to show variation. Measurement can be based on the fraction defective such as parts per million (PPM). Whilst variables' data follows

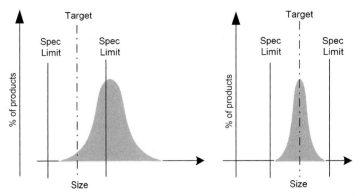

FIGURE 31-2   Difference between a process that is under control (left) and a capable process (right).

a distribution curve, attribute data varies in steps since you can't count a fraction. There will either be zero errors or a finite number of errors. Process capability studies can only commence once the process is under control, i.e., the results are predictable and exhibit a normal distribution. Under control does not necessarily mean that the process only delivers conforming product. A process that is under control is illustrated by the bell curve on the left in Fig. 31-2. A capable process is illustrated by the bell curve on the right in Fig. 31-2 where all results are within specification limits.

Achieving Best Practice

Each result requires resources and even when the required results are being achieved there may be better ways or more efficient ways of achieving them. Therefore, targets may be set for efficiency and these too need to be monitored and measured. One method is to monitor resource utilization. Another is to conduct benchmarking[①] against other processes to find the best practice.

Fulfilling Purpose

The planned results include organizational objectives and therefore the process objectives should be reviewed periodically to verify they are the right objectives. A 99.73% (3 sigma) yield may have been a real challenge five years ago but today your competitors achieve 99.999998% (6 sigma) and therefore the objective needs to change for the organization to maintain its position in the market.

Six Sigma

In a perfect world, we would like the range of variation to be well within the upper and lower specification limits for the characteristics being measured but invariably we produce defectives. If there were an 80% yield from each stage in a 10-stage process, the resultant output would be less than 11% and as indicated in Table 31-1 we would obtain only 107 374 good products from an initial batch of 1 million.

Even if the process stage yield was 99% we would still obtain 95 617 less products than we started with. It is therefore essential that multiple stage processes have a process

TABLE 31-1 10-Step Process Yield

Stage	Yield/stage	Total % yield	Initial population 1 million
1	0.80	80	800 000
2	0.80	64	640 000
3	0.80	51.2	512 000
4	0.80	41	409 600
5	0.80	32.8	327 680
6	0.80	26.2	262 144
7	0.80	21	209 715
8	0.80	16.8	167 772
9	0.80	13.4	134 218
10	0.80	10.7	107 374

stage yield well in excess of 99.99% and it is from this perspective that the concept of six sigma emerges.

The small Greek letter s is σ and is called sigma. It is the symbol used to represent the standard deviation in a population. We use s for standard deviation as measured by a sample of finite magnitude. Standard deviation is the square root of the variance in a population. In plain English, standard deviation tells us how tightly a set of values is clustered around the average of those same values. In statistical speak, standard deviation is a measure of dispersion (spread or variability) about a mean value (or average) of a population that exhibits a normal distribution and is expressed by the formula:

$$\sigma = \sqrt{\frac{\sum (x - \mu)^2}{n}}$$

where x is a value such as mass, length, time, money, μ is the mean of all values, Σ is capital sigma and is the mathematical shorthand for summation and n is the number in the population.

When the variance is computed in a sample, the formula is:

$$s = \sqrt{\frac{\sum (x - \bar{x})^2}{n}}$$

where x bar is the mean of the sample and n is the sample size.

However, small samples tend to underestimate the variance of the parent population and a better estimate of the population variance is obtained by dividing the sum of the squares by the number of degrees of freedom. Variance in a population thus becomes:

$$s = \sqrt{\frac{\sum (x - \bar{x})^2}{n - 1}}$$

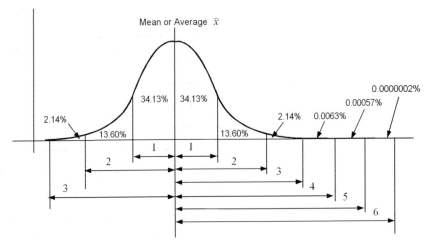

FIGURE 31-3 Normal distribution and standard deviation.

When the population of a variable is concentrated about the mean the standard deviation is small indicating stability and when the population of a variable is spread out from the mean the standard deviation is large indicating volatility.

When the frequency distribution of a set of values of a variable is symmetrical about a mean it may approximate to a normal distribution represented by the equation:

$$y = \frac{1}{\sqrt[\sigma]{2\pi}}e^{\frac{-(x-\bar{x})^2}{2\sigma^2}}$$

where y is the height of the curve at any point along the scale of x and e is the base of the Napierian logarithms (2.7183) and π the ratio of the circumference of a circle to its diameter which to three decimal places is 3.142.

The shape of the normal distribution embraces:

68.26% of the population within +/−1 standard deviation around the mean,

95.46% of the population within +/−2 standard deviations around the mean,

99.74% of the population within +/−3 standard deviations around the mean and,

99.9937 of the population within +/−4 standard deviations around the mean,

which is 100% for most applications. In industries producing millions of products even greater performance is deemed necessary especially when all output is being shipped to one customer. This is represented diagrammatically and extended to 6 standard deviations in Fig. 31-3.

Assuming a normal distribution of results at the six sigma level you would expect 0.002 parts per million or ppm but when expressing performance in ppm, it is common practice to assume that the process mean can drift 1.5 sigma in either direction. The area of a normal distribution beyond 4.5 sigma from the mean is 3.4 ppm. As control charts will detect any process shift of this magnitude in a single sample, the 3.4 ppm represents a very conservative upper boundary on the non-conformance rate.[1]

Although the concept of six sigma can be applied to non-manufacturing processes you cannot assume as was done in Table 31-1 that the nonconformities in a stage output

---

[1] Pyzdek, Thomas (2001). The Complete Guide to Six Sigma, McGraw-Hill.

TABLE 31-2 Process Yield at Various Sigma Values

Sigma	Product meeting requirements %	Number of errors per million products	
		Assuming normal distribution	Assuming 1.5 sigma drift
1	68.26	317 400.000	697 672.15
2	95.45	45 500.000	308 770.21
3	99.73	2700.000	66 810.63
4	99.9937	63.000	6209.70
5	99.999943	0.570	232.67
6	99.9999998	0.002	3.40

are rejected as unusable by the following stages. A person may pass through 10 stages in a hospital but you cannot aggregate the errors to produce a process yield based on stage errors. Patients don't drop out of the process simply because they were kept waiting longer than the specified maximum. You have to take the whole process and count the number of errors per 1 million opportunities.

Table 31-2 shows the number of products meeting requirements and the equivalent defects per million products for a range of standard deviations. In Table 31-2, six sigma (6σ) translates into two errors per billion opportunities but what Motorola found was that processes drift over time, what they call the Long-Term Dynamic Mean Variation. Motorola judged this variation as typically falling between 1.4 and 1.6 sigma, and will therefore account for special cause variation causing drift over the long term (years not months). Views do appear to differ on why the 1.5 sigma drift is applied but in any event in most situations whether six sigma is two errors per billion or 3.4 errors per million opportunities matters not to most people – it is only a target.

However, every measurement process, however complicated, has certain underlying assumptions that mean the results are valid only when certain conditions apply.

There are four assumptions that typically underlie all measurement processes; namely, that the data from the process at hand 'behave like' random samples:

- from a fixed distribution;
- with the distribution having fixed location; and
- with the distribution having fixed variation.

If the four underlying assumptions hold, then we have achieved probabilistic predictability – the ability to make probability statements not only about the process in the past, but also about the process in the future. In short, such processes are said to be 'in statistical control'. Thus, attempts at reaching six sigma levels in a process that is not in statistical control will be futile.

Having now acquired some understanding of measurement and variation, it is hoped that you can now approach the following requirements of ISO 9001 with greater confidence.

TAKING ACTION ON PROCESS VARIATION

The standard requires *correction and corrective action to be taken when planned results are not achieved.*

What Does this Mean?

When planned results are not achieved a process is exhibiting variation outside the acceptable limits. Correction is dealing with the unacceptable result: corrective action is preventing the unacceptable variation from recurring but there are dangers in adjusting a process every time it produces a nonconformity. These are dealt with below.

Why is this Necessary?

Variation is present in all systems. Nothing is absolutely stable. In fact in natural systems, variation is vital for survival as was observed by Charles Darwin in the 1830s. Darwin found that species living under one set of conditions had evolved differently to those living under another set of conditions and postulated that survival depended on the offspring inheriting attributes that equipped the parent to survive in a given environment. Whereas variation in natural systems is essential for survival of the species, variation in manmade systems tends to be detrimental to survival. Just imagine if variation was so unpredictable that you could not rely on the cleanliness in surgical operations, the safety of food, and drink, the stability of power distribution, the integrity of the air traffic control system or simply the handbrake in your car.

How is this Demonstrated?

If you monitor the difference between the measured value and the required value of a characteristic and plot the results on a horizontal timescale in the order the products were produced, you would notice that there is variation over time. There will be a natural scattering of the measured values about some central tendency value.[2] This scattering about a central value is known as a distribution. The distribution can be characterised by:

- Location (typical value),
- Spread (Span of values from smallest to largest),
- Shape (the pattern of variation; whether it is symmetrical, skewed etc.).

A process is deemed stable if it runs in a consistent and predictable manner. This means that the average process value is constant and the variability is controlled. If the variation is uncontrolled, then either the process average is changing or the process variation is changing or both.

If we can demonstrate that the process is stabilized about a constant location, with a constant variance and a known stable shape, then you have a process that is both predictable and controllable. This is required before you can set up control charts or conduct experiments. There are two type of variation, that which is attributable to an assignable or special cause and that which is attributable to a common cause.

---

[2] NIST/SEMATECH e-Handbook of Statistical Methods, http://www.itl.nist.gov/div898/handbook/, 2006

Special Cause

The cause of variations in the location, spread and shape of a distribution is considered special or assignable because the cause can be assigned to a specific or special condition that does not apply to other events. They are causes that are not always present. Wrong material, inaccurate measuring device, worn out tool, sick employee, weather conditions, accident, stage omitted are all one-off events that cannot be predicted. When they occur they make the shape, spread or location of the average change as shown in Fig. 31-4. The process is not predictable while special cause variation is present. Eliminating the special causes is part of quality control and many of these problems can be detected before they result in nonconforming product through preparatory measures and routine checks.

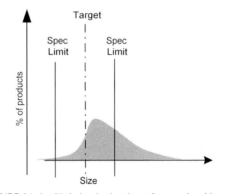

FIGURE 31-4   Variation in the shape from assignable cause.

Once all the special causes of variation have been eliminated, the shape and spread of the distribution and the location of the average become stable, the process is under control – the results are predictable. However, it may not be producing conforming product. You may be able to predict that the process could produce one defective product in every 10 produced. There may still be considerable variation but it is random. A stable process is one with no indication of a special cause of variation and can be said to be in *statistical control*. Special cause variation is not random – it is unpredictable. It occurs because something has happened that should not have happened, so you should search for the cause immediately and eliminate it. The person running the process should be responsible for removing special causes unless these causes originate in another area when the source should be isolated and eliminated.

Common Cause

Once the special cause of variation has been removed, the variation present is left to chance, it is random or what is referred to as common cause. This does not mean that no action should be taken but to treat each deviation from the average as a special cause will only lead to more problems. The random variation is caused by factors that are inherent in the system. The operators have done all they can to remove the special causes, the rest are down to management. This variation could be caused by poor design, working environment, equipment maintenance or inadequacy of information. Some of these events may be common to all processes, all machines, all materials of a particular type, all work performed in a particular location or environment, or all work performed using a particular method.

If the average value is the target value you will get a distribution similar to the curve as in Fig. 31-5 and this shows that the process is under control. If the average value is outside the upper specification limit you will get a distribution similar to the bell curve as on the left in Fig. 31-2. The goal is to get the average value on the target value with a spread of variation within the upper and lower limits as shown by the bell curve in Fig. 31-5 but this process still produces nonconforming product. As depicted by the edge of the bell curve going outside the spec limits.

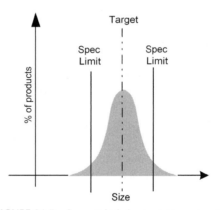

FIGURE 31-5   Symmetrical about a mean on target.

If every time you plot the results of a batch of product the distribution remains the same the process is under control regardless of the average being on the target value.

The factors causing these variations are referred to as 'common causes' and these need to be managed effectively.

By removing special causes, the process settles down and although nonconformities remain, performance becomes more predictable as shown in Fig. 31-6. Further improvement will not happen until the common causes are reduced and this requires action by management. However, the action management takes should not be to look for a scapegoat – the person whom they believe caused the error, but to look for the root cause – the inherent weakness in the system that causes this variation.

Common cause variation is random and therefore adjusting a process on detection of a common cause will destabilize the process. The cause has to be removed, not the process adjusted. When dealing with either common cause or special cause problems, the search for the root cause will indicate whether the cause is random and likely to occur again or a one-off event. If it is random, only action on the system will eliminate it. If it is a one-off event, no action on the system will prevent its recurrence, it just has to be fixed. Imposing rules will not prevent a nonconformity caused by a worn out tool that

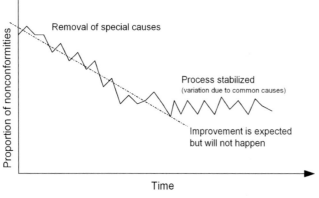

FIGURE 31-6 Inducing stability in processes.

someone forgot to replace. A good treatment of common cause and special cause variation is given in third footnote.[3]

With a stable process, the spread of common cause variation will be within certain limits. These limits are not the specification limits but are limits of natural variability of the process. These limits can be calculated and are referred to as the Upper and Lower Control Limits (UCL and LCL, respectively). The control limits may be outside the upper and lower specification limits to start with but as common causes are eliminated, they close in and eventually the spread of variation is all within the specification limits. Any variation outside the control limits will be rare and will signal the need for corrective action. This is illustrated in Fig. 31-7.

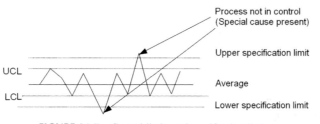

FIGURE 31-7 Control limits and specification limits.

Keeping the process under control is process control. Keeping the process within the limits of the customer specification is quality control. The action needed to make the transition from process control to quality control is an improvement action and this is dealt with next.

VERIFYING CONFORMITY WITH PRODUCT REQUIREMENTS (8.2.4)

The standard requires the organization *to monitor and measure the characteristics of the product to verify that product requirements have been met and that these activities be conducted at appropriate stages of the product realization process in accordance with the planned arrangements.*

---

[3] Deming, W. Edwards (1982). Out of the crisis, MITC.

What Does this Mean?

Activities that monitor and measure product are often referred to as inspection, test or verification activities. Appropriate stages of the product realization process means the stages at which:

- the achieved characteristics are accessible for measurement;
- an economic means of measurement can be performed;
- the correction of error is less costly than if the error is detected at later stages.

It may be possible to verify some characteristics on the final product just prior to shipment but it is costly to correct errors at this late stage resulting in delayed shipment. It is always more economic to verify product at the earliest opportunity.

The planned arrangements in this case are the plans made for verifying product in terms of what is to be verified, who is to verify it, when is it to be verified, how is to be verified, where is it to be verified and what criteria is to be used to judge conformity.

Product requirements are all the requirements for the product including customer, regulatory and the organization's requirements. Some of these may be met by inherent design features, others will be met in production, installation or service.

The forms of verification that are used in product and service development, should also be governed by these requirements as a means of ensuring that the product on which design verification is carried out conforms with the prescribed requirements. If the product is non-compliant it may invalidate the results of design verification. Product verification also applies to any measuring and monitoring devices that you design and manufacture to ensure that they are capable of verifying the acceptability of product as required. Product verification is part of product realization and not something separate from it although the way the requirements are structured may imply otherwise. Whenever a product is supplied, produced or repaired, rebuilt, modified or otherwise changed, it should be subject to verification that it conforms to the prescribed requirements and any deficiencies corrected before being released for use.

Why is this Necessary?

This requirement responds to the Factual Approach Principle.

One verifies product to establish that it meets requirements. If one could be certain that a product would be correct without it being verified, product verification would be unnecessary. However, most processes possess inherent variation due to common causes – variations that affect all values of process output and appear random. Although a process may be under statistical control, a special event could disturb performance and without checks on the output, its detection may go unnoticed. One can only check for those events we think might happen which is why our confidence in the 'system' is shaken when we discover a condition with a cause we had not predicted.

How is this Demonstrated?

As product requirements may include characteristics that are achieved by design, production, installation or service delivery, a high-level verification matrix is needed to provide traceability from requirement to the means of verification. This will undoubtedly lead to there being a few characteristics that need to be verified only once by design verification with many of the others being verified in production or service delivery. Characteristics that do not vary only need to be checked once. For example, a car designed with four wheels could not possibly be made with two, three or five wheels when put into production, but a body panel with screw inserts could emerge from the process without the inserts!

Having established that characteristics vary, the stage at which they need to be verified should be determined. This leaves three possibilities; on receipt, in-process or on completion. Receipt verification was addressed under *Purchasing* in Chapter 26. Here we address in-process and finished product verifications.

In-Process Verification

In-process verification is carried out in order to verify those features and characteristics that would not be accessible to verification by further processing or assembly. When producing a product that consists of several parts, sub-assemblies, assemblies, units, equipment and subsystems, each part, sub-assembly etc. needs to be subject to final verification but may also require in-process verification for the reasons given above. Your control plans should define all the in-process verification stages that are required for each part, sub-assembly, assembly etc. In establishing where to carry out the verification, a flow diagram may help. The verification needs to occur after a specified feature has been produced and before it becomes inaccessible for measurement. This doesn't mean that you should check features as soon as they are achieved. There may be natural breaks in the process where the product passes from one stage to another or stages at which several features can be verified at once. If product passes from the responsibility of one person to another, there should be a stage verification at the interface to protect the producer even if the features achieved are accessible later. Your verification plans should:

- identify the product to be verified;
- define the specification and acceptance criteria[①] to be used and the issue status which applies;
- define what is to be verified at each stage. (Is it all work between stages or only certain operations? The parameters to be verified should include those that are known to be varied by the manufacturing processes, those that remain constant from product to product, need verifying once only usually during design proving.);
- define the verification aids and test equipment to be used. (There may be jigs, fixtures, gauges and other aids needed for verification. Standard measuring equipment would not need to be specified because your verification staff should be trained to select the right tools for the job. Any special measuring devices should be identified.);
- define the environment for the measurements to be made if critical to the measurements to be made;
- identify the organization that is to perform the verification;
- make provision for the results of the verification to be recorded.

Finished Product Verification

Finished product verification is in fact the last verification of the product that you will perform before dispatch but it may not be the last verification before delivery if your contract includes installation. There are three definitions of finished product verification:

- The verification carried out on completion of the product; afterwards the product may be routed to stores rather than to a customer.
- The last verification carried out before dispatch; afterwards you may install the product and carry out further work.
- The last verification that you as a supplier carry out on the product before ownership passes to your customer; this is the stage when the product is accepted and consequently the term *product acceptance* is more appropriate and tends to convey the purpose of the verification rather than the stage at which it is performed.

There are two aspects to finished product verification. One is checking what has gone before and the other is accepting the product.

Final verification and test checks should detect whether:

- All previous verification activities have been performed;
- The product bears the correct identification, part numbers, serial numbers, modification status etc.;
- The as-built configuration is the same as the revision status of all the parts, sub-assemblies, assemblies etc. specified by the design standard. A Configuration Record containing this data would avoid argument later as to whether or not certain specification changes were embodied in the product;
- All recorded nonconformities have been resolved and verified;
- All concession applications have been approved;
- All verification results have been collected;
- Any result outside the stated limits is either subject to an approved concession, an approved specification change or a retest that shows conformance with the requirements;
- All documentation to be delivered with the product has been produced and conforms to the prescribed standards.

EVIDENCE OF CONFORMITY (8.2.4)

The standard requires *evidence of conformity with the acceptance criteria to be maintained*.

What Does this Mean?

Evidence of conformity is the information recorded during product verification that shows the product to have exhibited the characteristics required.

Why is this Necessary?

This requirement responds to the Factual Approach Principle.

At a point in the process, product will be presented for delivery to the next stage in the process or to a customer. At such stages a decision is made whether or not to release product and this decision needs to be made on the basis of facts substantiated by objective evidence.

How is this Demonstrated?

This requires that you produce something like an Acceptance Test Plan which contains, as appropriate, some or all of the following:

- Identity of the product to be verified.
- Definition of the specification and acceptance criteria[①] to be used and the issue status that applies.
- Definition of the verification aids and measuring devices to be used.
- Definition of the environment for the measurements to be made.
- Provision for the results of verification to be recorded – these need to be presented in a form that correlates with the specified requirements.

Having carried out these verification activities, it should be possible for you to declare that the product has been verified and objective evidence produced that will demonstrate that it meets the specified requirements. Any concessions given against requirements should also be identified. If you can't make such a declaration, you haven't done enough verification. Whether or not your customer requires a certificate from you testifying that you have met the requirements, you should be in a position to produce one. The requirement for a certificate of conformance should not alter your processes, your quality controls or your procedures. Your management system should give you the kind of evidence you need to assure your customers that your product meets their requirements without having to do anything special.

Your verification records should be of two forms – one which indicates what verification activities have been carried out and the other which indicates the results of such verification. They may be merged into one record but when parameters need to be recorded it is often cleaner to separate the progress record from the technical record. Your procedures, quality plan or product specifications should also indicate what measurements have to be recorded.

Don't assume that because a parameter is shown in a specification that an inspector or tester will record the result. A result can be a figure, a pass or fail or just a tick. Be specific in what you want recorded because you may get a surprise when gathering the data for analysis. If you use computers, you shouldn't have the same problems but beware, too much data is probably worse than too little! In choosing the method of recording measurements, you also need to consider whether you will have sufficient data to minimize recovery action in the event of the measuring device subsequently being found to be out of calibration. As a general rule, only gather that data you need to determine whether the product meets the requirements or whether the process is capable of producing a product that meets the requirements. You need to be selective so that you can spot the out of tolerance condition or variation in the measurement system. Sometimes, plotting the results as a histogram might indicate abnormalities in the results that are symptomatic of measurement errors. The acceptance criterion is therefore not simply specified upper and lower limits but evidence that results are located in a normal distribution that is centred on the nominal condition.

All verification records should define the acceptance criteria[1], the limits between which the product is acceptable and beyond which the product is unacceptable and therefore nonconforming.

IDENTIFYING THE PERSON(S) AUTHORIZING RELEASE (8.2.4)

The standard requires records *to indicate the person(s) authorizing release of product to the customer.*

What Does this Mean?

The person authorizing release of the product is the one whose permission is needed before product can pass beyond a defined stage. Release conditions may include conformity to specification and quantity. Product may be held until the required quantity has been produced. When a product is 'in-process', the operator is working on it. Product may be piling up in the output basket, but until the operator indicates it can be released, it remains under his or her control. This requirement means that such decisions are to be recorded in a manner that is traceable to the person who made them. With hardware, software and processed material this often means a signature or stamp on verification records or labels attached to the product. With documentation, it is an approval signature on a document or accompanying forms.

This requirement implies that there will always be a person in the process who decides when product should or should not be released. This may not always be so. With automated processes equipped with product verification instrumentation, the product may well pass straight into dispatch and onto the customer without any human intervention. The person releasing product in this case is the person controlling the process.

Why is this Necessary?

This requirement responds to the Factual Approach Principle.

Within an organization you may wish to identify the individual responsible so that you can go back to ask questions. This is more likely in the case of a reject decision as opposed to an acceptance decision. It is also necessary to be able to demonstrate (particularly with food and drugs) that product was released by someone who was aware of the consequences of their actions and was acting responsibly. An organization's reputation could be easily damaged if it emerged that the person releasing product into the food chain was unqualified.

How is this Demonstrated?

There are two parts to this requirement. The first is that the person who releases product is identified and the second that this person is authorized to release product. It is not enough for there to be a signature on a document, on a record or on a label. The signature has to be of a person who has the right to release the product. The signature therefore needs to be legible or at least traceable to the individual.

Some organizations maintain a list of authorized signatures as a means of being able to trace signatures to names of people who carry certain authority. If you have a large number of people signing documents and records and there is a possibility that the wrong person may sign a document, the list is a good tool for checking that there has been no abuse of authority. Otherwise, the name of the individual and his or her position below or alongside the signature should be adequate. The management system should have in-built provision for preventing the wrong people releasing product. When such provisions have been made, the authorized person could under certain circumstances be influenced to release bad product – it is only the strength of the shared values that would prevent such transgression.

PRODUCT RELEASE APPROVAL (8.2.4)

The standard requires *that unless otherwise approved by a relevant authority and where applicable, by the customer, the release of product and delivery of service to the customer is not to proceed until the planned arrangements have been satisfactorily completed.*

What Does this Mean?

This requirement can impose unnecessary constraints if taken literally. Many activities in planned arrangements are performed to give early warning of nonconformities. This is in order to avoid the losses that can be incurred if failure occurs in later tests and inspections. The earlier you confirm conformance the less costly any rework will be. One should therefore not hold shipment if later activities have verified the parameters, whether or not earlier activities have been performed. It is uneconomic for you to omit the earlier activities, but if you do, and the later activities can demonstrate that the end product meets the requirements, it is also uneconomic to go back and perform those activities that have not been completed. Your planned arrangements could cover installation and maintenance activities which are carried out after dispatch and so it would be unreasonable to insist that these activities were completed before dispatch or to insist on separate plans just to sanitize a point. A less ambiguous way of saying the same thing is to require *no product to be dispatched until objective evidence has been produced to demonstrate that it meets the product requirements and that authorization for its release has been given.*

If planned arrangements[⊕] cannot be achieved, a concession might be obtained from the recipient to permit release of product that did not fully meet the requirements. The recipient could be the owner of the process receiving product for processing or the external customer receiving product in response to an order.

Why is this Necessary?

This requirement responds to the Factual Approach Principle.

Having decided on the provisions needed to produce product that meets the needs and expectations of customers, regulators and the organization itself, it would be pretty foolish to permit release of product before confirming that all that was agreed to be done has been done. However, circumstances may arise where nonconforming product has been produced and instead of shipping such product without informing the intended recipient, an organization committed to quality would seek permission to do so.

How is this Demonstrated?

You need four things before you can release product whether it be to the stores, to the customer, to the site for installation or anywhere else:

1. Sight of the product.
2. Sight of the requirement with which the product is to conform including its packaging, labelling and other product related requirements.
3. Sight of the objective evidence that purports to demonstrate that the particular product meets the requirement.
4. Sight of an authorized signatory or the stamp of an approved stamp-holder who has checked that the particular product, the evidence and the requirement are in complete accord.

Once the evidence has been verified, the authorizedjk person can make the release decision and endorse the appropriate record indicating readiness for release. Should there be any discrepancies, they should be validated and if proven valid, the non-conforming product process should be initiated.

Chapter 32

Control of Monitoring and Measuring Equipment

CHAPTER PREVIEW

This chapter is aimed at those personnel responsible for using, verifying and calibrating measuring equipment. This may include laboratory managers, inspectors, testers, calibration engineers, metrologists, statisticians.

This requirement should strictly be located in Section 8 of ISO 9001 but has been included in Section 7 because it is believed that there are some applications where it does not apply. Clearly there are no applications where monitoring and measurement do not apply but there may be applications where physical calibration of measuring equipment may not be applicable in the traditional sense of the word *equipment*. The monitoring and measurements referred to are those required to carry out product verification rather than the measurements to calibrate a measuring device. The requirement is under product realization and not a sub-section, and therefore applies equally to product or service design, purchasing and production or service delivery. It should not be interpreted as only applying to the characteristics of a product that can be measured through examination. It applies equally to performance characteristics that are inherent in the product design such as durability, safety and security and to intangible characteristics such as courtesy, respect and integrity in service design.

In this chapter we examine the requirements in Clause 7.6 of ISO 9001:2008 and in particular:

- Determining monitoring and measurements to be undertaken;
- Determining monitoring and measuring equipment needed;
- Defining the monitoring and measuring processes;
- Calibrating and verifying measurements;
- Recording the basis for calibration;
- Adjustment of equipment;
- Indicating calibration status;
- Safeguarding monitoring and measuring equipment;
- Protection of monitoring and measuring equipment;
- Action on equipment found out of calibration;
- Calibration and verification records;
- Software validation.

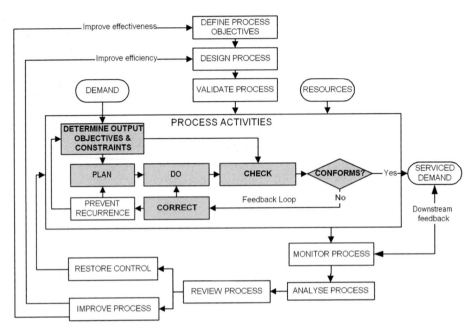

FIGURE 32-1 Where the requirements of Clause 7.6 apply in a managed process.

The position where requirements on the control of monitoring and measuring equipment feature in the managed process is shown in Fig. 32-1. In this case the PDCA in Process Operation (the bold rectangle) relates to controlling specific monitoring and measuring equipment. The input is a demand from an interfacing process for monitoring and measuring equipment and the output is monitoring and measuring equipment that is fit for purpose, i.e., calibrated. This process would be developed and managed as part of the resource management process.

DETERMINING MONITORING AND MEASUREMENTS TO BE UNDERTAKEN (7.6)

The standard requires the organization *to identify the monitoring and measurement to be undertaken to provide evidence of conformity of product to determined requirements.*

What Does this Mean?

Monitoring is the periodic or continual observation of operations to detect events before they occur so that action can be taken to prevent nonconformity. Measurement is the process of associating numbers with physical quantities and phenomena. Monitoring and measurement may be made by unaided human senses in which case they are often called estimates or, more usually, by the use of instruments, which may range in complexity from simple rules for measuring length to highly sophisticated systems designed to detect and measure quantities, entirely beyond the capabilities of human senses.

Determined requirements are those of the customer, the regulators and the organization as addressed previously in Clause 7.2.1 of the standard. It is therefore not only a question of measuring the characteristics of the product but also relating these characteristics to the defined requirements. For tangible product, the requirements may be expressed in performance terms (performance requirements) or in terms of form, fit and function (conformity requirements). Performance requirements may not be directly measurable. For instance safety requirements have to be translated into physical characteristics that are deemed to satisfy the requirement. You do not measure safety but the absence of hazard and therefore with a suspension bridge you would measure loading capacity and with food you would measure the absence of bugs. You can only search for what is known. Only when we know that a certain quantity of a substance is toxic can we search for it.

Where the product of a process is intangible, the range of measurements varies widely. Knowledge is measured as the output of a course of study, skill or competence is measured as an output of training. Perceptions and behaviours are measured where they are vital to service quality. Comfort in a vehicle may be function of ride quality measured by using a laser profilometer on a road surface of known characteristics.

Why is this Necessary?

This requirement responds to the Factual Approach Principle.

The reason for requiring monitoring and measurements to be identified is so that you have a means of relating the requirements to the characteristics to be measured in order to verify their achievement. If you did not identify the monitoring and measurements to be undertaken you would have no knowledge of whether or not requirements had been met before you delivered the product or service.

How is this Demonstrated?

The monitoring and measurements to be made should be derived from the characteristics defined in the product requirement. In some cases these may be directly measurable, such as size and weight, but often the characteristics need to be translated into measurable parameters. One way to identify what to monitor or measure is to produce a verification matrix showing the product characteristic, the measures, the units of measure, the target values and the level at which the measurements will be taken.

The product characteristic may be a specific dimension or an attribute. Length, mass and time are specific dimensions, response, safety, reliability, on-time delivery are attributes.

There are many different ways of expressing the units of measure as illustrated in Table 32-1.

With tangible products the level of measurement may be at system, sub-system, equipment, component or material level. With intangible products the measurement may be at the relationship, encounter or transaction level. The relationship is the long-term interaction between customer and supplier, the encounter tends to be the short-term interaction involving a single purchase and the transaction is a specific activity between representatives of the customer and supplier. A single encounter may involve several transactions and a relationship may involve years of encounters.

The measurements to be made should be identified in test specifications, process specifications and drawings etc. but often these documents will not define how to take the

TABLE 32-1 Units of Measure

Measure	Units
Length	Metres
Mass	Kilograms
Time	Seconds
Response	Time between receipt of call and engineer on site
Safety	Incidents per passenger mile
Reliability	Mean time between failures
Maintainability	Mean time to repair
On-time delivery	Percent delivered by agreed date
Conformity	Percent defective
Courtesy	Ratio of complaints to transactions

measurements. The method of measurement should be defined in verification procedures that take into account the measurement uncertainty, the devices used to perform the measurements, the competency of the personnel and the physical environment in which measurements are taken. There may be a tolerance on variable parameters so as to determine the accuracy required. You may use general tolerances to cover most dimensions and only apply specific tolerances where it is warranted by the application.

DETERMINING MONITORING AND MEASURING EQUIPMENT NEEDED (7.6)

The standard requires *the organization to identify the monitoring and measuring equipment needed to provide evidence of conformity of product to determined requirements.*

What Does this Mean?

The integrity of products depends on the quality of the equipment used to create and measure their characteristics. This part of the standard specifies requirements for ensuring the quality of such equipment.

Monitoring and measuring equipment comprises the monitoring or measuring instruments, software, measurement standard, reference material or auxiliary apparatus used to monitor and measure product characteristics. The measuring instrument itself could be software or a method or sensor that captures information. It includes equipment used during design and development for determining product characteristics and for design verification. Some characteristics cannot be determined by calculation and need to be derived by experiment. In such cases the accuracy of devices you use must be controlled, otherwise the parameters stated in the resultant product specification may not be achievable when the product reaches production.

Wherever there is a measurement to be made, a parameter or phenomena to be monitored, there is always a sensor employed to sense the variation. The key to effective measurement is plugging the right sensor into the right location.

Although the word equipment is often perceived to be hardware (see also Chapter 4 on *ISO 9001:2008 Commentary*), it can also include human senses (hearing, sight, touch, taste and smell), a physical instrument measuring mass, length, time, electric current, temperature etc. a perception obtained from surveys, interviews, questionnaires or behaviour obtained from psychometric tests or knowledge obtained from written examination. There may be others, but clearly measurement is a wide subject and beyond the scope of this book to explore in any detail.

The sensor senses what is taking place and registers it in a form suitable for transmission to a receiver. The measuring and monitoring equipment should encompass the sensor, the transmitter and the receiver because the purpose of measurement is to take decisions and without receipt of the information no decisions can be taken. Also, you need to be aware that the transmitter and receiver may degrade the accuracy and precision of the measurement.

Why is this Necessary?

This requirement responds to the Process Approach Principle.

It is necessary to identify the measuring and monitoring equipment so that a device capable of the appropriate accuracy and precision is used to take the measurements or monitor the parameters. If the equipment you use to create and measure characteristics are inaccurate, unstable, damaged or in any way defective the product will not possess the required characteristics and furthermore you will not know it. You know nothing about an object until you can measure it, but you must measure it accurately and precisely. The equipment you use therefore needs to be controlled.

How is this Demonstrated?

When identifying measuring and monitoring equipment you need to identify the characteristic, the unit of measure, the target value and then choose appropriate measuring or monitoring equipment. As the type of product may vary considerably, the range of measuring devices also varies widely. Considering that the term product in ISO 9000:2005 includes service, hardware, software and processed material, and that documents are considered to be software rather than information products, the range of characteristics for which measuring equipment are required is enormous. It is relatively easy to identify the measuring and monitoring equipment for hardware product and processed material but less easy for services, software and information.

In many cases there will be equipment available to do the measuring, but if you propose a new unit of measure, a new sensor may be required.

Physical Measurement

There are two categories of equipment that determine the selection of physical equipment – *general purpose* and *special to type equipment*. It should not be necessary to specify all the general-purpose equipment needed to perform basic measurements that should be known by competent personnel. You should not need to tell an inspector or tester which micrometer, vernier calliper, voltmeter or oscilloscope to use. These are the tools of the trade and they should select the tool that is capable of measuring the particular parameters with the accuracy and precision required. However, you will need to tell them which equipment to use if the measurement requires unusual equipment or the prevailing environmental conditions require that only equipment be selected that will operate in such an environment. In such cases the particular equipment to be used should be specified in the verification procedures. In order to demonstrate that you selected the appropriate equipment at some later date, you should consider recording the actual equipment used in the record of results. With mechanical equipment this may not be necessary because wear will be normally detected by periodic calibration well in advance of a problem with the operation of the equipment.

With electronic equipment subject to drift with time or handling, a record of the equipment used will enable you to identify suspect results in the event of the equipment being found to be outside the limits at the next calibration. A way of reducing the effect is to select equipments that are several orders of magnitude more accurate than is needed.

Service Measurements

In the service sector there are many measurements that cannot be made using physical equipment. The most common device is the customer survey either used directly by an interview with customers or by mail. Many service organizations develop metrics for monitoring service quality relative to the type of service they provide. Some examples are provided in Table 32-2.

TABLE 32-2 Service Quality Measures

Service provided	Measures
Laboratory	Turn round time Conformity with requirements Calibration accuracy Time to respond to complaints
Telephone	Line availability Call out response time Time to reply to complaints
Distribution	Time to respond to complaints Supply delivery time Received condition
Water	Time to reply to complaints Supply connection time Water quality
Data analysis	Report accuracy Conformity with requirements Time to respond to complaints
Education	Class size Percentage of pupils achieving pass grades

DEFINING THE MONITORING AND MEASURING PROCESSES (7.6)

The standard requires processes *to be established to ensure that monitoring and measuring can be carried out and are carried out in a manner consistent with the measuring and monitoring requirements.*

What Does this Mean?

A measurement process consists of the *operations* (i.e., the measurement tasks and the environment in which they are carried out), *procedures* (i.e., how the tasks are performed), *devices* (i.e., gauges, instruments, software etc. used to make the measurements), the *personnel* used to assign a quantity to the characteristics being measured and the measurement system (i.e., the units of measure and the process by which standards are developed and maintained).

Why is this Necessary?

This requirement responds to the Systems Approach Principle.

If the measurements of product and service are to have any meaning, they have to be performed in a manner that provides results of integrity – results others inside and outside the organization can respect and rely on as being accurate and precise. If the integrity of measurement is challenged and the organization cannot demonstrate the validity of the measurements, the quality of the product remains suspect.

How is this Demonstrated?

Controlling Measurements

A necessary measuring and monitoring requirement is that the measurements carried out are controlled, i.e., regulated in a manner that will ensure consistent results. Control in this instance means several things:

- Knowing what equipment is used for product verification purposes so that you can distinguish between controlled and uncontrolled equipment – you will need a maintained list of equipment for this purpose.
- Knowing where the equipment is located so that you can recall it for calibration and maintenance – you will need a recall notice for this purpose.
- Knowing who the current custodian is so that you have a name to contact.
- Knowing what condition it is in so that you can prohibit its use if the condition is unsatisfactory – you will need a defect report for this purpose.
- Knowing when its accuracy was last checked so that you can have confidence in its results – calibration records and labels fulfil this need.
- Knowing what checks have been made using the instrument since it was last checked so that you can repeat them, should the instrument be subsequently found out of calibration – this is only necessary for instruments whose accuracy drifts over time, i.e., electronic equipment. (It is not normally necessary for mechanical equipment. You will need a traceability system for this purpose.)

- Knowing that the measurements made using it are accurate so that you rely on the results – a valid calibration status label on the measuring device will fulfil this purpose.
- Knowing that it is only being used for measuring the parameters for which it was designed so that results are reliable and the equipment is not abused. The abuse of measuring equipment needs to be regulated primarily to protect the device but also if high pressures and voltages are involved, to protect the user. Specifying the equipment to be used for making measurements in your work instructions will serve to prevent the abuse of equipment.

You may not need to know all these things about every device used for product verification but you should know most of them. This knowledge can be gained by controlling:

- the selection of measuring equipment,
- the use of measuring equipment,
- the calibration of measuring equipment.

You may know where each equipment is supposed to be, but what do you do if equipment is not returned for calibration when due? The equipment maintenance process should track returns and make provision for tracking down any maverick equipment, because they could be being used on product acceptance. A model process flow for control of measuring and monitoring equipment is illustrated in Fig. 32-2.

Measurement Laboratories

In order to maintain the integrity of measurement, physical measurements need to be undertaken in a controlled environment often referred to as a 'laboratory'. The controlled environment consists of a workspace in which the temperature, humidity, pressure, cleanliness, access and the integrity of the measuring equipment and supporting equipment is controlled and the personnel competent. The systems in use in such areas can be assessed separately against international metrological standards. Some organizations are established solely for the purpose of providing measurement services to industry and therefore undertake certification of measurement equipment – testifying the equipment has been calibrated against standards traceable to national or international standards. Here we enter the world of accreditation$^{\odot}$ as opposed to certification.

Use of Laboratories

Wherever physical measurement is performed you need to be confident in the results and therefore whether the measurements are performed in-house or by external laboratories, both areas should meet the same standards in order that results are consistently accurate and precise. Wherever the calibration is performed, the same standards should therefore apply. Calibrating equipment in-house should not absolve you from complying with the same requirements that you would need to impose on an external test house.

The standard that applies to measurement laboratories is now ISO/IEC 17025. Laboratories meeting the requirements of ISO/IEC 17025, for calibration and testing activities, will comply with the relevant requirements of ISO 9001 when they are acting

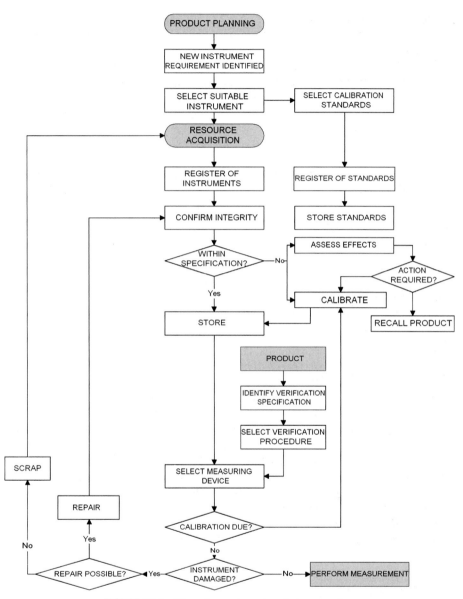

FIGURE 32-2 Measuring equipment control process flow.

as suppliers producing calibration and test results. However, laboratories meeting the requirements of ISO 9001 would not meet the requirements of ISO/IEC 17025. Therefore, in-house laboratories should be assessed against ISO/IEC 17025 to obtain the same level of confidence as obtained from external laboratories.

The requirements for measurement processes are duplicated in ISO 9001, therefore more information can be found in *Processes for demonstrating conformity* in Chapter 28.

CALIBRATING AND VERIFYING MEASUREMENTS (7.6a)

The standard requires *measuring and monitoring equipment to be calibrated or verified or both at specified intervals or prior to use, against measurement standards traceable to international or national standards.*

What Does this Mean?

In a measurement system the physical signal is compared with a reference signal of known quantity. The reference signal is derived from measures of known quantity by a process called calibration[①]. The known quantities are based on standards that in the majority of cases are agreed internationally.

Traceable to international or national standards means that an instrument will give the same reading when measuring a quantity under the same environmental conditions, wherever the measurement is taken.

Calibration applies to all measuring equipment used for providing evidence of conformity at any stage through product realization, not just at the product release stage as confirmed by published interpretation RFI 028.

With non-physical measurement systems, there is still a need for calibration but we tend to use the term integrity. The reference signal of known quantity is the standard and you can derive standards for anything – the only proviso is that they are agreed by those who benefit from the measurements. Therefore, if you set out to measure customer satisfaction, the standard used needs to be agreed with the customer. Standards that are not agreed by those they affect lack integrity.

The standard requires equipment to be calibrated or verified as though there is an option. Calibration is a process of comparing a physical signal with a reference signal of known quantity whereas verification is establishing the correctness of a quantity. Certain variables might be calibrated such as length or capacitance but attributes might be verified such as form and function. (The presence of a red light on the instrument panel is verified not calibrated.) Therefore, depending on the equipment being examined calibration, verification or both may be necessary. According to RFI 039, calibration and verification are not mutually exclusive.

Why is this Necessary?

This requirement responds to the Factual Approach Principle.

There are two systems used for maintaining the accuracy and integrity of measuring equipment – a calibration system and a verification system. The calibration system determines the accuracy of measurement and the verification system determines its integrity. If accuracy is important then the device should be included in the calibration system. If accuracy is not an issue but the form, properties or function of the equipment is important then it should be included in the verification system.

Variations can arise in measurements taken in different locations due to the measuring equipment not being calibrated to the same standards as other equipment. With the introduction of the SI system of units, this variation could be eliminated provided the quantity used to calibrate the measuring equipment was traceable to national or international standards.

How is this Demonstrated?

Calibration of Measuring and Monitoring Equipment

Calibration is concerned with determining the values of the errors of a measuring instrument and often involves its adjustment or scale graduation to the required accuracy (see *Accuracy and precision* in Chapter 28). You should not assume that just because equipment was once accurate it would remain so forever. Some equipment if well-treated and retained in a controlled environment will retain their accuracy for very long periods. Others if poorly treated and subjected to environmental extremes will lose their accuracy very quickly. Ideally you should calibrate measuring equipment before use in order to prevent an inaccurate equipment being used in the first place and after use to confirm that no changes have occurred during use. However, this is often not practical and so intervals of calibration are established which are set at such periods as will detect any adverse deterioration. These intervals should be varied with the nature of the equipment, the conditions of use and the seriousness of the consequences should it produce incorrect results.

It is not necessary to calibrate all measuring and monitoring equipment. Some equipment may be used solely as an indicator such as a thermometer, a clock or a tachometer – other equipment may be used for diagnostic purposes, to indicate if a fault exists. If such equipment is not used for determining the acceptability of products and services or process parameters, their calibration is not essential. However, you should identify such equipment as for 'Indication Purposes Only' if their use for measurement is possible. You don't need to identify all clocks and thermometers fixed to walls unless they are used for measurement.

There are two systems used for maintaining the accuracy and integrity of measuring equipment – a calibration system and a verification system. The calibration system determines the accuracy of measurement and the verification system determines the integrity of the equipment. If accuracy is important then the equipment should be included in the calibration system. If accuracy is not an issue but the device's form, properties or function is important then it should be included in the verification system. You need to decide the system in which your devices are to be placed under control and identify them accordingly.

There are two types of devices subject to calibration – those that are adjustable and those that are not. An adjustable device is one where the scale or the mechanism is capable of adjustment (e.g., micrometer, voltmeter and load cell). For non-adjustable devices, a record of the errors observed against a known standard can be produced which can be taken into account when using the device (e.g., slip gage, plug gage, surface table, thermometer).

Comparative references are not subject to calibration. They are, however, subject to verification. Such devices are those which have form or function where the criteria are either pass or fail, i.e., there is no room for error or where the magnitude of the errors does not need to be taken into account during usage. Such devices include software, steel rules or tapes, templates, forming and moulding tools. Devices in this category need to carry no indication of calibration due date. The devices should carry a reference number and verification records should be maintained showing when the device was last checked. Verification of such devices include checks for damage, loss of components, function etc.

Some electronic equipment has self-calibration routines built in to the start-up sequence. This should be taken as an indication of serviceability and not of absolute calibration. The device should still be subject to independent calibration at a defined frequency. (Note: **Use – not function – determines the need for calibration.**)

Traceability

If you calibrate your own devices, you will need (in addition to the 'working standards' which you use for measurement) calibration standards for checking the calibration of the working standards. The calibration standards should also be calibrated periodically against national standards held by your national measurement laboratory. This unbroken chain ensures that there is compatibility between measurements made in different locations using different measuring devices. By maintaining traceability you can rely on obtaining the same result (within the stated limits of accuracy) wherever and whenever you perform the measurement provided the dimensions you are measuring remain stable. The relationship between the various standards is illustrated in Fig. 32-3.

Determining Calibration Frequency

ISO 10012 requires that measuring equipment be confirmed at appropriate intervals established on the basis of stability, purpose and usage. With new equipment it is customary to set the frequency at 12-month intervals unless recommended otherwise by the manufacturer. Often this frequency remains despite evidence during calibration that accuracy and precision are no longer stable. Such action indicates that the calibration staff have not been properly trained or that cost rather than quality is driving calibration services. Calibrations should be performed prior to any significant change in accuracy that can be anticipated. The results of previous calibrations will indicate the amount of

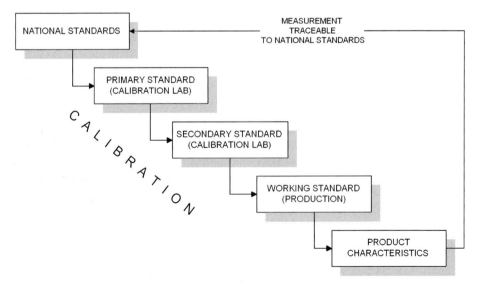

FIGURE 32-3   Traceability of standards.

drift – if drift is detected, the inte confirmation should be shortened. Conversely, if drift is not detected, the inter y be lengthened if two previous confirmations indicate such action would adve fect confidence in the accuracy of the device. Environment, handling, frequenc and wear are factors that can affect the stability of devices, therefore regardless calibration results, both previous and future conditions need to be taken into a In order to demonstrate you have reviewed the results and determined the appro llibration frequency, provision should be made on the calibration records for the ly to be decided at each calibration. Specifying a date is insufficient if the calib nstructions specify a frequency because it is unreasonable to expect the per sequently performing calibration to detect whether any change has been ma

Reference Materials

Comparative references are devic are used to verify that an item has the same properties as the reference. They n the form of colour charts or materials such as chemicals which are used in spe hic analysers or those used in tests for the presence of certain compounds in ure or they could be materials with certain finishes, textures etc. Certificates e produced and retained for such reference material so that their validity is o those who will use them. Materials that degrade over time should be date ven a use-by-date. Care should be taken to avoid cross-contamination and an ation due to sunlight (as can happen with colour charts). A specification for e ence material should be prepared so that its properties can be verified.

RECORDING THE BASIS FO RATION (7.6a)

The standard requires *that where no ional or national standards exist, the basis used for calibration or verification led.*

What Does this Mean?

For physical and chemical measure at are based on the fundamental units of measure (metre, kilogram, second, etc.) there are national or international standards but for other measures no or international standard may exist. Each industry has developed a series of by which the quality of its goods and services are measured and has acco veloped standards that represent agreed definitions of the measures. In the s tor involving interrelationships between people, standards become more diff fine in quantitative terms and therefore may be defined qualitatively. For the performance of a person handling customer calls may be defined by f results to be achieved. In setting the standard, the effort is focused on defi t a good job looks like.

Why is this Necessary?

This requirement responds to the Fac bach Principle.

   Without a sound basis for compari ffort of measurement is wasted.

How is this Demonstrated?

In some situations there may be no national standard against which to calibrate your devices. If you face this situation, you should get together a group of experts within your company or trade association and establish by investigation, experimentation and debate what constitutes the standard. Having done this you should document the basis of your decisions and produce a device or means devices that can be used to compare the product or result with the standard using visual, quantitative or other means. The device may be a physical instrument but be information such as a set of agreed criteria and the method of measurement.

Where you devise original solutions to the measurement of characteristics, the theory and development of the method should be documented and retained as evidence of the validity of the measurement method. Any measurement methods should be proven by rigorous experiment to detect the measurement uncertainty and cumulative effect of the errors in each measurement process. Samples used for proving the method should also be retained so as to provide a means of repeating the measurements should it prove necessary.

ADJUSTMENT OF DEVICES (7.6)

The standard requires measuring equipment be adjusted or re-adjusted as necessary.

What Does this Mean?

Adjustment is only possible with devices that have been designed to be adjustable.

Why is this Necessary?

This requirement responds to the Process Approach Principle.

When a measuring device is verified it may be found within specification and adjusted if the parameters have drifted towards the upper or lower limits. If the device is found outside specification it can be adjusted or re-adjusted (on subsequent occasions) within the specified limits.

How is this Demonstrated?

Mechanical devices are normally adjusted to the null position on calibration. Electronic devices should only be adjusted if found outside the limits. If you adjust the device at each calibration you will not be able to observe drift. Adjustments, if made very frequently, may also degrade the instrument. If the observed drift is such that the device may well be outside the specified limits by the next calibration, adjustment will be necessary.

In addition to calibrating the devices you will need to carry out preventive maintenance in order to keep them in good condition such as cleaning, testing, inspecting, replenishment of consumables etc. Corrective maintenance is concerned with restoring a device (after a failure has occurred) to a condition in which it can perform its required function. These activities may cover a range of skills and disciplines depending on the nature of the measuring device. It will include software development skills if you use test software for instance, or electronic engineering if you use electronic equipment. You can of course subcontract the complete task to a specialist who will not only maintain the equipment but also on request, carry out calibration. Take care to confirm that the supplier is qualified to perform the calibrations to national standards and to provide a valid certificate of calibration.

INDICATING CALIBRATION STATUS (7.6c)

The standard requires *measuring devices to have identification in order to determine its calibration status.*

What Does this Mean?

Calibration status is the position of a measuring device relative to the time period between calibrations. If the date when calibration is due is in the future, the device can be considered calibrated – if the current date is beyond the date when calibration is due, the device is not necessarily inaccurate but remains suspect until verified. However, devices can also be suspect if dropped or damaged even when the date of calibration is due is in the future. The requirement only applies to physical devices subject to wear, drift or variation with use or time.

Why is this Necessary?

This requirement responds to the Factual Approach Principle.

While a robust calibration system should ensure no invalid devices are in use, system failures are a possibility. As the consequences of failure are greater than the effort involved in checking the validity of devices before use, it is prudent to provide a means for checking calibration status.

How is this Demonstrated?

All devices subject to calibration should display an identification label that either directly or through traceable records, indicates the authority responsible for calibrating the device and the date when the next calibration is due. Don't state the actual calibration date because this would be no use without the user knowing the calibration frequency. Measuring equipment should indicate its calibration status to any potential user. Measuring instruments too small for calibration status labels showing the due date may be given other types of approved identification. It is not mandatory that users identify the due date solely from the instrument itself but they must be able to determine that the instrument has been calibrated. Serial numbers alone do not do this unless placed within a specially designed label that indicates that the item has been calibrated or you can fix special labels that show a circular calendar marked to show the due date. If you do use serial numbers or special labels then they need to be traceable to calibration records that indicate the calibration due date.

Devices used only for indication purposes or for diagnostic purposes should also display an identity that clearly distinguished them as not being subject to calibration. If devices are taken out of use for prolonged periods, it may be more practical to cease calibration and provide a means of preventing inadvertent use with labels indicating that the calibration is not being maintained. You may wish to use devices that do not fulfil their specification either because part of the device is unserviceable or because you were unable to perform a full calibration. In such cases, you should provide clear indication to the user of the limitation of such devices.

SAFEGUARDING MONITORING AND MEASURING EQUIPMENT (7.6d)

The standard requires measuring and monitoring equipment *to be safeguarded from adjustments that would invalidate the measurement result.*

What Does this Mean?

Once a device has been calibrated or verified there needs to be safeguards in place to prevent unauthorized or inadvertent adjustment.

Why is this Necessary?

This requirement responds to the Factual Approach Principle.

The purpose of this requirement is to ensure that the integrity of the measurements is maintained by precluding errors that can occur if measuring equipment is tampered with.

How is this Demonstrated?

To safeguard against any deliberate or inadvertent adjustment to measuring devices, seals should be applied to the adjustable parts or where appropriate to the fixings securing the container. The seals should be designed so that tampering will destroy them. Such safeguards may not be necessary for all devices. Certain devices are designed to be adjusted by the user without needing external reference standards, for example, zero adjustments on micrometers. If the container can be sealed then you don't need to protect all the adjustable parts inside.

Your procedures need to specify:

- those verification areas that have restricted access and how you control access;
- the methods used for applying integrity seals to equipment;
- the authority permitted to apply and break the seals;
- the action to be taken if the seals are found to be broken either during use or during calibration.

PROTECTION OF MONITORING AND MEASURING EQUIPMENT (7.6e)

The standard requires measuring and monitoring devices *to be protected from damage and deterioration during handling, maintenance and storage.*

What Does this Mean?

Each measuring and monitoring device has a range within which accuracy and precision remain stable – use the device outside this range and the readings are suspect.

Why is this Necessary?

This requirement responds to the Process Approach Principle.

Physical measuring and monitoring devices can be affected by inappropriate handling, maintenance and storage and thus jeopardize their integrity. Often measuring devices are very sensitive to vibration, dirt, shock and tampering and thus it is necessary to protect them so as to preserve their integrity.

How is this Demonstrated?

When not in use, measuring devices should always be stored in the special containers provided by the manufacturer. Handling instructions should be provided with the storage case where instruments may be fragile or prone to inadvertent damage by careless handling. Instruments prone to surface deterioration during use and exposure to the atmosphere should be protected and moisture absorbent or resistant materials used. When transporting measuring devices you should provide adequate protection. Should you employ itinerant service engineers, ensure that the instruments they carry are adequately protected as well as calibrated.

ACTION ON EQUIPMENT FOUND OUT OF CALIBRATION (7.6)

The standard requires *the validity of previous measuring results to be assessed and recorded when equipment is found not to conform to requirements and for appropriate action to be taken on the equipment and any product affected.*

What Does this Mean?

This is perhaps the most difficult of requirements to meet for some organizations. It is not always possible or practical to be able to trace product to the particular devices used to determine its acceptability. The requirements apply not only to your working standards but also to your calibration standards. When you send calibration standards away for calibration and they are subsequently found to be inaccurate, you may need a method of tracing the devices they were used to calibrate. A calibration standard that is found inaccurate within limits for a specified measurement may not be inaccurate for the range of measurement for which it is being use. Action would be needed only if the inaccuracies rendered the results obtained from previous use to be inaccurate.

If you have a small number of measuring devices and only one or two of each type, it may not be too difficult to determine which products were accepted using a particular device. In large organizations that own many pieces of equipment that is constantly being used in a variety of situations, meeting the requirements can be very difficult.

Why is this Necessary?

This requirement responds to the Process Approach Principle.

If a measurement has been taken with a device that is subsequently found inaccurate, the validity of the measurement is suspected and therefore an assessment is needed to establish the consequences. In most cases the devise used is accurate to an order of magnitude greater than that required, therefore, if found outside tolerance, it may not mean that the product measured is nonconforming. However, if measurements are taken at the extreme of device accuracy, the product may well be nonconforming if the device is found to be inaccurate.

How is this Demonstrated?

One way of meeting this requirement is to record the type and serial number of the devices used to conduct measurements but you will also need to record the actual measurements made. Some results may be made in the form of ticks or pass or fail and not by recording actual readings. In these cases you will have a problem in determining whether the amount by which the equipment is out of specification would be sufficient to reject the product. In extreme circumstances, if the product is no longer in the factory, this situation could result in your having to recall the product from your customer or distributor.

In order to reduce the effect, you can select measuring devices that are several orders of magnitude more accurate than your needs so that when the devices drift outside the tolerances, they are still well within the accuracy you require. There still remains a risk that the device may be wildly inaccurate due to damage or malfunction. In such cases you need to adopt the discipline of re-calibrating devices that have been dropped or are otherwise suspect before further use.

You need to carefully determine your policy in this area paying particular attention to what you are claiming to achieve. You will need a procedure for informing the custodians of unserviceable measuring devices and one for enabling the custodians to track down the products verified using the unserviceable device and assess the magnitude of the problem. You will need a means of ranking problems in order of severity so that you can resolve the minor problems at the working level and ensure that significant problems are brought to the attention of the management for resolution. It would be irresponsible for a junior technician to recall six months production from customers and distributors based on a report from the calibration laboratory. You need to assess what would have happened if you had used serviceable equipment to carry out the measurements. Would the product have been reworked, repaired, scrapped or the requirement merely waived. If you suspect previously shipped product to be nonconforming and now you have discovered that the measurements on which their acceptance was based were inaccurate, you certainly need to notify your customer. In your report to your customer, state the precise amount by which the product is outside specification so that the customer can decide whether to return the product – remember the product specification is but an interpretation of what constitutes fitness for use. Out of 'spec' doesn't mean unsafe, unusable, un-saleable etc.

CALIBRATION AND VERIFICATION RECORDS (7.6 AND 7.6.2)

The standard requires *records of the results of calibration and verification to be maintained*.

What Does this Mean?

Records of the results of calibration and verification are those records indicating the accuracy of the device prior to any adjustment and records after adjustment. These records apply not only to equipment designed and produced by the organization but also those owned by the organization and those owned by employees and customers when being used for product acceptance.

Why is this Necessary?

This requirement responds to the Factual Approach Principle.

It is important to record calibration and verification in order to determine whether the device was inside the prescribed limits when last used. It also permits trends to be monitored and the degree of drift to be predicted.

Calibration records are also required in order to notify the customer if suspect product or material has been shipped.

How is this Demonstrated?

Calibration and verification records are records of activities that have taken place. Records should be maintained not only for proprietary devices but also for devices you have produced and devices owned by customers and employees.

These records should include where appropriate:

- The precise identity of the device being calibrated or verified (type, name, serial number, configuration if it provides for various optional features);
- The modification status if relevant (applies to specially designed test equipment and gauges);
- The name and location of the owner or custodian;
- The date on which calibration was performed;
- Reference to the calibration or verification procedure, its number and issue status;
- The condition of the device on receipt;
- The results of the calibration or verification in terms of readings before adjustment and readings after adjustment for each designated parameter, e.g., any out of specification readings;
- An impact assessment of any out of specification conditions;
- The date fixed for the next calibration or verification;
- The permissible limits of error;
- The serial numbers of the standards used to calibrate the device;
- The environmental conditions prevailing at the time of calibration;
- A statement of measurement uncertainty (accuracy and precision);
- Details of any adjustments, servicing, repairs and modifications carried out;
- The name of the person performing the calibration or verification;
- Details of any limitation on its use;
- Notification to the customer if suspect product has been shipped.

Clearly not all this information would be presented on one record but the records should be indexed so that all this information is traceable both forwards and backwards. For example, the record containing the results of an assessment of out of specification conditions should carry a reference to the related calibration record and vice versa.

The records required are only for formal calibrations and verification and not for instances of self-calibration or zeroing using null adjustment mechanisms. Whilst calibration usually involves some adjustment to the device, non-adjustable devices are often verified rather than calibrated. However, it is not strictly correct to regard all calibration as involving some adjustment. Slip gauges and surface tables are calibrated but not adjusted. An error record is produced to enable users to determine the uncertainty of measurement in a particular range or location and compensate for the inaccuracies when recording the results.

SOFTWARE VALIDATION (7.6)

The standard requires *confirmation of the ability of software used for measuring and monitoring of specified requirements to satisfy intended application to be undertaken prior to initial use and reconfirmed as necessary.*

What Does this Mean?

These requirements apply not only to production, installation and servicing but also to design, development and operations.

Why is this Necessary?

This requirement responds to the Factual Approach Principle.

Software is used increasingly to drive equipment used for measurement or to interpret results. The integrity of the software therefore has a bearing on the integrity of the measurement and therefore needs to be verified prior to use. Although software does not degrade or wear out, it can be corrupted such that it no longer does the job it was intended to. In many cases software malfunction will be apparent by the absence of any result at all, but in some cases, a spurious result may be generated that appears to the observer as correct. Re-confirmation is necessary therefore after a period where the equipment may have been used in situations where intended or unintended changes to the configuration could have been made.

How is this Demonstrated?

The integrity of software is critical to the resultant product whether it be a deliverable product to a customer or a product being developed or in use. The design of software should be governed by the requirements of Clause 7.3 although this is not mandated by the standard. If these requirements are applied, the design verification requirements should adequately prove that the software is capable of verifying the acceptability of product. However, the design control requirements may be impractical for many minor verification devices and in many cases the software will not have been designed in-house but purchased. The hardware that provides the platform for the software should also be controlled if its malfunction could result in nonconforming product being accepted. Complex hardware of this nature should be governed by the design controls of Clause 7.3 if designed in-house because it ensures product quality. If bought-out, you should obtain all the necessary manuals for its operation and maintenance and it should be periodically checked to verify it is fully operational.

To control software you need to consider what it is that you need to control. As a minimum you should control its use, its configuration, modification, location (in terms of where it is installed), replication and disposal.

Use is controlled by specifying the software by type designation and version in the development and production test procedures or a register that relates products to the software that has to be used to verify its acceptability. It will be necessary to ensure that the correct version of the software is used with the associated equipment. Manufacturer's upgrades should be installed especially if they correct software errors, compatibility or security weaknesses. You should also provide procedures for running the software on the host computer or automatic test equipment. They may of course be menu driven from a display screen and keyboard rather than paper procedures.

Modifications should be controlled in a manner that complies with the requirements of Clause 7.3.7 of the standard.

The location could be controlled by index, register, inventory or other such means which enable you to identify on what machines particular versions of the software are installed, where copies and the master tapes or disks are stored.

Replication and disposal could be controlled by secure storage and prior authorization routines where only authorized personnel or organizations carry out replication and disposal.

Control of Nonconforming Product

CHAPTER PREVIEW

This chapter is aimed at all those personnel monitoring and measuring processes and products either as part of the demand fulfilment process or resource management process (see below for clarification). These will include operators, supervisors, inspectors and process engineers.

Nonconformities are caused by factors that should not be present in a process. There will always be variation but variation is not nonconformity. Nonconformity arises when the variation exceeds the permitted limits. The factors that cause nonconformity on one occasion will (unless removed) cause nonconformity again and again. As the objective of any process must be to produce conforming output, it follows therefore that it is necessary to eliminate the causes of nonconformity.

Nonconformity controls are one of the controls exercised over processes of all types. Although in principle a product is a result of a process, and nonconformity control applies to all processes, the products that are intended to be subject to the requirements of Clause 8.3 are destined for delivery to a customer in one form or another. They are not intended to apply to products of the mission management process or demand creation process or indeed the resource management process other than products used in the manufacture of products destined for delivery to a customer. They would therefore apply to processing materials.

They are included under measuring and monitoring rather than product realization because control of nonconforming product cannot be excluded from the management system as can other requirements of Section 7. Nonconforming product control represents a feedback loop within a process for handling products that failed to pass verification checks.

The position where the requirements on nonconformity control feature in the managed process is shown in Fig. 33-1. Note the action of removing the nonconformity is referred to as 'correct' meaning a correction.

In this chapter we examine the requirements in Clause 8.3 of ISO 9001:2008 and in particular:

- Preventing unintended use,
- Correction of nonconforming product,
- Authorizing use of nonconforming product,
- Action to preclude use or application,
- Records of nonconformity,

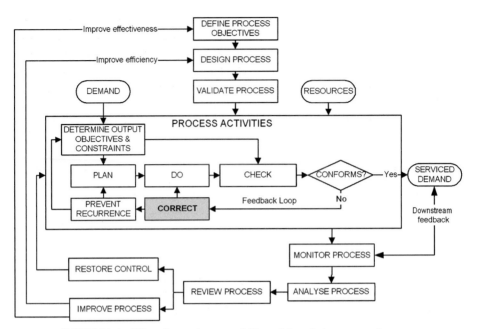

FIGURE 33-1 Where the requirements of Clause 8.3 apply in a managed process.

- Re-verification of corrected product,
- Consequences of nonconformity.

PREVENTING UNINTENDED USE (8.3)

The standard requires the organization *to ensure that product which does not conform to product requirements is identified and controlled to prevent unintended use or delivery.*

What Does this Mean?

Nonconforming product is product that does not conform to agreed product requirements when subject to either planned or unplanned verification. Product requirements are not limited to customer requirements (see also Chapter 24) therefore a nonconforming product is one that fails to meet the:

- specified customer requirements,
- intended usage requirements,
- stated or implied needs,
- organization's own requirements,
- customer expectations.

A product is judged either conforming or nonconforming at a verification stage. A product could also be judged nonconforming if it becomes damaged or fails at any other stage but the product is normally considered to be unserviceable in such cases.

Unserviceable products however are not necessarily nonconforming – they may simply lack lubrication or calibration. A piece of test equipment, the calibration date of which has expired, is not nonconforming, it is merely unserviceable. When checked against a standard it may be found to be out of calibration and then it is nonconforming, but it could be found to be within the specified calibration *limits*.

Suspect product (see Part 7 Introduction) should be treated in the same manner as nonconforming product and quarantined until dispositioned (i.e., dealt with). However, until nonconformity can be proven, the documentation of the nonconformity merely reveals the reason for the product being suspect.

This requirement relates to the controls exercised over the product itself whereas Clause 8.5.2 on corrective action addresses the measures needed to prevent recurrence of the nonconformity. The scope of procedures for control of nonconforming product should therefore focus on the product and its correction not the cause.

Why is this Necessary?

This requirement responds to the Process Approach Principle.

Nonconforming product needs to be prevented from use or delivery, simply because the organization should not knowingly supply nonconforming product.

How is this Demonstrated?

The only sure way of preventing inadvertent use of nonconforming product is to destroy it, but that may be a little drastic in some cases. It may be possible to eliminate the nonconformity by repair, completion of processing or rework. A more practical way of preventing the inadvertent use or installation of nonconforming or unserviceable product is to identify the product as *nonconforming* or *unserviceable* and then place it in an area where access to it is controlled. These two aspects are covered further below.

Identifying Nonconforming Product

The most common method is to apply labels to the product that are distinguishable from other labels. It is preferable to use red labels for nonconforming and unserviceable items and green labels for conforming and serviceable items. In this way you can determine product status at a distance and reduce the chance of confusion. You can use segregation as a means of identifying nonconforming product but if there is the possibility of mixing or confusion then this means alone should not be used.

On the labels themselves you should identify the product by name and reference number, specification and issue status if necessary and either a statement of the nonconformity or a reference to the service or nonconformity report containing full details of its condition. Finally the person or organization testifying the nonconformity should be identified either by name or inspection stamp. Unlike products, nonconforming services are usually rendered unavailable for use by notices such as 'Out of Order' or by announcements such as 'Normal service will be resumed as soon as possible'. Products are often capable of operation with nonconformities whereas services tend to be withdrawn once the nonconformity has been detected however trivial the fault.

Controlling Nonconforming Product

To control nonconforming product you need to:

(a) know when it became nonconforming;
(b) know who decided it was nonconforming;
(c) know of its condition;
(d) know where it is located;
(e) know that it is unable to be used.

On detection of a nonconformity, details of the product and the nonconformity should be recorded so as to address (a), (b) and (c) above.

Segregating a nonconforming product (or separating good from bad) places it in an area with restricted access and addresses (d) and (c) above. Such areas are called quarantine areas or quarantine stores. Products should remain in quarantine until disposal instructions have been issued. The store should be clearly marked and a register maintained of all items that enter and exit the store. Without a register you won't be able to account for the items in store, check whether any are missing, or track their movements. The quarantine store may be contained within another store providing there is adequate separation that prevents mixing of conforming and nonconforming articles. Where items are too large to be moved into a quarantine store or area, measures should be taken to signal to others that the item is not available for use and cordons or floor markings can achieve this. With services the simplest method is to render the service unavailable or inaccessible.

DOCUMENTING NONCONFORMITY CONTROL PROCEDURES

The standard requires *a documented procedure to be established to define the controls and related responsibilities and authorities for dealing with nonconforming product.*

What Does this Mean?

This means defining the provisions put in place to deal with the aspects addressed above (preventing unintended use) in a way that will ensure their consistent application.

Why is this Necessary?

The rationale for what should be documented is given in Chapter 11 but briefly, nonconformity controls need to be documented to communicate responsibility, ensure consistency in their application and to prevent nonconforming product being used and delivered to customers.

How is this Demonstrated?

Defining Controls

In order to demonstrate compliance with these requirements your nonconformity control procedures should include the following actions:

- Specify how product should be scrapped, or recycled, the forms to be used, the authorizations to be obtained;

- Specify the various repair procedures, how they should be produced, selected and implemented;
- Specify how modifications should be defined, identified and implemented;
- Specify how production permits (deviations) and concessions (waivers) should be requested, evaluated and approved or rejected;
- Specify how product should be returned to its supplier, the forms to be completed and any identification requirements in order that you can detect product on its return;
- Specify how regrading product is to be carried out, the product markings, prior authorization and acceptance criteria.

When making the disposition your correction needs to address:

- action on the nonconforming item to remove the nonconformity;
- a search for other similar items which may be nonconforming (i.e., suspect product);
- action to recall product containing suspect nonconforming product.

If you need to recall product that is suspected as being defective you will need to devise a *Recall Plan*, specify responsibilities and time-scales and put the plan into effect. Product recall is a *Correction* not a *Corrective Action* because it does not prevent a recurrence of the initial problem.

Defining Responsibility

Documented procedures should specify the authorities that make the disposition, where it is to be recorded and what information should be provided in order that it can be implemented and verified as having been implemented.

The decision on product acceptance is a relatively simple one because there is a specification against which to judge conformance. When product is found to be nonconforming, there are three decisions you need to make based on the following questions:

- Can the product be made to conform?
- If the product cannot be made to conform, is it fit for use?
- If the product is not fit for use, can it be made fit for use?

The authority for making these decisions will vary depending on the answer to the first question. If, regardless of the severity of the nonconformity, the product can be made to conform simply by rework or completing operations, these decisions can be taken by operators or inspectors, providing rework is economical. Decisions on scrap, rework and completion would be made by the fund-providing authority rather than the design authority. If the product cannot be made to conform by using existing specifications, decisions requiring a change or a waiver of a specification should be made by the authority responsible for drawing up or selecting the specification.

It may be sensible to engage investigators to review the options to be considered and propose actions for the authorities to consider. In your procedures you should identify the various bodies that need to be consulted for each type of specification. Departures from customer requirements will require customer approval; departures from design requirements will require design approval; departures from process requirements will require process engineering approval etc. The key lies in identifying who devised or selected the requirement in the first place. As all specifications are merely a substitute for knowledge of fitness for use – it follows that any departure from such specification must be referred back to the specification authors for a judgement.

CORRECTION OF NONCONFORMING PRODUCT (8.3a)

The standard requires the organization *to deal with nonconforming product by taking action to eliminate the detected nonconformity.*

What Does this Mean?

Action to remove the detected nonconformity is a correction; a term that is explained in more detail in the Introduction to this Part.

In some cases it may not be cost effective to attempt to eliminate the nonconformity and therefore such action would be inappropriate and the product should be disposed of.

Why is this Necessary?

This requirement responds to the Process Approach Principle.

Correction or remedial action is warranted when there are cost benefits from attempting to eliminate the nonconformity.

How is this Demonstrated?

To implement this requirement you will need a form or other such document in which to record the decision and to assign the responsibility for the correction or remedial action. These documents also need to stipulate the verification requirements to be implemented following rework. When deciding on repair or rework action, you may need to consider whether the result will be visible to the customer on the exterior of the product. Rework or repairs that may not be visible when a part is fitted into the final assembly might be visible when these same parts are sold as service spares. To prevent on-the-spot decisions being at variance each time, you could:

- identify in the specifications those products that are supplied for service applications, i.e., for servicing, maintenance and repair;
- provide the means for making rework invisible where there are cost savings over scrapping the item;
- stipulate on the specifications the approved rework techniques.

AUTHORIZING USE OF NONCONFORMING PRODUCT (8.3b)

The standard requires the organization when appropriate *to deal with nonconforming product by authorizing its use, release or acceptance under concession by a relevant authority and where applicable by the customer.*

What does this Mean?

If you choose to accept a nonconforming item as is without rework, repair etc. then you are in effect granting a *concession* or waiving the requirement *only* for that particular item. If the requirements cannot be achieved at all then this is not a situation for a concession but a case for a change in requirement. If you know in advance of producing the product or service that it will not conform to the requirements, you can then request a deviation from the requirements. This is often referred to as a *production permit* or *deviation*. Concessions apply *after* the product has been produced. Production permits or deviations apply *before* it has been produced. Both are requests that should be made to the acceptance authority for the product. The relevant authority is the authority that specified the requirement that has not been met. This authority could therefore be the customer, the regulator or the designer.

Why is this Necessary?

This requirement responds to the Process Approach Principle.

Product that does not conform to requirement may be fit for use. All specifications are but a substitute for knowledge of fitness for use. Any departure from such specification should be referred back to the specification authors for a judgment.

How is this Demonstrated?

In order to determine whether a nonconforming product could be used, an analysis of the conditions needs to be made by qualified personnel. There are two ways of doing this. Either you refer all such nonconformities to the relevant authority or the authority appoints representatives who are capable of making these decisions within prescribed limits. A traditional method is to classify nonconformities, assign authority for accepting concessions for each level and define the limits of their authority. These levels could be as follows:

- Critical Nonconformity. A departure from the requirements which renders the product or service unfit for use.
- Major Nonconformity. A departure from the requirements included in the contract or customer specification.
- Minor Nonconformity. A departure from the requirements not included in the contract or customer specification.

The only cases where you need to request concessions from your customer are when you have deviated from one of the customer requirements and cannot make the product conform (published interpretation RFI 018 confirms this). When you repair a product, provided it meets all of the customer requirements, there is generally no need to seek a concession from your customer. While it is generally believed that nonconformities indicate an out of control situation, provided that you detect and rectify them before release of the product, you have quality under control and have no need to report nonconformities to your customer. However, if the frequency of nonconformity exceeds process capability targets, the process has become unstable and requires corrective action.

In informing your customer when nonconforming product has been shipped you obviously need to do this immediately when you are certain that there is a nonconformity. If you are investigating a suspect nonconformity it only becomes a matter for reporting to your customer when the nonconformity remains suspect after you have concluded your investigations. Alerting your customer every time you think there is a problem will destroy confidence in your organization. Customers appreciate zeal but not paranoia!

Production permits or deviations are generally permitted for specific batches or a defined time period. This is to allow time for corrective action to be taken. It is therefore necessary to keep a log of the products and quantities produced that are subject to the production permit or authorized deviation. It is also necessary to ensure that when the batch or date when the corrective action becomes effective arrives, the production permit or deviation is withdrawn. Flags should be inserted into production schedules alerting planners to batches that are subject to authorized concession or production permit and when the date or batch beyond which authorization is invalid arrives.

When delivery subject to authorized concession or production permit commences the packaging should be duly annotated.

ACTION TO PRECLUDE USE OR APPLICATION (8.3c)

The standard requires *the organization when appropriate to deal with nonconforming product by taking action to preclude its original intended use or application.*

What Does this Mean?

Precluding intended use or application means either scrapping the product so no one can use it or regrading it so that it may be used in other applications. In some cases products and services are offered in several models, types or other designations but are basically of the same design. Those which meet the higher specification are graded as such and those which fail may meet a lower specification and can be *regraded*. The grading should be reflected in the product identity so that there is no confusion,

Why is this Necessary?

This requirement responds to the Process Approach Principle.

If a nonconforming product cannot be made conforming or accepted as is, some other action is needed to prevent inadvertent use and this leaves the two options stated.

How is this Demonstrated?

Regrading can be accomplished by assigning a new identity to the product. Scrapping an item should not be taken lightly – it could be an item of high value. Scrapping may be an economical decision with low cost items, whereas the scrapping of high value items may require prior authorization as salvage action may provide a possibility of yielding spares for alternative applications.

CONSEQUENCES OF NONCONFORMITY (8.3)

The standard requires the organization *to take action appropriate to the effects or potential effects of the nonconformity when nonconforming product is detected after delivery or use has started.*

What Does this Mean?

A nonconformity may be detected by a subsequent user of the product either within the organization or by the customer. Also a nonconformity might be detected prior to release and implicate products already in use such as when subsequent analysis reveals inaccurate measurements or when verification methods or acceptance criteria change. Such product may not have failed in service because it has not been used in a manner needed to cause failure but if part of the same batch or lot contains a common cause of nonconformity all the products are suspected. Action taken as a result of latent nonconformity may involve product recall, product alerts or the issue of instructions for correction.

Case Study – Nonconformity Control in Design Services

We employ a consultancy certified to ISO 9001:2000. They provide a broad range of services, not just the production of drawings. We found during an audit they do not have a documented process or procedure for control of nonconforming product (8.3). They say this is not a requirement in their organisation as they comply with Clause 7.3 – Design & Development and, their system is certified by an accredited certification body. Can this be correct?

Exclusions can only be granted for requirements in Section 7 and this is made clear in Section 1.2 therefore Clause 8.3 on control of nonconforming product applies. The consultancy provides at least two products. One appears to be a design and the other a design service, each having its own characteristics which need to be specified, planned, achieved, controlled and assured.

The quality of the design service is primarily governed by Clause 7.5 but a nonconforming service should be handled in compliance with the requirements of Clause 8.3.

The quality of design is governed by Clause 7.3 but a design that is found nonconforming following design verification or validation design would be passed through a design change process. However, this process would normally be designed to either cause change or prevent change. What the nonconformity controls do is to prevent use of nonconforming product, which is entirely different. While it may not be likely that a design that failed verification or validation would be delivered to a customer there does need to be controls in place that prevent use of nonconforming design and this might extend to requesting a concession. Although Clause 7.3.7 provides for design changes to be verified and validated it does not address nonconformities discovered after delivery.

If the design change procedures address all these issues and thus satisfies the requirements of Clause 8.3, the consultancy has not excluded Clause 8.3 relative to design but satisfied Clause 8.3 through design change procedures and not a separate procedure for control of nonconforming product.

However, there remains the issue of nonconforming services and Clause 8.2.3 invokes indirectly Clause 8.3 through the phrase "When planned results are not achieved correction...shall be taken".

It would appear that the certification body in question has misinterpreted the requirements of ISO 9001.

Why is this Necessary?

This requirement responds to the Customer Focus Principle.

The requirement acknowledges that problems may be detected after shipment or use that need action to prevent undesirable effects.

How is this Demonstrated?

Nonconformities detected by internal or external users indicate that the controls in place are not effective and should give cause for concern. Details should be recorded and an investigation conducted to establish why the planned verification did not detect the problem. Action should then be taken to improve the verification methods by changing procedures, acceptance criteria, equipment or retraining personnel.

When a nonconformity is detected by verification personnel in a product where products of the same type are in use, an analysis is needed to establish whether the nonconformity would previously have escaped detection. If not, there is no cause for alarm but if something has now changed to bring the nonconformity to light, an evaluation of the consequences needs to be conducted. It may only be a matter of time before the user detects the same nonconformity.

The procedures should cover:

- the rationale for notifying the customer and determining the appropriate course if action;
- the method of receiving and identifying returned product;
- the method of logging reports of nonconformities from customers and other users;
- the process of responding to customer requests for assistance;
- the process of dispatching service personnel to the customer's premises;
- a form on which to record details of the nonconformity, the date, customer name etc.;
- a process for acknowledging the report in order that the customer knows you care;
- a process for investigating the nature of the nonconformity;
- a process for replacing, or repairing nonconforming product and restoring customer equipment into service;
- a process for assessing all products in service that are nonconforming, determining and implementing recall action if necessary.

It is of course a matter for the organization's management to decide the appropriate action but it would not be conducive to strong customer relations if you were to neglect to inform the customer of any nonconformity with customer specified requirements or applicable regulatory or statutory requirements.

RECORDS OF NONCONFORMITY (8.3)

The standard requires *records of the nature of nonconformities and any subsequent actions taken, including concessions to be maintained.*

What Does this Mean?

The records of nonconformities are the documented details of the product (its identity), the specific deviations from requirements (what it is and what it should have been), the conditions under which the nonconformity was detected (the environmental or operating conditions – what was happening at the time when the nonconformity was detected), the time and date of detection, the name of the person detecting it and the actions taken with reference to any instructions, revised requirements and decisions.

Why is this Necessary?

This requirement responds to the Factual Approach Principle.

Records of nonconformities are needed for presentation to the authorities responsible for deciding on the action to be taken and for subsequent analysis. Without such records, decisions may be made on opinion resulting in the means for identifying opportunities for improvement being absent.

How is this Demonstrated?

There are several ways in which you can document the presence of a nonconformity. You can record the condition:

- on a label attached to the item;
- on a form unique to the item such as a nonconformity report;
- of functional failures on a failure report and physical errors on a defect report;
- in a logbook for the item such as an inspection history record or snag sheet;
- in a logbook for the workshop or area.

The detail you record depends on the severity of the nonconformity and to whom it needs to be communicated. In some cases a patrol inspector or quality engineer can deal with minor snags on a daily basis as can an itinerant designer. Where the problem is severe and the necessary action complicated, a panel of experts may need to meet. Rather than gather around the nonconforming item, it may be more practical to document the action on a form. In some cases the details may need to be conveyed to the customer off site and in such cases a logbook or label would be inappropriate. It is important when documenting the nonconformity that you record as many details as you can because they may be valuable to any subsequent investigation in order to help diagnose the cause and prevent its recurrence. An example of a Nonconformity Report is illustrated in Fig. 33-2. Note that in this example provision is made to record the impact of the nonconformity on both the customer and the business. This provision enables personnel to demonstrate compliance with Clause 6.2.2d.

RE-VERIFICATION OF CORRECTED PRODUCT (8.3)

The standard requires nonconforming product *to be subject to re-verification after correction to demonstrate conformity to the requirements.*

What Does this Mean?

Any rework, repair, modification or other action taken to correct the nonconformity will change the product and therefore it needs to be subject to re-verification. This may involve verification against different requirements to the original requirements.

Why is this Necessary?

This requirement responds to the Factual Approach Principle.

If a nonconforming product is accepted as is without correction, no re-verification is necessary, but if the product is changed the previous verification is no longer valid.

NONCONFORMITY REPORT

Part/Process			Qty	Used on		**NCR/**	
S/N		Location			Reported by	Date detected	

DESCRIPTION

	As relevant define:
	1. What is nonconforming
	2. Why it is nonconforming
	3. The criteria that states it is nonconforming
	4. The instruments used to detect the nonconformity
	5. The conditions at the time the nonconformity was detected

IMPACT

On Customer		On Business		As relevant define: The impact of the nonconformity on the customer and/or the business a) if no action is taken? b) If the item is scrapped?
Decision		Acceptance authority approval	Date	Acceptance authority approval required if decision is to Use-as-is

Date	CORRECTION	Responsibility	
	Rework, repair or modification		As relevant define: 1. Action to correct the specific nonconformity 2. Action to seek out and correct other examples that might be nonconforming
	Action on suspect items		3. Action to control the symptoms of the nonconformity until the root cause is located and a design or process change is completed 4. Verification action to be taken after remedial action is complete
	Containment action		
	Verification after correction		

EFFECTIVENESS OF ACTIONS

All actions taken and effective [Yes] [No]	Reviewed by:	Related NCRs	Date closed

FIGURE 33-2 Sample nonconformity report.

How is this Demonstrated?

Any product that has had work done to it should be re-verified prior to it being released to ensure the work has been carried out as planned and has not affected features that were previously found conforming. There may be cases where the amount of re-verification is limited and this should be stated as part of the recovery action plan. However, after rework or repair the re-verification should verify that the product meets the original requirement, otherwise it is not the same product and must be identified differently.

The verification records should indicate the original rejection, the disposition and the results of the re-verification in order that there is traceability of the decisions that were made.

Chapter 34

Analysis of Data

CHAPTER PREVIEW

This chapter is aimed at all those personnel monitoring and measuring processes and products at all levels therefore it would apply to the executives in the boardroom as much as the operators in the boiler room. In practice executives may employ personnel to collect and analyse data for them but in small companies that luxury may not be available.

In this chapter we examine the requirements in Clause 8.2.2 of ISO 9001:2008 and in particular:

- Collecting and analysing appropriate data;
- Evaluating where improvements in effectiveness can be made;
- Analysis of conformance to customer requirements;
- Analysis of product and process characteristics;
- Analysis of supplier data.

The analysis of customer satisfaction data is addressed in Chapter 29.

Unlike the control of nonconforming product, these requirement apply to all processes.

The position where the requirements on analysis feature in the managed process is shown in Fig. 34-1. Note that the analysis in the process cycle addresses both the process and the product of the process.

COLLECTING AND ANALYSING APPROPRIATE DATA (8.4)

The standard requires the organization *to determine, collect and analyse appropriate data to demonstrate the suitability and effectiveness of the quality management system including data generated as a result of monitoring and measurement and from other relevant sources.*

What Does this Mean?

The suitability of the management system as stated previously is whether it represents the best way of doing things rather than whether it is fit for purpose – this is expressed by the requirement for the management system to be adequate in Clause 5.6. (The omission of the word 'adequacy' in this clause is perhaps an oversight.) The effectiveness of the management system is judged by the extent it enables the organization to satisfy its stakeholders.

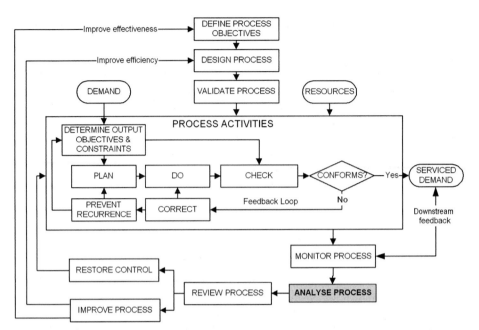

FIGURE 34-1 Where the requirements of Clause 8.5 apply in a managed process.

Appropriate data would be any data generated from the processes of the management system that assist in the determination of their suitability and effectiveness.

The standard does not indicate what data may be needed – the organization is required to determine the data needed, to collect and then analyse it in order to provide a basis for determining the performance of the management system.

Why is this Necessary?

This requirement responds to the Factual Approach Principle.

The requirements in Clause 5.6 indicate a need for data on which to judge whether the system is suitable, adequate and effective – this requirement addresses this need.

How is this Demonstrated?

Determining What Data to Collect

Suitability is concerned with doing things in the most appropriate way – perhaps the best way. If the system enabled staff to waste resources or under-utilize physical and human resources and under-utilize knowledge and capability, the output may conform to requirements but productivity would be down, bottlenecks would occur in processes and there would be a struggle to meet delivery targets without certain staff working flat out. System suitability cannot be examined if one perceives the system to be a set of documents – it is only possible if one perceives the system as a series of interacting processes encompassing everything the organization needs to fulfil its mission.

Effectiveness is concerned with doing the right things rather than with doing things right. So if the system enables management to stop the development of products for which there was no requirement, discover a potential safety problem, anticipate customer needs ahead of the competition, cut waste by 50%, successfully defend a product liability claim, meet all the delivery targets agreed with the customer, you would probably say that the system was pretty effective. If on the other hand a system allows the shipment of defective products every day, looses one in three customers, allows the development of unsafe products to reach the market, or the failure of a revolutionary power plant, you would probably say that system was pretty ineffective. So the first thing you need to do is establish what you want the quality system to do and this ought to be, to enable the organization to achieve its goals. Regretably many systems are only designed to meet the standard with the result that you can deliver defective product providing you also deliver some which are not defective.

As with many of these requirements, the place to start is to assess each process and determine the data needed to establish that:

- the process is achieving its objectives,
- the process is being run in the best way,
- the process objectives remain relevant to the organizational goals.

Data on achievement of objectives might include statistics on percent conforming, throughput, response time etc. Data on best practice may include measures of productivity, quality costs (i.e., the cost of prevention, appraisal and failure costs) or benchmarking[O] data from competitors or similar industries. Data on relevance of objectives may include contrasting the objectives and targets with the organization's objectives.

Data Collection

Plan the data requirements carefully so that you only:

- Collect data on events that you intend to analyse;
- Analyse data with the purpose of discovering problems;
- Provide solutions to the real problems;
- Implement solutions that will improve performance.

To analyse anything, you need data. Without data you cannot know for sure if your processes are under control and if your customers are satisfied. It is not sufficient to claim that you have had no problems unless you are confident that the processes in place will alert you to prob-

> **Food for Thought**
>
> Even though you are not aware of any problems will the processes in place alert you to problems should they arise?

lems should they arise. You also need to take care to avoid the 'garbage in–garbage out' syndrome. Your analysis will only be as good as the data with which you are provided. If you want to determine certain facts, you need to ensure that the means exist for the necessary information to be obtained. To do this you may need to change the input forms or provide new forms on which to collect the data. The data needed for corrective action is rarely of use to those providing it therefore design your forms with care. Reject any incomplete forms as a sign that you are serious about needing the data. A sure sign that forms have become obsolete is the number of blank boxes. It is also better to devise unique forms for specific uses rather than rely on general multipurpose forms because the latter have a tendency to degrade the reliability of the data.

Methods are needed to collect the data from the sensors and transmit it to the analysis stations. This may involve not only collecting data locally for immediate analysis for control purposes, but also transmission to central analysis stations. At such stations the data on all factors affecting performance may be aggregated and information produced for use in measuring overall performance relative to objectives. For example, a common target may be set for all processes relative to utilization, nonconformities, response time etc. Some processes may be better than others but when aggregated show that the corporate objectives are not being achieved. A computer network can aid data collection by enabling remote access and collection into interlinked databases. However, many organizations still rely on paper records and therefore you will need a means of enabling such records to be either submitted to the analysis points or collected from source. To achieve this you will need to insert submission or collection instructions in the relevant procedures that specify the records.

Many organizations use a Nonconformity Report to collect information on nonconformities. One report may deal with recovery action and another separate report may address corrective action in order to prevent the recurrence of one or more nonconformities. In this way you are not committed to taking action on every incident but on a group of incidents where the action and its cost can be more easily justified.

A general plan of action would include the following actions:

- Identify the key parameters to be measured.
- Locate where in the process they are achieved.
- Install data collection methods in relevant procedures.
- Collect and analyse the data.
- Use suitable presentation techniques to draw attention to the results.

In collecting the data, care should be taken to avoid data paralysis. The various quality tools can be used to prioritize the identified problems and corresponding decisions. As with all data collection tasks, you should show a direct correlation between what you are collecting and the goals to be achieved. All conclusions should lead to positive action otherwise the effort expended would be futile.

EVALUATING WHERE IMPROVEMENTS IN EFFECTIVENESS CAN BE MADE (8.4)

The standard requires the organization *to evaluate where continual improvement of the effectiveness of the quality management system can be made.*

What Does this Mean?

Juran writes on improvement thus *"Putting out fires is not improvement of the process – Neither is discovery and removal of a special cause detected by a point out of control. This only puts the process back to where it should have been in the first place".*[1] Continual improvement in the effectiveness of the management system is concerned with identifying where the objectives, standards or targets established for activities, tasks or processes are below those needed for the organization to accomplish its purpose or

---

[1] Deming, W. Edwards (1982). *Out of the crisis,* MITC.

mission. At one level this means changing the targets so that they are harder to meet and at another level it means changing the objectives so that work is driven in a new direction.

Why is this Necessary?

This requirement responds to the Continual Improvement Principle.

The analysis of the external environment (customers, markets, regulations, economy and society) will indicate whether the organization's objectives are relevant or whether they need to change. If the management system continues driving the organization against existing objectives, it will fail to satisfy customers and other stakeholders, therefore improvement in the effectiveness of the management system is needed. There may be cases where no specific objectives are set for some aspects that affect organizational performance. For example, managers may be desensitized to the level of rejects, the level of waste, scrap, delays, absenteeism and illness believing there is nothing that can be done to change it. By default, objectives have been set to maintain the status quo – if no objectives are set any level of performance is acceptable!

In evaluating the effectiveness of the management system, those areas where objectives have not been specified become opportunities for improvement.

How is this Demonstrated?

Implementation of this requirement needs a two-pronged approach. A review of established objectives and a review of performance attributes that have no objectives.

Reviewing Established Objectives

An approach to take is to:

- Identify the objectives and targets that have been established for each process and sub-process;
- Establish whether these targets are being achieved;
- Analyse the processes to determine the potential for raising the targets or eliminating targets on the basis that they no longer serve the organization's goals;
- Assess the feasibility of meeting the raised targets or the impact of eliminating inappropriate targets;
- Present the case to management for change.

Reviewing Uncontrolled Performance Attributes

One approach is to undertake the following actions:

- Analyse each process and identify all process outcomes – i.e., tangible and intangible outputs and effects on all stakeholders.
- Identify the outcomes for which no objectives and targets have been set. (Some outcomes may have objectives but no targets and are therefore not being measured.)
- Assess the significance of the outcome in terms of impact (both short and long term) on the organization's goals.
- Make proposals for setting objectives and plans for those outcomes that impact the organization's goals.

ANALYSIS OF CONFORMANCE TO CUSTOMER REQUIREMENTS (8.4b)

The standard requires *analysis of data to provide information relating to conformity to product requirements.*

What Does this Mean?

Data relating to conformity to product requirements is data generated from monitoring and measuring product characteristics. It comprises the results of all verification activities including those generated during design, purchasing, production, installation and operation. The data collected from customer feedback is also included where the cause of the complaint is product nonconformity. The requirement focuses on conformity and therefore the degree of variation is not relevant. A product either conforms or does not conform and it is this that is required to be analysed. The next requirement deals with variation.

Why is this Necessary?

This requirement responds to the Factual Approach Principle.

When setting your quality objectives, targets should be established for product conformity as a measure of achievement. Data needs to be collected and analysed for all products in order to determine whether these objectives are being achieved.

How is this Demonstrated?

Data on conformity and nonconformity can be collected from the product verification points in each process but it is also important to collect data on the size of the population involved. In general terms it is important to know the overall ratio of conforming product to nonconforming product, i.e., of a quantity of products, how many were conforming and how many nonconforming. The data could also relate to specific product characteristics such as reliability, safety, power output, strength etc.

A concept that has become very popular is the sigma value – which is a measure of the capability of a process to produce conforming product (see Chapter 31 under *Six sigma*).

ANALYSIS OF PRODUCT AND PROCESS CHARACTERISTICS (8.4c)

The standard requires *analysis of data to provide information relating to characteristics and trends of processes and product including opportunities for preventive action.*

What Does this Mean?

The data that can provide information relating to characteristics and trends of processes and product comprises all the product and process measurements taken. The measurements provide useful data for indicating variation in product conformity and process capability. Variation is measured when the characteristics are variable such as dimensions, voltage, power output and strength.

Opportunities for preventive action may arise when the trend in a series of measured values indicates deterioration in performance and if the deterioration was allowed to continue, nonconformity would result.

Why is this Necessary?

This requirement responds to the Factual Approach Principle.

The purpose behind the requirement is to generate information that can be used to bring about an improvement in product and process quality. It is only by analysing trends that the opportunity for improvement is revealed. A series of figures may appear random at first glance but on closer analysis proves the process to be out of control. The points on a graph may indicate measurements are within limits but further analysis may reveal that a change has occurred that if ignored, may result in the process generating nonconforming product in the future. Data analysis is therefore useful for assessing current performance and predicting future performance.

How is this Demonstrated?

Measurements of product and process characteristics should be collected but remember to collect only data that is useful in improving conformity or capability. Just because it can be measured does not mean that it should be! In automated processes, the machine performs the analysis and adjusts the process. In operator controlled processes, the operator takes the measurements, conducts the analysis and makes the adjustments often using control charts such as Average and Range Charts, Median Charts, p Charts, np Charts etc. The characteristics measured are taken from the product specification. In management controlled processes, the data needs to be collected, reduced, analysed, interpreted and presented in a suitable form before any meaningful information can be transmitted to decision makers. The data may come from many processes and locations. Analysing the Staff Development process for instance may require data from all managers on staff numbers, development needs, training programmes etc. to be consolidated and interpreted before meaningful results will emerge. The output may be in the form of histograms showing the levels of competence in each department, the cost of training per employee, the proportion of turnover spent on training etc. All information should relate to specific product or process characteristics to prevent the management being inundated with reports that add no value.

A methodical approach would be to generate a list for each product and process that includes:

- The product or process characteristics;
- The location(s) of the data capture point(s);
- The method of capturing measurement data (automated, operator or manager);
- The form in which the data is transmitted to the analysis point (raw data, reports, surveys etc.);
- The unit responsible for data analysis;
- The methods used to extract meaningful information from the data;
- The form in which the resultant analysis is transmitted to decision makers
- The frequency of reporting;

- The levels through which the information must pass before reaching the decision makers;
- The unit responsible for making decisions.

Data on product characteristics may be analysed and acted on by the operator and therefore do not warrant separate indication of each characteristic. However, some characteristics such as reliability and availability are not measured by operators but from field data or service centres. For processes the characteristics will be key performance measures against the process objectives. Where a process involves many departments, extracting and analysing the data can be a complex process in itself. Juran uses the analogy of telephone transmission to illustrate the problems with communication.[2]

- The amplifier: A device for restoring intensity of a signal that has become weak over long distances.
- The filter: A device for admitting a selected part of the spectrum and to reject all else.
- Redundancy: Multiple transmission channels to increase reliability.
- Shielding: A device to keep noise from invading the message, keeps the message secure and keeps out cross-talk.

Depending on the significance of the information, all of these devices may be active in the channels transmitting information of product and process performance. Some of the devices may be installed deliberately to safeguard information from outsiders. Other devices may be used covertly to prevent information from reaching the top management. In the list described above, the penultimate bullet is inserted to bring such ploys out into the open. If the information is based on fact and is relevant to the product and process objectives, it should be allowed free passage. Redundancy may be warranted when the information is of vital importance to the business and some means of validating the facts is needed.

ANALYSIS OF SUPPLIER DATA (8.4d)

The standard requires *analysis of data to provide information relating to suppliers.*

What Does this Mean?

Information relating to suppliers includes that related to their performance regarding product and service quality, delivery and cost.

Why is this Necessary?

This requirement responds to the Factual Approach Principle.

Suppliers are a key contributor to the performance of an organization and therefore information on the performance of suppliers is necessary to determine the adequacy, suitability and effectiveness of the management system.

---

[2] Juran, J. M. (1995). *Managerial Breakthrough Second Edition,* McGraw-Hill.

How is this Demonstrated?

There are several aspects of the purchasing process that can be analysed and used to reveal information about suppliers in order to determine their performance and opportunities for improvement. However, the resources allocated for the analysis need to be appropriate to the potential risks to the organization and its customers. It is therefore necessary to focus on those suppliers that indicate the greatest risk to the organization's performance.

Order Value

One group of suppliers at risk are those that provide the highest value of products and services. High order value implies significant investment by the organization because such decisions are not taken lightly. Should the supplier fail, the organization may not be able to recover sufficiently to avoid dissatisfying its customers.

The supplier database should identify the value of orders with suppliers and from an examination of these data; a Pareto analysis may reveal the proportion of suppliers that receive the highest value of orders. The performance of those in the top 20% in the order value list will obviously have more effect on the organization than the performance of the other 80% and may warrant closer attention. If you establish the effort spent on developing these suppliers as opposed to the others, it may reveal that the priorities are wrong and need adjustment.

Order Quantity

Another group at risk are those suppliers that process the greatest number of orders. If there is a systemic fault in their processes, many deliveries may contain the same fault. It is possible that some of the high order value suppliers are the same as the high order quantity suppliers such as those supplying consumables. The performance of the top 20% in the order quantity list may affect all your products especially if the product is a raw material, fasteners, adhesives or any item that forms the basis of the product's physical nature.

Quality Risk

The third group at risk comprises suppliers that supply products or services that a product failure modes analysis has shown are mission critical regardless of value or quantity. You may only need a few of these and their cost may be trivial, but their failure may result in immediate customer dissatisfaction. The FMEA[1] should show the probability of failure and therefore the Pareto analysis could reveal the top 20% of products that are critical to the organization in terms of quality.

Delivery Risk

The fourth group at risk are those suppliers that must meet delivery targets. Some items are on a long lead time with plenty of slack, others are ordered when stocks are low and others are ordered against a schedule that is designed to place product on the production line just-in-time to be used. It is the latter that are most critical although a JIT scheme does not have to be place for delivery to be critical. The top 20% of these suppliers deserve special attention, regardless of value, quantity or product quality risk. A late delivery may have ramifications throughout the supply chain.

Suppliers per Item

The fifth group of suppliers is not necessarily a group at risk. Many organizations insist on having more than one qualified supplier for a given item or service just in case a supplier under-performs. As Deming points out "*A second source for protection in case of ill luck puts one vendor out of business temporarily or forever, is a costly policy. There is lower inventory and a lower total investment with a single supplier than with two*".[3] Remember you don't have to be ordering from different suppliers at the same time for a second source to be a costly policy. There are the costs associated with the evaluation and approval which are double for maintaining a viable second source. An analysis of purchased items by supplier will reveal how many items are sourced from more than one supplier. Those items sourced from the most number of suppliers are therefore candidates for a supplier reduction programme.

Costs

Cost is also a factor but often only measured when there is a target for suppliers to reduce costs year-on-year. An analysis of these suppliers may reveal the top 20% that miss the target by the greatest amount.

Once the top 20% have been identified in each group, further analysis should be carried out to establish how each of these suppliers perform on quality, cost and delivery, the amount of effort spent in developing these suppliers and the degree to which these suppliers respond to requests for action.

A common method of assessing suppliers was to send out questionnaires that gathered data about the supplier. These add little value apart from gathering data. A measure of how many of your suppliers have ISO 9001 certification does not reveal anything of value because it does not indicate their performance. Analysis of supplier data should only be performed to obtain facts from which decisions are to be made to develop the supplier or terminate supply.

---

[3] Deming, W. Edwards (1982). *Out of the crisis*, MITC.

Chapter 35

Continual Improvement

CHAPTER PREVIEW

This chapter is aimed at all those personnel engaged in managing, measuring or monitoring products and processes for there should be no measurement or monitoring without analysis and no analysis without recommendations for improvement.

The positions where the requirements on continual improvement feature in the managed process are shown in Fig. 35-1. Improvements can arise through better control of the process, through better product planning, better product requirements, better process design and more challenging process objectives.

In this chapter we examine the requirements in Clause 8.5.1 of ISO 9001:2008 and in particular what they mean, why they are necessary and how conformity is demonstrated.

But first we take a look at the concepts and principles of improvement so that we may then take a more informed approach to the requirements.

IMPROVEMENT

There are three things that are certain in this life, death, taxes and change! We cannot improve anything unless we know its present condition and this requires measurement and analysis to tell us whether improvement is both desirable and feasible. Improvement is always relative. Change is improvement if it is beneficial and a retrograde step if it is undesirable but there is a middle ground where change is neither desirable (beneficial) nor undesirable, it is inevitable and there is nothing we could or should do about it. Change is a constant. It exists in everything and is caused by physical, social or economic forces. Its effects can be desirable, tolerable or undesirable. Desirable change is change that brings positive benefits to the organization. Tolerable change is change that is inevitable and yields no benefit or may have undesirable effects when improperly controlled. The challenge is to cause desirable change and to eliminate, reduce or control undesirable change so that it becomes tolerable change.

Deming cited two mistakes frequently made in attempts to improve results, both of them being costly. The first was to react to an outcome as if it came from a special cause, when actually it came from a common and thus random cause of variation. The second was to treat an outcome as if it came from common causes of variation, when it actually came from a special cause.

Juran writes on improvement thus "Putting out fires is not improvement of the process – Neither is discovery and removal of a special cause detected by a point out of

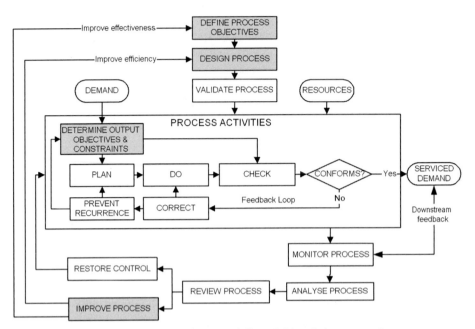

FIGURE 35-1 Where the requirements of Clause 8.5.1 apply in a managed process.

control. This only puts the process back to where it should have been in the first place".[1] This we call *restoring the status quo*. If eliminating special causes is not improvement but maintaining the status quo, that leaves two areas where the improvement is desirable – the reduction of common cause variation and the raising of standards.

Figure 35-2 illustrates the continuing cycle of events between periods of maintaining performance and periods of change. The transition from one target to another may be gradual on one scale but is considered a breakthrough on another scale. The variation

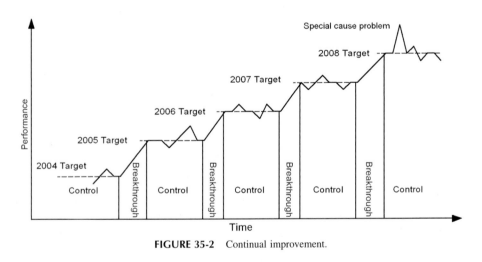

FIGURE 35-2 Continual improvement.

[1] Deming, W. Edwards (1982). Out of the crisis, MITC. Juran as observed by Edwards Deming.

around the target value is due to common causes that are inherent in the system. This represents the expected performance of the process. The spike outside the average variation is due to a special cause, a one-off event that can be eliminated. These can be regarded as fires and is commonly called *fire fighting*. Once removed, the process continues with the average variation due to common causes.

When considering improvement by raising standards, there are two types of standards: one for results achieved and another for the manner in which the results are achieved. We could improve on the standards we aim for, the level of performance, the target or the goal but use the same methods. There may come a point when the existing methods won't allow us to achieve the standard, then we need to devise a new method, a more efficient or effective method or due to the constraints on us, we may choose to improve our methods simply to meet existing standards.

This leads us to ask four key questions:

- Are we doing it right?
- Can we keep on doing it right?
- Are we doing it in the best way?
- Is it the right thing to do?

The ISO 9000:2005 definition of *quality improvement* states that it is that part of quality management focused on increasing the ability to fulfil quality requirements. If we want to reduce the common cause variation we have to act on the system. If we want to improve efficiency and effectiveness we also have to act on the system and both are not concerned with correcting errors but concerned with doing things better and doing different things.

There is a second dimension to improvement – it is the rate of change. We could improve 'gradually' or by a 'step change'. Gradual change is also referred to as incremental improvement, continual improvement or *kaizen*. 'Step change' is also referred to as 'breakthrough' or a 'quantum leap'. Gradual change arises out of refining the existing methods, modifying processes to yield more and more by consuming less and less. Breakthroughs often require innovation, new methods, techniques, technologies and new processes.

Are We Doing it Right?

Would the answer be this?	No, we are not because every time we do it we get it wrong and have to do it again.
Or would it be this?	Yes, we are because every time we do it we get it right and we never have to do it over again.

Quality improvement in this context is for better control and is about improving the rate at which an agreed standard is achieved. It is therefore a process for reducing the spread of common cause variation so that all products meet agreed standards as shown in Fig. 35-3 until it is all within the specification limits. It is not about removing special cause variation, i.e., this requires the corrective action process.

This type of improvement is only about reducing variation about a mean value or closing the gap between actual performance and the target. This is improvement by better control and in some sectors is not regarded as improvement at all. In the automotive sector, continual improvement is implemented once manufacturing processes are

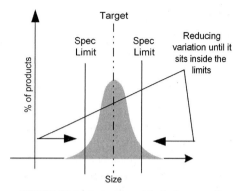

FIGURE 35-3   Improvement by better control.

capable and stable or product characteristics are predictable and meet customer requirements.[2] The target remains static and the organization gets better and better until all output meets the target or falls between the acceptance limits.

When a process is stable the variation present is only due to random causes. There may still be unpredictable excursions beyond the target due to a change in the process but this is special cause variation. Investigating the symptoms of failure, determining the root cause and taking action to prevent recurrence can eliminate the special cause and reduce random causes. A typical quality improvement of this type might be to reduce the spread of variation in a parameter so that the average value coincides with the nominal value. Another example might be to reduce the defect rate from three sigma to six sigma. The changes that needed to meet this objective might be simply changes in working practices or perhaps complex changes that demand a redesign of the process or a change in working conditions. These might be achieved using existing methods or technology but it may require innovation in management or technology to accomplish.

Can We Keep on Doing it Right?

Would the answer be this?   No, we can't because the supply of resource is unpredictable, the equipment is wearing out and we can't afford to replace it.

Or would it be this?   Yes, we can because we have secured a continual supply of resources and have in place measures that will provide early warning of impending changes.

This question is about continuity or sustainability. It is not enough to do it right first time once, i.e., you have to keep on doing it right and this is where a further question helps to clarify the issue.

What affects our ability to maintain this performance?

It could be resources as in the example, but to maintain the status quo might mean innovative marketing in order to keep the flow of customer orders of the type that the process can handle. Regulations change, staff leave, emergencies do happen: Can you keep on doing it right under these conditions?

---

[2] ISO/TS 16949:2002.

Are We Doing it in the Best Way?

Would the answer be this?	We have always done it this way and if it isn't broken why fix it?
Or would it be this?	Yes, we think so because we have compared our performance with the best in class and we are as good as they are.

One might argue that any target can be met providing we remove the constraints and throw lots of money at it. Although the targets may be achieved, the achievement may consume too much resource; time and materials may be wasted – there may be a better way of doing it. By finding a better way you release resources to be used more productively and therefore bring about improvement through better utilization of resources as illustrated in Fig. 35-4.

Over 21 years since the introduction of ISO 9001, it is strange that more organizations did not question if there was a better way than writing all those procedures, filling in all those forms, insisting on all those signatures. Although ISO 9001 did not require these things there was more than one way of interpreting the requirements.

The search for a better way is often more effective when in the hands of those doing the job and you must therefore embrace the 'leadership' and 'involvement of people' principles in conjunction with continual improvement.

Much of the improvement potential does not require large injections of cash. It requires the right attitude – seizing opportunities from observed weaknesses – it requires a change in culture. When your management has the right attitude then and only then can you declare your policy on continual improvement. Cultural changes arise through awareness, experiences and training that become embedded in the behavioural patterns of the organization. Cultural change requires an understanding of human psychology and requires you to overcome the inertia of the human mind.

Is it the Right Thing to Do?

Would the answer be this?	I don't know – we always measure customer satisfaction by the number of complaints.
Or would it be this?	Yes, I believe it is because these targets relate very well to the organization's objectives.

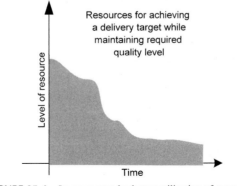

FIGURE 35-4 Improvement by better utilization of resources.

Quality improvement in this context is accomplished by raising standards and is about setting a new level of performance, a new target that brings additional benefits for the stakeholders. These targets are performance targets for products, processes and the system. They are not targets established for the level of errors, such as nonconformities, scrap, and customer complaints. Such targets are not in fact targets at all; they are simply historical standards or performance.

One needs to question whether the targets are still valid so we ask, 'How do we know this is the right thing to do?' These new targets have to be planned targets as exceeding targets sporadically are a symptom of out-of-control situations. Targets need to be derived from the organization's goals but as these change the targets may become disconnected. Targets that were once suitable are now obsolete – they are not the right things to do any longer. We need to ask, "Are these targets still relevant to the stakeholder needs?"

Functions are often measured by their performance against budget. We need to ask whether this is the right thing to do – does it lead to optimizing organizational performance? You may have been desensitized to the level of nonconformities or customer complaints – they have become the norm – is this the right level of performance to maintain or should there be an improvement programme to reach much a lower level of rejects?

New standards are created through a process that starts with an analysis of stakeholder needs and expectations followed by the identification of opportunities for change, then a feasibility stage, progressing through research and development to result in a new standard, proven for repeatable applications. Such standards result from innovations in technology, marketing and management. This is improvement by better understanding of stakeholder needs. A typical quality improvement of this type might be to redesign a range of products to increase the achieved reliability from one failure every 5000 hours to one failure every 100,000 hours. Another example might be to improve the efficiency of the service organization so as to reduce the guaranteed call-out time from the specified 36 to 12 hours or improve the throughput of a process from 1000 to 10,000 components per week. Once again, the changes needed may be simple or complex and might be achieved using existing technology but it may require innovation in technology to accomplish.

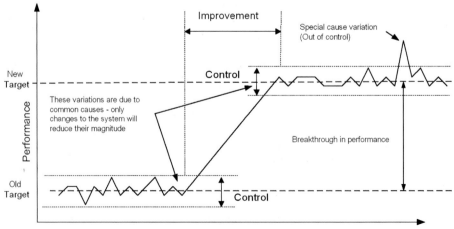

FIGURE 35-5   Improvement by reaching higher standards.

The transition between where quality improvement stops and quality control begins is where the level has been set and the mechanisms are in place to keep quality on or above the set level. In simple terms, if quality improvement reduces quality costs from 25% of turnover to 10% of turnover, the objective of quality control is to prevent the quality costs rising above 10% of turnover. This is illustrated in Fig. 35-5.

The Improvement Process

Improving quality by better control or raising standards can be accomplished by the following these 10 steps advocated by Juran.

1. Determine the objective to be achieved, e.g., new markets, products or technologies, or new levels of organizational efficiency or managerial effectiveness, new national standards or government legislation. These provide the reasons for needing change.
2. Determine the policies needed for improvement, i.e., the broad guidelines to enable management to cause or stimulate the improvement.
3. Conduct a feasibility study. This should discover whether accomplishment of the objective is feasible and propose several strategies or conceptual solutions for consideration. If feasible, approval to proceed should be secured.
4. Produce plans for the improvement that specifies the means by which the objective will be achieved.
5. Organize the resources to implement the plan.
6. Carry out research, analysis and design to define a possible solution and credible alternatives.
7. Model and develop the best solution and carry out tests to prove it fulfils the objective.
8. Identify and overcome any resistance to the change in standards.
9. Implement the change, i.e., put new products into production and new services into operation.
10. Put in place the controls to hold the new level of performance.

This improvement process will require controls to keep improvement projects on course towards their objectives. The controls applied should be designed in the manner described previously.

Deming's improvement cycle is more similar but more concise.

Plan	Identify the need for change aimed at improvement, propose solutions, evaluate solutions and select the most appropriate.
Do	Carry out the change, preferably on a small scale according to the plan.
Study	Study the results. If they do not meet expectations, diagnose the root cause and identify improvements to the plan.
Act	Adopt the change if expectations are met or abandon it and run through the cycle again with a revised plan.

CONTINUALLY IMPROVING SYSTEM EFFECTIVENESS (8.5.1)

The standard requires the organization *to continually improve the effectiveness of the quality management system through use of the quality policy, quality objectives, audit results, analysis of data, corrective and preventive actions and management review.*

What Does this Mean?

There are 10 requirements for improvement in the standard including the one above:

1. The organization shall establish, document, implement and maintain a management system and continually improve its effectiveness (Clause 4.1).
2. The organization shall implement actions necessary to achieve continual improvement of the processes needed for the management system (Clause 4.1f).
3. Top management shall provide evidence of its commitment to continually improving the effectiveness of the management system (Clause 5.1).
4. Top management shall ensure that the quality policy includes a commitment to continually improve the effectiveness of the management system (Clause 5.3).
5. The output from the management review shall include any decisions and actions related to improvement of the effectiveness of the quality management system and its processes (Clause 5.6.3a).

> **Food for Thought**
>
> Confirm where the 'Goal Posts' are located for they may have moved since the last time you checked.

6. The output from the management review shall include any decisions and actions related to improvement of product related to customer requirements (Clause 5.6.3b).
7. The organization shall determine and provide the resources needed to continually improve the effectiveness of the management system (Clause 6.1).
8. The organization shall plan and implement the monitoring, measurement, analysis and improvement processes needed to continually improve the effectiveness of the management system (Clause 8.1).
9. The organization shall determine, collect and analyse appropriate data to evaluate where continual improvement of the management system can be made (Clause 8.4).
10. The organization shall continually improve the effectiveness of the management system through the use of the quality policy, quality objectives, audit results, analysis of data, corrective and preventive actions and management review (Clause 8.5.1).

Requirement 1 is duplicated by requirement 10. Requirement 2 omits reference to *effectiveness* therefore improvement can be interpreted as applying to improved efficiency as well as effectiveness. Requirements 3 and 4 are linked in that they relate to policy. Requirement 8 serves to identify opportunities for improvement of which requirement 9 forms a part and requirement 7 provides the means by which improvement are to be made. Requirement 6 is the only one that refers specifically to product as all the others refer to improvement in the management system. They are therefore not separate requirements but are all derivatives of the first requirement, each focusing on either the whole or a part of the improvement process.

Published interpretations RFI 025 states that the realization of a new product to improve an old one could be one of the results of the management review but this is not considered to be a continual improvement by RFI 024 which states that the improvement addressed in Clause 8.5.1 does not include product improvement. Some clarity is obviously needed.

As previously stated there are three types of improvement:

- improvement by better control,
- improvement by better utilization of resources,
- improvement by better understanding of stakeholder needs.

The quality management system is not simply a tool that is used to produce product. It produces the products and services as illustrated in Fig. 7-15. (Even Fig. 1 in ISO 9001 shows the output of the QMS to be product.) Therefore, when the effectiveness of the system is improved, it will result in:

- More products meeting customer requirements. For example, variations outside specified limits reduced and processes brought under statistical control.
- More processes utilizing fewer resources. For example, outputs will be produced more quickly with less waste, lower labour costs and fewer accidents.
- More products satisfying the needs of stakeholders. For example, products that set new standards of performance providing greater benefits and less impact on the environment.

Why is this Necessary?

This requirement responds to the Continual Improvement Principle.

The policies, objectives and targets maybe those that are set on the basis of existing capability and performance. In other words, what was achieved last year? They may be those that were set last time customer needs and expectations were evaluated or the prevailing regulations were determined. These may no longer be the policies, objectives and targets that are required to keep the organization focused on its purpose and mission. The *goal posts* may have moved! The competition may have pushed forward the frontiers of technology, innovation, and performance. The market may have changed, the economic and social climate changed and in order to stay ahead of the competition, the organization has to aim for different objectives and targets, to set new policies that will form different behaviours. On the premise that for a system to be effective it has to fulfil its purpose and in the environment in which the system operates, the route towards that fulfilment is always changing; the system will never be totally effective – there will always be a new goal to aim for. It is therefore necessary to continually improve the effectiveness of the management system even though by the time the system is effective the goal posts will have moved. There is nothing more certain in life than death, taxes and change!

How is this Demonstrated?

This particular requirement duplicates that contained in the first paragraph of Clause 5.4.2 because Clause 5.4.1 required objectives consistent with policy and Clause 5.3 required policy to address continual improvement, therefore plans for continual improvement would be required. Planning for improvement is also addressed by Clauses 8.1 and 8.5.1 so there is no shortage of clauses related to improvement planning. They all amount to the same thing. There is not a lot of difference between them but the most informative reference is that in Annex B of ISO 9004 where an improvement methodology is described.

Case Study – Demonstrating Continual Improvement

In an organization at the forefront of technology we are always changing our way of working and our technologies just to stay in the race – if we don't we will fall behind. We are not in mass production and therefore don't produce large quantities where continuous improvement can easily be demonstrated. As soon as the bugs are fixed, we launch a new product and if we were to hold production until we fixed all the bugs we would be too late to market. So when change is an inherent attribute of the business how can we demonstrate continual improvement?

Continual improvement is defined in ISO 9000:2005 as a recurring activity to increase the ability to fulfil requirements. But the standard does not indicate which requirements, so the place to start is with identifying the requirements. If the product specification identifies 400 requirements, you have at least 400 opportunities for improvement.

One area is that of fulfilling customer requirements. One of these requirements might be product integrity so continual improvement in this context is the activity of increasing the probability of your products or services being free of bugs or security risks. This could be demonstrated by showing that the number of software bugs is declining or that you are improving the speed with which you detect the bugs and upgrade products already shipped. (Microsoft is a good example of this. At one time you had to obtain a new version of the software from Microsoft and install it yourself. A few years later you could go onto their web site and download the updates thereby improving accessibility. Now you can configure Windows XP to download updates automatically so that you don't have to intervene at all – thereby demonstrating continual improvement in product integrity.)

Another area might be that of fulfilling market requirements (What you call changes to stay in the race). Here you would be concerned with whether your products fulfilled market needs rather than with the extent of product conformity to a specification.

Another area might be that of fulfilling process efficiency requirements. The hours taken to code the software might be cut dramatically by installing new coding tools. The time to market might be improved by streamlining the process steps. The time spent testing the product might be cut by introducing new technology or simply retraining your staff. So by measuring time at various points in the product development process you might be able to demonstrate continual improvement after installing new techniques or training staff.

The clue here is to firstly identify the specific requirement where there is potential for improvement. It could be a product requirement or a process requirement. Wherever requirements, objectives, goals, targets etc. have been set, current performance should be measured and the difference between actual and target identifies opportunities for improvement. Plot the change in performance over time and provided the trend is positive you are demonstrating continual improvement.

Continual improvement was addressed briefly in Chapter 10 under the same heading but to understand the implications of this requirement it is necessary to understand:

- the composition of a management system;
- the types of improvement;
- the difference between random and continuous improvement.

We introduced the notion of a management system in Chapter 1 under *A quest for capability* then analysed it in more detail in Chapter 7 in A *systems approach* so you would have learnt that a management system is much more than a set of documents. A management system consists of the processes required to deliver the organization's products and services as well as the resources, behaviours and environment on which they depend. It follows therefore that in continually improving the management system, you need to continually improve the processes, resources, behaviours and the physical and human environment within the organization.

Improvement can be random and unstructured – arising from fire fighting measures, reactions to situations that have got out of control or to threats that appear on the horizon that must be dealt with in order to survive – but this is not strictly improvement, it merely restores performance to where it should have been in the first place. ISO 9001 requires a continual quest for improvement and this will only come about if you are continually measuring performance and acting on the results so as to improve conformity and improve capability. The frequency by which performance is measured should be appropriate to the parameter concerned. Some parameters change by the second, others by the year. The frequency of measurement should provide factual data that if acted on will move the organization forward incrementally or in great steps. What should not happen is that the measurements are taken too late to halt a significant decline in performance.

Improvement plans do not need to be consolidated into documents with the title Continual Improvement Plan. Separate plans may exist, focused on general or specific improvements. For example, there may be:

- new product development plans;
- new process development plans;
- new system development plans;
- corrective action plans;
- preventive action plans;
- staff development plans;
- plant development plans.

All these plans aim to provide the organization with an improved capability and for data management purposes, it may help to catalogue these plans under *Improvement* as well as product, process, department, division or corporate headings etc.

A process for continual improvement would be a series of interrelated activities, resources and behaviours that bring about an improvement in performance. Such a process would operate at several different levels in the organization rather than at one level. It would be impractical to have a single continual improvement process through which all improvements are channelled.

Each managed process has three improvement mechanisms. As shown in Fig. 35-6.

- One that brings about improvement by better control – reducing variation about a mean;
- One that brings about improvement by greater efficiency – reducing resources, doing more for less (The seven Wastes and five Ss are two examples – see Glossary);
- One that brings about improvement by greater effectiveness – raising standards, doing the right things, being ahead of stakeholder needs and expectations.

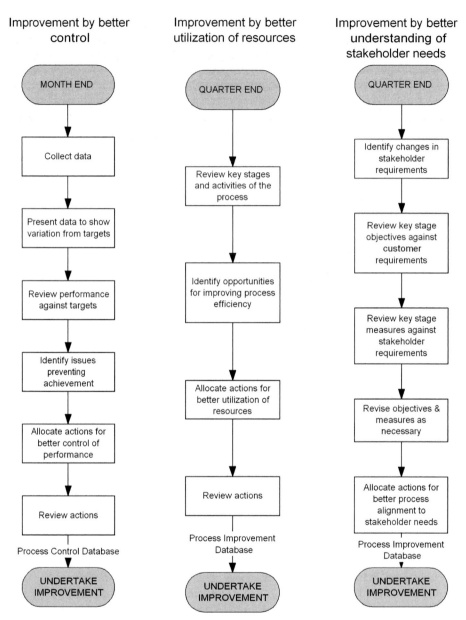

FIGURE 35-6 Three improvement processes.

With improvement programme there will be spin offs.

- A programme for reducing variation reduces rework time and waste;
- A reduction in waste leads to less storage space required;
- A reduction in storage space might lead to a reduction insurance premiums, rent etc.;
- A reduction in insurance and rent might enable a reduction in price;
- A reduction in price might draw in more customers.

There are many ways of running through an impact analysis and it is useful to express your improvement programmes in these terms so as to make a persuasive case that will influence the decision makers. By quantifying the level of reduction you can make it more convincing still.

Some customers require or expect a reduction in price year on year in return for giving you their business. On the face of it, this might appear an unreasonable requirement but when thought of in terms of a share of the bounty gained from a well managed improvement programme, it is equivalent to giving employees a share of the profits in good years.

Chapter 36

Corrective Action

CHAPTER PREVIEW

This chapter is aimed at all those personnel monitoring and measuring processes and their outputs.

The position where the requirements on corrective action feature in the managed process is shown in Fig. 36-1. Note that the outputs from corrective action are changes to either plans or requirements.

In this chapter we examine the requirements in Clause 8.5.2 of ISO 9001:2008 and in particular:

- Eliminating the cause of actual nonconformities;
- Reviewing nonconformities;
- Determining the cause of nonconformities;
- Evaluating the need for action;
- Determining and implementing actions;
- Evaluating the need for action;
- Determining and implementing actions;
- Recording results of actions taken;
- Reviewing corrective action taken.

ELIMINATING THE CAUSES OF ACTUAL NONCONFORMITIES (8.5.2)

The standard requires the organization *to take action to eliminate the causes of nonconformities in order to prevent recurrence and requires the actions to be appropriate to the effects of the nonconformities encountered.*

What Does this Mean?

The term corrective action was explained in the Introduction to this Part of Handbook.

Although the natural inclination is to think of nonconformities in the context of product and manufacturing process, any departure from a requirement is a nonconformity. It follows therefore that any failure to meet the organization's objectives is a nonconformity and its cause should be eliminated.

Action appropriate to the effects of the nonconformities encountered, means that if the nonconformity is a random occurrence and of insignificant consequence, no corrective action might be needed, e.g., if a person makes a mistake and knows a mistake has been made – there is no pattern of behaviour that would suggest it would occur

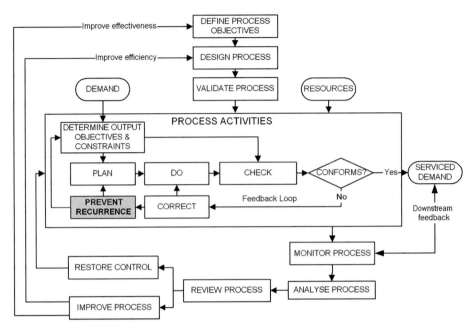

FIGURE 36-1   Where the requirements of Clause 8.5.2 apply in a managed process.

again. If there is frequent occurrence of nonconformity this indicates a systemic problem that can be prevented from recurrence by retraining, changing work practices or modifying product or process design. It is therefore not practical to attempt action to prevent recurrence for each nonconformity.

The requirement also acknowledges that there may be more than one cause of a nonconformity, e.g., several mechanisms in a chain of actions may have failed for the nonconformity to have occurred.

Why is this Necessary?

This requirement responds to the Process Approach Principle.

Nonconformities are caused by factors that should not be present in a process. There will always be variation but variation is not nonconformity. Nonconformity arises when the variation exceeds the permitted limits. The factors that cause nonconformity on one occasion will (unless removed) cause nonconformity again and again. As the objective of any process must be to produce conforming output, it follows therefore that it is necessary to eliminate the causes of nonconformity. This does not simply apply to products of the production process but to products of all processes – mission management, demand creation and demand fulfilment.

How is this Demonstrated?

The Actions

The other requirements in this clause identify most of the steps needed to eliminate the cause of nonconformity. There are one or two stages that have been omitted. A more complete list of steps is shown in Table 36-1. These will be addressed later in this chapter.

TABLE 36-1 Steps in the Corrective Action Process

Step	Action	Clause 8.5.2 requirement
1	Collect the nonconformity data and classify	Reviewing
2	Conduct Pareto analysis to identify the vital few and trivial many	nonconformities
3	Organize a diagnostic team	Determining
4	Postulate causes and test theories	the cause
5	Determine the root cause of nonconformity	
6	Determine the effects of nonconformity and the need for action	Evaluating the
7	Determine the action needed to prevent nonconformity recurring	action needed
8	Organize an implementation team	Implementing the
9	Create or choose the conditions which will ensure effective implementation	action needed
10	Implement the agreed action	
11	Record the results of Pareto analysis	Records of results
12	Record the causes of nonconformity	
13	Record the criteria for determining severity or priority	
14	Record the proposed actions to be taken	
15	Record the actions actually taken	
16	Record the results of actions taken	
17	Assess the actions taken	Reviewing
18	Determine whether the actions were those required to be taken	corrective actions
19	Determine whether the actions were performed in the best possible way	
20	Determine whether the nonconformity has recurred	
21	If nonconformity has recurred repeat steps 1–20	

The Procedures

As the sources of nonconformity are so varied, it may not be practical to have a single corrective action procedure. It may be more practical to embody corrective action provisions in the following procedures:

- Failure investigation procedure.
- Nonconforming material review procedure.
- Customer complaints procedure.
- Document change procedure.
- Specification change procedure.
- Maintenance procedures.

The Responsibilities

Corrective actions should be determined and implemented relative to the process of which the activities belong rather than in isolation to avoid destabilizing effects on the

TABLE 36-2 Corrective Action Responsibilities

Action	Responsibility
Reviewing variations	Process owner
Determining the cause	Diagnostic team
Evaluating the action needed	Diagnostic team
Implementing the action needed	Implementation team
Reviewing corrective actions	Diagnostic team

process. Those responsible for the process outputs should either conduct or initiate a review of the variations (the nonconformities) and should establish a diagnostic action to determine their root cause and an implementation team to restore the status quo. These teams may comprise any competent personnel and the same personnel may be in each team but it is useful to distinguish between diagnostic and implementation actions because their purpose is distinctly different. The allocation of responsibilities is illustrated in Table 36-2, but with all allocated responsibilities the Process Owner remains responsible for the performance of the process.

REVIEWING NONCONFORMITIES (8.5.2a)

The standard requires *the documented procedure for corrective action to define requirements for reviewing nonconformities (including customer complaints).*

What Does this Mean?

As indicated above, the review of nonconformities means that nonconformity data should be collected, classified and analyzed. The reference to customer complaints is that every customer complaint is a nonconformity with some requirement. They may not all be product requirements. Some may relate to delivery, to the attitude of staff or to false claims in advertising literature. Any complaint implies that a requirement (expectation, obligation or implied need) has not been met even if that requirement had not been determined previously. We have to accept that we could have overlooked something. Just because it was not written in the contract does not mean that the customer is wrong.

Why is this Necessary?

This requirement responds to the Continual Improvement Principle.

A review is another look at something therefore the first view of the nonconformity was when it was detected and recorded. The second view of it should aim to:

1. establish if the nonconformity had been predicted in the planning phase;
2. establish why the preventive action measures were not effective;
3. prevent it from happening again where possible.

How is this Demonstrated?

Your corrective action procedures need to cover the collection and analysis of product nonconformity reports and the collection and analysis of process data to reveal process nonconformities. The standard does not require you to take corrective action on every nonconformity. Here it is suggested that the decision to act should be *appropriate to the effects of the nonconformities encountered*. It is therefore implying that you only need act on the vital few. To find the vital few nonconformities out of the total population that provide the bulk of improvement potential an analysis should be conducted. A Pareto analysis is a management tool that finds a few needles in a haystack of trivia, e.g., most of the wealth is concentrated in few hands or 10% of customers account for 60% of sales.[1] When dealing with nonconformity, the question we need to ask is 'what are the few sources of nonconformities that comprise the bulk of all nonconformities?' If we can find these nonconformities and eliminate their cause, we will reduce variation significantly.

The first step is to assign a short description to the nonconformity such as dry solder joint, hole not plated, broken track, incorrect part number, dirty terminal etc. The next step is to sort the nonconformities by product and process. Then rank the nonconformities in order of occurrence so that the nonconformity having the most occurrences would appear at the top of the list. The result might be that for a particular product or process a few types of nonconformity would account for the greatest proportion of nonconformities. An example is given in Table 36-3. Here we see there are 15 types of nonconformity with just four types accounting for 76% of the total. It follows therefore that if we eliminate the bottom four causes of nonconformity, productivity would increase by a mere 3%. However, if we eliminate the four most dominant types of nonconformity, productivity would increase by staggering 76%.

Another way of ranking the nonconformities is by seriousness. Not all nonconformities will have the same effect on product quality. Some may be critical and others may be insignificant. By classification of nonconformities in terms of criticality a list of those most serious nonconformities can be revealed using the Pareto analysis. Even though the frequency of occurrence of a particular nonconformity may be high, it may not affect any characteristic that impacts customer requirements. This is not to say the cause should not be eliminated but there may be other more significant problems to eliminate first.

Before managers will take action, they need to know:

- What is the problem or potential problem?
- Has the problem been confirmed?
- What are the consequences of doing nothing, i.e., what effect is it having?
- What is the preferred solution?
- How much will the solution cost?
- How much will the solution save?
- What are the alternatives and their relative costs?
- If I need to act, how long have I got before the effects damage the business?

---

[1] For an explanation of the Pareto principle and its origins see Juran on quality by design (1992). The Free Press, Division of Macmillan Inc.

TABLE 36-3 Pareto Analysis

Nonconformity type	Frequency	%	Cumulative %
Too much solder	400	34.19	34.19
Lifted tracks	230	19.66	53.85
Solder bridge	180	15.38	69.23
Dirty terminals	90	7.69	76.92
Component not flat on board	70	5.98	82.91
Cracked insulation	40	3.42	86.32
Uncropped component legs	30	2.56	88.89
Under spec plating thickness	30	2.56	91.45
Too little solder	20	1.71	93.16
Dry joints	20	1.71	94.87
Unplated holes	20	1.71	96.58
Broken tracks	10	0.85	97.44
Incorrect component fitted	10	0.85	98.29
Incorrect part number	10	0.85	99.15
Damaged edge connector	10	0.85	100.00
Total nonconformities	1170		

Whatever you do, don't act on suspicion, always confirm that a problem exists or that there is a certain chance that a problem will exist if the current trend continues. Validate causes before proclaiming action!

Customer Complaints, Rejected and Returned Products

The customer can be mistaken and customer complaints therefore need to be validated as genuine nonconformities before entering the corrective action process. Parts returned from dealers, customer manufacturing plants etc. might not be nonconforming. They may be obsolete, surplus to requirements, have suffered damage in handling or have been used in trials etc. Products may have failed under warranty and not be logged as a complaint but nonetheless they are nonconforming. Whatever the reason for return, you need to record all returns and perform an analysis to reveal opportunities for corrective action when appropriate. You should process these items as indicated previously but prior to expending effort on investigations, you should establish your liability and then investigate the cause of any nonconformities for which you are liable.

When parts are rejected subsequent to delivery it is indicative that your processes are not under control. Rejected parts analysis should be focused on determining the reason why the process failed to detect nonconformity. There could be some weakness in the process that if not corrected, further nonconforming parts might be shipped.

Nonconformity Reduction

Previously it was suggested that action be taken on the vital few nonconformities that dominated the population. If this plan is successful these nonconformities will no longer appear in the list the next time the analysis is repeated. As the vital few nonconformities are tackled, the frequency of occurrence will begin to decline until there are no nonconformities left to deal with. This is nonconformity reduction (or special cause removal) and can be applied to specific products or processes. If you were to aggregate the nonconformities for all products and processes you would observe that it is quite possible to take corrective action continuously and still not reduce the number of nonconformities – no matter how hard you try, you cannot seem to reduce the number. This is because the objectives and targets keep changing. They rarely remain constant long enough to make valid comparisons from year to year. There is always some new process, practice or technology being introduced that triggers the learning cycle all over again.

DETERMINING THE CAUSE OF NONCONFORMITIES (8.5.2b)

The standard requires *the documented procedure for corrective action to define requirements for determining the causes of nonconformity.*

What Does this Mean?

The cause of nonconformity is the reason it occurred. What you observe when detecting nonconformity is a symptom of the cause. A product is damaged, the immediate cause might be poor packaging or poor handling but these too have their causes and although they may appear unconnected there maybe a common cause which, if eliminated, resolves two problems of concern.

There are three types of corrective action, product related, process related and system related. Product-related nonconformities can be either internal or external and you will have nonconformity reports to analyze. Process-related nonconformities may arise out of product nonconformity but if you expect something less than 100% yield from the process, the reject items may not be considered *nonconformities*. They may be regarded as *waste*. Unlike products, process nonconformities are often not recorded in the same way and therefore the data is not as readily accessible. By analyzing the process you can find the cause of low yield and improve performance of the process. Product and process nonconformities may be detected at planned verification stages and may also be detected during product and process audits. System-related nonconformities could arise out of internal and external system audits but also arise as a result of tracing the root cause of a problem to a system inadequacy.

Why is this Necessary?

This requirement responds to the Continual Improvement Principle.

All nonconformities are caused – all causes within your control can be avoided – all that is needed is concerted action to prevent recurrence. Nonconformity costs money and wastes resources. The fewer the nonconformities, the more resources available for producing productive output.

How is this Demonstrated?

Discovering the Cause (Five Why's)

To eliminate the cause of nonconformity the cause needs to be known and therefore the first step is to conduct an analysis of the symptoms to determine their cause. Simply asking why an event occurred might reveal a cause but don't accept the first reason given because there is usually a reason why this previous event occurred. Toyota discovered that asking *why* successively five times would invariably discover the root cause. There may be more or less than five steps to the root cause but it is critical to stop only when you can't go any further. The following example illustrates the technique:

A trainer arrives to conduct a training course to discover that the materials have not been delivered from head office as he expected them to be.

1. Why were the materials not delivered? – Answer: Because the administrators thought the trainer would bring his own materials.
2. Why did they think the trainer would bring his own materials? – Answer: Because they had not been informed otherwise.
3. Why weren't the administrators given the correct information? – Answer: Because the office manager had not communicated the agreed division of responsibility when setting up training courses.
4. Why had the office manager not communicated the agreed division of responsibility? Answer: Because the office manager had put other matters before internal communication in his order of priorities.
5. Why had the office manager not got his priorities right? Answer: Because he was not yet competent.
6. Why was the office manager not yet competent? Answer: Because the top management had made the appointment in haste.
7. Why had top management made the appointment in haste? Answer: Because they were not applying leadership.

Therefore, a lack of leadership (the second quality management principle) is the root cause. It took seven questions to get there but if we had stopped at question 3, and made the assumption that giving the administrators the correct instructions would prevent recurrence of the problem, we would be wrong. It might well prevent recurrence of the specific

> **Food for Thought**
>
> The root cause of most problems can be traced to lack of application of one or more of the eight quality management principles.

problem with that particular office manager but not similar problems with other managers. If the office manager forgot to issue the instructions, it indicates that he did not complete the process that commenced when the division of responsibility was agreed. This is quite typical. A meeting is held and agreements reached and when everyone departs they get on with what they were doing before the meeting, not realizing that a process has been initiated that needs to continue and be completed outside the meeting. If the staff were competent, they would complete the process before moving on.

Another technique is the cause and effect diagram (also known as the Ishikawa Diagram or fishbone diagram). This is a graphical method of showing the relationship between cause and effect. Each type of nonconformity (an effect) would be analyzed to postulate the causes so, for example, the question would be put to a diagnostic team – What could cause too much solder on a joint? The team would come up with a number of possibilities. Each one would be tested either by experiment or further examination of the soldering process and a root cause will be established.

Some nonconformities appear random but often have a common cause. In order to detect these causes, statistical analysis may need to be carried out. The causes of such nonconformities are generally due to noncompliance with (or inadequate) working methods and standards. Other nonconformities have a clearly defined special or unique-cause that has to be corrected before the process can continue. Special cause problems generally require the changing of unsatisfactory designs or working methods. They may well be significant or even catastrophic. These rapidly result in unsatisfied customers and loss of profits. In order to investigate the cause of nonconformities you will need to:

1. identify the requirements which have not been achieved;
2. collect data on nonconforming items, the quantity, frequency and distribution;
3. identify when, where and under what conditions the nonconformities occurred;
4. identify who was carrying out what operations at the time.

The investigation of parts rejected subsequent to delivery needs to probe not only into the root cause but also establish

- why the controls in place did not detect the nonconformity;
- why the Process FMEA failed to identify the potential failure mode;
- why the process capability studies concluded the process was capable when clearly it wasn't.

It could be of course that the rejected parts are simply within the 3.4 ppm that were expected from a process with six-sigma capability. However, if the number of parts rejected already exceeds this limit, process improvement action will be necessary.

Common Problem Solving Tools (8.5.2.1)

There are many tools you can use to help you determine the root cause of problems. These are known as *disciplined problem solving methods*. A common method in the automotive industry is known as 8D meaning eight disciplined methods. Originally conceived by the Ford TOPS (Team Oriented Problem Solving) programme in 1987[2] and upgraded and renamed as 'Prevent Recurrence' in 1992.

D1 – Establish the team.
D2 – Describe the problem.
D3 – Develop an interim containment action.
D4 – Define or verify root cause.
D5 – Choose or verify permanent corrective action.
D6 – Implement or validate permanent corrective action.
D7 – Prevent recurrence.
D8 – Recognize the team.

---

[2] Six sigma dictionary at http://www.isixsigma.com.

CORRECTIVE ACTION REPORT			
Problem name			**CAR/**
Product/Process	Location	Reported by	Date observed

PROBLEM DESCRIPTION

As relevant define:

1. Nature of problem

2. Acceptance criteria

3. Number of ocurrences

4. Significance of problem (% of total ocurrences for this product)

DIAGNOSTIC TEAM

Role	Name	Department	Location

IMPACT & PRIORITY

On Quality	On Cost	On Delivery	Other impact	Priority Urgent ☐ High ☐ Medium ☐ Low ☐	Required Date

ROOT CAUSE

Cause	% Contribution	Verified	As relevant state the results of applying the 5 Why test to the problem Reference relevant diagnostic reports
Cause 1			
Cause 2			
Cause 3			
Cause 4			
Cause 5			

RESOLUTION

Solution for eliminating root cause	Responsibility	Date validated	As relevant define: 1. Product design changes & validation method 2. Process design changes & validation method 3. FMEA Update 4. Control Plan Update

EFFECTIVENESS OF ACTIONS

All actions taken Yes No	Number of potential recurrence opportunities since last action	Result	
Related NCRs/CARs	Reviewed by:		Date closed

FIGURE 36-2   Sample corrective action report (CAR) or 8D report.

Another version is shown in the boxed section indicating that the labels and interpretations vary. In both examples the terms do not quite align with those in ISO 9001. D5, 6 and 7 are all part of the same action to prevent recurrence which ISO 9000:2005 defines as 'Corrective Action'. The notion of a permanent corrective action implies that there is a temporary corrective action. Corrective action either removes the root cause or it doesn't. If the 'permanent corrective action' were effective it would also deal with the issues raised under D7 like modifying specifications but other interpretations indicate that this would form part of D6. However, it is not too important what the steps are called provided all the steps are completed. A typical 8D Form or correction action report (CAR) for collecting and reporting the basic data in illustrated in Fig. 36-2.

The 8D Approach to Problem Solving

1. Use team approach

Establish a small group of people with the knowledge, time, authority and skill to solve the problem and implement corrective actions. The group must select a team leader.

2. Describe the problem

Describe the problem in measurable terms. Specify the internal or external customer problem by describing it in specific terms.

3. Implement and verify short-term corrective actions

Define and implement those intermediate actions that will protect the customer from the problem until permanent corrective action is implemented. Verify with data the effectiveness of these actions.

4. Define and verify root causes

Identify all potential causes which could explain why the problem occurred. Test each potential cause against the problem description and data. Identify alternative corrective actions to eliminate root cause.

5. Verify corrective actions

Confirm that the selected corrective actions will resolve the problem for the customer and will not cause undesirable side effects. Define other actions, if necessary, based on potential severity of problem.

6. Implement permanent corrective actions

Define and implement the permanent corrective actions needed. Choose on-going controls to ensure the root cause is eliminated. Once in production, monitor the long-term effects and implement additional controls as necessary.

7. Prevent recurrence

Modify specifications, update training, review work flow, improve practices and procedures to prevent recurrence of this and all similar problems.

8. Congratulate your team

Recognize the collective efforts of your team. Publicize your achievement. Share your knowledge and learning.

By courtesy of National Semiconductor 2004

Whilst 8D has a certain meaning, disciplined methods are simply those proven methods that employ fundamental principles to reveal information. There are two different approaches to problem solving. The first is used when data are available as is the case when dealing with nonconformities. The second approach is when not all the data needed are available.

The seven quality tools in common use are as follows:

1. *Pareto diagrams* – used to classify problems according to cause and phenomenon.
2. *Cause and effect diagrams* – used to analyze the characteristics of a process or situation.
3. *Histograms* – used to reveal the variation of characteristics or frequency distribution obtained from measurement.
4. *Control charts* – used to detect abnormal trends around control limits.
5. *Scatter diagrams* – used to illustrate the association between two pieces of corresponding data.
6. *Graphs* – used to display data for comparative purposes.
7. *Check-sheets* – used to tabulate results through routine checks of a situation.

The further seven quality tools for use when not all data are available are as follows:

8. *Relations diagram* – used to clarify interrelations in a complex situation.
9. *Affinity diagram* – used to pull ideas from a group of people and group them into natural relationships.
10. *Tree diagram* – used to show the interrelations among goals and measures.
11. *Matrix diagram* – used to clarify the relations between two different factors (e.g., QFD).
12. *Matrix data-analysis diagram* – used when the matrix chart does not provide information in sufficient detail.
13. *Process decision program* chart – used in operations research.
14. *Arrow diagram* – used to show steps necessary to implement a plan (e.g., PERT).

The source of causes is not unlimited. Nonconformities are caused by one or more deficiencies in:

- communication,
- documentation,
- personnel training and motivation,
- materials,
- tools and equipment,
- the operating environment.

Each of these is probably caused by not applying one or more of the eight quality management principles.

Once you have identified the root cause of the nonconformity you can propose corrective action to prevent its recurrence. Eliminating the cause of nonconformity and preventing the recurrence of nonconformity are essentially the same thing.

EVALUATING THE NEED FOR ACTION (8.5.2c)

The standard requires *the documented procedure for corrective action to define requirements for evaluating the need for action to ensure that nonconformities do not recur.*

What Does this Mean?

All nonconformities are the result of something not going to plan no matter how insignificant the problem is. Whether action is taken depends on the effects of the nonconformity. The plan may not be right, the deviation from plan may have no effect at all, it may be a one-off and unlikely to recur but on the other hand it may be disastrous and likely to recur unless something is done about it. An evaluation of the need for action is therefore necessary to determine if action should be taken and if so when that action should be taken.

Why is this Necessary?

This requirement responds to the Continual Improvement Principle.

With a multitude of problems to resolve, some method of determining priorities is needed and rather than require the recurrence of all nonconformities to be prevented, the standard requires quite sensibly an evaluation to determine the need for action.

How is this Demonstrated?

In reviewing the nonconformity data, you can rank nonconformities by class or cost so that you reveal the most important problems to tackle. A simple classification is to classify nonconformities on the basis of affecting form, fit or function and this is sufficient for most purposes. However, there are various degrees of fit, form and function. In the automotive industry a severity ranking as shown in Table 36-4 is used when conducting an FMEA.[3]

Although used in the design phase, these criteria can be applied in the production and operational phases to rank nonconformities – those detected before and after shipment. Clearly a nonconformity that falls into any category above Low, (6–10) is a candidate for immediate corrective action. However, many nonconformities may not have these effects and would therefore receive a ranking of 1 but this does not mean that such nonconformities are not important to other processes.

DETERMINING AND IMPLEMENTING ACTIONS (8.5.2d)

The standard requires *the documented procedure for corrective action to define requirements for determining and implementing the corrective action needed.*

What Does this Mean?

The action needed is the action that will eliminate the cause of the nonconformity and therefore prevent its recurrence. You would think that after many years eliminating causes on nonconformity, there would be no nonconformities left, but you would be wrong, primarily because most of what purports to be corrective action is little more than a recovery action. There are countless corrective action procedures being implemented that do not get close to eliminating the cause of nonconformity. They focus on the immediate cause and not the root cause.

Why is this Necessary?

This requirement responds to the Continual Improvement Principle.

---

[3] Potential failure mode and effects analysis, Chrysler, Ford & General Motors (1995).

TABLE 36-4 Failure Severity Ranking

Effect	Severity	Ranking
Hazardous without warning	Very high severity ranking when a potential failure mode affects safe vehicle operation and/or involves noncompliance with government regulation without warning	10
Hazardous with warning	Very high severity ranking when a potential failure mode affects safe vehicle operation and/or involves noncompliance with government regulation with warning	9
Very high	Vehicle or item inoperable, with loss of primary function	8
High	Vehicle or item operable, but at reduced level of performance. Customer dissatisfied	7
Moderate	Vehicle or item operable, but comfort or convenience item(s) inoperable. Customer experiences discomfort	6
Low	Vehicle or item operable, but comfort or convenience item(s) operate at reduced level of performance. Customer experiences some dissatisfaction	5
Very low	Fit and finish/squeak and rattle item do not conform. Defect noticed by most customers	4
Minor	Fit and finish/squeak and rattle item do not conform. Defect noticed by *average* customers	3
Very minor	Fit and finish/squeak and rattle item do not conform. Defect noticed by discriminating customers	2
None	No effect	1

Getting at the root of the problem is crucial to corrective action. Action on the immediate cause is only a palliative – which is a temporary measure. Fixing the immediate cause will result in another nonconformity eventually appearing somewhere else. It is also important to minimize the time taken to isolate and eliminated the root cause in order to minimize the impact on the customer of further deliveries. While you are contemplating the cause of the nonconformity, batches of nonconforming product might be on their way to already dissatisfied customers.

How is this Demonstrated?

It is important to distinguish between four separate actions when dealing with nonconformity:

- Action to remove the specific nonconformity in the nonconforming item (this is addressed by Clause 8.3 covering the control of nonconforming product and referred to as correction or recovery action).

- Action to discover other occurrences of the nonconformity (this should also be covered by the provisions to meet Clause 8.3).
- Action to prevent recurrence in the short term (this is the local action taken on the immediate cause and often referred to as 'containment action').
- Action to prevent recurrence in the long term (this is the action taken on the root cause).

At the time you review a nonconforming product you should consider whether other similar products could be affected. Some of these products might already exist; others may be in the process of being produced and others in the process of being designed. For example, a nonconformity might be that a component was fitted the wrong way round.

1. The first action is to remove the component and fit a new one, the right way round.
2. The second action is to search for other assemblies and replace the component. Some regard this as recovery action because it is searching for like items. Others regard it as corrective action because it is preventing a recurrence of a problem in the same and similar products. There are also some people who regard it as preventive action because although the nonconformity has occurred in a specific product or process, it is only a potential nonconformity for other products or processes.
3. The third action is to display warning notices to alert operators and show them how to identify correct component orientation.
4. The fourth action is to install error-proofing measures and update the generic process FMEA.
5. The fifth action is to introduce into the design process a Process Failure Mode Effects Analysis that will detect when mistake-proofing is needed.

When corrective actions require interdepartmental action, it may be necessary to set up a corrective action team to introduce the changes. Each target area should be designated to a person with responsibility in that area and who reports to a team leader. In this way the task becomes a project with a project manager equipped with the authority to make the changes through the department representatives.

Take care not to degrade other processes by your actions. The corrective action plan should detail the action to be taken to eliminate the cause and the date by which a specified reduction in nonconformity is to be achieved. You should also monitor the reduction therefore the appropriate data collection measures need to be in place to gather the data at a rate commensurate with the production schedule. Monthly analysis may be too infrequent. Analysis by shift may be more appropriate.

Your management system needs to accommodate various corrective action strategies, from simple intradepartmental analysis with solutions that affect only one area, procedure, process or product, to projects that involve many departments, occasionally including suppliers and customers. Your corrective action procedures need to address these situations in order that when the time comes you are adequately equipped to respond promptly.

Before we leave this topic, let us not overlook the most obvious corrective action; that of changing the requirement. It does not always follow that the requirement is correct but it rather depends firstly on when the nonconformity is detected and with whose requirement the product is nonconforming.

If the nonconformity arises from a process the capability of which has been proven, it is highly likely that the nonconformity is due to special cause variation induced by a change in an otherwise stable parameter. However, one should not overlook the possibility that the standards/limits/targets imposed might be far tighter than necessary. If the nonconformity arises in a process, the capability of which has not been formally determined, such as a design process, a purchasing process or recruitment process etc. and it is not an isolated case, the requirement might be too stringent and a relaxation might be the most sensible solution.

If the nonconformity arises from a failure to meet a customer requirement after product approval, it is likely that no change to the requirement would be permitted. However, if arising during development and again it is not an isolated case, there might be a valid argument for changing the requirement. Requirements are sometimes ambiguous, inconsistent or simply not achieveable. Designers sometimes make assumptions and impose limits that are far tighter than needed.

RECORDING RESULTS OF ACTIONS TAKEN (8.5.2e)

The standard requires the documented procedure for corrective action *to define requirements for recording results of action taken.*

What Does this Mean?

If taken literally this requirement means that one need to only record whether the action had the desired effect. But we are interested in not only doing what the standard prescribes but also doing what is necessary for effective process management. Records of all the intentions and actions relative to the elimination of nonconformity should be generated.

Why is this Necessary?

This requirement responds to the Factual Approach Principle.

Records are necessary in this case to chart the trail from nonconformity to root cause and back again to effective action. This is so that you can check through the logic of the analysis and subsequent actions and verify that the planned action was implemented.

How is this Demonstrated?

Through the corrective action process there are several things that should be recorded:

- the results of Pareto analysis;
- the likely causes;
- the root cause;
- the criteria for determining severity or priority;
- the tests conducted to validate the root cause;
- the actions proposed to eliminate the cause;
- the actions taken;
- the results of the actions taken.

In order for it to be possible to verify the actions taken, records need to exist to provide traceability. For example, if your Corrective Action Report (e.g., CAR023) indicates that procedure XYZ requires a change, a reference to the Document Change Request (e.g., DCR134) initiating a change to procedure XYZ will provide the necessary link. The Change Request can reference the Corrective Action Report as the reason for change. If you don't use formal change requests, the Amendment Instructions can cross-reference the Corrective Action Report. Alternatively, if your procedures carry a change record, the reason for change can be added. There are several methods to choose from, but whatever the method you will need some means of tracking the implementation of corrective actions. This use of forms illustrates one of the many advantages of form serial numbers.

REVIEWING CORRECTIVE ACTION TAKEN (8.5.2f)

The standard requires the documented procedure for corrective action *to define requirements for reviewing the effectives of corrective action taken.*

What Does this Mean?

The review of the effectiveness of corrective actions means establishing that the actions have been effective in eliminating the cause of the nonconformity.

Why is this Necessary?

This requirement responds to the Factual Approach Principle.

Every process should include verification and review stages not only to confirm that the required actions have been taken but also that the desired results have been achieved. It is only after a reasonable time has elapsed without a recurrence of a particular nonconformity that you can be sure that the corrective action has been effective.

How is this Demonstrated?

This requirement implies four separate actions:

- A review to establish what actions were taken;
- An assessment to determine whether the actions were those required to be taken;
- An evaluation of whether the actions were performed in the best possible way;
- An investigation to determine whether the nonconformity has recurred.

The effectiveness of some actions can be verified at the time they are taken but quite often the effectiveness can only be checked after a considerable lapse of time. Remember it took an analysis to detect the nonconformity therefore it may take further analysis to detect that the nonconformity has been eliminated. In such cases the Corrective Action Report should indicate when the checks for effectiveness are to be carried out and provision made for indicating that the corrective action has or has not been effective.

Some corrective actions may be multidimensional in that they may require training, changes to procedures, changes to specifications, changes in the organization, changes to equipment and processes – in fact so many changes that the corrective action becomes more like an improvement programme. Checking the effectiveness becomes a test of the system carried out over many months. Removing the old controls completely and committing yourselves to an untested solution may be disastrous therefore it is often prudent to leave the existing controls in place until your solution has been proven to be effective.

The nonconformity data should be collected and quantified using one of the seven quality tools, preferably the Pareto analysis. You can then devise a plan to reduce the 20% of causes that account for 80% of the nonconformities.

Chapter 37

Preventive Action

CHAPTER PREVIEW

This chapter is aimed at all those personnel planning, designing and managing products, processes and systems for it is in these operations that opportunities for preventing problems are identified and their occurrence prevented.

Preventive action mainly takes place during the planning, preparatory and designing operations in the organization but it will also arise in process monitoring. Potential nonconformities might arise due to the inherent nature of the product or the process or its design, production, installation or operation. The way a product has been designed or the way it is intended to produce, install or maintain a product might lead to a failure either during production, installation or maintenance or a failure when in service. (For an explanation of the differences between corrective and preventive action see Part 7 Introduction.)

In this chapter we examine the requirements in Clause 8.5.3 of ISO 9001:2008 and in particular:

- Eliminating the cause of potential nonconformities;
- Determining potential nonconformities;
- Evaluating the need for action;
- Determining and implementing preventive action;
- Recording the results of preventive action;
- Reviewing preventive action.

The positions where the requirements on preventive action feature in the managed process are shown in Fig. 37-1. In the case of products preventive action takes place during planning and in processes it takes place during process design.

ELIMINATING THE CAUSE OF POTENTIAL NONCONFORMITIES (8.5.3)

The standard requires the organization to determine action *to eliminate the causes of potential nonconformities in order to prevent their occurrence and for such actions to be appropriate to the effects of the potential problems.*

What Does this Mean?

The term preventive action was explained in the Introduction to this Part of the Handbook. Action to eliminate a potential nonconformity could be any action taken to anticipate failure in products or processes and remove the cause. If one cannot conceive

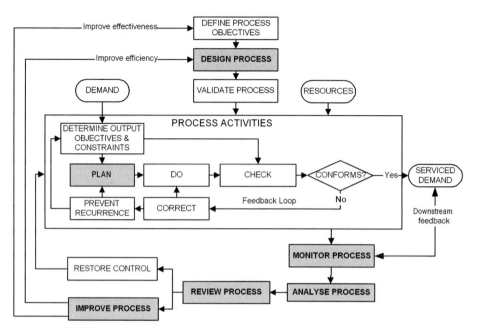

FIGURE 37-1   Where the requirements of Clause 8.5.3 apply in a managed process.

Precautionary Principle

Not having the evidence that something might be a problem is not a reason for not taking action as if it were a problem.

a mode of failure, no action could possibly be taken. This is often regarded as operating within the 'state of the art'. There may be a time in the future when the risk of failure is known but one cannot plan for events about which one has no knowledge. This means of course that if there is evidence in the organization or in the public domain of a mode of failure that could occur with the organization's products or processes, preventive action needs to be taken. The grey areas arise where such evidence remains unproven. Many organizations are reluctant to take action on the basis of opinion especially when the action may be costly. It is an area where the organization's values can be tested and where moral or ethical judgements may need to be made. In the environmental arena there is growing support for the precautionary principle (see box) but you need to apply this principle with caution. A vivid imagination could result in very costly measures being taken to prevent an event that has never happened and is unlikely to happen – but having said that, we didn't think that aluminium would burn until the Royal Navy's 3322 tonnes LSL, Sir Galahad was destroyed by fire in 1982 in the Falklands.

Actions appropriate to the effects of the potential problems mean that the probability of occurrence, the significance of the effects and the certainty of eliminating the cause should be taken into account when deciding on the action required.

Why is this Necessary?

This requirement responds to the Continual Improvement Principle.

Regardless of the attitude of people towards quality, they all desire to be successful. No one really wants to fail but without foresight brought about by instinct, experience, diligence or luck, they will invariably fail. Organizations cannot put their results at risk and expect success to be achieved by chance. They have to take preventive action of the type described above if they are to succeed.

How is this Demonstrated?

The other requirements in this clause identify most of the steps needed to eliminate the cause of a potential nonconformity. However, there are one or two stages that have been omitted. A more complete list of steps is given in Table 37-1: these will be addressed later in this chapter.

> **Food for Thought**
>
> One cannot plan for events about which one has no knowledge.

It makes no sense to have a procedure with the title Preventive Action Procedure because the sequence of actions in Table 37-1 does not occur in isolation. They are part of and embedded within other actions and processes. The saying 'Look before you leap' is a preventive action but developed into a habit by the time we are teenagers. We sometimes forget or are distracted and don't look before we leap but it is not something where we stop and say to ourselves, 'Now what's the next step – oh yes; I must now look before I leap', we just do it without a conscious thought. The first 12 actions in Table 37-1 might form the basis of an FMEA procedure with the other 12 actions being implemented by other processes. For example, the records will be generated from planned verification activities, not special preventive action activities. The diagnostic and implementation teams will not be special teams but the normal design and production staff acting in their normal roles. There may be a need on occasions for a special task force to resolve a particularly difficult problem but most of the time preventive action will be performed as a routine part of your job.

Preventive action provisions may therefore be embodied in the following types of procedures:

- Business planning procedures,
- Design planning procedures,
- Production planning procedures,
- Resource planning procedures,
- Risk assessment procedures,
- Hazard analysis procedures,
- Training procedures,
- Performance analysis procedures,
- Design review procedures,
- Design analysis procedures (reliability, safety, maintainability etc.),
- Supplier or subcontractor performance review procedures,
- Management review procedures.

The actions in Table 37-1 could be included in a general planning guide. It would be unwise to make them policies because they describe a general methodology and there will be occasions when not all apply.

TABLE 37-1 Steps in the Preventive Action Process

Step	Action	Clause 8.5.3 requirement
1	Determine the objectives of the product, process, task or activity	Determining potential nonconformities and their causes
2	Organize a diagnostic team	
3	Perform an analysis to determine the factors critical to the achievement of these objectives	
4	Determine how the factors might act to adversely affect the product, process, task or activity (the mode of failure)	
5	Determine the potential effect of such condition on the achievement of the objectives	
6	Determine the severity of the effect on meeting the objectives	
7	Assess the probability of this condition occurring	
8	Postulate causes and test theories	
9	Determine the root cause of potential nonconformity	
10	Identify the provisions currently in place that will prevent this adverse condition occurring or detect it before it has a detrimental effect on performance	Evaluating the action needed
11	Assess the probability that these provisions will prevent the occurrence of this condition or of detecting it before it has a detrimental effect on performance	
12	Determine any additional action needed to prevent the occurrence of the potential nonconformity	
13	Organize an implementation team	Implementing the action needed
14	Create or choose the conditions which will ensure effective implementation	
15	Implement the agreed action	
16	Record the results of all the analysis	Records of results
17	Record the causes of potential nonconformity	
18	Record the criteria for determining severity or priority	
19	Record the proposed actions to be taken	
20	Record the actions actually taken	
21	Determine whether the actions actually taken were those required to be taken	Reviewing preventive actions
22	Determine whether the actions were performed in the best possible way	
23	Determine whether the nonconformity has occurred	
24	If nonconformity has occurred, undertake corrective action and review the preventive action methods	

DETERMINING POTENTIAL NONCONFORMITIES (8.5.3a)

The standard requires a documented procedure that defines requirements *for determining potential nonconformities and their causes.*

What Does this Mean?

Potential nonconformities are those that have not occurred, therefore the determination of potential nonconformities is a quest to discover inherent characteristics of products and processes that if not changed will eventually result in actual nonconformities. As a nonconformity is non-fulfilment of a requirement, it follows that any requirement placed on or within the organization that may not be fulfilled provides the opportunity for a potential nonconformity. As indicated above, the determination of potential nonconformities involves the analysis and evaluation of risk. Once risks are known and the effect on the product, process or organization established, action could be taken to eliminate, reduce or control these risks.

Why is this Necessary?

This requirement responds to the Continual Improvement Principle.

The prevention of potential nonconformities cannot be relied on to occur by chance – it has come about by systematic analysis of products and processes and their interrelationships both inside and outside the organization.

How is this Demonstrated?

As indicated in Table 37-1, there are nine steps to determining potential nonconformities and their causes. Each step will now be addressed briefly.

Determining Objectives, Requirements and Constraints

As nonconformity is non-fulfilment of a requirement, the first step is to determine what those requirements are. In some cases they will be product requirements, in others they will be process objectives or at a higher level the corporate objectives or quality objectives. They may also be requirements that act as constraints on other activities. For example, environmental requirements apply only if the operations you undertake impact the environment. If the operations do impact the environment, the requirements constrain what you can or cannot do.

It is necessary when considering the requirements for preventive action to avoid limiting your imagination to products because the potential for failure is just as present in the organization and its processes as in products – in fact it is these very weaknesses that are likely to be the cause of potential product nonconformities. Such objectives therefore will include the following:

- Corporate objectives (Clause 5.4.1),
- Product requirements (Clause 7.2.1),
- Process objectives (Clause 4.1c).

As the product will comprise units, components, materials etc. each of these will be defined by requirements. As each process will comprise sub-processes, tasks and activities, each of these will be defined by objectives and input or entry requirements.

Organizing Diagnostic Teams

As indicated previously, specially organized diagnostic teams may not be needed because the prevention of nonconformity is the job of the product or process design team. It is usually a one-off activity. Once the provisions are put in place to prevent nonconformity, the team moves on to other things. At the corporate level, a team of senior managers or a research team may be needed to analyse the organization as a whole to discover risks to achievement of its objectives. At the sub-process and task level a team approach may also be necessary but at the activity level, it is the individual who should assess the risks, anticipate problems and either put in place provisions to prevent failure or take precautions.

Analysis of Critical Success Factors

With every organization, process or product there are some critical factors on which their success depends. The success of a book is not whether there is an unwanted printing mark on a page – it is whether the book lives up to the reader's expectations and that is more to do with the substance than with materials. The success of electronic equipment is dependent on appearance, function and reliability. The success of an automobile depends on appearance, function, safety, reliability and maintainability. In most cases these factors will be defined in the product specification and will be the functions that the product is required to perform. With processes, success may depend on throughput, resource consumption, traceability and/or response. With the organization, success may depend on market intelligence, retaining competent people, short product development timescales, the quality of conformity or service standards – these will be the organization's objectives.

Answering the key question – 'What affects our ability to fulfil our mission or to achieve our objectives?' reveals the critical success factors.

Failure Modes

By taking each factor, one at a time and asking one or more of the following questions you develop a series of failure modes – the outward appearance of a specific failure effect[1]:

- How might this part or process fail to meet the requirements?
- What could happen which would adversely affect performance?
- What would a stakeholder consider to be unacceptable?

Previous experience is a good starting point for determining what could go wrong but often these failures only arise under certain conditions. When a product is stressed by being subject to extreme environmental conditions, it may fail. When a process is overloaded or under-resourced or the operators are put under pressure, certain failures might occur. This can be presented as a Fault Tree diagram that describes the combination of events leading to a defined product or process or organization failure. For a description of this analysis see footnote 1.

Failure Effects

For each failure mode, determine the effect of the failure through the hierarchy to the end product, the system and the external interfaces. The effect should be described in terms of what the customer might experience or the impact on overall organizational performance. A component may distort under load resulting in a failure of the assembly of which it is a part and consequently cause customer dissatisfaction. A process may deliver output late that in turn injects delays into other processes, which ultimately causes the organization

---

[1] Smith, David (1997). *Reliability, Maintainability and Risk Fifth Edition,* Butterworth Heinemann.

TABLE 37-2 Failure Occurrence Ranking

Rank	Cpk	Failure rate	Probability of failure
10	>0.33	>1 in 2	Very high:
9	>0.33	1 in 3	failure almost inevitable
8	>0.51	1 in 8	High: repeated failures
7	>0.67	1 in 20	
6	>0.83	1 in 80	Moderate: occasional failures
5	>1.00	1 in 400	
4	>1.17	1 in 2000	
3	>1.33	1 in 15 000	Low: relatively
2	> 1.50	1 in 150 000	few failures
1	> 1.67	<1 in 1 500 000	Remote: failure is unlikely

to lose a customer. The omission of safety checks could endanger operators and subsequently put the line out of service or in breach of government regulations. The omission of records could leave a process down stream without necessary input data. This in turn could result in decisions being made without the full facts being available thus impacting work priorities that divert effort away from key projects.

Severity

Once the effects of failure have been identified, it does not follow that you have to act on all of them. Some will have more impact than others and the next step is therefore to assess the severity of the effect. A convention for ranking failure severity in the automotive industry is shown in Table 8-5.

Probability of Occurrence

A potential nonconformity is one that could occur but the probability of occurrence might be as low as 1 in 1,000,000 or as high as 1 in 2. With electronic components, standard failure rate data can be used to predict the probability of failure. With other failure effects you will have to rely on past experience and this is where the records from previous projects are useful. Simply asking the question – 'When did we last experience this type of problem?' might bring forth useful information.

A convention for ranking the probability of occurrence in the automotive industry is shown in Table 37-2.[2]

Determining Cause

The diagnostic effort needed to determine the potential cause of the identified mode of failure can be considerable and therefore it is prudent to determine the probability of occurrence first so that the diagnostic effort can be focused on the failures with highest probability of occurrence.

---

[2] From Potential failure mode and effects analysis, Chrysler, Ford & General Motors (1995).

Determining the Root Cause

Here once again the '5 Why's' technique is useful to get at the root cause. In some cases it may be necessary to experiment using Design of Experiments (DOE) technique in order to determine which causes are the major contributors.

EVALUATING THE NEED FOR ACTION (8.5.3b)

The standard requires a documented procedure that defines requirements *for evaluating the need for action to prevent occurrence of nonconformities.*

What Does this Mean?

The likelihood that an event might occur is not a command to take action for provisions may already be in place to prevent its occurrence, reduce or control its effects. The requirement therefore means that in evaluating the need for action, the existing provisions should be evaluated and on this basis, requirements for action to prevent the occurrence of nonconformity should be determined.

Case study – Preventive action procedures

During a recent audit to ISO 9001:2008, we explained to the auditor that we use risk assessment and failure modes analysis in our planning processes and have documented procedures covering the conduct of these activities. However, when shown the procedures he did accept these as complying with the planning requirements of Clauses 5.4.2 and 7.1 but not the requirement for a procedure for preventive action. Can this be right?

It rather depends upon the scope of your planning processes. Risk assessment and Failure Modes and Effects Analysis are indeed techniques that if used properly will prevent potential nonconformity. The auditor might have found evidence that you were not using these techniques in all processes.

Planning is itself a preventive action if you anticipate what could go wrong and put in place measures to eliminate, reduce or control such events. However, there are several types of planning – strategic planning, process design planning, product design planning, manufacturing planning, installation planning, maintenance planning etc. If these techniques are only used in product design the auditor was right to question your preventive action procedures.

However, simply having one preventive action procedure would not be the right answer unless it is plugged into every planning, design and review activity at strategic, process and product level. But a procedure that had this degree of flexibility would be more like a set of guidelines. If you are thinking that a formal procedure might be required that involves completing a form identifying the problem, the cause and the action taken, you will have great difficulty getting this accepted in all areas. Those organizations that have gone down this route limit the use of the form to specific processes. For processes where the formality is inappropriate, specific guidance can be written into the procedures or instructions associated with planning, design and review.

If you can show that each process is periodically reviewed to seek better ways to achieve the process objectives and in doing so you carry out a risk assessment, you would have adequate preventive action procedures in place.

Why is this Necessary?

This requirement responds to the Continual Improvement Principle.

The existing provisions may not adequately prevent the occurrence of nonconformity and therefore it is necessary to determine the additional actions that are needed. Events are not always mutually exclusive and therefore provisions designed to eliminate, reduce or control one failure mode may well eliminate, reduce or control others.

How is this Demonstrated?

Adequacy of Current Provisions

All existing processes should include provisions for verifying output and monitoring progress even if they are informal. However, if the processes have evolved over a long period it is also likely that there will be many more controls in place. Every requirement of previous versions of the standard will, if implemented, impose control over the process. Such controls should have been installed to prevent failure but many were often installed just to meet the requirements of the standard and get the *badge on the wall*. If you now re-examine these controls from the perspective of establishing the failure modes they prevent, you will at least give justification to those that serve a useful purpose and provide an action list for those that don't.

The inherent characteristics of a product or process that will reduce the probability of occurrence should be identified. For a product these might be safety factors, component redundancy, error-proofing or specifying high reliability components. For processes these might be error-proofing, inspections, tests, audits, identification techniques, skill training, peer reviews etc.

Existing controls can be classified as those that prevent occurrence and those that detect occurrence. Table 37-3 illustrates the difference. There are two types of detection measures – those that detect a failure and lead to a permanent change and those that detect a failure and lead to a restoration action such as to reset the product or process or to replenish consumables.

TABLE 37-3 Contrasting Preventive and Detection Measures

Preventive measures	Detection measures
Planning	Material testing
Training	Component Inspection
Safety factors	Assembly testing
Component derating	Prototype testing
Component redundancy	Receipt inspection
Product identification	Design review
Error-proofing	Peer review
Warning notices	Alarms
	Alerts

TABLE 37-4 Ranking the Probability of Detection

Detection	Criteria	Rank
Absolute uncertainty	Design Control will not and/or cannot detect a potential cause/mechanism and subsequent failure mode; or there is no Design Control.	10
Very remote	Very remote chance that the Design Control will detect a potential cause/mechanism and subsequent failure mode.	9
Remote	Remote chance that the Design Control will detect a potential cause/ mechanism and subsequent failure mode.	8
Very low	Very Low chance that the Design Control will detect a potential cause/ mechanism and subsequent failure mode.	7
Low	Low chance that the Design Control will detect a potential cause/mechanism and subsequent failure mode.	6
Moderate	Moderate chance that the Design Control will detect a potential cause/mechanism and subsequent failure mode.	5
Moderately high	Moderately High chance that the Design Control will detect a potential cause/mechanism and subsequent failure mode.	4
High	High chance that the Design Control will detect a potential cause/mechanism and subsequent failure mode.	3
Very high	Very High chance that the Design Control will detect a potential cause/mechanism and subsequent failure mode.	2
Almost certain	Design Controls will almost certainly detect a potential cause/mechanism and subsequent failure mode.	1

Probability of Detection

You can go a step further and rank the probability that the existing provisions will detect the failure. The ranking could range from 'Certainty of detection' to 'Absolute uncertainty of detection' with variations in between. Examine the existing provisions to see whether there are any alarms, alerts or reporting arrangements that would bring potential problems to the attention of management.

What may appear trivial on a case-by-case basis may well be significant when taken over a longer period or a larger population. Determining this deterioration requires some detective work that focuses on processes and not on specific products. Managers have a habit of reacting to events particularly if they are serious nonconformities in the form of a customer complaint. What we are all poor at is perceiving the underlying trends that occur daily and gradually eat away at our profits. If there is no means of alerting management to these trends clearly something is missing.

A convention for ranking the probability of detection in the automotive industry is shown in Table 37-4.[3]

There are some problems with this method as it is subjective. It comes down to one person's guess as to the likelihood of detection. Where a product is designed with sensors, the chances to detection can be calculated more precisely using component failure rates. But with business and work processes it is often up to the personnel running the process. For example, currently there is no process that will detect swarf inside an oil channel unless an endoscope can travel every channel and into every crevice. We rely on cleaning processes and manual observation of particles on a filter to determine whether the channels are free of swarf.

Risk Priority Number (RPN)

The RPN is the product of three numbers, the severity (S), occurrence (O) and detection (D). The ideal situation is where the product of $S \times O \times D = 1$. Any severity ranking of 10 should be eliminated by design. Any result above 50 should require action and therefore the priority needs to be given to those failures ranked above 50. However, RPN is a qualitative assessment of risk and takes no account of cost.

Financial Impact

There is merit in extending the FMEA to include financial impact because there is not a linear relationship between RPN and life cycle cost. There might be some failures that are less costly to prevent or detect in process than to design out of the product. There might also be other failures that are more costly to prevent or detect in process due to the initial cost of equipment and recurring maintenance costs. A third scenario is where one solution to the reduction in RPN has a lower impact on product cost but higher impact on life cycle cost.[4] Traditional FMEA identifies a failure mode and its effect and postulates a cause and its probability of occurrence. The weakness with this approach is that it does not permit the investigation of several causes and effect chains each having a different probability of occurrence and thus a different solution in terms of cost.

DETERMINING AND IMPLEMENTING PREVENTIVE ACTION (8.5.3c)

The standard requires a documented procedure that *defines requirements for determining and implementing action needed to prevent occurrence of nonconformities.*

What Does this Mean?

Action to prevent a potential nonconformity may be in the form of a redesign of a product or process, the introduction of new routines, precautions, procedures, techniques or

---

[3] From Potential failure mode and effects analysis, Chrysler, Ford & General Motors (1995).
[4] Scenario based FMEA: Life cycle cost perspective. Kemta and Ishii, 2000 ASME Design Engineering Conference.

methods or involve changing the behaviour of personnel including management. The work environment may be the cause of some problems – changing it should not be ruled out.

Why is this Necessary?

This requirement responds to the Continual Improvement Principle.

The most cost-effective action one can take in any organization is an action designed to prevent problems from occurring. It is always cheaper to do a job right the first time rather than do it over.[5] Preventive action therefore saves money even though there is a price to pay for the discovery of potential nonconformities. Once the analysis has been performed a few times, a pattern will begin to emerge where provisions made to handle one situation will prove suitable for other situations and so it will get easier until the analysis almost becomes a habit.

How is this Demonstrated?

The action necessary to eliminate, reduce or control the effects of a potential noncon-formity may be as simple as applying existing techniques or methods to a new product or process. In other cases it might involve designing new techniques and methods – something that may require additional resources and a development team. If the solution requires the involvement of more than one function, the formation of a multidisciplinary team may be necessary. This is not the same as a multifunctional team where representatives from each function meet to discuss a problem then go back to their departments and get on with what they were doing. A multidisciplinary team comprises people of different disciplines, brought together to pool their skills and knowledge. They may all be from the same function but the focus is on getting the problem solved, not playing departmental politics.

The steps you need to take to deal with specific problems will vary depending on the nature of the problem. The part that can be proceduralized is the planning process for determining the preventive action needed. A typical process may be as follows:

- Devise a strategy for eliminating the cause together with alternative strategies, their limitations and consequences.
- Gain agreement on the strategy.
- Prepare an improvement plan which if implemented would eliminate the potential problem and not cause any others.
- Prepare a timetable and estimate resources for implementing the plan.
- Gain agreement of the improvement plan, timetable and resources before going ahead.
- Calculate the Potential RPN for the failure mode.

Some plans may be very simple and require no more than an instruction to implement an existing procedure. Others may be more involved and require additional resources. By incorporating the actions into a formal improvement plan you are seen to operate a coherent and co-ordinated improvement strategy rather than a random and unguided strategy. While those on the firing line are best equipped to notice the trends, any

---

[5] Crosby, Philip (1979). *Quality Is Free*, McGraw-Hill.

preventive action should be co-ordinated in order that the company's resources are targeted at the problems that are most significant.

Recalculating the RPN

When the improvement plan has been devised the RPN needs to be recalculated. Redesign of the product or the process may well reduce the probability of occurrence but the severity of failure might remain the same. By increasing the probability of detection the RPN may be reduced well below the original RPN indicating that the planned action will bring about the necessary improvement. However, the important thing is that the planned action is taken and provision made with the associated documentation to indicate as much.

Error-proofing

Error-proofing is a means to prevent the manufacture or assembly of nonconforming product. Errors are unintentional unplanned events in the design, planning or production of a product or delivery of a service. Machines don't run forever without attention. We occasionally forget things and we can either make machines and actions error-proof or we can provide signals to alert us of unintentional events that are about to happen, are happening or have happened.

Error-proofing can be accomplished by product design features in order that the possibility of incorrect assembly, operation or handling is avoided. In such cases the requirements for error-proofing need to form part of the design input requirements for the part. The design FMEA should be analysed to reveal features that present a certain risk that can be contained by redesign with error-proofing features.

Error-proofing can also be accomplished by process design features such as sensors to check the set-up before processing, stop the machine if an abnormal situation arises (jammed mechanism or defect part), trigger audible signals to remind operators to do various things and stop the machine when the correct number of items have been produced (see Autonomation in the Glossary). However, signals to operators are not exactly error-proof, only a mechanism that prevents operations commencing until the right conditions have been set is proof against errors. In cases where computer data entry routines are used, error-proofing can be built into the software such that the operator cannot bypass a stage.

RECORDING THE RESULTS OF PREVENTIVE ACTION (8.5.3d)

The standard requires a documented procedure that *defines requirements for recoding of results of action taken to prevent occurrence of nonconformities.*

What Does this Mean?

If taken literally this requirement means that one need only record whether the action had the desired effect and as we are addressing potential nonconformities, a record of the absence of nonconformities would seem to address this issue. However, in addition (and not in place of), other records are necessary for effective process management. Records of all the intentions and actions relative to the elimination of potential nonconformity should be generated.

Why is this Necessary?

This requirement responds to the Factual Approach Principle.

Records of the actions taken en route to discover and eliminate potential nonconformities serve a number of important uses:

- They show due diligence and consideration to matters that affect customer satisfaction and compliance with regulations.
- They can be used as evidence in any prosecution against the organization.
- They provide a basis for comparison of actual nonconformity against predicted nonconformity and therefore a means to improve product and process design techniques.

How is this Demonstrated?

Through the preventive action process there are several things that should be recorded:

- the objectives or requirements of the product, process or organization;
- the critical success factors;
- the modes of failure, their cause and effect, their severity and probability of occurrence (an example of a design FMEA is shown in Fig. 37-2.[6] Note that the same form can be used for both design and process FMEA);
- the criteria for determining severity;
- current provisions to detect nonconformity;
- the probability that current provisions will detect the nonconformity;
- the actions proposed to eliminate the cause;
- the actions taken;
- the results of an analysis of nonconformity data showing the effectiveness of actions taken.

REVIEWING PREVENTIVE ACTIONS (8.5.3e)

The standard requires a documented procedure that *defines requirements for reviewing the effectiveness of the preventive actions taken.*

What Does this Mean?

The review of preventive actions means establishing that the actions have been effective in preventing the occurrence of the nonconformity.

Why is this Necessary?

This requirement responds to the Process Approach Principle.

A lot of effort goes into preventing problems and therefore it is necessary to periodically review results to establish whether this effort is being effectively applied. Is the

---

[6] From Potential failure mode and effects analysis, Chrysler, Ford & General Motors (1995).

FAILURE MODE & EFFECTS ANALYSIS - DESIGN PROCESS

Subsystem Passenger airbag system	Part/AssemblyProcess Airbag	Specification/Revision GL-50-265-1789 G	Supplier ASSL	FMEA Ref/Issue DF2986-B
Function Crash Protection	Core Team G Hunt, R Tuesday, J May, S Samadi	Design Responsibility R Tuesday	Project Grampian	Date 20/10/2000 — Page 1 of 3
				Prepared by G Hunt

No.	Item function or Process stage/ operation	Potential failure mode	Effect of failure	Cause of Failure	Current Controls	Current Status				Recommended Action	Actionee	Revised Status				
						O	S	D	RPN			Action Taken	O	S	D	RPN
1	Inflate Airbag	Bag does not open on impact	Passenger injured	Malfuntion of sensor	Sensor failure warning light	2	8	6	96	Add redundant sensor in parallel	Designer	2nd sensor circuitry added 7/01/2001	1	8	2	16
2	Restrain passenger	Occupant unable to withstnd inflation pressure	Injure lightweight occupant	Occupant not wearing seat belt	Seat belt warning light	4	8	10	320	Install switch to deactivate airbag unless seat belt is fastened	Designer	Switch and circuitry added 7/01/2001	1	8	3	24
			Bruise medium weight occupant	Force regulator malfunction	Regulator life testing	2	3	3	18	None	None	None				0

FIGURE 37-2 Example of a design FMEA.

effort focusing on the right things or does it repeatedly fail to prevent any significant problems occurring?

How is this Demonstrated?

This requirement implies four separate actions:

- A review to establish what actions were taken.
- An assessment to determine whether the actions were those required to be taken.
- An evaluation of whether the actions were performed in the best possible way.
- An investigation to determine whether the nonconformity has occurred.

The first action is to trace forward from the failure analysis in order to locate evidence that the planned action was taken. The action taken may not be the same as planned simply because a better solution emerged and the FMEA was not updated (a common weakness but not drastic). The review should cause the FMEA to be updated but a more important issue is whether the *better solution* had the desired effect. An analysis of performance over a set time interval may reveal the evidence. It will show whether there are nonconformities occurring and if so, the preventive action has not been 100% effective. However, a study of actual nonconformities will reveal whether:

- The root cause matches that in the FMEA;
- The nonconformity is one that had been anticipated;
- The solution merely created another problem.

The fact that nonconformities occur should not necessarily be cause for despair. If a process is very mature and the organization stable, there perhaps should be no nonconformities. But life is never thus. There is always change and some of it we can predict and some we cannot. A lot of what is needed to prevent potential nonconformity is to do with imagination, knowledge and commitment. You need imagination to postulate the modes of failure, knowledge to confirm your suspicions and isolate the causes and commitment to do something about it, especially where cost is involved and it is uncertain as to whether nonconformity will occur. There are costs versus benefits and often the benefits are external to the organization rather than internal such as protection of the environment and safety of people. The management has to balance the costs and make value judgements which if they share good values, the decisions will always fall in favour of the external parties (employees are external parties in this situation).

Key Messages from Part 7

1. You know nothing about an object until you can measure it, but you must measure it accurately and precisely.
2. You can derive standards for anything - the only proviso is that they are agreed by those who benefit from the measurements. Therefore if you set out to measure customer satisfaction the standard used needs to be agreed with the customer.
3. Use - not function - determines need for calibration.
4. The product specification is but an interpretation of what constitutes fitness for use. Out of 'spec' doesn't mean unsafe, unusable, un-saleable etc.
5. Both customers and managers have a need for an assurance of quality because they are not in a position to oversee operations for themselves. They need to place trust in the producing operations, thus avoiding constant intervention.
6. The control of quality depends on an ability to measure quality be it the quality of products, services, processes, systems, organizations or simply the quality of actions and decisions. As quality, like beauty is abstract we need to translate it into measurable characteristics to control it.
7. Before we measure we need standards, targets, requirements etc. we can use to judge the results of measurement, in fact there is little point measuring anything unless we have a clear idea of what we are looking for.
8. Invariably, managers monitor and measure outputs, react to the results and start tinkering with the process and make thing worse, when what they observed was not random but the effect of a specific problem with an assignable cause.
9. Stable conditions exist when all variation is under statistical control. This means that all variation is then due to common causes only and none due to assignable or special causes.
10. The most important part of a customer survey is to ask the right questions and these need to be derived from what customers expect from your products not what you want from them.
11. It is a common practice for management to set goals like undertaking a commitment to meet a customer requirement without knowing whether the systems in place are capable of fulfilling those goals.
12. If we begin with the end in mind, we would start an audit by wanting to know if the system was effective, and only then establish that it conformed to ISO 9001.
13. Many of the one to two day auditing courses are no more than tutored reading. An in-depth understanding is what is required and this comes from application followed by examination.

14. It is processes that produce products and therefore measuring products tells us whether the products are correct but does not tell us whether the process is achieving its objectives.

15. A person may pass through 10 stages in a hospital but you cannot aggregate the errors to produce a process yield based on stage errors. Patients don't drop out of the process simply because they were kept waiting longer than the specified maximum.

16. Keeping the process under control is process control. Keeping the process within the limits of the customer specification is quality control.

17. Characteristics that do not vary only need to be checked once. For example a car designed with four wheels could not possibly emerge with two, three or five wheels when put into production, but a body panel with screw inserts could emerge from the process without the inserts!

18. Wherever there is a measurement to be made, a parameter or phenomenon to be monitored, there is always a sensor employed to sense the variation. The key to effective measurement is plugging the right sensor into the right location.

19. The only cases where you need to request concessions from your customer are when you have deviated from one of the customer requirements and cannot make the product conform.

20. If a system allows the shipment of defective products every day, loses one in three customers, allows the development of unsafe products to reach the market, or the failure of a revolutionary power plant, you would probably say that system was pretty ineffective. But if a system enables management to stop the development of products for which there was no requirement, discover a potential safety problem, anticipate customer needs ahead of the competition, cut waste by 50%, successfully defend a product liability claim, meet all the delivery targets agreed with the customer, you would probably say that system was pretty effective.

21. We cannot improve anything unless we know its present condition and this requires measurement and analysis to tell us whether improvement is both desirable and feasible.

22. If the product specification identifies 400 requirements, you have at least 400 opportunities for improvement.

23. Nonconformities are caused by factors that should not be present in a process.

24. The saying 'Look before you leap' is a preventive action but developed into a habit by the time we are teenagers. We sometimes forget or are distracted and don't look before we leap but it is not something where we stop and say to ourselves, 'Now what's the next step?'

System Assessment Certification and Continuing Development

INTRODUCTION TO PART 8

Almost every organization that uses standards in the ISO 9000 family will either be intending to pursue ISO 9001 certification or be an organization assisting or auditing such organizations. This part of the handbook provides some self-assessment tools that can be used prior to the external assessment and for continual improvement. There is also an overview of the certification process together with suggestions for developing the system beyond certification.

Chapter 38

Self Assessment

CHAPTER PREVIEW

This chapter is aimed at everyone in the organization to help them assess their own objectives, processes and activities and identify opportunities for improvement.

This is a chapter of lists and tables designed to check conformity, test understanding, stimulate thought and provoke improvement. We include:

- A simple maturity grid based on the quality management principles;
- A set of questions to test to verify that all system and process elements are in place;
- A list of all requirements in ISO 9001 by section that can be used to check that all have been addressed;
- A series of questions by section of ISO 9001 to test understanding and stimulate debate.

SYSTEM MATURITY

Table 38-1 is a maturity grid to test where your organization is relative to the eight quality management principles. An organization fully committed to quality would be at maturity level III. Simply score your organization against the criteria placing your score at the most appropriate level. For example, if you believe that there is no proactive process for understanding customer needs, you would place a 1 in Level I. There is a range (e.g., 4–6) to allow for 'not sure, might be, some but not all' responses etc.

SYSTEM ASSESSMENT

Table 38-2 contains a series of questions that will provoke an analysis of various aspects of the organization into finding answers. They can be used to test the robustness of processes or for audit. There are no right or wrong answers only answers that work for your organization except the answer 'don't know'. If you truly don't know the answer, make it your job to find out.

SYSTEM REQUIREMENT CHECKLIST

In this section are lists of requirements numbered from 1 to 263 representing the individual topics addressed by the clauses of the standard. This list could be used to build an ISO 9001 Exposition or Compliance Table.

TABLE 38-1 Quality Management Principles – Self Assessment

PRINCIPLE	MATURITY LEVEL		
	Level I (1–3)	Level II (4–6)	Level III (7–10)
1. **Customer focused organization** Understanding customer needs and expectations	No proactive process for understanding customer needs ☐	A proactive process exists but not in the QMS ☐	The process is fully integrated into the QMS ☐
2. **Leadership** Creating a unity of purpose and a quality culture	No clearly defined and communicated organization purpose, values and objectives ☐	We know where we are going but we are not all pulling in the same direction ☐	Everyone understands the organization's purpose and objectives and are motivated and supported to achieve them ☐
3. **Involvement of people** Developing and motivating the people	People are just another esource to be used to achieve our results ☐	We involve everyone in decisions that affect them ☐	We value our people and achieve our results through team work ☐
4. **Process approach** Managing processes effectively	We have a set of random task-based procedures that are independent of the business objectives ☐	We have departmental processes that serve departmental goals ☐	We design our processes to meet the organization's objectives and continually measure, review and improve their performance ☐

#			
5.	**System approach to management** Understanding interactions and interdependencies	Our system for achieving quality is organized around the clauses of ISO 9001 ☐	We have formalized our operational processes so that they deliver conforming product ☐	We have a system of processes that enables us to manage the interactions between processes and so deliver the desired organizational outcomes. ☐
6.	**Continual improvement** Continually seeking better ways of doing things	Continual Improvement is perceived as correcting mistakes only ☐	Continual Improvement is perceived as responding to problems ☐	Continual Improvement is perceived as proactively seeking opportunities to improve performance in everything we do ☐
7.	**Factual approach to decision making** Basing decisions on facts	We don't use any data generated by the QMS to make business decisions ☐	We mainly use audit data, customer complaints and nonconformity data as inputs to decision making ☐	We base our decisions on process performance data generated by the management system ☐
8.	**Mutually beneficial supplier relationships** Realizing that you need others to succeed	We treat our suppliers as adversaries and keep them at arms length ☐	We work with our suppliers to improve our overall performance ☐	We involve our key suppliers in our future strategy ☐
	Total			

TABLE 38-2 System Assessment Questions

To discover the:	Ask:	Notes
Mission/Purpose	What is the organization trying to do?	Also involves asking What is our business? What have we been set up to do? Who is our customer? Where is our customer? What does the customer buy? What is value to the customer? Which customer needs are unfulfilled?
Vision	What the organization should look like as it successfully implements and fulfils its mission?	A vision of success allows people to take action without constant oversight by leaders and managers.
Values	What principles will guide us on our journey?	These are corporate values not personal values – these values characterize the culture in the organization and answer the questions: What is it like to work in this organization? What do we stand for? How do we want to conduct our business?
Stakeholders	For whom does the organization exist?	Primarily this will be customers and investors but ultimately if employees, suppliers and society do not want it, the organization will not survive
Business outcomes	What do the stakeholders want from the organization?	These are the stakeholder needs and expectations
Stakeholder success measures	What will the stakeholders look for to assess if their needs have been met?	Obtained by filtering the mission through each stakeholder These are the Key performance indicators
Critical Success Factors	What factors affect our ability to accomplish our goals?	Ask this question of the mission to reveal factors that might be turned into a set of corporate values and strategic objectives. Also ask: What things must we do to be successful? What criteria must we meet to be successful?
Business outputs	What outputs will deliver successful outcomes?	You can't control a process by measuring outcomes as they arise long after the process has delivered its output.
Performance Targets	What criteria will indicate whether our performance is acceptable?	These are the standards you need to achieve
Business Processes	What processes deliver the business outputs?	There are probably only four of these so if you discover more, it is likely that two or more have a common objective

Continued

TABLE 38-2 System Assessment Questions — cont'd

To discover the:	Ask:	Notes
Process purpose	What is the main function of this process?	Derived from an assessment of what the process outputs have in common – the essence
Process activities	What affects our ability to deliver process outputs?	If you are doing something that does not affect your ability to deliver a successful output – stop doing it
Risks	How could this process fail to achieve its objectives?	These are the failure modes or hazards inherent in the process that need to be eliminated, reduced or controlled
Process output	What outputs would we look for as evidence that the process objectives have been achieved?	Must match a stakeholder need
Process measures	What parameters will we control to ensure delivery of the process outputs?	These are the subjects for which you apply the universal steps of quality control
Process constraints	How do the expectations of the other stakeholders impact (influence or constrain) the process for achieving customer expectations?	The other stakeholders are investors, employees, suppliers and society
Competence	What skills, knowledge and behaviours are needed to produce these outputs?	The results that people can deliver under the stipulated conditions are more important than what people say they know or can do or once did
Distinctive competences	What abilities enable us to perform well against the critical success factors or KPIs?	A competency that is difficult for other organizations to replicate and so is a source of enduring advantage
Performance review	How do we know we are doing things right?	Process outputs are reviewed resulting in improvement by better control
Efficiency review	How do we know we are doing things in the best way?	Practices are reviewed resulting in improvement by better utilization of resources
Effectiveness review	How do we know we are doing the right things?	Objectives are reviewed resulting in improvement by better understanding of stakeholder needs

QUALITY MANAGEMENT SYSTEM REQUIREMENTS

General Requirements

1. Establishing and implementing a documented quality management system
2. Implementing a documented quality management system

3. Maintaining a documented quality management system
4. Continually improving the effectiveness of the QMS
5. Determining processes
6. Determining process sequence and interaction
7. Determining criteria and methods for operation and control of processes
8. Availability of resources necessary to support the operation and control of processes
9. Availability of information necessary to support the operation and control of processes
10. Monitoring, measurement and analysis of processes
11. Restoring the status quo
12. Continual process improvement
13. Process management
14. Outsourcing processes
15. Responsibility for outsourced processes

Documentation Requirements

General

16. Documenting the quality policy and objectives
17. Documenting the quality manual
18. Documenting the quality management system procedures
19. Documenting the information needed for the effective operation and control of processes
20. Documenting records

Quality Manual

21. Establishing and maintaining a quality manual

Control of Documents

22. Controlling documents required by the QMS
23. Establishing document control procedures
24. Approval of documents
25. Review and revision of documents
26. Re-approval of documents after revision
27. Identifying revision status
28. Availability of documents
29. Identification and legibility of documents
30. Control of documents of external origin
31. Control of obsolete documents
32. Identifying obsolete documents retained for use

Control of Records

33. Establishing and maintaining records of conformity
34. Establishing and maintaining records of effectiveness
35. Ensuring legibility of records

36. Ensuring identification of records
37. Retrieval of records
38. Establishing a records control procedure

MANAGEMENT RESPONSIBILITY REQUIREMENT

Management Commitment

39. Evidence of management commitment to developing a QMS
40. Evidence of management commitment to continually improving the effectiveness of the QMS
41. Evidence of communicating the importance of meeting customer and regulatory requirements
42. Evidence of management establishing the quality policy
43. Evidence of management establishing quality objectives
44. Evidence of management commitment by conducting management reviews.
45. Evidence of management commitment by ensuring the availability of necessary resources.

Customer Focus

46. Determining customer requirements
47. Meeting customer requirements

Quality Policy

48. Purpose of organization
49. Commitment to comply with requirements
50. Commitment to continual improvement
51. Framework for quality objectives
52. Communication of quality policy
53. Review of quality policy

Planning

Quality Objectives

54. Establishing quality objectives
55. Measurement of quality objectives
56. Consistency of quality objectives
57. Objectives for meeting product requirements
58. Inclusion of quality objectives in business plan

Quality Management System Planning

59. Planning of quality management system in line with process management principles
60. Planning of quality management system to meet quality objectives
61. Maintaining integrity of quality management system

Responsibility, Authority and Communication

Responsibility and Authority

62. Defining and communicating responsibility and authority

Management Representative

63. Appointment of management representative
64. Responsibility and authority of management representative

Customer Representative

65. Designation of customer representatives

Internal Communication

66. Establishing communication processes
67. Communicating the effectiveness of the QMS

Management Review

General

68. Top management review of quality management system
69. Assessing opportunities for improvement
70. Records of management review

Review Input

71. Performance information for input to management review
72. Changes affecting the QMS
73. Analysis of field failures

Review Output

74. Decisions and action arising from management review

RESOURCE MANAGEMENT REQUIREMENTS

Provision of Resources

75. Determination and provision of resources to implement the QMS
76. Determination and provision of resources to enhance customer satisfaction

Human Resources

General

77. Competence of personnel

Competence, Awareness and Training

78. Determination of competence
79. Provision of training
80. Evaluating the effectiveness of training
81. Awareness of impact on the achievement of quality objectives
82. Records of education, training, skills and experience

Training

83. Establishing procedures for identifying training needs
84. Qualification of personnel
85. Training in specific customer requirements

Infrastructure

86. Provision and maintenance of infrastructure

Work Environment

87. Determination and management of work environment

PRODUCT REALIZATION REQUIREMENTS

Planning of Product Realization

88. Planning product realization processes
89. Consistency of process planning
90. Determining product quality objectives and requirements
91. Establishing product specific processes
92. Providing product specific documents
93. Providing product specific resources
94. Determining product specific verification and validation activities and methods
95. Determining product specific acceptance criteria
96. Determining product specific records
97. Determining process specific records
98. Ensuring suitability of planning output

Customer-Related Processes

Determination of Requirements Related to the Product

99. Determining customer specified requirements
100. Determining requirements for intended use
101. Determining statutory and regulatory requirements
102. Determining organizational requirements

Review of Requirements Related to the Product

103. Reviewing product requirements
104. Timing of product requirements review

105. Ensuring product requirements are defined
106. Resolving differing requirements
107. Ensuring the organization has the ability to meet requirements
108. Maintaining records of the results of product reviews
109. Confirming undocumented requirements
110. Amending documents affected by changed product requirements
111. Informing personnel of changed requirements

Customer Communication

112. Arrangements for communicating product information to customers
113. Arrangements for dealing with customer enquiries
114. Arrangements for dealing with contracts and orders
115. Arrangements for dealing with amendments to contracts and orders
116. Arrangements for dealing with customer feedback

Design and Development

Design and Development Planning

117. Planning product design and development
118. Controlling product design and development
119. Determining design and development stages
120. Determining review, verification and validation for each stage
121. Determining design and development responsibilities and authority
122. Managing design and development interfaces
123. Updating design and development planning

Design and Development Inputs

124. Determining design and development inputs
125. Determining functional and performance requirements
126. Determining statutory and regulatory requirements
127. Determining information from previous design
128. Determining other design and development requirements
129. Reviewing design and development inputs
130. Quality of design and development inputs

Product Design Input

131. Identification of customer requirements
132. Product design information deployment process
133. Product quality targets

Design and Development Outputs

134. Provision of design and development outputs
135. Approval of design and development outputs
136. Ensuring design output meets requirements

137. Provision of purchasing, production and service information
138. Referencing product acceptance criteria
139. Specifying product operation and safety characteristics

Design and Development Review

140. Performing design reviews
141. Evaluation of design
142. Identifying problems and correcting errors
143. Participants at design reviews
144. Records of design reviews

Design and Development Verification

145. Conducting design and development verification
146. Records of design and development verification

Design and Development Validation

147. Conducting design and development validation
148. Completing validation prior to the delivery or implementation of product
149. Records of design and development validation

Control of Design and Development Changes

150. Identification of design and development changes
151. Recording of design and development changes
152. Review, verification and validation of design changes
153. Evaluating the effects of change
154. Recording the results of design change reviews

Purchasing

Purchasing Process

155. Ensuring purchased product conforms to requirements
156. Determining supplier controls
157. Supplier evaluation
158. Establishing selection and evaluation criteria
159. Results of supplier evaluations

Purchasing Information

160. Description of product
161. Product approval requirements
162. Qualification of personnel
163. Quality management system requirements
164. Adequacy of purchasing requirements

Verification of Purchased Product

165. Ensuring purchased product meets requirements
166. Verification at supplier's premises

Production and Service Provision

Control of Production and Service Provision

167. Planning production and service provision
168. Controlling production and service provision
169. Ensuring the availability of information describing the product
170. Ensuring the availability of work instructions
171. Using suitable equipment
172. Availability and use of monitoring and measuring devices
173. Implementing monitoring and measurement activities
174. Implementing of release activities
175. Implementing of delivery activities
176. Implementing of post-delivery activities

Validation of Processes for Production and Service Provision

177. Validating special processes
178. Defining criteria for review and approval
179. Approving process equipment
180. Qualifying process personnel
181. Establishing process methods and procedures
182. Defining recording requirements
183. Defining revalidation requirements
184. Scope of process validation

Identification and Traceability

185. Identifying product
186. Identifying product status
187. Controlling unique identification of product

Customer Property

188. Exercising care of customer property
189. Identifying customer property
190. Verifying customer property
191. Protecting customer property
192. Reporting lost or damaged customer property

Preservation of Product

193. Preserving conformity of product
194. Identifying, handling, packing storage and protection of product
195. Scope of product preservation

Control of Monitoring and Measuring Equipment

196. Determining monitoring and measurements to be undertaken
197. Determining the monitoring and measurement equipment required
198. Establishing monitoring and measurement processes
199. Calibrating and verifying measuring equipment
200. Recording the basis for calibration and verification
201. Adjusting measuring equipment
202. Determining calibration status
203. Safeguarding adjustments
204. Protecting measuring equipment
205. Assessing nonconforming equipment
206. Action on equipment found out of calibration
207. Maintaining records of calibration and verification
208. Confirming integrity of computer software used for measurement

MEASUREMENT, ANALYSIS AND IMPROVEMENT REQUIREMENTS

General

209. Establishing processes necessary to demonstrate product conformity
210. Establishing processes necessary to ensure system conformity
211. Establishing processes necessary to improve system effectiveness
212. Determining monitoring, measurement and analysis methods

Monitoring and Measurement

Customer Satisfaction

213. Monitoring customer perceptions
214. Determining customer satisfaction monitoring methods

Internal Audit

215. Conducting internal audits for conformity with planned arrangements
216. Conducting internal audits for conformity with ISO 9001:2000
217. Conducting internal audits for conformity with the organization requirements
218. Determining effective implementation and maintenance of QMS
219. Planning the internal audit program
220. Defining audit criteria, scope, frequency and methods
221. Selecting auditors
222. Documenting audit procedures
223. Ensuring prompt action on audit findings
224. Following up audit actions

Monitoring and Measurement of Processes

225. Monitoring QMS processes
226. Demonstrating processes achieve planned results
227. Taking action on process measurements

Monitoring and Measurement of Product

228. Monitoring and measurement of product characteristics
229. Determination of product monitoring and measurement stages
230. Maintaining evidence of conformity
231. Indicating product release authority
232. Holding product release

Control of Nonconforming Product

233. Preventing unintended use of delivery of nonconforming product
234. Documenting nonconforming product controls
235. Maintaining records of the nature of nonconformities and actions taken
236. Re-verification of nonconforming product
237. Detecting product nonconformity subsequent to delivery

Analysis of Data

238. Collecting and analysing data on the effectiveness of the QMS
239. Collecting data to identify improvements in system effectiveness
240. Providing information on customer satisfaction
241. Providing information on product conformity
242. Providing information on trends and characteristics
243. Providing information on suppliers
244. Supporting decision making and long-term planning

Improvement

Continual Improvement

245. Improving the effectiveness of the QMS

Corrective Action

246. Taking action to eliminate the cause of nonconformity
247. Appropriateness of corrective actions
248. Documenting corrective action procedures
249. Reviewing nonconformities including customer complaints
250. Determining the causes of nonconformities
251. Evaluating the needs for action
252. Determining and implementing action needed
253. Recording the results of corrective actions
254. Reviewing corrective actions taken

Preventive Action

255. Determining action to eliminate potential nonconformities
256. Appropriateness of preventive actions
257. Documenting preventive action procedures
258. Determining the cause of potential nonconformities

259. Evaluating the needs for action
260. Determining and implementing action
261. Recording the results of preventive actions
262. Reviewing preventive actions taken

SYSTEM DEVELOPMENT – FOOD FOR THOUGHT

In this section are a number of questions grouped by section of ISO 9001 that will cause the reader to reflect on the previous chapters, perhaps even change perceptions but mostly confirm understanding, aid system development and provoke discussion or a search for answers.

Quality Management System

1. Is your system thought of as a set of documents or a set of interacting processes that deliver the organization's objectives?
2. Is your system integrated into the organization so that people do the right things right without having to be told?
3. Is your system a collection of interacting processes rather than a series of interconnected functions?
4. Does every process in the chain of processes from requirements to their satisfaction add value?
5. Are your business objectives functionally oriented driving a functional oriented organization or are they process oriented?
6. If you have simply changed the names of your procedures to reflect processes where have you defined the process objectives, the resources and the behaviours required to cause these objective to be achieved and the measures required to determine process adequacy, efficiency and effectiveness?
7. Are your processes described simply as a series of transactions or are there provisions in the process to manage its performance?
8. Do the outputs from one process connect with other processes?
9. Do all the inputs to a process have their origin in other processes or external organizations?
10. Do you have measures that enable you to determine how well each process is performing and are these known to those controlling the process?
11. Is there any activity, task or process that exists only to meet the requirements of ISO 9001?
12. Are you sure that all the documentation in place is needed for the effective operation and control of your processes?
13. There are six ways of conveying information. Before you document, have you eliminated the other five ways as being unsuitable for the particular situation?
14. If a person moves onto another job, how much of what is removed from the process is essential for the process to maintain its capability?
15. How much of what affects your ability to achieve results is dependent on staff following documented procedures?
16. How much of what affects your ability to achieve results is dependent on the physical and human environment in which your staff work?

17. How much of what affects your ability to achieve results is dependent on your staff's ability to do the right things right?
18. How much of what affects your ability to achieve results is dependent on your staff's motivation to do the right things right?
19. Do you question the effectiveness of established policies and procedures first when the customer complains or do look for someone to blame?
20. What causes the actions and decisions for which there are no documented policies or procedures?
21. Are you managing a set of functions or a series of processes and do you know the difference?
22. Do you know what each process aims to achieve?
23. Do you know how each process causes the observed results?
24. Do you know whether the process is producing outcomes that satisfy the process objectives?
25. Do you know how to change process performance to bring it into line with the objectives?

Management Responsibility – Food for Thought

26. Does your quality management system make the right things happen or is it just a set of procedures?
27. Does your management perceive the quality management system as the means by which the organization's objective are achieved?
28. Does your organization exist to make profit or to create and retain satisfied customers?
29. Does your quality policy affect how people behave in your organization or is it simply a slogan?
30. Do all your quality objectives relate to the organization's mission or are they focused only on what the quality department will achieve?
31. Does your management review examine the way the organization is managed or does it simply focus on conformity issues?
32. Do you struggle to obtain the necessary resources to do your job, or have you designed your job so that you get all the resources necessary for you to achieve your objectives?
33. Do you wait to receive a customer enquiry before identifying customer needs and expectations or have you researched the market in which you operate so that your offerings respond to customer needs?
34. What made you think that by simply publishing your quality policy, anything would change?
35. If you didn't know your current performance, how did you manage to set meaningful objectives?
36. When you set your objectives, was there any discussion on how they might be achieved and did it result in changing the way you do things?
37. If you don't think you need to change, how come you didn't meet these objectives last year?
38. How will you know when you have achieved your objectives?

39. Are you sure that none of the managers' objectives relate to extracting more performance from unstable processes?

40. Are you sure that none of the managers are tasked with meeting objectives for which no plans have been agreed for their achievement?

41. How many plans are the managers working on that have an objective that is not derived from the business plan?

42. Are you sure that managers are not pursing objectives that will cause conflict with those of others managers?

43. Are you sure you have not imposed a target on a member of staff for performance improvement when it is the system that requires improvement?

44. Do you always consider the impact of change on other processes before you proceed?

45. Does your staff know of the results for which they are accountable and are their job descriptions limited to such responsibilities?

46. If your management representative is unable to influence the other managers to implement, maintain and improve the management system, are you sure you have appointed the right person?

47. How often do you check that messages conveyed from management are actually understood by those they are intended to affect?

48. How often do you check that messages conveyed by workers are actually given due consideration by management and interpreted as the consequence of their own actions and decisions?

49. If the person with the most interest in the effectiveness of the management system is not your CEO, is there not something wrong with the way the management system is perceived in the organization?

Resource Management – Food for Thought

50. Are there any activities in an organization that require no human or financial resources?

51. What business benefit is derived from excluding particular resources from the management system?

52. Why are the costs of maintaining the management system different from those of maintaining the organization?

53. Are functional budgets justified when in reality the organization's objectives are achieved through a series of interacting processes?

54. If your staff are your most important asset, why are staff development programmes a prime target for cost reduction?

55. Do you select people on what they demonstrate they know or what they demonstrate they can do?

56. How do you know that your staff are competent to achieve the objectives you have agreed with them?

57. From where does the evidence come from to assess staff competence?

58. There are seven ways of developing staff. Before you send them on a training course have you eliminated the other six ways as being unsuitable for the particular situation?

59. What action was taken the last time your staff returned from an external training course that resulted in an improvement in their performance?

60. How do you ensure that staff work on those things that add value to the organization?
61. What's the second thing you do on receipt of a customer complaint?
62. How much information is contained in your personnel training records to help you make decisions on staff development needs?
63. Do you know of all the ways in which an infrastructure failure might impact customer satisfaction?
64. How much of what affects individual performance depends on the relationship between management and staff?
65. In describing what it is like to work in your organization, would you identify any factors that were detrimental to overall performance?
66. Would anyone in your organization take an action or make a decision that might be considered unethical in your society?
67. What actions do managers take to create conditions in which their staff are motivated to achieve their objectives?
68. When did you last examine the organization's culture for its relevance to the current conditions in which the organization operates?
69. When did you last recognize or appreciate the efforts of your staff in contributing towards improved performance?
70. When did you last reprimand a member of staff for something that was symptomatic of the natural variation inherent in the process?
71. If every employee were to follow the examples set by the management, in what way would the organization's performance change?
72. Do you wait until you have no option but to take drastic action to restore financial stability or do you involve your workforce in seeking ways to lessen the impact?
73. When did you last calculate the time lost by the inappropriate location of tools, information, equipment and facilities?
74. If an employee was dissatisfied with the working environment, could you be sure that he or she would approach the manager with confidence that the matter would be dealt with objectively and sympathetically?

Product Realization – Food for Thought

75. Do your marketing, sales and design processes include provisions for measuring the extent to which the process objectives are being achieved?
76. Are you confident that your sales personnel will not commit the organization beyond the capability of its processes?
77. When preparing plans for product realization how do you know you have taken account of all the factors that will affect successful implementation?
78. How do you know that the plans made for fulfilling product requirements will reach those who will create the product and process features necessary for successful implementation?
79. Do you assume customers will define the characteristics necessary to give satisfaction or do you recognize they are not experts and endeavour to find out what their expectations really are?
80. How do you know you have identified all the relevant regulations that apply to the customer transaction?

81. Would your customer expect you to proceed knowing there are issues to be resolved or to wait while you sought resolution?
82. However, simple the order, do you always confirm understanding before proceeding?
83. How confident are you that the sales literature does not lead potential customers to expect more than you are prepared to provide?
84. Is customer feedback collected from all the points of contact with customers or only through the mailbox?
85. At what point do you bring design under control, before the design is released or before you spend money?
86. Are you confident that you won't make the same mistakes on the next new design as you did on the last design?
87. How do you stop your designers reinventing solutions to problems solved previously?
88. How do you ensure that design weaknesses revealed through risk analysis techniques are eliminated, reduced or at least controlled before the design is released?
89. What research is performed to discover the probability of success with new designs?
90. Is your purchasing process sufficiently flexible to prevent inappropriate conditions being placed on your suppliers?
91. Does your supplier selection procedure permit value-based decisions to be made or is it one size that fits all?
92. How do you know that the products you purchase for incorporation into supplies will satisfy the regulatory conditions that apply to the item you deliver to your customer?
93. How do your receipt inspectors know of the decisions your supplier verification personnel have made on the shipments received?
94. How do you know the processes are capable of achieving the required product or service features before commencing production or service delivery?
95. Does your system include all the distribution channels that have been established for delivering product to customers?
96. If a product was returned because of a failure, are you confident that you could find out what work has been done on it as it passed through the production process?
97. If an item of customer property was recalled, could you find it and return it in its original condition if requested to do so?
98. Have you put in place methods that will ensure the integrity of all of the devices used for monitoring and measurement?

Measurement, Analysis and Improvement – Food for Thought

99. Do you manage the system as a series of processes or as a series of functions?
100. If no objectives are set for a process, will any level of performance be acceptable?
101. Do you spend more time putting out the fires than on improving the process?
102. Do you act on suspicion, or always confirm that a problem exists or might exist before taking action?
103. Have you discovered any root cause of a problem that cannot be traced to lack of application of one or more of the eight quality management principles?

104. Do you always undertake data analysis with the intent of taking action on the results?
105. Are you confident that you are not expecting success to be achieved by chance?
106. Have you re-examined existing controls from the perspective of establishing the failure modes they prevent?
107. Have you equipped every process with provisions for measuring its performance?
108. How often do you check that your objectives and targets are still relevant to the organization's goals?
109. If the questions in your customer satisfaction questionnaires were generated internally, how do you expect to obtain unbiased results?
110. When was the last time your internal audit programme found something that led to improved performance?
111. If you discontinued your internal audit programme, would anyone other than the internal or external auditors demand its reinstatement?
112. If your auditing approach has been to verify compliance with procedures, what approaches are you intending to take now that the system has to enable the organization to achieve its objectives?
113. Do you consider that the system is effectively implemented if people are following the documented procedures or would you also verify that the planned results are being achieved?
114. Would you accept a box of 1000 components by simply checking one sample, if not why would you base your audit conclusions on a few unrepresentative samples?
115. Why shouldn't the manager perform the internal audits, and if you should think he or she is not competent to do so, why do you trust him or her to manage the function?
116. How do you know that each of the processes is achieving the planned results?
117. When was the last time you changed your operating methods in order to increase resource utilization?
118. Do you continue with the current level of product verification regardless of detecting no nonconformities?
119. Are you sure that those examining products or services for conformity apply the same criteria as those using them?
120. Is the data used by management to make decisions generated from the processes of the management system and if not why not?
121. Are there any data collection routines that are not triggered by a process in the management system?
122. How continuous is your continual improvement process?
123. When was the last time a problem recurred?
124. When did asking the question? What if … become a habit?
125. What we remember we can avoid, what we forget we can repeat therefore what have you done lately to help you remember the mistakes of the past?

Chapter 39

System Certification

CHAPTER PREVIEW

This chapter is aimed at all those personnel who will be involved in the certification process including top management and below.

In this chapter we examine:

- The reasons for seeking certification.
- How to choose a certification body.
- Factors to consider when preparing for certification.
- How to deal with permitted exclusions from the requirements.
- How to define the scope of registration.
- How to handle the assessment.
- How to deal with nonconformities.
- How to use the certification to best advantage and avoid bad practices.

REASONS FOR SEEKING CERTIFICATION

Being ISO 9001 certificated should make no difference to the way the organization is managed. If your organization operates in a sector in which it is well-known, certification will add no value. But if your organization wants to win business from outside its community, from customers that have no knowledge of its capability, ISO 9001 certification might add significant value. It is therefore imperative that the approach you take to ISO 9001 is one that does not force you to do things that add no value.

If ISO 9001 is perceived as a requirement it is more than likely that it is your customers that demand certification in order that your organization may remain on an approved suppliers list or retain eligibility for receiving invitations to tender.

If ISO 9001 certification is not a customer requirement, there are perhaps a few benefits for some organizations.

- The value of an independent audit of your management system;
- The pressure to formalize the management system;
- The recognition that certification brings in the market place.

There is no valid evidence that ISO 9001 certification provides a guarantee of product and service quality or that those organizations that obtain ISO 9001 certification produce better quality products and services than others. Therefore, if preparing

To the Critics of ISO 9000

Until such time as every child leaving secondary education is equipped with the skills and knowledge to apply the principles of quality management, we have to make do with the crude tools we have and one of these is the ISO 9000 family of standards. They can help those people who don't have the time or the inclination to work out for themselves the right things to do by providing a prescription that experience has shown will reduce failure and improve quality when applied intelligently.

Every manager has the right to decline to do anything that cannot be shown to add value and it is the responsibility of consultants and auditors to behave professionally at all times by providing such justification.

There is no requirement in ISO 9001 that is mandatory. It is the complexity of an organization and the market in which it operates that will determine which requirements apply.

Where proven to apply, there will be few situations where an organization will not be applying or will not gain from applying the prescribed principles and practices to some extent.

The extent of application is at the discretion of the management. If they choose not to constrain the application of requirements it will more than likely result in an over bureaucratic system of documentation that stifles initiative and innovation. This is not the intent of the ISO 9000 family of standards and never has been. When used intelligently, it can help organizations towards sustained success.

ISO 9001 does not yet address all of the factors upon which the achievement of quality depends therefore its application is not a guarantee of product quality. However, the intelligent manager will realize this and not place too much reliance on the certification process.

a case for certification and the consequential costs without a customer mandate, the arguments are weak but it rather depends on the type of organization and its management style. Some organizations won't make any progress without external pressure.

CHOOSING A CERTIFICATION BODY

An issue of concern to organizations choosing certification to ISO 9001:2008 is the variability of the certification process and the validity of the resultant certificate. There are certification bodies that are not accredited by a recognized accreditation body[1]. There is a wide variation in the competence of certification body auditors meaning that there is no guarantee that an auditor from an accredited certification body will be any better than one from an unaccredited certification body. The only difference is that one can show evidence of nationally recognized independent review and the other can't. Therefore, if you choose an accredited certification body and have problems they are unwilling to resolve then you can appeal to the accreditation body. You have no redress with an unaccredited certification body except for litigation.

Unfortunately auditors are more likely to certify ineffective systems than fail robust systems and most managers will accept minor corrective actions rather than create a fuss. But you need to believe you are getting value for money.

Questions addressed to the certification body might include:

1. What experiences do your auditors have in our type of business?
2. Which organizations of our type have you registered?
3. What continuity will be maintained between auditors chosen for the initial audit and the surveillance visits?
4. What criteria will be employed to judge the effectiveness of our quality management system?
5. What are the conditions for issuing a certificate?
6. How soon after the assessment will we receive the written report and what will it contain?
7. What approach will the auditor take to the audit – will he/she be checking element by element or process by process or department by department?
8. Will the auditor wish to visit every site?
9. How much notice do we have to give for the initial assessment?
10. When will the auditor want to review the documentation?
11. Which documents will the auditor wish to examine in the documentation review?
12. Will the auditor want to review the documentation on or off site?
13. How will the results of the documentation review be conveyed?
14. Will the auditor require objective evidence of compliance with every requirement at the time of the audit or can some requirements be checked at a later stage?
15. Will you permit us to vet any new auditors before their assignment?
16. What guidance will be provided to help us improve the system?
17. Should we wish to lodge a complaint about the audit what is the procedure?

Some of these questions may already be addressed by the literature you are sent, but a discussion to clarify them is often useful for both parties.

PREPARING FOR CERTIFICATION

In order that you are properly prepared for the external assessment, you need to be fully aware of how your system meets the requirements of ISO 9001:2008. Although the auditors will seek evidence to demonstrate conformity to the requirements, it is also necessary that you are able to show the auditor how your system meets the requirements. The more skilled you are at predicting what auditors will look at and what they are looking for when they find it, the more successful you will be. But if you know the system is noncompliant then don't try to bluff as it only makes things worse when you are found out. It is better to declare you don't know the answer than pretend you do.

Everyone should know where the ISO 9001 compliance table (see Chapter 30) is located and have access to it during the external assessment. In addition all personnel should be able to explain:

- The objectives of what they are doing;
- The process they are using to achieve these objectives;

- How they know they are achieving these objectives?
- How they know they are doing this in the best way?
- How they know these are the right objectives?

Personnel should also be informed to only answer the questions asked but if the auditor looks as though they are going down the wrong path through ignorance of the system, personnel should know the system well enough to offer redirections.

DEALING WITH EXCLUSIONS

Clause 1.2 of ISO 9001 states that where any requirement cannot be applied due to the nature of an organization and its product, this can be considered for exclusion. In the past organizations were able to register parts of their system or parts of their organization and product ranges. Certification is applied to some product lines but not to others. This is no longer applicable. However, *exclusion* is a certification issue – not an issue with using ISO 9001 as a basis for designing a management system.

Exclusions are limited to requirements within Clause 7 meaning that you cannot exclude requirements contained in Clauses 4, 5 and 8. If you want certification against ISO 9001, Clauses 4, 5 and 8 are mandatory and any exclusion in Clause 7 is only acceptable if it does not affect the organization's ability or responsibility to provide product that meets customer and applicable regulatory requirements. You cannot exclude requirements of Clause 7 simply because you do not want to meet them.

Examining each sub-clause in Clause 7 may serve to explain the impact of this requirement.

7.1 – Planning of Product Realization

Every organization provides outputs of one kind or another and therefore needs processes to deliver these outputs that obviously need to be planned. It is therefore inconceivable that this clause could be excluded.

7.2 – Customer Related Processes

Every organization has customers, people or organizations they serve with their outputs. Organizations do not have to be profit-making to have customers because the term customer is not used in ISO 9001 to imply only those that make purchases. It is therefore inconceivable that this clause could be excluded.

7.3 – Design and Development

Every organization provides outputs that can be in the form of tangible or intangible products (computers, materials, software or advice), services that may process product supplied from elsewhere or services that develop, distribute, evaluate or manipulate information such as in finance, education or government.

With every product there is a service. If the product is provided from elsewhere, a service still needs to be designed to process it. While not every organization may provide a tangible product, every organization does provide a service. Some organizations manufacture products designed by their customers and therefore product design could be excluded. However, they still provide a service possibly consisting of sales, production and distribution and so these processes need to be designed. In franchised operations, the service is designed at corporate headquarters and deployed to the outlets but it is not common for an outlet alone to seek ISO 9001 certification – it would be more common for ISO 9001 certification to be a corporate policy and therefore the scope of certification would include service design. The IAF guidance[1] is that if you are not provided with the product or service characteristics necessary to plan product or service realization and have to define those characteristics; this is product design and development. There is also a note in ISO 9001:2008 Clause 7.3 stating that the requirements given in 7.3 can also be applied to the development of product realization processes.

It is therefore inconceivable that this clause could be excluded for anything other than for product design meaning that it must be included for service design.

7.4 – Purchasing

Every organization purchases products and services because no organization is totally self-sufficient except perhaps a monastery and it is doubtful that ISO 9001 would even enter the thoughts of a monk! Some purchases may be incorporated into products supplied to customers or simply passed onto customers without any further processing. Other purchases may contribute to the processes that supply product or deliver services to customers and there are perhaps some purchases that have little or no effect on the product or service supplied to customers but they may affect other stakeholders. It is therefore inconceivable that this clause could be excluded in an organization that was considering certification to ISO 9001.

7.5 – Production and Service Provision

Every organization provides outputs and these outputs need to be either produced or distributed. If the organization only designs products, it provides a design service and therefore this clause would apply. If the organization only moves product, it provides a distribution/shipment/transportation service and therefore this clause would apply. If the organization provides a service that some other organization designed, this clause *would* apply. There are however sub-clauses within this clause that could be excluded.

7.5.2 – Validation of Processes

This clause applies to processes where the resulting output cannot be verified by subsequent monitoring or measurement. Firstly, if we apply the provisions of clause 7.3 to all processes, this clause is redundant. Secondly, there is nothing within these requirements that should not apply to all processes.

Defined criteria for review and approval of processes are addressed by Clause 4.1c for all processes where it requires criteria and methods needed to ensure operation and control of processes.

Approval of equipment is addressed by Clause 4.1d for all processes where it requires resources necessary to support the operation and monitoring of processes to be available. The qualification of personnel is addressed by Clause 6.2.1 for all processes where it

---

[1] International Accreditation Forum, (2001). www.iaf.nu.

requires personnel to be competent on the basis of appropriate education, training, skills and experience.

Use of specific methods and procedures is addressed by Clause 4.1d for all processes where it requires information necessary to support the operation and monitoring of processes to be available.

Requirements for records are addressed by Clause 4.2.4 for all processes where it requires records to be established and maintained to provide evidence of conformity to requirements.

Revalidation is addressed by Clause 8.2.3 for all processes where it requires methods to demonstrate the ability of processes to achieve planned results.

It is therefore inconceivable that this clause could be excluded.

7.5.3 – Identification and Traceability

This clause only applies where product identification is appropriate and where traceability is a requirement – there is therefore no need to exclude it.

7.5.4 – Customer Property

Not every organization receives customer property but such property does take a variety of forms. It is not only product supplied for use in a job or for incorporation into supplies, but also can be intellectual property, personal data or effects. Even in a retail outlet where the customer purchases goods, customer property is handed over in the sales transaction perhaps in the form of a credit card where obviously there is a need to treat the card with care and in confidence. In other situations the customer supplies information in the form of requirements and receives a product or a service without other property belonging to the customer being supplied. Information about the customer obtainable from public sources is neither customer property nor is information given freely but there may be constraints on its use.

It is therefore conceivable that this clause could be excluded.

7.5.5 – Preservation of Product

This clause applies to organizations handling tangible product and would include documentation shipped to customers. It applies in service organizations that handle product, serve food and transport product or people. It does not apply to organizations that deal in intangible product such as advisers although if the advice is documented and the documents are transmitted by post or electronic means, preservation requirements would apply.

Although in theory it is conceivable that this clause could be excluded, in practice it is unlikely.

7.6 – Control of Monitoring and Measuring Equipment

Every organization measures the performance of its products, services and processes. It uses instruments for such measurement. They may be physical, financial or human and while calibration may not be appropriate, validation certainly is. Whatever the method of measurement, it is important that the integrity of measurement is sound. It is therefore inconceivable that this clause could be excluded. If however, we take individual requirements within this clause, the specific requirements for calibration could be excluded if no physical measuring devices are employed.

It is inconceivable therefore that this clause could be excluded but conceivable that certain requirements within it could be excluded.

Summary

This analysis has shown that in only one instance could a clause in Section 7 of ISO 9001 be excluded (Customer property) and in only two other instances could requirements within a clause be excluded. It is therefore not the big loophole that it may have appeared to be at first glance.

SCOPE OF REGISTRATION

The scope describes the products and service for which you require your quality management system to be certified. In assessing the quality management system, auditors are looking to see that the system is capable of ensuring you have the capability of supplying the products and services specified in the 'scope'. For example, if you register your system for the manufacture of washing machines then add to your business the manufacture of electronic components, you cannot claim that you are certificated to ISO 9001:2008 for the manufacture of electronic components. The registration is limited only to the scope. When selecting suppliers you cannot rely on the fact that they are registered to ISO 9001:2008. You need to know for what products and services they are registered. You also need to know which accreditation body issued the certificate since not all are registered with the national accreditation agency.

The scope of registration is not the same as the scope of your quality management system (see Chapter 11). You may include many functions and processes in the quality management system that are not addressed by ISO 9001 or which do not affect product or service quality directly. If you choose to design a management system which reflects how you conduct your business then you may include, legal, medical, catering,

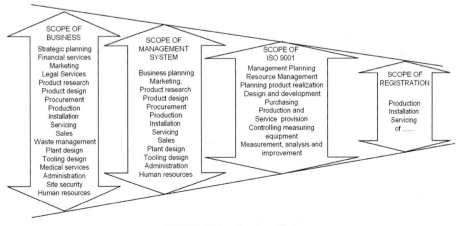

FIGURE 39-1 Scoping effect.

personnel, health and safety, environmental and other management processes. If you do not intend to sell or supply these services to your customer then you don't have to include them in the scope of registration. However, if you do, they will be assessed. This scoping effect is illustrated in Fig. 39-1.

HANDLING THE ASSESSMENT

The Assessment Process

It is important to commence a dialogue with your certification body early in the programme so as to obtain their interpretation of the standard to your business. The assessment itself is only as good as the auditor who conducts the assessment. Auditors are subject to annual appraisal by an independent council but the process in no way guarantees that high standards are maintained. You need to determine whether the auditors appointed will be fair and reasonable and that they have an adequate understanding of your business. Further information on how the assessment will be carried out is given in ISO 17021. The certification process is illustrated in Fig. 39-2. ISO 17021 requires the audit programmes to include a two-stage initial audit, surveillance audits in the first and second years and a re-certification audit in the third year prior to expiration of the certificate.

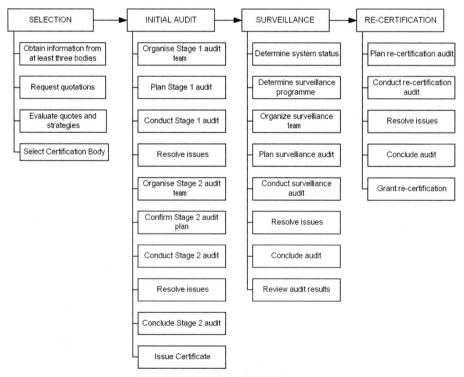

FIGURE 39-2 The certification process.

The Initial Audit – Stage 1

The purpose of the stage 1 audit is to establish both parties' readiness for the stage 2 audit and will:

- Evaluate the locations and site-specific conditions and establish their level of preparedness for the stage 2 audit;
- Establish the extent to which the organization understands the requirements of ISO 9001 particularly with respect to identifying objectives, processes and key performance parameters;
- Verify the scope of system to be audited, the locations involved, to identify the customer specific, statutory and regulatory requirements that apply and to identify associated risk factors;
- Examine the management system documentation and establish the extent to which it reflects a system that conforms with the requirements of ISO 9001;
- Determine the resources required for the stage 2 audit and agree with the detail of the stage 2 audit;
- Evaluate the internal audit and management review processes and establish that sufficient evidence exists to substantiate that the organization is ready for stage 2 audit.

It will be clear from this that it will be necessary for the third party auditor to visit the site during the stage 1 audit. This used to be undertaken unofficially as part of an initial visit or preliminary assessment but ISO 17021 now rolls these into the formal audit.

The Initial Audit – Stage 2

The purpose of stage 2 audit is to verify the implementation and effectiveness of the organization's management system and will:

- Gather evidence of conformity with all requirements of ISO 9001;
- Gather evidence that the organization is achieving the declared objectives and key performance targets;
- Gather evidence that the organization's processes are being managed effectively;
- Gather evidence of a clear alignment between what the organization is under an obligation to achieve, what is being reported internally it is achieving and what actual performance data demonstrate it is achieving.

Documentation Audit in Stage 1

During stage 1 of the initial audit the quality management system documentation will be examined to gather evidence that the requirements ISO 9001 have been addressed in one form or another.

As the auditor is now required to visit the site, there is no need to send a copy of your documentation to the certification body. In fact many documented systems are now intranet based, it is probably more desirable that the auditor sees the documentation in the same form as the users. Many of the requirements in the standard cannot be addressed at the policy level. Some will only be addressed in the process descriptions and others much lower down in the tiers of documentation, in operating procedures, instructions and forms. If the relationship between policies and procedures is vague then the auditor may well want to look at the detail. Writing the quality manual around the requirements of the standard makes the auditor's job easier but may well not provide a system which reflects the way you conduct your business. A manual written around the standard is often not user friendly and hence is likely to been seen as only serving the needs of the auditor. It is also not what the auditor either wants or needs. A better approach is for your Quality Manual to describe your system and for a Compliance Table to show the relationship between your manual and the requirements of the ISO 9001 (see Chapter 30).

In assessing the documentation the auditor is looking for conformance, although some auditors may appear to be only concerned with nonconformity. It is therefore of no interest that you may have included aspects outside the scope of the standard. The documentation should reflect a system with elements which fit together. The outputs from one process should be inputs to other processes. There should be no loose ends, conflict, gaps, unnecessary overlaps and ambiguities. The auditor should therefore look for coherence. The auditor may well request further information in order to gain an adequate understanding of your system. Should an element of the standard have not been addressed the auditor will proceed no further until it has been resolved. Should a clause of the standard[①] have not been addressed this may be resolved by providing documented procedures for review. Should a requirement of a clause have not been addressed the auditor may add this to the check list and establish whether the practice is compliant during the implementation audit.

Implementation Audit in Stage 2

In assessing the practice, the auditor is looking for evidence that you are implementing the system you have documented. If you are not implementing those parts of the system that are outside the scope of the standard or the scope of registration, the auditor may or may not regard this as a nonconformity by virtue of the requirements of Clause 4.1. (*The organization shall implement the quality management system.*) If your practices are compliant with the requirements of the standard but the practices are not addressed in your documented system, this should be deemed as a minor nonconformity (see later). Table 39-1 should help to distinguish between valid and invalid nonconformities. The table clearly shows that the only nonconformities that are valid are those where the requirements of ISO 9001 and the scope of registration match.

Audit Process

The site visit will take the following form:

There will be an opening meeting to introduce the assessment team, confirm the scope and timetable, outline the assessment process and reporting method and clarify any unclear aspects.

During the assessment the auditors interview members of your staff to determine how work is carried out in certain areas, establish if it conforms with your documented processes, seek objective evidence of the facts and compare the facts with the requirements of the standard. Any observations will be documented and the auditor may seek confirmation of the facts and request you to endorse the observation report before proceeding further.

At the end of the assessment the auditor will prepare a report detailing the observations and identifying those which are nonconformities with the requirements of the standard. The Lead Auditor will draw conclusions from the results and formulate the recommendations.

TABLE 39-1 Determining Validity of Nonconformities

Condition	Requirement of ISO 9001	Provision of registration	Process documented	Process implemented	Example	Result
Activity outside scope of business	Yes	No	No	No	Design or servicing	No nonconformity
Activity outside scope of quality system	No	No	No	Yes	Finance, Security, Medical	No nonconformity
Activity outside scope of quality system	Yes	No	No	Yes	Design or servicing	Nonconformity*
Activity outside scope of registration	No	No	Yes	No	Advertising, Public relations	No nonconformity
Activity outside scope of registration	Yes	No	Yes	No	Marine products	No nonconformity
Pertinent activity	Yes	Yes	No	No		Nonconformity
Pertinent activity	Yes	Yes	No	Yes		Nonconformity
Pertinent activity	Yes	Yes	Yes	No		Nonconformity

*If the organization carries out an activity upon which its ability to meet customer requirements depend it has to be included within the scope of the quality system (Clause 1.2)

There will be a closing meeting to thank the participants, emphasize the good points, explain the nonconformities and observations and make recommendations as to whether or not the company will be recommended for registration. The auditor may leave the assessment report with you or it may be issued later following a review by the certification body.

If there is more than one auditor there will be a Lead Auditor who will manage the assessment. If the assessment takes more than one day the Lead Auditor may call a daily meeting with the company to convey the results thus far.

DEALING WITH NONCONFORMITIES

Results

If the auditors use the Pass/Fail method to summarize the findings there are only three possible results of the assessment: Pass, Open or Fail.

- A pass verdict means that no major or minor nonconformities were detected.
- An open verdict means that the auditors found no more than one major and several minor nonconformities and that you have 90 days in which to eliminate them.
- A fail verdict means that the auditors found two or more major nonconformities.

Should you fail the assessment or fail to correct the nonconformities to the auditor's satisfaction within the 90 days, you will have to make a new application for registration and go through the whole process again. With an open verdict, the auditors return to conduct a follow-up audit on the specific areas where nonconformities were detected. Certification is not granted until all outstanding nonconformities have been closed.

Nonconformities

The definitions of major and minor nonconformities are given in the Glossary. The definitions given rely on there being unified understanding on what a clause of the standard[①] is and there is currently no ISO definition of this or of major and minor nonconformities. All nonconformities should be identified with the element and clause of the standard which has not been met. The nonconformity statement should be concise, accurate and supported with objective evidence. It should enable you to correct the problem to eliminate the nonconformity. Vague statements should be challenged. Nonconformity statements should therefore specify:

- The object of the nonconformity;
- The location of the object;
- The requirement of the standard which has not been met.

Here are some typical nonconformity statements:

- No measures had been taken with Avometer S/N 3568 located in the final test area to safeguard the measuring instrument from adjustments which would invalidate the measurement result as required by Clause 7.6d of ISO 9001:2008.
- No records could be found for latch mechanism JC 478 held in despatch area which provide evidence of conformity with acceptance criteria as required by Clause 8.2.4 of ISO 9001:2008.

- There was no evidence that the interfaces between latch mechanism JC 034 designed by the company and boom arm JC 021 had been effectively communicated to the subcontractor of the boom arm as required by Clause 7.3.1 of ISO 9001:2008.

You will note that in all these examples the exact wording from the standard has been used. This is to ensure objectivity.

As the detection of a major nonconformity can be cause for refusal to award certification, it is extremely important that there is agreement on the nature of such nonconformity. One instance of failing to meet one requirement or a clause of the standard[①] is not a major nonconformity. To be a major there has to be no provision in place to meet a clause of the standard or the provisions in place are not working as intended. If the provisions in place have not been followed in one instance then the auditor should look for more objective evidence. If compliance is established in some cases but not in others then a judgment needs to be made as to whether it signifies that the system has broken down. The auditor needs to establish whether the nonconformity is the result of random error or indicative of operations being out of control. In any case, the auditor should establish if the system is incapable of stopping the supply of nonconforming product or service before classifying a nonconformity as major.

Existing Nonconformities

If you know of nonconformities and have in fact put in place corrective action plans which have not yet been implemented, whether the nonconformities would be deemed sufficient cause for refusing certification will depend on their magnitude. Failure to address an element or clause of the standard, regardless of your plans to remedy the situation, will be cause for refusal of certification, simply because your system is incomplete. You may however, have commenced a programme of change which is only partially complete. Providing there is evidence of conformance to the standard in some cases, then you will be deemed compliant. The auditor will want to check progress at the subsequent surveillance visit. If no progress has been made, this may indicate that the system is not in place.

Challenging Nonconformities

If you believe the nonconformity to be invalid, challenge the auditor to demonstrate its validity by showing you the requirement of the standard that has not been met and showing you the evidence of nonconformity. If you are still not satisfied, ask the auditor to explain how the nonconformity can effect quality. A useful question to put to external auditors when they report they have found nonconformity that you disagree with is '*How does confidence in the quality of the product diminished if this requirement is not met?*'

Remember you are paying for the assessment, although if you withdraw on the basis that you are dissatisfied, you may have to pay another certification body to repeat the assessment. It is prudent only to challenge the auditor when you are on firm ground and when the corrective measures may well be costly, and in your view, add no value. Minor nonconformities are best accepted if their correction is trivial. You do not want to give the impression that you are not committed to quality by dismissing errors in the paperwork. The smallest error in paperwork has been known to result in severe penalties. Auditors should note that "You are only taken as seriously as your most insignificant nonconformity".

Correcting Nonconformities

The standard requires that you take corrective action without undue delay for nonconformities found during internal quality audits and the same is expected (but not required except in the contract with the Registrar) of external audits. The auditor will request proposals from you to correct the reported problems and to prevent them recurring together with dates by which these actions will be completed. The auditor will not normally agree to timescales in excess of three months to correct the nonconformities as it indicates that the assessment was premature. In some cases the nonconformity can be closed by letter or submission of changed pages to the system documentation. In other cases the new provisions need to be in place for several months before sufficient evidence has been generated to show that the system is effective.

USING AND ABUSING CERTIFICATION STATUS

There is some very useful guidance in the ISO publication; Publicizing your ISO 9001:2000 or ISO 14001:2004 certification. The guidance will help you to apply good practice in publicizing, communicating and promoting your certification to all stakeholders. There are a number of issues to bear in mind.

Scope

When you are awarded an ISO 9001 certificate it is for the organization's quality management system and as such has a scope of registration that defines the extent to which the certificate applies. You should be accurate and precise about the scope of your organization's ISO 9001:2000 certification, as far as both the activities, the products and geographical locations covered by the certification are concerned.

Logos

When you receive your certificate you will be given instructions by your certification body on what you can and cannot do but these sometimes get mislaid. You should not use, adapt or modify ISO's logo. In fact ISO will take whatever actions it considers necessary to prevent the misuse of its logo. If you want to use a logo, ask your certification body for permission.

What You Can and Cannot Claim

If your organization is certified to ISO 9001:2008, use the full designation (not just 'ISO 9001') because your certificate is valid only for a specific version of ISO 9001. You should replace use of the generic term 'ISO 9000 certification' by the specific term 'ISO 9001 certification'.

In the context of ISO 9001:2008, 'certified' (and certificated), registered (and registration) are equivalent in meaning and you can use either term. However, you should not say your organization has been 'accredited' or 'ISO certified', or has 'ISO certification'. You should use instead 'ISO 9001:2008 certified' or 'ISO 9001:2008 certification'. Only certification bodies are accredited. This means they are authorized to conduct certification of conformity to prescribed standards and ISO 9001 certification does not authorize your organization to do this.

You cannot display ISO 9001:2008 certification marks of conformity on products, product labels, or product packaging, or in any way that may be interpreted as denoting product conformity. You should also not give the impression in any context that ISO 9001:2008 certification is a product certification or product guarantee.

Chapter 40

Beyond ISO 9001 Certification

CHAPTER PREVIEW

This chapter is aimed at those managers who were involved in making the decision to pursue ISO 9001 certification.

After registration to ISO 9001, you have made a major achievement, but you may have just started on the road to a quality culture. Meeting the requirements at the assessment is like passing a school exam. You know the syllabus and could be asked any questions. You did your homework and you are fortunate that you could give the right answers to the questions or show the auditor acceptable evidence of conformity. But passing the exam doesn't mean you have become educated. You weren't tested on the whole syllabus, only a sample. You could have failed if other questions had been asked. And so it is in the ISO 9001 assessment. ISO 9001 certification implies that you have the capability to satisfy the requirements of your current customers but this may not be sufficient to win business from your competitors. You will need to do three things: maintain, improve and innovate. Maintain your standards, Improve on the efficiency and effectiveness with which you meet these standards and Innovate occasionally to set new standards. The MII cycle can be illustrated as shown in Fig. 40-1. In this cycle, routine activities should be moving between maintenance and improvement with periodic excursions into innovation when the routines have exhausted improvement potential.

In this chapter we look at:

- The payback from the investment in a certified quality management system;
- The ongoing development of the system;
- The type of improvement that arise from the requirements of ISO 9001;
- The type of improvements that arise beyond the requirements of ISO 9001;
- Extending the envelope beyond current perceptions of quality.

THE PAYBACK

The investment to achieve ISO 9001 certification may be considerable and therefore top management will be looking for the payback. When should you expect to reap the dividends? Well, it depends on the state of your organization before you started on the road to ISO 9001 certification. It was for this reason that we recommended you measure your performance in key areas before commencing quality

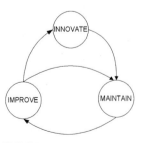

FIGURE 40-1 The MII cycle.

system development (see Chapter 1). If you have done this and installed the necessary provisions to capture the relevant performance data, you will be able to detect when performance starts to improve. It is not unusual for performance to decline slightly during system development as effort is diverted to building an effective management system and more efficient means of collecting data emerge. It should therefore be expected that improvement will be slow on installing the system. If improvement is not attained by the first surveillance visit, you will be noncompliant with the standard. However, the degree of improvement will vary depending on the level of performance at the start.

One of the most significant findings as a result of installing a quality management system should be that performance becomes less erratic and less sensitive to changes in organization or customer requirements. A factor which will distort the picture is the effect of any new technology that you have introduced in the same period.

There will of course be short-term gains such as:

- The decline in reactive management as the processes begin to be managed more effectively;
- The decline in overtime as fewer products are found nonconforming before delivery;
- The decline in design modifications as customer requirements are established before commencing design;
- The increase in orders as customers apply selection policies based on ISO 9001.

If performance remains persistently poor even though you remain compliant with ISO 9001, it is a sign that the problem is much deep-routed than can be detected by conventional auditing. What is often needed is a breakthrough in attitudes, management behaviour and staff motivation and a management system that adopts all three approaches to quality as expressed in Part 2 of this book.

CONTINUING DEVELOPMENT

There may be many areas where development of the system needs to be completed. If your organization is one that is constantly changing to respond to market forces, quality system development never ceases. The initial assessment only took a sample of your operations to test for conformity against the standard. Subsequent audits will reveal more nonconformities and if you do not continue with the development of your system, these audits may find major nonconformities which if not corrected promptly, will lead to the withdrawal of your certificate.

It is important that you continue with any system development activities. Every policy and requirement of every process needs to be tested over the range of operations to which it applies. This may well yield opportunities for improvement which should be followed. Some work processes may have yet to be exercised such as contingency plans for recovery from disasters although where possible all processes should be validated even if by simulation.

The certification body will conduct periodic audits to verify that you remain compliant with the requirements of ISO 9001. These audits only establish that you have retained the capability of meeting your customer's requirements and that the system is being implemented effectively. The audits are not intended to address standards other than the one against which you were certificated. However, there several areas in which improvement is addressed in ISO 9001.

BECOMING MORE COMPETITIVE AND PROFITABLE

You don't have to have a successful or profitable business to gain ISO 9001 certification. You need a capable quality management system but even with an ISO 9001 certification you might be loosing customers. There are four key factors on where action can be taken to improve competitiveness and profitability.

Reduction Measures

In any organization there is always too much of something (except profit!) and so there are usually ample opportunities for reduction. There follows some pointers to areas where reduction may yield considerable benefits.

- Reduce complexity so that there are fewer ways of doing something, fewer interfaces and fewer things to go wrong.
- Reduce variation so that processes produce results of consistent quality.
- Reduce waste so that resource costs can be kept low.
- Reduce time through the process so that bottle-necks are avoided.
- Reduce error in products, services, documents, decisions and communication so that costs decline.
- Reduce job classifications so that you are not reliant on particular individuals.
- Reduce inspection so that costs decline.
- Reduce anti-quality attitudes which will jeopardize the programme.

Increasing Measures

For some factors any reduction would only make things worse and so you have to increase the amount of it to gain any benefit. Here are some which can yield considerable benefits.

- Increase utilization of material, machines, tools, equipment, personnel and facilities.
- Increase training of management and staff.
- Increase discipline and adherence to policy and practices.
- Increase tidiness and cleanliness.
- Increase availability and retrievability of information.
- Increase motivation of management and staff.

Stabilizing Measures

Before you had a quality system performance that may have been erratic, now it should be less erratic but you may need to apply further effort to make it stable. Aspects you can address are as follows:

- Stabilizing the controls so that less correction is needed to maintain quality standards.
- Stabilizing methods so that once a good method is found it is used throughout the organization.
- Stabilizing materials so that the effects of variation in materials are reduced.
- Stabilizing suppliers so that you depend on fewer but more reliable suppliers.
- Stabilizing processes so that variation is predictable.
- Stabilizing the environment to reduce the effect that variations might have on operations.

Keeping Measures

Whilst the introduction of a formal quality management system may have resulted in you throwing away many obsolete documents, there are several things you need to keep and they are not all documents. Here are some of the more important things to keep.

- Keep commitments to signal to the workforce that you are serious about quality.
- Keep records so that you don't have to rely on opinions.
- Keep measuring performance so that you know where you are at any time.
- Keep analysing results so that you know what to fix, what is about to happen and what has happened.
- Keep auditing to determine the health of the organization.
- Keep questioning the status quo as nothing stands still.
- Keep reducing, increasing and stabilizing.
- Keep maintaining, improving and innovating.

Improvement on Price

The price charged for products is a function of cost, profit and what the market will pay. Sometimes price is much higher than cost and in other cases only slightly higher.

In your particular business, it may be profitable to sell some products below cost in order to capture other business where you can make more profit. This will create a force to drive down costs. Remember that if you control change you control cost so the more stable your processes the less they cost. If you find that you cannot absorb increases in labour and raw material costs, then you may either have to look for alternative approved sources, alternative materials, alternative methods or consider alternative designs. By including price in the improvement formula, it will act as a driving force. You can offer price reductions to customers as a result of making your processes more efficient.

Improvement on Timing

With a developed quality management system you can monitor the cycle time for each process and not just the production processes. Often the other processes are a source rich in cycle time improvements such as the time taken to change a document, a design, a policy etc. The time taken to place an order, arrange a training course, authorize budgets and expenditure etc. Reaction time is also important as in servicing, maintenance, customer support etc. You will collect masses of data, do the analysis, show that there is a problem to be solved. How long does it take to get management to react to a situation that requires their attention? There are priorities of course, but question these priorities if you believe they hinder continuous improvement.

Improvements in Productivity

Your general aim should be to improve product quality, increase productivity and reduce the cost of development and manufacture or service delivery. However, productivity is not easy to measure unless you have one product on one production line. With multiple products on multiple lines each at a different stage of maturity, it makes comparisons to detect changes in productivity difficult, if not impossible.

Key Messages from Part 8

1. The more skilled you are at predicting what auditors will look at and what they are looking for when they find it, the more successful you will be.
2. All personnel should be able to explain:
   - The objectives of what they are doing;
   - The process they are using to achieve these objectives;
   - How they know they are achieving these objectives;
   - How they know they are doing this in the best way; and
   - How they know their efforts are focussed on the right objectives.
3. If you want certification against ISO 9001, Clauses 4, 5 and 8 are mandatory and any exclusion in Clause 7 is only acceptable if it does not affect the organization's ability or responsibility to provide product that meets customer and applicable regulatory requirements.
4. With every product there is a service. If the product is provided from elsewhere, a service still needs to be designed to process it.
5. Whatever the method of measurement, it is important that the integrity of measurement is sound. It is therefore inconceivable that Clause 7.6 on the control of measuring equipment could be excluded.
6. If you register your system for the manufacture of washing machines then add to your business the manufacture of electronic components, you cannot claim that you are certificated to ISO 9001:2008 for the manufacture of electronic components.
7. The scope of registration is not the same as the scope of your quality management system. If you do not intend to sell or supply services that are outside the scope of ISO 9001 to your customer (legal, medical, catering etc.) then you don't have to include them in the scope of registration.
8. The assessment itself is only as good as the auditor who conducts the assessment.
9. A quality manual written around the standard is often not user friendly and hence is likely to be seen as only serving the needs of the auditor.
10. Only certification bodies are accredited. This means they are authorized to conduct certification of conformity to prescribed standards and ISO 9001 certification does not authorize your organization to do this.
11. You should not give the impression in any context that ISO 9001:2008 certification is a product certification or product guarantee.
12. ISO 9001 certification implies that you have the capability to satisfy the requirements of your current customers but this may not be sufficient to win business from your competitors.

13. One of the most significant findings as a result of installing a quality management system should be that performance becomes less erratic and less sensitive to changes in organization or customer requirements.
14. Another benefit of an effective quality management system is that you can offer price reductions to customers as a result of making your processes more efficient.

Appendices

Related Web Sites

Auditing

Information on auditing practices at www.iqnet-certification.com
The official line from the International Accreditation Forum on the transition at http://www.iaf.nu
Books on auditing and auditor training at http://www.transition-support.com

Benchmarking

Information on benchmarking from the American Productivity Centre http://www.apqc.org/

Competence

The Institute of Personnel Development at http://www.ipd.co.uk/

Customer Satisfaction

The American customer satisfaction index at http://www.theacsi.org/

Customer Service Standards

U.S. Bureau of the Census customer service standards at http://www.census.gov/mso/www/custstd.html
Nottinghamshire Police http://www.nottinghamshire.police.uk/about/our_customer_service_standards/

Process Management

Articles on process management at http://www.transition-support.com

Product Recall

Recalls notified by UK Trading Standards at http://www.tradingstandards.gov.uk/cgi-bin/newslist.cgi?area=safe

Recall notices from the UK Vehicle & Operator Services Agency at http://www.vosa. gov.uk/vosa/apps/recalls/default.asp
Recall notices from USA at http://www.recalls.gov/nhtsa.html and http://www.cpsc. gov/

Quality – General

Chartered Quality Institute Quality Resources http://www.thecqi.org/resources/
Purchase past articles from ASQ Journals at http://qic.asq.org/infosearch.html
Articles on quality management at http://www.transition-support.com

Quality Tools

Descriptions and examples of the 19 tools used in management and operations at http://www.skymark.com/resources/tools/management_tools.asp
A wide range of tools on leadership, problem solving, decision making, project management, and many more at http://www.mindtools.com/
Quality tools, decision making and process analysis tools at http://www.asq.org/learn-about-quality/quality-tools.html
Handbook of statistical methods http://www.itl.nist.gov/div898/handbook/index.htm
Integrated Definition Methods. Tools for enterprise modelling and analysis at www.idef. com

Related Initiatives

UK's manufacturing industry initiative at http://www.fitforthefuture.co.uk/
Tomorrows company at http://www.tomorrowscompany.co.uk/
Information from the European Organization for Quality at http://www.eoq.org/

Scientific Management

Information on Frederick Winslow Taylor at http://en.wikipedia.org/wiki/Frederick_Winslow_Taylor

Sector Schemes

Telecommunications sector scheme at http://www.tl9000.org/
Automotive sector scheme at http://www.iaob.org/showPage.php
Aerospace http://www.iaqg.sae.org/iaqg/

Six Sigma

A good range of articles on six sigma at http://www.qualityamerica.com/six_sigma.html
An online statistics handbook http://davidmlane.com/hyperstat/index.html

Standards

Articles on ISO 9001 at http://www.transition-support.com

Standards and articles at http://www.iso.ch

Access news from the technical committee that created the ISO 9000 family at http://www.tc176.org/

Official interpretations on ISO 9001:2000 at http://www.tc176.org/Interpre.asp

Mil-Q-9858 can be obtained at http://www.quality-control-plan.com/mil-q-9858-spec.htm

ISO standards can be obtained at http://www.iso.ch

British standards can be obtained at http://www.bsi-global.com/

Standards on Investors in People at http://www.iipuk.co.uk/

Articles on ISO 9000 at http://www.iso.org/iso/iso_catalogue/management_standards/iso_9000_iso_14000.htm

Guidance on the concept and use of the process approach for management systems at http://www.iso.org/iso/iso_catalogue/management_standards/iso_9000_iso_14000/iso_9001_2008/concept_and_use_of_the_process_approach_for_management_systems.htm

Supply Chain

Supply Chain Council. Tools and resources at www.supply-chain.org

Related Standards

ISO 13485:2003	Medical devices – Quality management systems – Requirements for regulatory purposes
ISO/TR 14969:2004	Medical devices – Quality management systems – Guidance on the application of ISO 13485: 2003
ISO/TS 16949:2002	Quality management systems – Particular requirements for the application of ISO 9001:2000 for automotive production and relevant service part organizations
ISO 14964:2000	Mechanical vibration and shock – Vibration of stationary structures – Specific requirements for quality management in measurement and evaluation of vibration
ISO 15161:2001	Guidelines on the application of ISO 9001:2000 for the food and drink industry
ISO 15189:2007	Medical laboratories – Particular requirements for quality and competence
ISO/CD 15189	Medical laboratories – Particular requirements for quality and competence
ISO 15378:2006	Primary packaging materials for medicinal products – Particular requirements for the application of ISO 9001:2000, with reference to Good Manufacturing Practice (GMP)
ISO/TS 19218:2005	Medical devices – Coding structure for adverse event type and cause
ISO 22000:2005	Food safety management systems – Requirements for any organization in the food chain
ISO/DIS 22006	Quality management systems – Guidelines for the application of ISO 9001:2000 in crop production
ISO 22870:2006	Point-of-care testing (POCT) – Requirements for quality and competence
ISO 28000:2007	Specification for security management systems for the supply chain
ISO/TS 29001:2007	Petroleum, petrochemical and natural gas industries – Sector-specific quality management systems – Requirements for product and service supply organizations

ISO/IEC 90003:2004	Software engineering – Guidelines for the application of ISO 9001:2000 to computer software
ISO/IEC TR 90005:2008	Systems engineering – Guidelines for the application of ISO 9001 to system life cycle processes
ISO/PAS 30000:2008	Ships and marine technology – Ship recycling management systems – Specifications for management systems for safe and environmentally sound ship recycling facilities
IWA 2:2007	Quality management systems – Guidelines for the application of ISO 9001:2000 in education
IWA 4:2005	Quality management systems – Guidelines for the application of ISO 9001:2000 in local government
BS 7850-1:1992	Total quality management Part 1 guide to management principles.
BS OHSAS 18001:2007	Occupational health and safety management systems

Glossary

This list includes some of the terms and common words that have acquired a special meaning in the field of quality management.

3 Ms These are three Japanese words associated with lean production
M1 Muda (waste)
M2 Muri (overburden)
M3 Mura (unevenness)
5 Why's. These typically refer to the practice of asking, five times, why a failure has occurred in order to get to the root cause.
5 Ss These Japanese words apply to the visual management of a workspace

1. **S1** Seiri (straighten up). Differentiate between the necessary and unnecessary and discard the unnecessary
2. **S2** Seiton (put things in order). Put things in order
3. **S3** Seido (clean up). Keep the workplace clean
4. **S4** Seiketsu (personal cleanliness). Make it a habit to be tidy
5. **S5** Shitsuke (discipline). Follow the procedures

6 Ms These are the six words that are used to title the arms in a fishbone diagram. The words vary but the most commonly used Ms are:
M1 machines
M2 methods
M3 materials
M4 measurements
M5 milieu (surrounding environment)
M6 manpower

One could also add money and management.

7 Wastes. These are:
W1 overproduction
W2 excess inventory
W3 waiting time
W4 unnecessary transportation
W5 unnecessary movement
W6 over processing or incorrect processing
W7 defects

8D. A problem solving method that is structured into eight disciplined steps. The eight basic steps are:

D 1 establish the team

D 2 describe the problem

D 3 develop an interim containment action

D 4 define or verify root cause

D 5 choose or verify permanent corrective action

D 6 implement or validate permanent corrective action

D 7 prevent recurrence

D 8 recognize and reward the team

(Note: some of these terms are not consistent with ISO 9000 definitions for corrective and preventive action.)

Acceptance authority. An organization with the right to decide on the acceptability of something, typically products, services, designs, projects or proposals for changing a design or project. Also referred to as Design Authority and Project Authority.

Acceptance criteria. The standard against which a comparison is made to judge conformance.

Accreditation. A process by which organizations are authorized to conduct certification of conformity to prescribed standards. Laboratory accreditation is defined by ISO as formal recognition that a laboratory is competent to carry out specific tests or specific types of tests. An accredited organization is authorized to issue certificates of conformity to national or international standards. Like certification, accreditation is awarded for a specific scope of service or range of products, except that for laboratory accreditation they are accredited for very specific tests or measurements – usually within specified ranges of measurement – with associated information on uncertainty of measurement and for particular product and test specifications. An ISO 9001 certificate for a laboratory does not accurately specify the performance characteristics of the product that the certificated organization is capable of supplying.

Activity. An element of work that produces an output required by a process. Activities comprise tasks or operations.

Adequate. Suitable for the purpose.

Applicable. In the context of documents, applicable means capable of being applied to the activities to be undertaken. In the context of activities applicable means where it applies, e.g if there is a requirement for all electronic circuits to be grounded and the product in question contains no electronic circuits, the requirement cannot be applied – it is therefore not applicable.

Applicability. It is a technical issue unlike appropriateness which can be subjective.

Appropriate. Means suitable for its purpose or to the circumstances and required knowledge of this purpose or circumstances. Without criteria, an auditor is left to decide what is or is not appropriate based on personal experience.

Approved. Something that has been confirmed as meeting the requirements.

Assessment. The act of determining the extent of compliance with requirements.

Assurance. Evidence (verbal or written) that gives confidence that something will or will not happen or has or has not happened.

Attribute data. Qualitative data that can be counted for recording and analysis, e.g. presence or absence of a required characteristic, number of failures in a production run, number of people eating in the cafeteria on a given day, etc.

Audit. An examination of results to verify their accuracy by someone other than the person responsible for producing them. (See also ISO 9000:2005 clause 3.9.1.)

Authority. The right to take actions and make decisions.

Authorized. A permit to do something or use something that may not necessarily be approved.

Autonomation. Automation with the human touch. The purpose is to free equipment from the necessity of constant human attention, separate people from machines and allow workers to staff multiple operations. In Japanese the word is Jidoka. (See also Error proofing.)

Balanced scorecard. A strategic planning and review methodology that enables organizations to clarify their vision and strategy and translate them into action. It provides feedback around both the internal business processes and the external outcomes in order to continuously improve strategic performance and results.

Benchmarking. A technique for measuring an organization's products, services and operations against those of its competitors resulting in a search for best practice that will lead to superior performance.

Business management system. The set of interacting and managed processes that function together to achieve the business objectives.

Business objectives. Objectives the business needs to achieve in order to accomplish its mission. These are usually derived from an analysis of stakeholder needs and expectations.

Business plan. Provisions made to fulfil the organization's mission, and vision and apply its values in terms of the strategy, objectives, measures, targets and enabling processes.

Business process. A process that is designed to deliver outputs that satisfy business objectives.

Calibrate. To standardize the quantities of a measuring instrument.

Capability index C_p. The capability index for a stable process is defined as the quotient of tolerance width and process capability where process capability is the 6σ range of a process' inherent variation.

Capability index C_{pk} The capability index which account for process centering for a stable process using the minimum upper or lower capability index.

Certification body. See Registrar.

Certification. A process by which a product, process, person or organization is deemed to meet specified requirements.

Class. A group of entities having at least one attribute in common or a group of entities having the same generic purpose but different functional use.

Clause of the standard. A numbered paragraph or subsection of the standard containing one or more related requirements such as 7.2.2. Note: each item in a list is also a clause.

Codes. A systematically arranged and comprehensive collection of rules, regulations or principles.

Commitment. An obligation a person or an organization undertakes to fulfil i.e. doing what you say will do.

Common cause variation. Random variation caused by factors that are inherent in the system.

Competence. The ability to demonstrate the use of education, skills and behaviours to achieve the results required for the job.

Competence-based assessment. A technique for collecting sufficient evidence that individuals can perform or behave to the specified standards in a specific role (Shirley Fletcher).

Competent. An assessment decision that confirms a person has achieved the prescribed standard of competence.

Concession. Permission granted by an acceptance authority to supply product or service that does not meet the prescribed requirements. (See also ISO 9000:2005 clause 3.6.11.)

Concurrent engineering. See also simultaneous engineering.

Configuration control. Systematic evaluation, co-ordination, approval or disapproval of all changes to the baseline configuration (NASA SP 6001).

Configuration management. A discipline applying technical and administrative direction and surveillance to the identity, documentation, control and recording of the functional and physical characteristics of a product taking into account system interfaces (DEF STAN 05-57).

Conformity assessment. Any activity concerned with determining directly or indirectly that relevant requirements are fulfilled (ISO/IEC Guide 2).

Conformity control. Ensuring that products remain conforming once they have been certified as conforming.

Continual improvement. A recurring activity to increase the ability to fulfil requirements (ISO 9000:2005).

Contract loan. An item of customer-supplied property provided for use in connection with a contract that is subsequently returned to the customer.

Contract. An agreement formally executed by both customer and supplier (enforceable by law) which requires performance of services or delivery of products at a cost to the customer in accordance with stated terms and conditions. Also agreed requirements between an organization and a customer transmitted by any means.

Contractual requirements. Requirements specified in a contract.

Control charts. A graphical comparison of process performance data to computed control limits drawn as limit lines on the chart.

Control methods. Particular ways of providing control which do not constrain the sequence of steps in which the methods are carried out.

Control procedure. A procedure that controls product or information as it passes through a process.

Control. The act of preventing or regulating change in parameters, situations or conditions.

Controlled conditions. Arrangements that provide control over all factors that influence the result.

Core competence. A specific set of capabilities including knowledge, skills, behaviours and technology that generate performance differentials.

Corrective action. Action planned or taken to stop something from recurring. (See also ISO 9000:2005 clause 3.6.5.)

Corrective maintenance. Maintenance carried out after a failure has occurred that is intended to restore an item to a state in which it can perform its required function.

Covey's 7 habits of highly effective people:

1. Be proactive – the principle of personal vision
2. Begin with the end in mind – the principles of personal leadership
3. Put first things first – the principle of personal management
4. Think win/win – the principle of interpersonal leadership
5. Seek first to understand then to be understood – the principle of empathetic communication
6. Synergize – the principle of creative cooperation
7. Sharpen the saw – the principle of balanced self-renewal

Critical success factors (CSFs). Those factors on which the achievement of specified objectives depend.

Critical to Quality (CTQ). This relates to the key measurable product or process characteristics that must meet the agreed performance standards in order to satisfy the customer.

Cross-functional team. See multidisciplinary team.

Customer complaints. Any adverse report (verbal or written) received by an organization from a customer.

Customer feedback. Any comment on the organization's performance provided by a customer.

Customer-supplied product. Hardware, software, documentation or information owned by the customer which is provided to an organization for use in connection with a contract and which is returned to the customer either incorporated in the supplies or at the end of the contract.

Customer. Organization that receives a product or service – includes, purchaser, consumer, client, end user, retailer or beneficiary (ISO 9000:2005).

Cusum chart. A type of control chart (cumulative sum control chart) used to detect small changes between 0 and 0.5σ. Cusum charts plot the cumulative sum of the deviations between each data point (a sample average) and a reference value, T. Unlike other control charts, one studying a cusum chart will be concerned with the slope of the plotted line, not just the distance between plotted points and the centreline.

Data. Information that is organized in a form suitable for manual or computer analysis.

Define and document. To state in written form, the precise meaning, nature or characteristics of something.

Demand creation process. A key business process that penetrates new markets and exploits existing markets with products and a promotional strategy that influences decision makers and attracts potential customers to the organization.

Demand fulfilment process. A key business process that converts customer requirements into products and services in a manner that satisfies all stakeholders.

Deming's 14 points of management

1. Create constancy of purpose

2. Adopt the new philosophy

3. Cease dependence on inspection

4. End the practice of awarding business on the basis of price tag

5. Improve constantly and forever the system of production and service

6. Institute training on the job

7. Institute leadership

8. Drive out fear

9. Break down barriers between departments

10. Eliminate slogans, exhortations, and targets

11. Eliminate quotas and management by objectives and by numbers

12. Remove barriers that rob the hourly worker of his right to pride of workmanship

13. Institute a vigorous program of education and self-improvement

14. Put everybody in the company to work to accomplish the transformation

Demonstrate. To prove by reasoning, objective evidence, experiment or practical application.

Department. A unit of an organization that may perform one or more functions. Units of organization regardless of their names are also referred to as functions (see Functions).

Design and development. Design creates the conceptual solution and development transforms the solution into a fully working model (See also ISO 9000:2005 3.4.4.)

Design of experiments. A technique for improving the quality of both processes and products by effectively investigating several sources of variation at the same time using statistically planned experiments.

Design review. A formal documented and systematic critical study of a design by people other than the designer.

Design. A process of originating a conceptual solution to a requirement and expressing it in a form from which a product may be produced or a service delivered.

Disposition. The act or manner of disposing of something.

DMAIC. Define, Measure, Analyse, Improve and Control – the problem solving technique at the heart of Six Sigma programmes.

Documented procedures. Procedures that are formally laid down in a reproducible medium such as paper or magnetic disk.

DPMO. Defects per million opportunities – the units of measure for process capability.

Effectiveness of the system. The extent to which the system fulfils its purpose.

Embodiment loan. An item of customer-supplied property provided for incorporation into product that is subsequently supplied back to the customer or a party designated by the customer.

Employee empowerment. An environment in which employees are free (within defined limits) to take action to operate, maintain and improve the processes for which they are responsible using their own expertise and judgement.

EMS. Environmental management system.

Ensure. To make certain that something will happen.

Establish and maintain. To set-up an entity on a permanent basis and retain or restore it in a state in which it can fulfil its purpose or required function.

Evaluation. To ascertain the relative goodness, quality or usefulness of an entity with respect to a specific purpose.

Evidence of conformance. Documents which testify that an entity conforms to certain prescribed requirements.

Executive responsibility. Responsibility vested in those personnel who are responsible for the whole organization's performance. Often referred to as top management.

Fagan inspection. A software inspection technique in which someone other than the creator of a product examines it with the specific intent of finding errors. Software Inspections were introduced in the 1970s at IBM, which pioneered their early adoption and later evolution. Michael Fagan helped to develop the formal software inspection process at IBM, hence the term 'Fagan inspection'.

Failure mode effects analysis (FMEA). A technique for identifying potential failure modes and assessing existing and planned provisions to detect, contain or eliminate the occurrence of failure. (See also Risk assessment.)

FIFO. First in first out. A term used to describe a method of inventory control.

Final inspection and testing. The last inspection or test carried out by the organization before ownership passes to the customer.

Finite element analysis. A technique for modelling a complex structure.

First party audits. Audits of a company or parts thereof by personnel employed by the company. These audits are also called Internal Audits.

Flinching. The tendency of inspectors to falsify the results of measurements that are borderline.

Follow-up audit. An audit carried out following and as a direct consequence of a previous audit to determine whether agreed actions have been taken and are effective.

Force majeure. An event, circumstance or effect that cannot be reasonably anticipated or controlled.

Function. In the organizational sense, a function is a special or major activity (often unique in the organization) which is needed in order for the organization to fulfil its purpose and mission. Examples of functions are design, procurement, personnel, manufacture, marketing, maintenance, etc.

Geometric dimensioning and tolerancing. A method of dimensioning the shape of parts that provides appropriate limits and fits for their application and facilitates manufacturability and interchangeability.

Grade. Category or rank given to entities having the same functional use but different requirements for quality, e.g. hotels are graded by star rating and automobiles are graded by model. (See also ISO 9000:2005 clause 3.1.3.)

Hazard. Anything that may cause harm to people, product, property or the natural environment.

Hazard analysis. The process of collecting and evaluating information on hazards and conditions leading to their presence to decide which are significant for product safety and therefore should be addressed in the HACCP plan (ISO 15161).

HACCP. Hazard Analysis and Critical Control Point. A technique used particularly in the food industry for the identification of hazards and control of risks. The CCP is a step at which control can be applied and is essential to prevent or eliminate a food safety hazard or reduce it to an acceptable level (ISO 15161).

Hoshin kanri. A Japanese term for a systems' approach to goal achievement. Hoshin means a course, a policy, a plan or an aim. Kanri means administration, management, control, charge of or care for. Also known as policy deployment but it goes further than this.

IAF. International Accreditation Forum. The world association of Conformity Assessment Accreditation Bodies in the fields of management systems, products, services, personnel and other similar programmes of conformity assessment.

Identification. The act of identifying an entity, i.e. giving it a set of characteristics by which it is recognizable as a member of a group.

Implement. To carry out a directive.

Implementation audit. An audit carried out to establish whether actual practices conform to the documented quality system. Note: also referred to as a conformance audit or compliance audit.

Indexing. A means of enabling information to be located.

In-process. Between the beginning and the end of a process.

Inspection authority. The person or organization that has been given the right to perform inspections.

Inspection, measuring and test equipment. Devices used to perform inspections, measurements and tests.

Inspection. The examination of an entity to determine whether it conforms to prescribed requirements. (See also ISO 9000:2005 clause 3.8.2.)

Installation. The process by which an entity is fitted into larger entity.

Integrated management. The understanding and effective direction of every aspect of an organisation so that the needs and expectations of all stakeholders are justly satisfied by the best use of all resources (CQI – Integrated Management Special Interest Group).

Integrated management system. A management system that enables the organization to achieve all its objectives in a manner that satisfies the needs and expectations of all stakeholders. Synonymous with Business management System. Often perceived to be the amalgamation of quality, environmental, health and safety management systems and other similar systems.

Intellectual property. Creations of the mind: inventions, literary and artistic works, and symbols, names, images, and designs used in commerce. Intellectual property is divided into two categories: industrial property and copyright.

Interested party. Person or group having an interest in the performance or success of an organization which normally includes: customers, owners, employees, contractors, suppliers, investors, unions, partners or society (ISO 9000:2005). Interested parties can be benevolent or malevolent and the latter group might include terrorists, criminals and competitors whose only interest is to harm the organization (See also Stakeholder.)

ISO. International Organization for Standardization.

Issues of documents. The revision state of a document.

Just-in-time. A method of lean production where the demand comes from the end of the process through to the beginning so that the only parts that are delivered are those that are needed at the time they are needed.

Kanban. A Japanese work for 'tag' or 'ticket' or 'sign board'. These tickets are used as a means of picking up and receiving the right quantity of parts required by a process thus ensuring parts are delivered just-in-time by preceding processes.

Kaizen. Continuing improvement in personal life, home life, social life and working life. When applied to the workplace it means continuing improvement involving everyone – managers and workers alike (Masaaki Imai).

Key performance indicators (KPI). The quantifiable characteristics that indicate the extent by which an objective is being achieved. (See also Stakeholder success measures.)

Lagging measures. Measures that indicate an aspect of performance long after the conditions that created it have changed (e.g. profit and return on capital).

Leading measures. Measures that indicate an aspect of performance while the conditions that created it still prevail (e.g. response time, conformity).

Lean production. A method of production that is demand driven (pull) rather than supply driven (push) as with mass production. There is zero waiting time, zero inventory, line balancing and reduction in process time with less space required for materials and finished product. This results in product being produced only to satisfy a demand. In lean production the person goes to the job and performs multiple tasks.

Line balance. Balancing the resources in a process or number of processes by optimizing speeds, feeds, batch size, number of workstations, operators, idle time, changeover time, cycle time and process yield.

Manage work. To manage work means to plan, organize and control the resources (personnel, financial and material) and the tasks required to achieve the objective for which the work is needed.

Management representative. The person management appoints to act on their behalf to manage the quality management system.

Management system. The set of interacting processes by which an organization defines and achieves its goals. A qualifying prefix would describe a management system that achieves the organization's goals relative to this qualifying prefix (e.g. a quality management system achieves the organization's quality goals).

Mass production. A method of production that is supply driven based on sales forecasts rather than firm orders. It produces large amounts of standardized products on parallel production lines that stretch from raw materials to finished product (vertical integration). In mass production the job comes to the worker who passes it on to the next worker to perform the next operation on the line.

Master list. An original list from which copies can be made.

Measures. The characteristics by which performance is judged. They are the characteristics that need to be controlled in order than an objective will be achieved. They are the response to the question 'What will we look for to reveal whether the objective has been achieved?'

Measurement. The act of measuring. It is a process of associating numbers with physical quantities and phenomena.

Measurement capability. The ability of a measuring system (device, person and environment) to measure true values to the accuracy and precision required.

Measuring equipment. The monitoring or measuring instruments, software, measurement standard, reference material or auxiliary apparatus used to monitor and measure product characteristics (ISO 9000:2005 clause 3.10.4).

Measurement process. Activities, measuring devices, personnel, operating environment and the measurement system to determine the value of a quantity.

Measurement system. The units of measure and the process by which standards for these units of measure are developed and maintained.

Measurement uncertainty. The variation observed when repeated measurements of the same parameter on the same specimen are taken with the same device.

Mission. An expression of the purpose of an organization, why it exists, what it is being mobilized to accomplish in the long term.

Mission management. A key business process that determines the direction of the business, continually confirms that the business is proceeding in the right direction and makes course corrections to keep the business focussed on its mission.

Modifications. Entities altered or reworked to incorporate design changes.

Monitoring. To check periodically and systematically. It does not imply that any action will be taken.

Motivation. An inner mental state that prompts a direction, intensity and persistence in behaviour.

Muda. The Japanese term for waste.

Multidisciplinary team. A team comprising representatives from various functions or departments in an organization, formed to execute a project on behalf of that organization.

Nationally recognized standards. Standards of measure that have been authenticated by a national body.

Nature of change. The intrinsic characteristics of the change (what has changed and why).

Objective evidence. Information that can be proven true based on facts obtained through observation, measurement, test or other means. (See also ISO 9000:2005 clause 3.8.1.)

Objective. A result to be achieved usually by a given time.

Obsolete documents. Documents that are no longer required for operational use. They may be useful as historic documents.

OEM. Original Equipment Manufacturer.

Operating procedure. A procedure that describes how specific tasks are to be performed (might be called a work instruction).

Organizational goals. Where the organization desires to be, in markets, in innovation, in social and environmental matters, in competition and in financial health.

Organizational interfaces. The boundary at which organizations meet and affect each other expressed by the passage of information, people, equipment, materials and the agreement to operational conditions.

Performance index P_{pk}. The performance index which accounts for process centering and defined as the minimum of the upper or lower specification limit minus the average value divided by 3σ.

Performance indicators. Quantifiable measures of performance related to specific objectives. They respond to the question 'What would we expect to see happening if this objective had been achieved?' (see also Measures).

PEST Analysis. Political, Economic, Social and Technological Analysis. A tool used in scanning the environment for changes affecting an organization's success in fulfilling its mission.

Plan. Provisions made to achieve an objective.

Planned arrangements. All the arrangements made by the organization to achieve the customer's requirements. They include the documented policies, objectives, plans, specifications and processes and the documents derived from such requirements.

Planned maintenance. The maintenance carried out with forethought as to what is to be checked, adjusted, replaced, etc.

Poka-yoke. Japanese term that means 'mistake proofing', a concept introduced by Shigeo Shingo to Toyota in 1961. It is a device that prevents incorrect parts from being made or assembled, or prevents correct parts being assembled incorrectly. Previously the term baka-yoke was used but as this means fool proofing and is rather offensive it was discontinued. Even mistake proofing have evolved into 'error proofing' to avoid the personal implications. Error proofing is one of the two pillars of the Toyota Production System (TPS).

Policy. A guide to thinking, action and decision.

Positive recall. A means of recovering an entity by giving it a unique identity.

Positively identified. An identification given to an entity for a specific purpose which is both unique and readily visible.

Potential nonconformity. A situation that if left alone will in time result in nonconformity.

Predictive maintenance. Work scheduled to monitor machine condition, predict pending failure and make repairs on an as-needed basis.

Pre-launch. A phase in the development of a product between design validation and full production (sometimes called pre-production) during which the production processes are validated.

Prevent. To stop something from occurring by a deliberate planned action.

Preventive action. Action proposed or taken to stop something from occurring (See also ISO 9000:2005 clause 3.6.4.)

Preventive maintenance. Maintenance carried out at predetermined intervals to reduce the probability of failure or performance degradation, e.g. replacing oil filters ate defined intervals. Also referred to as Planned maintenance.

Procedure. A sequence of steps to execute a routine activity. (See also ISO 9000:2005 clause 3.4.5.)

Process. A series of activities that use resources to produce a result. An effective process would be one in which the activities use resources to achieve a prescribed objective. The activities may be interrelated, interdependent and may interact. (See also ISO 9000:2005 clause 3.4.1.)

Process approach. An approach to managing work in which the activities and resources (including behaviours) function together in such a relationship as to produce results consistent with the process objectives.

Process capability. The inherent ability of a process to reproduce its results consistently during multiple cycles of operation.

Process description. A set of information that describe the characteristics of a process in terms of its purpose, objectives, measures, design features, inputs, activities, resources, behaviours, outputs, constraints, measurements and reviews.

PFMEA. Process Failure Mode and Effects Analysis.

Process management. The planning, operation and control of interrelated and interacting activities to produce a desired result.

Process measures. Measures used to judge the performance of processes. They are generally a response to the question 'What will we look for to reveal whether the process objectives have been met?'

Process parameters. Those variables, boundaries or constants of a process that restrict or determine the results.

Product realization. All those processes and resources necessary to transform a set of requirements into a product or service that fulfil the requirements.

Product. Anything produced by human effort, natural or man-made processes. Result of activities or processes (ISO 9000:2005).

Production. The creation of products.

Proprietary designs. Designs exclusively owned by the organization and not sponsored by an external customer.

Prototype. A model of a design that is both physically and functionally representative of the design standard for production and used to verify and validate the design.

Purchaser. One who buys from another.

Purchasing documents. Documents that contain the organization's purchasing requirements.

Qualification test. Determination by a series of tests and examinations of a product, and its related documents and processes that the product meets all the specified performance capability requirements under operational conditions.

Quality. The degree to which a set of inherent characteristics fulfil a need or expectation that is stated, generally implied or obligatory (ISO 9000:2005).

Quality assurance. Part of quality management focused on providing confidence that quality requirements will be fulfilled (ISO 9000:2005).

Quality characteristics. Any characteristic of a product or service that is needed to satisfy customer needs or achieve fitness for use.

Quality circles (or QC circles). A group of volunteers who perform activities within a process participating continuously together for the purpose of self-development, mutual development, control and improvement of the process (derived from the texts of Kaoru Ishikawa).

Quality conformance. The extent to which the product or service conforms to the specified requirements.

Quality control. A process for maintaining standards of quality that prevents and corrects change in such standards so that the resultant output meets customer needs and expectations. (See also ISO 9000:2005 clause 3.2.10.)

Quality costs. Costs incurred because failure is possible. The actual cost of producing an entity is the no failure cost plus the quality cost. The no failure cost is the cost of doing the right things right first time. The quality costs are the prevention, appraisal and failure costs.

Quality function deployment (QFD). A technique to deploy customer requirements (the true quality characteristics) into design characteristics (the substitute characteristics) and deploy them into subsystems, components, materials and production processes. The result is a grid or matrix that shows how and where customer requirements are met.

Quality improvement. Part of quality management focused on increasing the ability to fulfil quality requirements (ISO 9000:2005).

Quality management system requirements. Requirements pertaining to the design, development, operation, maintenance and improvement of quality management systems.

Quality management system. The set of interacting processes used by the organization to achieve its quality objectives. (See also ISO 9000:2005 3.2.3.)

Quality objectives. Those results which the organization needs to achieve in order to improve its ability to meet needs and expectations of all the stakeholders.

Quality planning. Provisions made to achieve the needs and expectations of organization's stakeholders and prevent failure.

Quality plans. Plans produced to define how specified quality requirements will be achieved, controlled, assured and managed for specific contracts or projects.

Quality problems. The difference between the achieved quality and the required quality.

Quality requirements. Those requirements which pertain to the features and characteristics of a product or service which are required to be fulfilled in order to satisfy a given need.

Quarantine area. A secure space provided for containing product pending a decision on its disposal.

Reductionism. A way of using logic and causal thinking to separate the individual parts of what is being studied and draw conclusions about a group based on the analysis of its constituent parts. It is not always possible to predict the behaviour of systems as any changes can lead to unintended consequences. Reductionism tends to ignore the influence that individual parts have on each other.

Registrar. An organization that is authorized to certify organizations. The body may be accredited or non-accredited.

Registration. A process of recording details of organizations of assessed capability that have satisfied prescribed standards.

Regulator. A legal body authorized to enforce compliance with the laws and statutes of a national government.

Regulatory requirements. Requirements established by law pertaining to products or services.

Remedial action. Action proposed or taken to remove a nonconformity in a product previously deemed conforming (see also Corrective and preventive action).

Representative sample. A sample of product or service that possesses all the characteristics of the batch from which it was taken.

Resources. Something of which there is an available supply that can be called on when needed. Resources include time, personnel, skill, machines, materials, money, plant, facilities, space, information, knowledge, etc. Resources are used by processes resulting in some being reusable and others changed, lost or depleted by the process.

Resource management. A key business process that specifies, acquires and maintains the resources required by the business to fulfil the mission and disposes of any resources that are no longer required.

Responsibility. An area in which one is entitled to act on one's own accord or able to respond by virtue of having caused an event.

Review. Another look at something.

Rework. Continuation of work on a product to make it conform to the specified requirements without additional procedures or techniques.

Risk. The likelihood of something happening that could have a positive or negative effect. Also the combination of the probability of an event and its consequences (ISO/IEC Guide 73).

Risk assessment. A study performed to quantify potential risks associated with a particular event or situation. It identifies hazards or failure modes, their effect on people, product, property or natural environment, the probability of their occurrence and detection and the severity of their effect in order to identify provisions taken or needed to eliminate, control or reduce the root cause. (See also FMEA, HACCP.)

Risk management. The process whereby organizations methodically address the risks attaching to their activities with the goal of achieving sustained benefit within each activity and across the portfolio of all activities (a Risk Management Standard IRM 2002).

Scheduled maintenance. Work performed at a time specifically planned to minimize interruptions in machine availability, e.g. changing a gearbox when machine is not required for use (includes predictive and preventive maintenance).

Shall. A provision that is binding.

Should. A provision that is optional.

Simultaneous engineering. A method of reducing the time taken to achieve objectives by developing the resources needed to support and sustain the production of a product in parallel with the development of the product itself. It involves customers, suppliers and each of the organization's functions working together to achieve common objectives.

Six sigma. Six standard deviations.

SMART. An acronym used for testing objectives as follows:

S **Specific.** Objectives should be *specific* actions completed while executing a strategy or delivering an output. They should be derived from the mission and relevant to the process or task to which they are being applied. Objectives should be specified to a level of detail that those involved in their implementation fully understand what is required for their completion – not vague or ambiguous and defining precisely what is required.

M **Measurable.** Objectives should be *measurable* actions that have a specific end condition. Objectives should be expressed in terms that can be measured using available technology. When setting objectives you need to know how achievement will be indicated, the conditions or performance levels that will indicate success.

A **Achievable.** Objectives should be *achievable* with resources that can be made available – they should be achievable by average people applying average effort.

R **Realistic.** Objectives should be *realistic* in the context of the current climate and the current and projected workload. Account needs to be taken of the demands from elsewhere that could jeopardize achievement of the objective.

T **Timely.** Objectives should be *time-phased* actions that have a specific start and completion date. Time-phased objectives facilitate periodic review of progress and tracking of revisions. The specific date or time does not need to be expressed in the objective unless it is relevant – in other cases the timing for all objectives might be constrained by their inclusion in the 2005 Business Plan, implying all the objectives will be achieved in 2005. The Business Plan for 2005–2008 implies all objectives will be achieved by 2008.

SMS Safety management system.

Special cause variation. A cause of variation that can be assigned to a specific or special condition that does not apply to other events, e.g. weather, power failure, tool breakage, etc.

Specified requirements. Requirements prescribed by the customer and agreed by the organization or requirements prescribed by the organization that are perceived as satisfying a market need. Such requirements may or may not be documented.

Stakeholder. The individuals and constituencies that contribute, either voluntarily or involuntarily, to an organization's wealth-creating capacity and activities, and that are therefore its potential beneficiaries and/or risk bearers (Post, Preston and Sachs). They are the response to the question 'Who are we working for?' (See also Interested party.)

Stakeholder measures. Measures used to judge the performance of an organization. They are generally a response to the question, 'What measures will the stakeholders use to reveal whether their needs and expectations have been met?' (See also Key performance indicators.)

Statistical control. A condition of a process in which there is no indication of a special cause of variation.

Status. The relative condition, maturity or quality of something.

Strategy. The broad priorities adopted by an organization in recognition of its operating environment in pursuit of its mission (Paul Niven).

Subcontract requirements. Requirements placed on a subcontractor that are derived from requirements of the main contract.

Subcontractor. A person or company that enters into a subcontract and assumes some of the obligations of the prime contractor.

SWOT Analysis. Strengths, Weaknesses, Opportunities and Threats Analysis. A tool for determining the capability of an organization to achieve prescribed objectives.

System.

(a) An ordered set of ideas, principles and theories that fulfil a purpose.

(b) A set of interacting processes that function together to achieve a specified objective.

(c) A set of variables that influence one another (Senge).

Systems approach. An approach to managing an organization that recognises its performance results from the interaction of interrelated processes and cannot be predicted by analysing each process, activity or function separately.

System audit. An audit carried out to establish whether the quality system conforms to a prescribed standard in both its design and its implementation.

System effectiveness. The ability of a system to achieve its stated purpose and objectives.

System management. The management of an organization as a system of interacting processes that function together to achieve the goals of the organization.

Systems thinking. The understanding of dynamic relationships between interacting variables or process outputs within the context of a larger whole. A scientific field of knowledge for understanding change and complexity through the study of dynamic cause and effect over time. (Kambiz E Maani)

Targets. The level of performance to be achieved, e.g. standard, specification, requirement, budget, quota, plan.

Task. The smallest component of work. A group of tasks comprise an activity.

Technical interfaces. The physical and functional boundary between products or services.

Tender. A written offer to supply products or services at a stated cost.

Theory of constraints. A thinking process optimizing system performance. It examines the system and focuses on the constraints that limit overall system performance. It looks for the weakest link in the chain of processes that produce organizational performance and seeks to eliminate it and optimize system performance.

Theory X. A label given to a belief that workers inherently dislike and avoid work and must be driven to it (Douglas McGregor (1906–1964)).

Theory Y. A label given to a belief that work is natural and can be a source of satisfaction when aimed at higher order human psychological needs (Douglas McGregor (1906–1964)).

Theory Z. A label given to a belief that organizations should focus on increasing employee loyalty to the company by providing a job for life with a strong focus on the well being of the employee, both on and off the job (William Ouchi).

TQC. Total Quality Control.

  (a) An effective system for integrating the quality-development, quality-maintenance and quality-improvement efforts of the various groups in an organization so as to enable product and service at the most economical levels which allow for full customer satisfaction (A.V. Feigenbaum).

  (b) The total participation of everyone in an organization to develop, design, produce and service a quality product or service which is most economical, most useful and always satisfactory to the customer (from I. Ishikawa's What is total quality control?)

TQM. Total Quality Management. A management philosophy and company practices that aim to harness the human and material resources of an organization in the most effective way to achieve the objectives of the organization (BS 7850: 1992).

Traceability. The ability to trace the history, application, use and location of an individual article or its characteristics through recorded identification numbers. (See also ISO 9000:2005 3.5.4.)

Unique identification. An identification that has no equal.

Validation. A process for establishing whether an entity will fulfil the purpose for which it has been selected or designed. (See also ISO 9000:2005 3.8.5.)

Values. The fundamental principles that guide the organization in accomplishing its goals. They are what it stands for such as integrity, excellence, innovation, inclusion, reliability, responsibility, equality, fairness, confidentiality, safety of personnel and property, etc. These values characterize the culture in the organization.

Value engineering. A technique for assessing the functions of a product and determining whether the same functions can be achieved with fewer types of components and materials and the product produced with fewer resources. Variety reduction is an element of value engineering.

Value stream mapping. A tool that helps visualise and understand the flow of material and information as a product or service makes its way through the value stream from customer to suppliers. The maps show the current steps, delays, and information flows required to deliver the target product or service. The objective is to increase value by reducing delay and waste.

Variables data. Quantitative data of measurable characteristics, e.g. diameter of a bearing, weight of a component, viscosity of a liquid, etc.

Verification activities. A special investigation, test, inspection, demonstration, analysis or comparison of data to verify that a product or service or process complies with prescribed requirements.

Verification requirements. Requirements for establishing conformance of a product or service with specified requirements by certain methods and techniques.

Verification. The act of establishing the truth or correctness of a fact, theory, statement or condition. (See also ISO 9000:2005 clause 3.8.4.)

Vision. An expression of the aspirations of an organization; what success will look like as it fulfils its mission.

Waiver. See Concession.

Work Break down structure. A structure in which elements of work for a particular project are placed in a hierarchy.

Work environment A set of conditions under which people operate and include physical, social and psychological environmental factors (ISO 9000:2005:2000).

Work instructions. Instructions that prescribe work to be executed, who will do it, when it is to start and be complete and if necessary how, it is to be carried out.

Work packages. An assembly of related work elements.

Workflow system. A method of manufacture whereby value is added to the product in each process as it moves along a production line. Invented in 1910 by Charles Sorensen, first President of Ford Motor Company.

Workmanship criteria. Standards on which to base the acceptability of characteristics created by human manipulation of materials by hand or with the aid of hand tools.

Zero defects. The performance standard achieved when every task is performed right first time with no errors being detected downstream.